AF526818

Water, Environment & Society

Chief Patron

Prof. A. Venugopal Reddy

Vice-Chancellor, J.N.T. University Hyderabad

Patron

Dr. A. Govardhan

Rector, J.N.T. University Hyderabad

Chairman

Dr. N. Yadaiah

Registrar, J.N.T. University Hyderabad

Co-Chairman

Dr. B. Venkateswara Rao

Director, IST, J.N.T.University Hyderabad

Convener

Dr. M.V.S.S. Giridhar

Professor, CWR, IST, J.N.T. University Hyderabad

Organized by

Centre for Water Resources

Institute of Science and Technology

Jawaharlal Nehru Technological University Hyderabad

Kukatpally, Hyderabad - 500 085

Disclaimer

The Publisher and Organisers do not claim any responsibility for the accuracy of the data, statements made, and opinions expressed by authors. The authors are solely responsible for the contents published in the paper.

Published by

BSP BS Publications

A unit of **BSP Books Pvt. Ltd.**

4-4-309/316, Giriraj Lane, Sultan Bazar,
Hyderabad - 500 095
Phone : 040 - 23445600, 23445688
e-mail : info@bspbooks.net
www.bspbooks.net

ISBN: 978-93-87593-23-7 (Hardback)

Preface

Water, Environment and Society are inextricably linked so that actions in one area commonly have impacts on the others. Of all natural resources, energy, water and food are most needed to sustain life on earth. They are so familiar in our daily life that we often forget its importance and uniqueness. Water is an essential resource that underpins human societies and is a requirement for ecosystem health. Clean, safe drinking water is scarce. India is bestowed with good number of rivers, ponds and lakes but still, far too many people spend their entire day searching for it. It's hard for most of us to imagine that clean, safe water is not something that can be taken for granted. But, in the developing world, finding a reliable source of safe water is often time consuming and expensive.

The scarcity of clean water is from the result of multiple reasons like over usage of water, pollution, increasing demand of population, climate change etc. Water scarcity involves water crisis, water shortage, water deficit or water stress. Water scarcity can be due to physical water scarcity and economic water scarcity. Physical water scarcity refers to a situation where natural water resources are unable to meet a region's demand and economic water scarcity is a result of poor water management resources.

It is in this context and backdrop that the Centre for Water Resources, Institute of Science and Technology, JNTUH felt the need to organize a three day 6th national conference on Water, Environment and Society (NCWES-2019) to take stock of the current status of applications in water resources development and management and also to identify areas most relevant to ensure sustainable development of water resources and environment to benefit the society at large.

Researchers, engineers, site managers, regulatory agents, policy makers, Consultants, NGO's, academicians and vendors will all benefit from the opportunity to exchange information on recent research trends and to examine ongoing research programs in the areas of water and environment. The conference is expected to recommend suitable strategies and policy guidelines to operationalize the initiatives and dovetail them into various watershed development programmes appropriately. Keeping in view the importance and need of the hour, this issue of proceedings is brought out to coincide with the conduct of the national conference. The high value contributions by eminent speakers, Research scholars and participants have been overwhelming and encouraging.

The three day national conference on NCWES will focus its attention on various themes in the form of technical sessions such as

1. Hydrologic parameter estimation & modeling
2. Climate change and environment
3. Urbanisation, Bio-diversity and EIA
4. Groundwater Exploration, Development, Recharge, Modeling and Quality
5. Water quality, Water treatment, Pollution and Society
6. Water Conservation and Irrigation management
7. Water Management, Rainfall and Rainwater Harvesting
8. Geospatial Applications in Water resources

More than 150 delegates and about 60 technical papers are being presented in these eight technical sessions.

I hope the present conference would serve as a link between technology, policy, practice and decision making in the quest for synergetic solutions for sustainable development of water resources and environment.

I wish and expect that the participants will find this conference useful and give their total participation to make it a grand success.

It is with this great pleasure; I extend a warm welcome to all the delegates, speakers and participants to NCWES- 2019.

Dr. M.V. S.S Giridhar
-Editor

Contents

KEYNOTES

Urban Flooding:
[The Role of Urbanization & Climate Change]

S. Jeevananda Reddy
Chief Technical Advisor – WMO/UN & Expert – FAO/UN
Fellow, Telangana Academy of Sciences [Founder Member]
Convenor, Forum for a Sustainable Environment
Jeevananda_reddy@yahoo.com

ABSTRACT

Urbanization means un-planned growth of the city. Slum culture is the principal character of unplanned growth in urban areas, more particularly in capital cities of states. However, the proportion of such character varies significantly with city to city. Extremes in rainfall also vary with space and time in association with the rhythmic variation in rainfall, which is the principal component of climate change. River floods follow this natural rhythmic pattern in rainfall. Both these scenarios are aggravated multi-fold by human interference through the destruction of nature/hydrology. These impacts change with location of urban areas, namely coastal, hilly & inland. The wastewater generated in urban areas also follows the population but major share of this water is imported from far away sources [ex-situ] and taken from groundwater. As a result even with not much rain, urban floods became a regular phenomenon. These issues are discussed in brief in the article with few examples.

Keywords: Urbanization, Hydrology, Flood, Population.

Introduction

Urban flooding is significantly different from rural flooding, as urbanization leads to developed catchments, which increases the flood peaks and flood volume and consequently flooding occurs very quickly due to faster flow. Our governments look at urban flood management, which is like post-mortem report. We rarely follow the precautionary principle, i.e., prevention is better than cure. In the case of GO111, the Supreme Court put forth this principle but governments rarely follow this. Urban floods can be classified as:

- Natural floods
- Human induced floods
- Natural + human induced floods

However, the impact varies with the place, namely coastal, hilly and interior.

Role of Urbanization on Urban Flooding

Urbanization means unplanned growth. Slum culture is part of unplanned growth in urban areas, more particularly capital cities of the states. Hut dwellers directly dump solid and liquid wastes in to nalas and water bodies wherein they encroached and built huts. Around 75% of the water used in urban areas, practically turns in to wastewater. This component has been steadily increasing with the population growth. This water includes groundwater and water brought from faraway places. In Hyderabad currently more than 2000 mld of wastewater is generated. It may go up further with the time unless urban growth is contained. Urban floods are caused by joining this wastewater with rainwater.

Urban drainage system is to take the responsibility of storm water runoff gathering, transport and discharge. Emergency discharge system is used to deal with the extra runoff that could not be discharged by urban drainage system under the extreme storm events. The emergency discharge system may include the natural water bodies (lakes or reservoirs), multiple function detention ponds, spillways, storm water tunnels, etc. These three systems are not independent of each other but they can be integrated to work together for comprehensive urban storm water management.

Historically, urban areas are located along the coasts. With the bifurcation of states and tourism created urban areas in hilly regions and inland areas. With the population growth, the farm sizes have been coming down drastically year by year per farmer. Input costs in agriculture are also going up with poor agriculture systems. Both states and central governments think this will be solved by incentives/subsidies. This has lead rural to urban migration. The rulers instead of improving the quality of life in rural India, developmental activities have been concentrating in urban areas. Thus, the meaning of urbanization changed as "unplanned growth". In this process the major casualty is natural resources. Water bodies turned in to concrete jungle and the remaining have been turned in to cesspool of poison with the polluted water. Rainwater channels/nalas turned in to shelters for hut dwellers. In this process rivers-rivulets and storm water drains disappeared or encroached.

Human induced floods, therefore, are associated with human actions in changing the hydrology of the place such as:

(i) Reduction in infiltration amount of rainfall due to roads & buildings, destruction of greenbelts, destruction of hillocks, etc.;
(ii) Reduction in surface water storage capacity due to destruction of water bodies [lakes/tanks, rivers/rivulets, etc.];
(iii) Destruction of storm water/rainwater drains;
(iv) Increased load of wastewater due to water from ex-situ sources use, groundwater, etc.

All these are associated with urban population and slum population.

Growth of Urban Population

Water and oxygen are the two nature's gifts to human survival on the Earth. With the non-linear growth in human population and associated infrastructure built to meet their needs under changing lifestyles lead imbalance on these two nature's gifts.

Rural population is static but urban population is dynamic. Urban population is growing steadily with abrupt jumps. At global level, Indian population constituted 18.0% of global population occupying 2.7% of the land area and uses only 4.6% of world's fresh water but uses 25% of global groundwater. At all-India level currently with urban population of 440 million, around 25% are living in slums. Such people are vulnerable for urban floods but at the same time the very same people are generating wastewater & water bodies; and thus increasing flood impact.

The population changed from 1950 to 2014 of China and India respectively is 543 million to 1.39 billion and 376 million to 1.27 billion. Though in terms of geographical area China is around "three times" to India but population is more or less the same. Also urban population in 2017 of China and India respectively are 57.9% and 33.5%. Though in China, area available per person is around three times to that of India, yet urban concentration of population in China is nearly double to that of India in 2017; and greenery is in opposite direction – recently NASA released data based on satellite images.

The urban population in India went up from 27.81% in 2001 to 31.16% in 2011. A total of 42.6 million people were living in slums according to 2001 census and they are 95 million in 2011, doubled. However, at state-wise these are quite different. Table 1 presents urban and slum population in % for few states in India in 2011: Andhra Pradesh they are 33.5% & 32.69%; Maharashtra 45.23% & 31.66%; Tamil Nadu 48.45% & 17.85%; and Karnataka 38.57% & 11.51%.

Table 1 Urban Population & Slum Population

Urban Area	Population Total (Million)	Urban	Slum
China [2017]		57.9%	
India [2017]		33.5%	
2011 census data			
Tamil Nadu	14.1	48.45%	17.85%
Kerala	02.5	47.72%	17.93%
Maharashtra	33.6	45.23%	31.66%
Gujarat	11.4	42.58%	11.78%
Karnataka	11.0	38.57%	11.51%
AP	15.7	33.50%	32.69%
WB	14.2	31.89%	26.82%
UP	18.8	22.28%	22.12%
MP	09.8	27.63%	24.32%
Bihar	04.8	11.30%	10.53%
Delhi	11.0		18.45%

River Encroachment

The Bangladesh High Court (HC) delivered on 30/31 January 2019 a historic judgment on river encroachments. In a petition, the organisation cited a report published in "The Daily Star" on November 6, 2016, headlined "Time to Declare Turag Dead". The court directed:

- The authorities to remove all structures from it in 30 days
- If any person, whose name is included in the list of river land grabbing, he or she will be disqualified from contesting elections and directed the Election Commission to take steps in this regard
- The Bangladesh Bank to make sure no river grabbers get bank loans
- The government to make a list of every grabber in the country and publish the list in the media to expose them to the public
- The education ministry to take steps for holding an hour-long class every two months at all public and private academic institutions, including school, madrasa, college and university, to build awareness among students about the importance of rivers
- The industries ministry to take measures for arranging an hour-long meeting every two months with factory workers across the country, also to create awareness

The grabbers include powerful individuals, businesses and, ironically, government offices. The Gazipur City Corporation is among the grabbers, a judicial inquiry has found – unfortunately, same is the case in India and it is worse than that in Bangladesh. A classic example is river Musi in Hyderabad and several hundreds of chained lakes in and around Hyderabad. Also, the remaining parts are cesspools of poison. A 2009 landmark HC judgment had detailed measures on how to recover the ailing rivers from land grabbers and save them from pollution. After the verdict, the Bangladesh government now will have to amend the National River Protection Commission (NRPC) Act 2013 with provisions for punishment and fine for river grabbing. The current NRPC Act does not have provisions for punishment. The government must report to the court in six months on its action in this regard. The HC also declared the NRPC as the legal guardian of all rivers and act like their "parents". The landmark verdict comes when river grabbing by influential groups seems unstoppable. Often, grabbers return to steal river land soon after being evicted. India needs one such a judgment but there is no scope as the judges themselves are corrupt to the core. Just after retirement they expect plum posts and thus follow the rulers in delivering the judgments. See the case of KWDT-II, a highly "Technical Fraud" Award on Krishna River sharing among the three riparian states in favour of Karnataka state. As a reward, the government goes on extending the term of the tribunal and governments are spending huge sum on advocates for years for defending their side versions, though it is a technical issue and not a judicial issue. The reality in India is different.

Role of Climate Change on Urban Flooding

It has become a fashion to man on the street to people occupying the responsible positions in the society or governance or institutions to attribute every event to global warming – also they masque their ignorance by referring it as climate change, which they mean it as de-facto global warming. The fraudulent groups are also poisoning the young minds with such false alarming messages. The fact is that climate change is a vast subject.

To meet the greed, humans have been destroying the natural flow systems and now governments are putting the blame on global warming as it can't defend against such onslaught by politicians and bureaucrats to protect themselves for wrong doings. For all ills [heat waves, cold waves, floods, cyclonic fury, etc.] the easy prey is global warming. Natural variability in rainfall is the main component of climate change that plays the crucial role in urban flooding. However, the impact is controlled with through human interference.

The other important component of urban flooding relates to human interference on nature. Majority of such cases are attributed to weather, as it cannot defend against such attacks on it both by media and by the government agencies. Under natural weather conditions scenario, many a time human greed plays vital role in the loss of human lives and destruction of property and natural resources. A classic case is that of Uttarkhand floods of 2013. However, the urban flooding clears pollution by washing away and by diluting both surface water and groundwater.

December 2015 floods in England and October 2015 floods in Nile Delta in Egypt were attributed to climate change. Even in India the scenario is the same. In all such human greed the easy scapegoat is climate change. On 6th December 2015 received 9-10" rain in three days due to a huge storm "Goliath" in a belt across the central United States, centred just Southwest of St. Louis in the Mississippi River Basin. 1982 flood was similar to 2015 flood. Also, both were winter floods during an El Nino event. They should have been similar but it is not so. After 1982 there were large scale changes in the catchment area with construction activities and thus water level rise by several feet over 1982 flood level, resulting in the damage of 7000 buildings and several other damages and few people died. It is a manmade disaster. February 27, 2019 Mississippi River zone is facing flood fury.

A popular opinion is that the amount of available freshwater is decreasing because of climate change. Climate change has caused receding glaciers, reduced stream and river flow, and shrinking lakes and ponds. Many aquifers have been over-pumped and are not recharging quickly. Although the total fresh water supply is not used up, much has become polluted, salted, unsuitable or otherwise unavailable for drinking, industry and agriculture. To avoid a global water crisis, farmers will have to strive to increase productivity to meet growing demands for food, while industry and cities find ways to use water more efficiently.

A New York Times article, "Southeast Drought Study Ties Water Shortage to Population, Not Global Warming", summarizes the findings of Columbia University researcher on the subject of the droughts in the American Southeast between 2005 and 2007. The findings published in the *Journal of Climate* say that the water shortages resulted from population size more than rainfall. Census figures show that Georgia's population rose from 6.48 to 9.54 million between 1990 and 2007. After studying data from weather instruments, computer models, and tree ring measurements, they found that the droughts were "not unprecedented and result from normal climate patterns" and random weather events. "Similar droughts unfolded over the last thousand years", the researchers wrote, "Regardless of climate change, they added, similar weather patterns can be expected regularly in the future, with similar results." As the temperature increases, rainfall in the Southeast will increase but because of evaporation the area may get even drier. The researchers concluded with a statement saying that any rainfall comes from complicated internal processes in the atmosphere and are very hard to predict because of the large amount of variables.

Climate Change Vulnerability Index

Instead of presenting the impact of climate change in the right perspective, groups of researchers/groups used lump models such as Vulnerability Index, Environmental Vulnerability Index, Climate Change Vulnerability index, etc. by integrating several factors [up to 50 factors]. All these referred Climate Change, but in reality it is not realistically climate change.

The Environmental Vulnerability Index (EVI) is a measurement developed by the South Pacific Allied Geoscience Commission (SOPAC), the United Nations Environmental Program (UNDP) and others to characterize the relative severity of various types of Environmental Issues affected by 243 enumerated Individual nations. Environmental Vulnerability data for the 50 indicators are also divided up in the issue categories for use as required: climate change, biodiversity, water, agriculture and fishing, human health aspects, desertification, and exposure to natural disasters.

Climate Change Vulnerability Index (CCVI) in 2011 was rereleased by global risks advisory firm Maplecroft enables organizations to identify areas of risk within their operations, supply chains in investments. It evaluates 42 social, economic and environmental factors to assess natural vulnerability across three core areas: (i) exposure to climate related natural disasters and sea-level rise; (ii) human sensitivity in terms of population patterns, development, natural resources, agricultural dependency and complexities; and (iii) the index assesses future vulnerability by considering the adaptive capacity of country's government and infrastructure to combat climate change. 84% of the world's fastest growing cities are going to face extreme climate change risks.

In many countries, cities are located in coastal areas, beside rivers, on steep slopes or other risk-prone areas. Infrastructure such as roads, water networks, transmission lines, schools and hospitals providing basic services for urban populations, are vulnerable to extreme climatic events such as floods, storms or landslides. Cities located in tropical coastal areas are particularly vulnerable to cyclones or rising sea levels, the frequency and intensity of which have been on the increase over the past three decades. In addition, salt water intrusion restricts the availability of fresh water in coastal areas, jeopardizing food security as once fertile land becomes barren due to high salt content. Cities located in the hinterland or along rivers may be vulnerable to flooding. Conversely, areas where climate change is expected to reduce rainfall may be affected by drought, shrinking water tables and food scarcity. In urban areas, the poor are the most vulnerable to the effects of climate change, and particularly slum dwellers in developing countries.

Existing climate change models indicate that in warmer climates, overall average rainfall would increase by about one to two per cent per degree of warming (IPCC, 2007). This is caused by increased evaporation leading to increases in rainfall. A warmer atmosphere as a result of global warming can hold more moisture before becoming saturated.

Here, we must remember one important fact: severity, intensity, frequency of occurrence of extreme weather events occurred in the past, occurring now and will occur in future. For this one need to understand the natural variability in climate, more particularly in rainfall. Without such study, harping on rhetoric will not serve the real problem related to climate. Let us the case of Maputo in Mozambique:

Maplecroft observed that the over the past 25 years, Mozambique has suffered from an uninterrupted succession of droughts and floods, with damaging consequences for social and economic development. The most severe drought periods were recorded in 1981-1984, 1991-1992 and 1994-1995; while floods were observed in 1977-1978, 1985, 1988, 1999-2000 and more recently in 2007- 2008. Floods are often exacerbated by cyclones. Since 1970, Mozambique has been hit by 34 significant cyclones or tropical depressions and four major flood events (2000, 2001, 2007 and 2008). In particular, the number of recorded cyclones during the 1999-2000 wet seasons was extraordinarily high and flooding had terrible consequences. In February and March 2000, a combination of torrential rains and tropical cyclones caused the most devastating floods in the history of Mozambique, killing 700 and causing US$600 million worth in damages. These postulations are inaccurate.

Reddy (1987) presented estimation of global solar radiation and evaporation through rainfall over northeast Brazil. Historical carbon dioxide followed ocean surface temperature but carbon dioxide is lagging by temperature. That is when the ocean surface temperature is warmer carbon dioxide from the ocean is released in to the atmosphere and when the ocean temperature is low carbon dioxide from the atmosphere is absorbed by the ocean. Vice-versa is not true. That is ocean temperature is not rising due to carbon dioxide [Reddy, 2019a – page 279, Fig. 15a].

Figure 1 presents the annual temperature variation with annual rainfall over India. 2002 and 2009 are drought years with 0.81 and 0.79 % of average raised the temperature by 0.7 and 0.9 ^{o}c. Formation of cyclonic activity is not related to temperature as they occur in pre-monsoon, monsoon post-monsoon seasons. However, irrespective of temperature, temperature gradient plays important role in the formation of cyclonic storms.

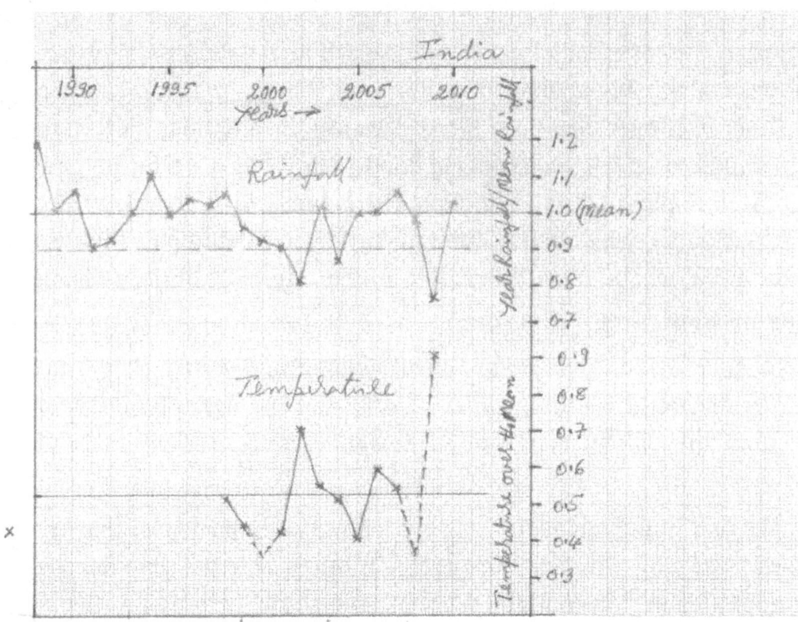

Figure 1 Annual march of rainfall and temperature over India

Figure 2 presents the annual march of southwest monsoon (SWM) and northeast monsoon (NEM) coastal Andhra met sub-division rainfall for 1871 to 1994. They both follow 56-year cycle but in opposite direction [[see dotted lines pattern]. The cyclonic activity follows the NEM rainfall pattern.

I studied the agro-climate scenario of Mozambique and the reports are available in the National Institute of Agriculture Research [INIA] under the Ministry of Agriculture, Maputo. The study of Reddy (1986) presented natural variability in Mozambique rainfall – Reddy & Mersha (1990) presented the same for Ethiopia. However, there is phase shift with coast to inland and with latitude. These are clearly evident from Catuane, Maputo & Beira rainfall with 54 year cycle superposed on it the sub-multiple of 18 years. Table 2 presents information of these three Mozambique met stations rainfall

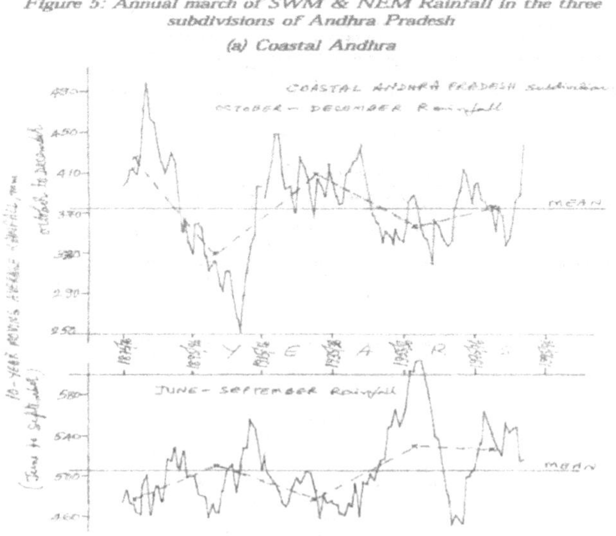

Figure 2 Annual march of SWM & NEM rainfall of Coastal Andhra met sub-division

Figures 3 (a&b) presents the annual march of rainfall over Catuane in Mozambique and Durban in South Africa [Reddy 2019a – pages73 & 74, Figs. 4.1 & 4.2]. Table 2 presents the starting and ending years of 54 year cycle superposed on it the sub-multiple of 18-years for Catuane, Maputo & Beira in Mozambique and Durban of 66-year cycle superposed on it the sub-multiple of 22 years in South Africa. Cyclone Idai caused severe blow to Sofala-Beira reaching Zimbabwe and Malawi on 14th March 2019 killing around 300 people. The flood and drought years mentioned above follow the 54 year cycle of Beira [Table 2]. These publications are available in INIA, Maputo/Mozambique. However, the droughts and floods are not uniform over different parts of the country as the natural variability changes with coast to inland and with latitude. The recent droughts in southern South Africa are clearly seen in Durban cyclic pattern, wherein 2010 to 2076 of 66 years, the 1st part is W and the second part is M. That is, in 2010 to 2032 of W, 14-5-14 years present dry-wet-dry years in 33 year period.

Table 2 Catuane, Maputo & Beira information

Station	Average Rainfall (mm)	Cycle 1	Cycle 2	Cycle 3
Catuane	low rainfall (620)	1943-1996	1997-2050	2051-2104
Maputo	medium rainfall (900)	1925-1978	1979-2032	2033-2086
Beira	high rainfall (1480)	1931-1984	1985-2038	2039-2092
W =	1985-1995(-)	1996-2000(+)		2001-2011(-)
M =	2012-2022(+)	2023-2027(-)		2028-2038(+)
Durban	medium (1050)	1876-1942	1943-2009	2010-2075
W =	2010-2023(-)	2024-2028(+)	2029-2042(-)	
W =	2043-2056(+)	2057-2061(-)	2062-2075(+)	

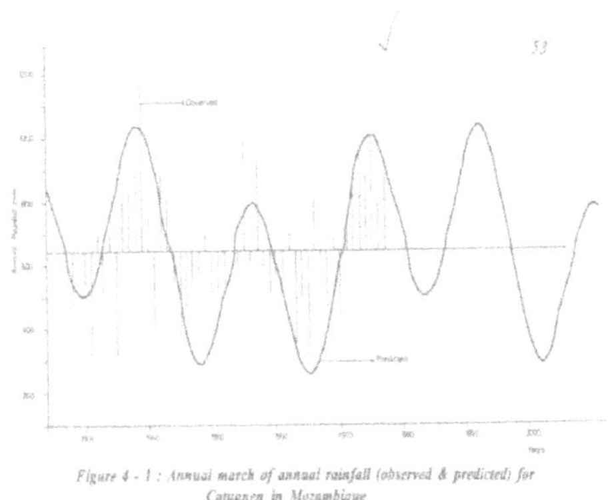

(a) Catuane/Mozambique

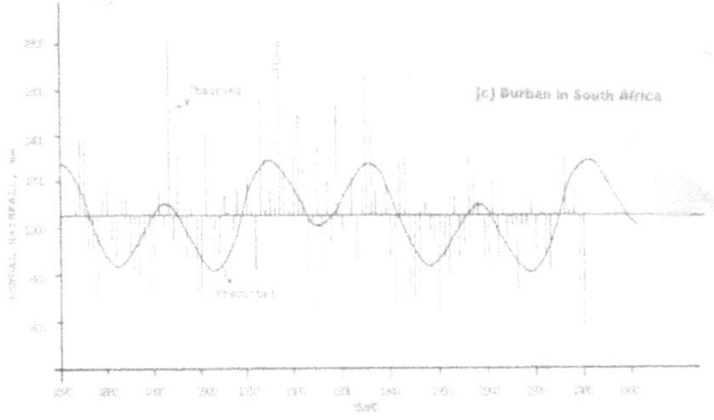

(b) Durban/South Africa

Figure 3 (a&b) Annual march of Rainfall [observed and predicted] (a) Catuane and (b) Durban

Vulnerability of Indian Agriculture to Climate Change

When the global warming studies are in the birth stage, I carried out agro-climatic and agrometeorological studies for India, northern Australia, Mozambique, Ethiopia, Upper Volta [now Burkina Faso] and Senegal countries. All these I compiled in a book form [earlier they were published in Journals of international repute] (Reddy, 1993). Now the second edition is out from the publisher (Reddy, 2019) with the same title. Scientists from Pune University from India carried out the analysis for Maharashtra State in India in 2009 [Akumunchi Anand, et al., 2009]; a student did Ph.D. from Pune University on Bhīma Basin in Maharashtra state earlier to this under my guidance. Such analysis provides the level of sustainability – drought risk. These are linked to local/regional rhythmic pattern in rainfall.

Droughts and floods are part of natural variability in precipitation and are modified by local conditions and method used to define them. Rao, et al. (2013) presented elaborate study and prepared an Atlas on "Vulnerability of Indian Agriculture to Climate Change". The study has not really studied the climate change but used it as an adjective. Same is the case with Raja, et al. (2014) study though they made some efforts in the rainfall variability [referred my agro-climatic model]. Rao, et al. (2009) report presented a map of drought proneness. It showed except coastal Gujarat, all parts of India present less than 25%. In my work cited above [Reddy, 1993, 2019] they reach as high as 60% in the Eastern side of Western Ghats – Anantapur-Bellary-Sangly belt. Figure 4 [a, b, c] presents the drought proneness maps from the three publications.

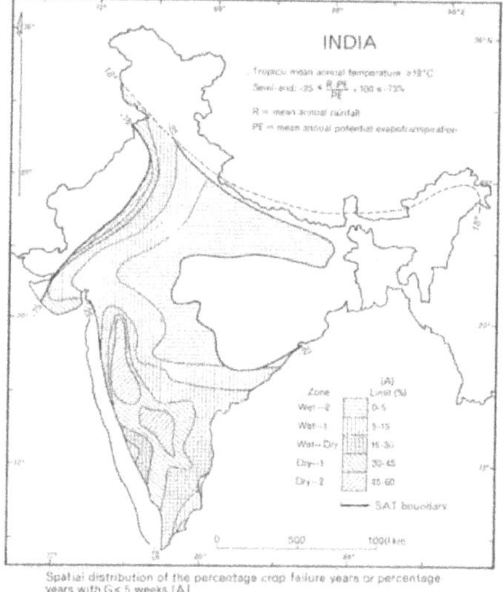

Spatial distribution of the percentage crop failure years or percentage years with G≤5 weeks [A]

(a) India by Reddy [1993]

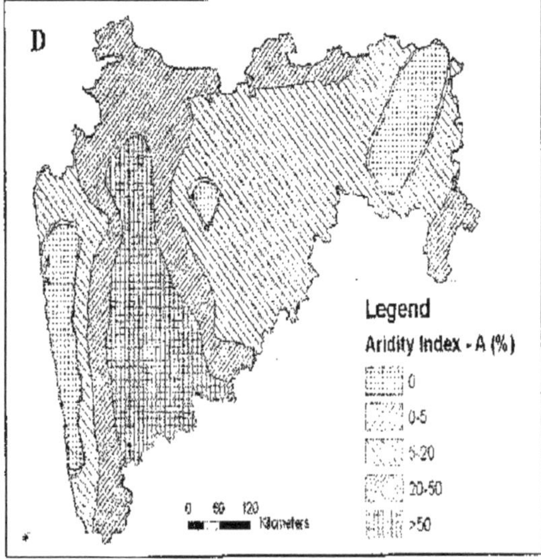

(b) Maharashtra by Akumunchi Anand, et al. [2009].

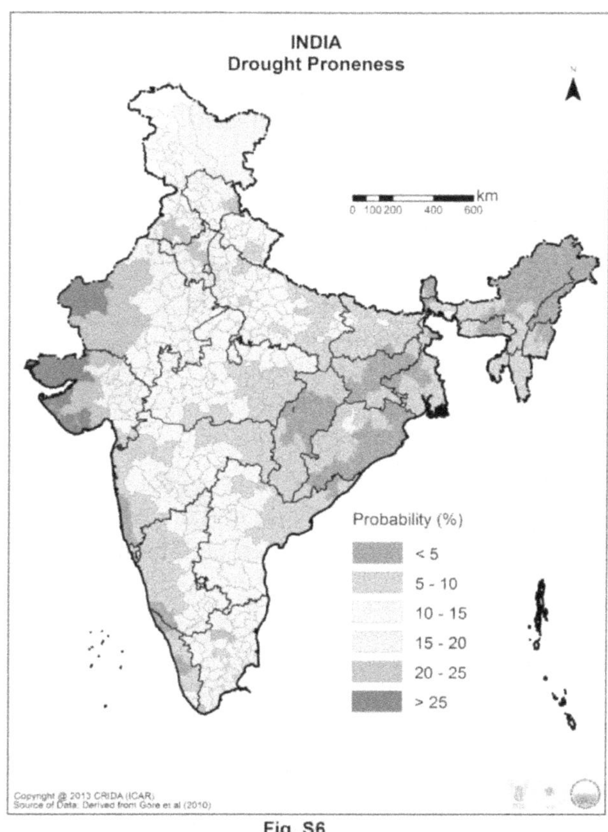

(c) India by Rao, et al. [2013]

Figure 4(a,b&c) Drought proneness maps

Figure 5 presents the drought proneness variation with 56 year cycle for Kurnool in AP [S = week of sowing rains, G = available effective rainy period in weeks]. The drought proneness is on an average is 45% with 35% and 70% respectively during the above and below the average cycle parts.

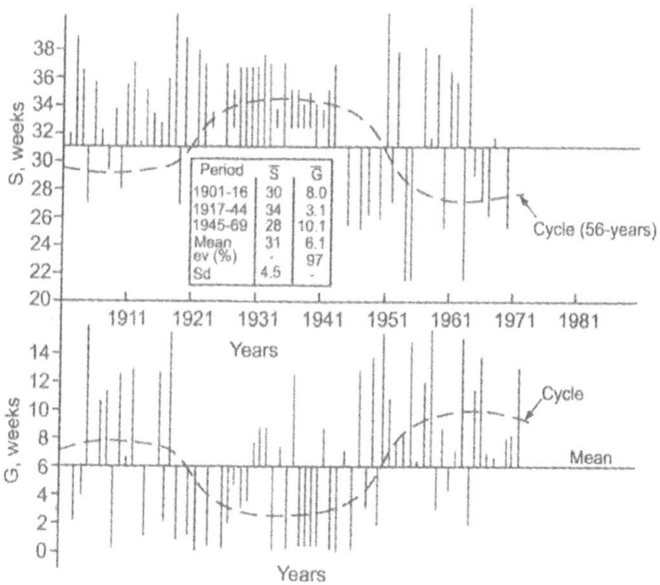

Figure 5 Annual march of agroclimatic variables: G & S

In agriculture sense, climate change relates to water availability from rainfall and snowfall; and destruction of production through floods and cyclonic activity – they are region specific. However, they are the major sources for water availability at local and regional level. Normal rains rarely work for sustainability of dams/reservoirs, tanks, etc. In the case of human interference, the major issues are the sharing of water and pollution – domestic, industrial and agricultural.

Natural Variability part of Climate Change

Climate Change is a vast subject. It includes both natural variability and human induced variations. Natural variability in rainfall form part of it. This is the principal component of climate change. However, agencies like World Bank harping on the global warming carbon credit even after IPCC in its AR5 report categorically said no. Also World Bank and other agencies [national & international] giving hype to such reports and sensationalizing. Here modellers playing spoil sport as such studies are far from realities.

We have seen now a day media bombarding with all sorts of stories on temperature. In fact if we look at the highest temperature recorded in February to June in Hyderabad they are given as: 37.2, 42.2, 43.3, 44.4 & 43.9 °C. Then highest average for the February month is 35.3 °C. The current weather scenario in Hyderabad has not deviated from these past occurrences. Also, there is a theory on high temperature built up in north western parts of India, "monsoon will be early".

In fact Earth's climate is dynamic and it is always changing through the natural cycles. What we are experiencing now is part of this system only. They are highly region specific systems. Floods and droughts are part of this natural variability component of climate change in rainfall. Extreme rain spells are also part of this. Figure 6 presents the annual rainfall (June to May) at all-India from 1871-72 to 2014-15. This data series presents the 60-year cycle with zero trend which shows no human induced component. Floods in rivers follow the natural rhythm in rainfall. Table 3 presents the frequency of occurrence of high magnitude floods in north western Rivers following the 60-year cycle presented in Figure 6. To a question raised in Indian Parliament, IITM & IMD scientists prepared a reply saying that Indian rainfall is decreasing. They arrived at this answer based on the data of one 60-year cycle [sine curve – one up and then one down]. If they would have shifted 30 years forward, then the conclusion would have been that "Indian rainfall is increasing".

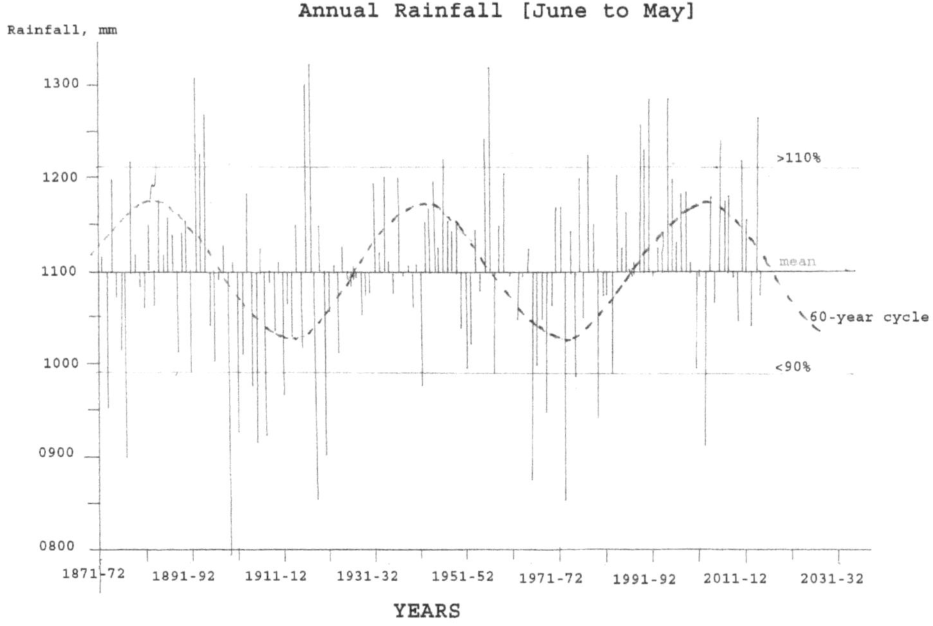

Figure 6 All India Annual Rainfall [Observed, vertical lines & Predicted, dotted curve]

Table 3 Frequency of occurrence of high magnitude floods in few northwest Indian Rivers

	Frequency of high magnitude floods*		
River	**Period**	**Frequency**	**Climatic cycle**
Chenab1962-1990	1 in 9 years	(a) below the average	
1990-1998	1 in 3 years	(b) above the average	
Ravi	1963-1990	1 in 14 years	(a)
1990-1998	1 in 3 years	(b)	
Beas	1941-1990	1 in 8 years	(a)
1990-1995	1 in 2 years	(b)	

*State of Environment Report, India – 2009, MoEF/GoI : The frequency of floods in India is largely due to deforestation in the catchment area, destruction of surface vegetation, changes in land use, increased urbanization and other developmental activities – this is a false statement but it is more in association of cyclic variation in rainfall.

Figure 7 presents [a - upper] the 132 year cycle in annual rainfall of AP. Bangalore was under severe drought during 1876-78 [Figure 8] which is clearly seen in annual rainfall [Figure 8a] wherein the rainfall was less than 50% of the average with four successive years with below the average. Figure 7 also presents [b - lower] the water flows in Krishna River. It follows the rainfall pattern. The current below the average part started in 2001. This can be seen from the water reached Srisailam dam during 2009-10 to 2018-19 in tmc ft. They are: 1222, 1028, 736, 197, 848, 614, 59, 345, 489, & 562. This presents the character of low rainfall after 2001 [Figure 7a].

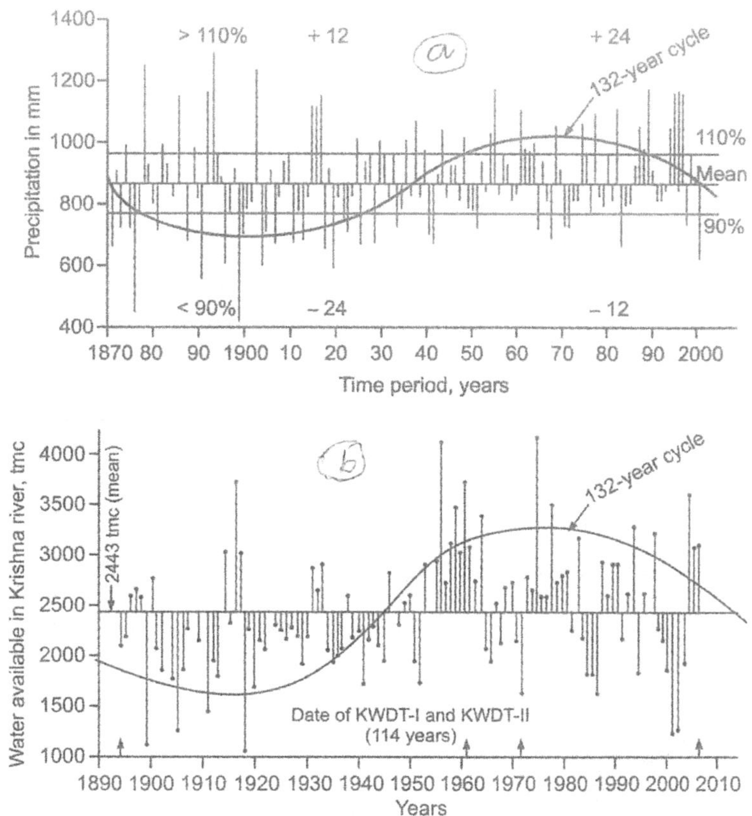

Figure 7 (a) Andhra Pradesh Annual Rainfall & (b) Annual Water Availability in Krishna River

Figure 8 1876-78 severe drought impacts on Bangalore

Natural Floods

Areas with high rainfall, cyclonic activity, sloppy terrain, etc. are affected by floods. A classic example of natural floods is the floods prior to 1930 in Musi River in Hyderabad. On September 26, 1908, due to cyclone in the Bay of Bengal Musi River catchment area received over 48 hours 98.57 cm of rainfall. Of the 788 tanks and lakes in the river basin, 221 breached due to cloudburst. At 11 a.m. on September 28, an estimated 4.25 lakh cusecs of water raced through the city swirling away over 19,000 homes, and killing about 15,000 people. However, some other reports put the death toll at 55,000. Figure 9 presents a scene of floods in Hyderabad. The arched gateway of the then British Residency, now OU Women's College, was partially in water, which never again saw a flood of such proportion. A 200 year old Tamarind Tree in Osmania Hospital premises saved 150 people who climbed on it.

Figure 9 A Picture of 1908 September Floods in Hyderabad

Human Induced Floods

Natural floods associated with heavy rainfall are contained with building of reservoirs in many river basins. For example, after 1908 September devastating floods to Musi River, the then ruler built two reservoirs, namely

Osmansagar on Musi River and Himayatsagar on Musi's tributary Esi to contain Musi floods. With these two reservoirs Hyderabad has not experienced natural floods after 1930. Now inflows also have comedown drastically from upstream catchment area due to human actions. In the case of Hyderabad after 1930 the floods are associated with human actions on nature. They are caused by:

- Hussainsagar Lake was built in1562 [Hyderabad was built in 1591]. The extent of the lake covered 5.7 sq. km. "Save Lakes for a Better Future" released on 4-5-2000 showed the present area was 1356 acres; "Restoration and Management of Hussainsagar waters" showed it as 1066 acres; and now it is around 700 acres. The depth of the lake was 32' and now it is less than 25 ft due to silting and immersion of different kinds of idols over years;
- In 2001 "Water Conservation Committee" got a report prepared by NRSA, Hyderabad on lakes in Huda area. According to this report in Huda area there were 932 tanks spread over 51,480 acres. Now, more than half of them were encroached and this process of encroachment is continuing. The remaining are partially encroached and are now filthy lakes;
- More than 80% of the rainwater drains/nalas were encroached;
- More than 50% of Musi River was encroached and still continued the encroachments;
- With the growing population, water has been brought from faraway places. Around 75% of them along with the groundwater used is turning in to wastewater. Thus we add daily around 2000 mld of water in to drains/waterbodies and finally in to Musi River. This will steadily rise with the time. This amount adds to rainfall.

The huts are built on drains and in water bodies; and solid and liquid wastes generated by those hut dwellers have been dumping in to the nalas and lakes. Domestic solid waste that includes plastics has been dumping in to nalas and water bodies. After desilting, the waste is not moved away and thus finally joins the same water bodies/nalas. Sewage treatment plants generate huge solid waste. During rainy season the large part joins the water bodies, nalas/drains and Musi. Industrial solid waste is another part of it. In India produced around 15,000 t/day of plastic in 2015. 70% is going as waste. Figure 10 presents plastic waste in drains in Hyderabad & Delhi and similar scenarios can be seen in all urban areas in India.

(a) Hyderabad [Lingampalli to BHEL]

(b) Delhi

Figure 10 Nalas choking with the plastic waste in (a) Hyderabad & (b) Delhi

With drastic reduction in the water holding capacity of Hussainsagar & drains that carry water from Hussainsagar Lake to Musi, in 2000 the flood water reached 2nd floor in Ashok Nagar. Flood water carried cars with it.

Figure 11 presents the floods in 2016 in Hyderabad. In Durgam Cheruvu area, where drainage system non-existing & tanks/lakes were enriched in upstream areas and thus sceptic tank waste was carried by flood water to the 1st floor of the houses that were built illegally in Durgam Cheruvu FTL and Buffer zone. Military personal helped stranded people in flood water were rescued through boats. During 2000 floods ICRISAT Colony-Phase-I was under flood water as the three lakes were encroached. Boats were used to rescue residents.

Figure 11 Hyderabad floods in September 2016

Figure 12 presents a case of Chennai floods in November 2015. There is no unusual scenario with the November 2015 rains. The main problem here is that governments over the years allowed the people as well government agencies to destroy the rainwater carrying channels and water bodies/rivers. This resulted the aggravating the impact for the same flood capacity. Added to the rains is huge quantity of the domestic sewage, which is increasing year by year. We have seen above the population increase in urban Tamil Nadu. Similar scenario was the case with Nellore floods.

Figure 12 Chennai Floods in 2015

Tamil Nadu government appointed a committee to evaluate Chennai 2015 floods. The report highlighted the encroachments of three rivers, namely Gooum, Adiyar and Kasasthalayar and flood channels/river beds, release of water from overflowing reservoirs, etc. It received 1200 mm in November 2015. More than 500 people dead and more than 1.8 million displaced due to flooding of Chennai.

Nellore to Chennai receives rains during the Northeast Monsoon Season [October to December] that coincides with severe cyclonic activity. The highest rainfall in 24 hours (mm)/year and the highest per month (mm)/year for Nellore and Chennai from 1931-60 Normal (Red Book of IMD) are given below:

Nellore – 444/1950, 357/1936 & 189/1902; 647/1920, 982/1915, 494/1946

Chennai – 234/1888, 236/1922 & 262/1901; 892/1943, 1088/1918, 699/1946

Natural + Human Induced Floods

The severity of destruction changes with the time of the year, the terrain, with the population growth, and growth in infrastructure. However, with the violation of existing local, state and national laws the rainfall based destruction is aggravated. Here human factor plays crucial role on the death toll and destruction. Let us see few examples.

Kurnool Floods: Natural floods are associated with heavy rainfalls in the catchment areas of the rivers flowing through urban areas. 2009 floods in Krishna-Tungabhadra-Hundri Rivers that inundated Kurnool town and Mantralayam in Kurnool district causing heavy toll on humans and property. These are associated with human negligence. Even with the early warning, water from Srisailam dam [reached 896.5 ft, with capacity 885 ft] was not released to Nagarjunasagar Dam in the downstream. This is basically because of poor water use practices in Nagarjunasagar dam. This dam was full, which includes carryover water for the next deficit year [if any] but they reduced 590 to 536 ft by the flood time. This led animosity between the people of these two dams that caused severe damage to humans and property in the upstream of Srisailam dam. Also, government sanctioned money to build karakattas at Kurnool along the Tungabhadra River bank. The officials did not do their job and on the contrary they showed the compound wall at Saibaba Temple as karakatta [I was on TV channels discussions from morning to night]. Figure 13 presents Kurnool Floods around Kondareddy Burju.

Figure 13 Kurnool Floods in 2009

Uttarkhand Floods: Figure 14 presents the classic example of human greed wherein in violation of laws built commercial complexes in the river bed by cutting the bank. June 2013 a multi-day cloudburst centered on the north Indian state of Uttarkhand caused devastating floods and landslides. The main day of the flood was June 16, 2013. The rainfall recorded was 375%.

Figure 14 Uttarkhand Floods in 2013

However as seen in Figure 14, it is a manmade disaster with the government's apathy. Even the disaster management was taking rest until everything was over. People illegally filled with rubbles the sloping river bed on bankside and built commercial buildings on this. With the heavy upstream rains, the guessing river water washed away these commercial complexes along with 10,000 people. Though in the upper regions pilgrims were also affected by the heavy rains, but the casualty was not that severe and thus deaths are not that high.

Srinagar Floods: It is also a manmade disaster in the presence of heavy rains. On the 5th September, the Jhelum River in Srinagar was reported to be flowing at 22.4 ft which was 4.40 ft above the danger mark. Ironically the entire flood basin that saved Srinagar in 1893 from greater devastation was vandalized through mindless planning for urbanization. Most of the southern waters of the Dal Lake were drained through "Nallar Mar" (Serpentine Canal) that went round the city and drained all its surplus water in to the Jhelum. The impact of rainfall is aggravated with human interference in both urban areas and catchment area. In fact when we fill the river banks, the width of the river reduces and thus with heavy upstream rains, water overflows on either side. This is clearly seen in Figure 15 wherein flood water entering the city on either side.

Figure 15 Srinagar Floods in 2014 – Jhelum River

Mumbai Floods: Mithi River is the main river system that used to carry rainwater in to the Sea. This has been filled with rubbles and waste around 62% and converted in to concrete jungle. The protective Mangroves were destroyed and thus water that enters the river moves in to urban sprawl in Mumbai. Also several lakes were converted in to real estate ventures. Added to this is the infiltration has been gradually come down with urbanization roads & buildings. With high percent of urban population & slum dwellers, sewage component gradually increased. Figure 16 presents an aerial scenario of flooding disaster in Mumbai during July 2018 rains.

Kerala Floods: It is not unusual to Kerala. In 2013, June 1 to August 15 received 2087 mm with normal of 1606 mm – 30% excess. In 1924, received the highest is the recorded history. A total of 3368 mm rainfall has been received and most parts of Kerala submerged. Kerala received heavy monsoon rainfall on the mid evening of August 8, resulting in dams filling to capacity were opened. In the first 24 hours received 310 mm. 80 reservoirs were full. According to district-wise IMD data for rainfall the 2018 monsoon, 45% of the districts of the country are facing deficit rain. But Kerala 12 out of 14 districts received heavy rainfalls and floods. Figures 17 presents the aerial Photo of flood impact. Table 3 presents the rainfall figures.

Figure 16 Aerial scenario of flood devastation in Mumbai during July 2018

Figure 17 Kerala Floods Aerial Photo [Periyar River floods in Aluva]

Table 3 Actual and observed rainfall

Period (mm)	Actual Rainfall (mm)	Normal Rainfall (% normal)	Actual Rainfall
June 2018	750	650	15
July 2018	857	726	18
1-19 August	759	288	164
Total	2347	1649	42

Excessive mining, quarrying, use of land for non-forest purposes and construction of high rises in the catchment area of rivers & river beds created landslides and heavy flows and as a result reservoirs were filled with debris and water. This is on one side. Around 10% of the area of Kerala is below the sea level and thus on the other side sea water entered the rivers mouths and pushed back the flood waters from the reservoirs overflows. 80 reservoirs gates raised and water released.

Summary & Conclusions

The rainfall extremes are not new under natural variability [climate change] in rainfall. The cyclic nature of rainfall presents periods of droughts and floods. However they present high variation with space and time in association with local/region specific general circulation patterns in different seasons. The severity of floods impact changes with location [coastal, hilly and inland]. The severity of urban floods is aggravated with human interference on nature to meet their greed at the cost of environment. The wastewater [polluted water] generated in urban areas that relates to population and slum population create flood condition even with little rains. That is, even if we contain the flood fury related to natural variability, this will help in creating local floods.

Humans are the major source of urban flooding in Indian cities. To minimize this, as a major first step, governments must put a break on rural to urban migration of population and thus slum population. This can be achieved by bring out new agricultural policy & developing appropriate food processing units, completing smaller irrigation projects that give immediate results, developing education & health care systems. Etc. Around 50% of total subsidies/incentives in India has been going in to the pockets of unscrupulous people. By saving such money, storage facilities could be built at village/mandal level.

To reduce the urban flooding treat and re-use wastewater locally & protect water resources, start treating wastewater from upstream side and not from downstream side, see more rainfall infiltrates in to the ground, bring out a law to punish the encroachers of natural water resources, etc. Unplanned urbanization raises temperature and thus increases power consumption and water demand.

References

Akumanchi Annand, et al., 2009. Agro-climatic Zonation of Maharashtra State Using GIS. Trans. Inst. Indian Geographers, 31(1).

Raja, R., Panda, B.B. & Nayak, A.K., 2014. 21. Analysis of meteorological variables for understanding climate change/variability and planning of adaptation practices. (Eds.)Mohammad Shahid, Pratap Bhattacharya and A.K. Nayak, "Advanced Technologies for Assessment of soil health, GHG emissions and Carbon sequestration in Rice under Changing Climatic scenario and mitigation strategies", 11Nov. – 1 Dec. 2014, Crop Production Division, ICAR-CRRI, Cuttack, pp.203-212 [Training Manual, ICAR Sponsored Winter School].

Rao, C.A.R., Raju, B.M.K., Rao, A.V.M.S., Rao, K.V, Rao, V.UVV.M., Kausalya Ramachandran, Venkateswarlu, B. & Sikka, A.K., 2013. Atlas on Vulnerability of Indian Agriculture to Climate Change, National Initiative on Climate Resilient Agriculture (NICRA), Central Research Institute for Dryland Agriculture (CRIDA/ICAR), Hyderabad, India.

Reddy, S.J., 1986: 'Climatic fluctuations in the precipitation data of Mozambique during meteorological record', Comm. No. 39, Series Terra e Agua, INIA, Maputo, Mozambique, pp.40.

Reddy, S.J., 1987: 'The estimation of global solar radiation and evaporation through precipitation', Solar Energy, 38:97-104, Ethiopia, pp.28.

Reddy, S. J., 1993: 'Agroclimatic/Agrometeorological Techniques: As Applicable to Dry-land Agriculture in Developing Countries', www.scribd.com/Google Books, 205pp. [Book Review: Agricultural and Forest Meteorology, 67: 325-27, 1994].

Reddy, S.J., 2019a: 'Agroclimatic/Agrometeorological Techniques: As applicable to Dry-land Agriculture in Developing Countries [2nd Edition]', "Brillion Publishing", New Delhi, 372p.

Reddy, S.J., 2019b: 'Workable "*Green*" Green Revolution: A Framework [Agriculture in the perspective of Climate Change]', "Brillion Publishing", New Delhi, ---p.

Estimation of Runoff and Soil Loss in Conservation Agriculture Treatments under Rainfed Condition

R. S. Patode, M. S. Khakare, M. B. Nagdeve, M. M. Ganvir and V.V. Gabhane
All India Coordinated Research Project for Dryland Agriculture
Dr. Panjabrao Deshmukh Krishi Vidyapeeth, Akola (M.S.)

ABSTRACT

Conservation agriculture is characterized by three principles which are linked to each other, namely; continuous minimum mechanical soil disturbance, Continuous organic soil cover, Diversified crop rotations in the case of annual crops or plant associations in case of perennial crops. Excessive tillage of agricultural soils may result in short term increases in fertility, but will degrade soils in the medium term. Soil erosion resulting from soil tillage has forced us to look for alternatives and to reverse the process of soil degradation. The logical approach to this has been to reduce tillage. The experiment on Conservation Agriculture was undertaken at AICRP for Dryland Agriculture, Dr. PDKV, Akola. The results obtained during 2018-19 are presented in this paper. The highest total runoff of 41.6mm was observed in conventional tillage (CT) without crop residue mulch treatment (T_2) and lowest total runoff of 25.2mm was observed in Zero tillage + crop residue treatment (T_4) followed by permanent BBF furrow after every 4 rows + crop residue mulch treatment (T_5), Reduced tillage (T_3) and conventional tillage (CT) with crop residue mulch treatment (T_1) . Runoff in conventional tillage without crop residue mulch treatment (T_2) was 65.10% more than that of Zero tillage + crop residue treatment (T_4). Also the Zero tillage + crop residue treatment (T_4), permanent BBF furrow after every 4 rows + crop residue mulch treatment (T_5) and Reduced tillage (T_3) has less soil loss (0.8, 0.9 and 1.4tons ha^{-1}) as compared to conventional tillage without crop residue mulch (1.9tons ha^{-1}) and with crop residue mulch treatments (1.7tons ha^{-1}).

Keywords: Runoff, tillage, mulch, soil loss.

Introduction

Many regions in the country are characterized by variable and low rainfall and the soils have low productivity. The fragile ecosystems in the dry areas are prone to degradation. Minimum tillage is an ecological and science-based approach for resource conservation and sustainable production. Tillage is one of the important components to improve soil conditions and conserve water. Sub-optimal soil conditions have significant impact on soil-water and nutrient regime, gaseous exchange between soil and atmosphere and crop growth. The term minimum tillage is often used synonymously with conservation tillage which retains productive amount of mulch on the surface. Minimum tillage includes reduced frequency and intensity of tillage operation, use of those implements that loosen the soil without turning over and do not excessively pulverize it and perform the needed tillage operations when soil conditions are within the optimum soil condition range to produce the desired tilth. It facilitates intensive cultivation with minimum risk of degradation. Conservation Agriculture is a concept for resource-saving agricultural crop production that strives to achieve acceptable profits together with high and sustained production levels while concurrently conserving the environment and maintaining soil health. It is based on enhancing natural biological processes above and below the ground.

Methods

In *Kharif* - Soybean

T_1: Conventional tillage (CT) - Ploughing once in 3 years + 2 pre-sowing harrowings + One hand weeding + Opening of furrow with hoe in each row at 30-35 DAS + Crop residue mulch

T_2: Conventional tillage (CT) - Ploughing once in 3 years + 2 pre-sowing harrowings + One hand weeding + Opening of furrow with hoe in each row at 30-35 DAS without crop residue mulch

T_3: Reduced tillage (RT) – Broad bed and furrow every year + Pre and post emergence herbicide application + crop residue

T_4: Zero tillage + crop residue

T_5: Permanent BBF furrow after every 4 rows + crop residue mulch

In *Rabi* –Chickpea

T_1- Conventional tillage (CT)- Pre sowing harrowing + One hoeing + One hand weeding + Crop residue mulch

T_2- Conventional tillage (CT) - Pre sowing harrowing + One hoeing + One hand weeding + No crop residue mulch

T_3- Reduced tillage (RT) – Pre sowing harrowing + Broad bed and furrow every year + Pre-emergence herbicide application + Crop residue mulch

T_4- Zero tillage + crop residue

T_5- Permanent Broad bed and furrow + Pre-emergence herbicide application + crop residue mulch

Operations

- Sowing of soybean (JS-9305) and RDF application has been done in all treatment plots.
- Pre emergence herbicide application was done as per treatments.
- Runoff and soil loss monitoring Gadgets has been installed and monitoring was done.
- Post emergence herbicide application in T_3, T_4, T_5 was done.
- Hoeing as per treatment has been given.
- Treatment wise opening of furrow (T_1 & T_2) in each row has been done.
- Harvesting of soybean was done on 12/10/2018.

Results

The experiment on Conservation Agriculture was undertaken during 2018-19 at AICRP for Dryland Agriculture, Dr. PDKV, Akola. The results obtained are presented here.

Runoff and soil loss

The runoff and soil loss observed in different treatments is given in Table 1. During the season total 6 runoff events were occurred out of which only two events were major. The highest total runoff of 41.6mm was observed in conventional tillage (CT) without crop residue mulch treatment (T_2) and lowest total runoff of 25.2mm was observed in Zero tillage + crop residue treatment (T_4) followed by permanent BBF furrow after every 4 rows + crop residue mulch treatment (T_5), Reduced tillage (T_3) and conventional tillage (CT) with crop residue mulch treatment (T_1) . Runoff in conventional tillage without crop residue mulch treatment (T_2) was 65.10% more than that of Zero tillage + crop residue treatment (T_4). Also the Zero tillage + crop residue treatment (T_4), permanent BBF furrow after every 4 rows + crop residue mulch treatment (T_5) and Reduced tillage (T_3) has less soil loss (0.8, 0.9 and 1.4tons ha^{-1}) as compared to conventional tillage without crop residue mulch (1.9tons ha^{-1}) and with crop residue mulch treatments (1.7tons ha^{-1}).

Table 1 Runoff and soil loss during the *Kharif* season 2018 for soybean crop

Date	Rainfall, mm	Runoff, mm					Soil loss, tons ha^{-1}				
		T_1	T_2	T_3	T_4	T_5	T_1	T_2	T_3	T_4	T_5
12/07/2018	27.7	3.4	3.9	3.3	2.8	3.0	0.17	0.20	0.15	0.11	0.12
13/07/2018	35.8	5.4	6.0	4.7	3.5	4.2	0.38	0.43	0.30	0.15	0.18
22/07/2018	25.3	3.3	3.7	3.2	2.4	2.8	0.30	0.34	0.28	0.18	0.22
17/08/2018	78.7	10.5	12.0	9.8	8.1	9.4	0.40	0.47	0.35	0.22	0.26
20/08/2018	37.2	6.0	6.6	4.9	3.4	4.6	0.27	0.31	0.20	0.08	0.11
21/08/2018	51	8.4	9.3	6.3	5.0	5.9	0.15	0.17	0.09	0.05	0.06
Total	255.7	37.1	41.6	32.2	25.2	29.8	1.7	1.9	1.4	0.8	0.9

Soil Moisture

The soil moisture at the depths 0-15 and 15-30cm is given in Table 2. The soil moisture status observed was good during vegetative and flowering stage and very less during maturity stage of crop growth. It was better in all treatment combinations during flowering stage of crop growth at the depths 0-15 and 15-30cm. The soybean crop was under mild moisture stress initially at pod formation/ seed initiation due to dry spell of 23 days but later on it recovers.

Table 2 Soil moisture content at different crop growth stages recorded at 0-15 and 15-30cm depth

Treatments	Depth (cm)	Soil Moisture Content (%)		
		Vegetative stage (30/07/18)	Flowering stage (28/08/18)	Maturity stage (29/09/18)
T_1	0-15	30.45	30.80	25.42
	15-30	32.30	32.70	26.26
T_2	0-15	30.40	30.55	25.14
	15-30	31.23	32.05	25.94
T_3	0-15	32.15	33.12	27.10
	15-30	33.30	34.06	28.54
T_4	0-15	31.80	32.00	26.05
	15-30	32.70	33.75	27.15
T_5	0-15	31.85	32.11	26.12
	15-30	33.10	33.86	27.20

Yield of *Kharif* soybean

Treatment wise data in respect of crop growth and productivity was given in Table 3. It was observed that the treatment effects on growth and yield attributes was significant. In terms of grain yield of soybean the treatment T_3 was found significantly superior over other treatments and is at par with treatment T_1. In terms of straw yield of soybean the treatment T_3 was found significantly superior over other treatments and is at par with treatment T_1. The rainwater use efficiency as influenced by different treatments is given in Table 4. The rainwater use efficiency was observed to be higher in the treatments T_3 and T_1 as compared to other treatments.

Table 3 Growth, yield attributes and productivity of soybean as influenced by different treatment combinations

Treatments	Plant height (cm)	Number of pods/plant	Grain weight (g plant^{-1})	Grain yield (kg ha^{-1})	Straw yield (kg ha^{-1})
T_1	30.20	29.2	7.81	1982	2457
T_2	28.35	28.1	7.69	1900	2302
T_3	29.88	30.6	8.07	2140	2673
T_4	27.60	26.6	7.51	1879	2229
T_5	27.05	26.3	7.12	1806	2141
S. E. (m)	0.299	0.377	0.169	63.10	70.67
C.D. at 5%	0.931	1.17	0.527	196	220

Table 4 Rain water use efficiency as influenced by different treatments

Treatments	Grain yield (kg ha^{-1})	Rain water use efficiency (kg ha^{-1}mm^{-1})
T_1	1982	2.50
T_2	1900	2.41
T_3	2140	2.68
T_4	1879	2.33
T_5	1806	2.24

Inferences

Runoff in conventional tillage without crop residue mulch treatment (T_2) was 65.10% more than that of Zero tillage + crop residue treatment (T_4). Also the Zero tillage + crop residue treatment (T_4), permanent BBF furrow after every 4 rows + crop residue mulch treatment (T_5) and Reduced tillage (T_3) has less soil loss (0.8, 0.9 and 1.4tons ha^{-1}) as compared to conventional tillage without crop residue mulch (1.9tons ha^{-1}) and with crop residue mulch treatments (1.7tons ha^{-1})

Acknowledgments

The authors are thankful to the Director and Principal Investigators, ICAR-CRIDA, Hyderabad, Government of India for their financial support under Consortia Research Platform (CRP) on Conservation Agriculture Project.

References

Ghazanfar, M., M.A. Asoodar and K.A. Saeed. 2010. The effect of conservation tillage, planting and irrigation methods on water use efficiency and wheat grain yield in North of Ahwaz. *Journal of Engineering Apllied Science*, 5(2), pp. 101-105.

Gilley, J.E. and L.M. Risse. 2000. Runoff and soil loss as affected by application of manure. *Trans ASAE* 43(6), pp. 1583 - 1588

Goamez, J.A., J.V. Giraldez, M. Pastor and E. Ferere. 1999. Effect of tillage method on physical properties, infiltration and yield in an olive orchard. *Soil & Tillage Research,* 52:167-175.

Saha, S., D. Chakraborty, A.R. Sharma, R.K. Tomar, S. Bhadraray, U. Sen, U.K. Behera, T.J. Purakayastha, R.N. Garg and N. Kalra. 2010. Effect of tillage and residue management on soil physical properties and crop productivity in maize (*Zea mays*) Indian mustard (*Brassica juncea*) system. *Indian Journal of Agricultural Sciences* 80 (8): 679-685.

Wang, X.B., D.X. Cai, W. B. Hoogmoed , O. Oenema, U.D. Perdok. 2007. Development in conservation tillage in rainfed regions of north china. *Soil & Tillage Research,* (93), pp. 239-250.

Wiedenfeld, B. 2006. Conservation vs conventional tillage, fall double cropping and cover crop effects on crop water use in subtropical south texas. *Conservation Systems Conference*, Amarillo TX, June 26-28.

Yao, Y. 2004. Influence of tillage practices on yield, water conservation and soil loss results of field experiments in the Eastern Loess Plateau. *International Conserving Soil and Water for Society.*

Preserve: A Novel Product of Nagpur Mandarin

S. R. Patil[1], Anjali Khade[2], S. D. Deshmukh[3] and Premlata M. Chandan[4]

[1]Professor, [2]M. Sc. Student, [3]Ph.D. Scholar, [4]Assistant Professor

Department of Fruit Science, Faculty of Horticulture

Dr. Panjabrao Deshmukh Krishi Vidyapeeth, Akola

srpatil1812@rediffmail.com

ABSTRACT

An experiment entitled "Studies on preservation of Nagpur mandarin segments in sugar syrup" was carried out during the year 2017-18 at Post Harvest Technology Laboratory, Department of Horticulture, Post Graduate Institute, Dr. Panjabrao Deshmukh Krishi Vidyapeeth, Akola with the objectives to assess the effect of sugar concentrations on preservation of Nagpur mandarin segments and to study the physico-chemical changes in Nagpur mandarin segments during the storage. The experiment was conducted in RBD design consisting of eight treatments and three replications with using seedless Nagpur mandarin segments. Observations of Nagpur mandarin segments viz., physical, chemical and sensory evaluation were recorded periodically at an 15th days of interval. From the findings it was observed that, there was gradual increase in length, breadth and weight of Nagpur mandarin segments in sugar syrup concentrations during storage period of 120 days. In chemical analysis there was a gradual increase in TSS (%), reducing sugar (%), total sugar (%), non reducing sugar (%) and there was no microbial infestation found up to four month of storage period. Further, the experimental data recorded on sensory qualities of Nagpur mandarin segments in sugar syrup at an interval of 15th days. The overall scores of Taste, colour, flavour, and texture found decrease in trend with the advancement of storage period.

Keywords: Nagpur mandarin segments, Preservation, Sugar syrup, Sensory quality.

Introduction

Mandarin orange (Citrus *reticulata* Blanco) is one of the most popular citrus fruits having attractive, bright colour, appealing taste and flavour. Citrus is grown in 114 countries around the world. Out of these, 53 countries grow citrus commercially with a total production of more than 115 million tones. On the production basis, China tops the list with 22.9 million tones, followed by Brazil with 22.7 million tones and USA with 10.4 million tones, India with 10.48 million tones is in 4th position (Anonymous, 2015). At present mandarin cultivation has assumed great importance among north Indian growers and a large acreage of land is being brought under cultivation , particularly in Punjab, Madhya Pradesh, Andhra Pradesh, Maharashtra, Rajasthan. Total mandarin production in India is 3.70 million ha area and 9.3 t/ha as productivity (Ladaniya, 2015). In Maharashtra, mandarin is grown in an area of 1.35 Lakh ha area with the production of 7.425 Lakh MT with the productivity of 5.5 MT per hectares (Anonymous, 2015). Citrus fruits have numerous therapeutic properties like anticancer, anti-tumor and anti- inflammatory reported by Etebu *et al.* (2014).

Citrus fruits have the attractive shape, size, colours, fragrant, and appetizing with high nutritional values. These are one of the richest sources of vitamin C along with vitamin A and B and contain 5-8% sugar and minerals such as calcium, phosphorous and iron in appreciable amounts, which are essential for proper human health and vigour.

Indian fruit processing industries are established only on a few important fruits like aonla, mango and pineapple. Production of new processed products, from various fruits is immediate necessity for the ultimate survival and economical growth of processing industries in India. This would also helps to meet out new product taste and demand at home as well as capturing the foreign export market. Hence, there is a great scope to develop most feasible and suitable processing technology for the preparation and preservation of Nagpur mandarin.

Methods and Material

The present research work entitled "Studies on preservation of Nagpur mandarin segments in sugar syrup" was carried out in Post- Harvest Technology Laboratory, Department of Horticulture, Dr. P.D.K.V., Akola during the year 2017-18. The details of material used and methods adopted during the course of investigation are presented in this chapter under appropriate headings and sub headings.

Treatment Details

T1 - Sugar syrup – 20° Brix

T2 - Sugar Syrup - 25° Brix

T3 - Sugar Syrup - 30°Brix

T4 - Sugar Syrup - 35°Brix

T5 - Sugar syrup - 40° Brix

T6 - Sugar syrup - 45°Brix

T7 - Sugar Syrup – 50°Brix

T8 - Sugar Syrup - 60°Brix

Procedure of preservation of Nagpur mandarin segments in sugar syrup

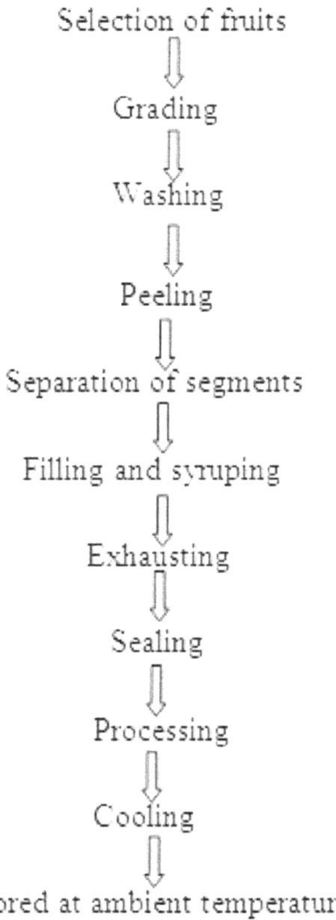

Nagpur mandarin segments were analyzed for various physico- chemical composition such as TSS, acidity, ascorbic acid, reducing sugar, non reducing sugar, total sugar (AOAC, 1990). The sensory quality such as taste, colour, flavour, texture, and overall acceptability were also by five expert panels using hedonic scale and the data were statistically analyzed as described by Panse and Sukhatme (1985).

Results and Discussion

The data pertaining to these observations of Nagpur mandarin segments and chemical changes took place during storage period or presented under appropriate headings and subheadings.

1. **Length of segment**

   In general, the length of Nagpur mandarin segments was gradually increased in all sugar concentrations up to 45 days. Then it remains unchanged up to 120 days of storage period. Maximum change in length of segment was found to be more from 8.81 to 8.98 cm in 60° Brix solution. Whereas, minimum change in length of Nagpur mandarin segments was observed 8.38 to 8.54 cm in 40° Brix sugar solution up to 45 days storage period. After 45 days there were no change found in length of Nagpur mandarin segments up to 120 days of storage.

The increase in length of Nagpur mandarin segments in various sugar concentrations at ambient conditions because low concentrations of sugar syrup may get diluted and rich near saturation point , which would not help in removal of more water during osmosis process. The similar results were observed by Hope and Vitale (1972) in apple, Novakunda *et al.* (2004) in banana and Singh et al. (2008) in pineapple.

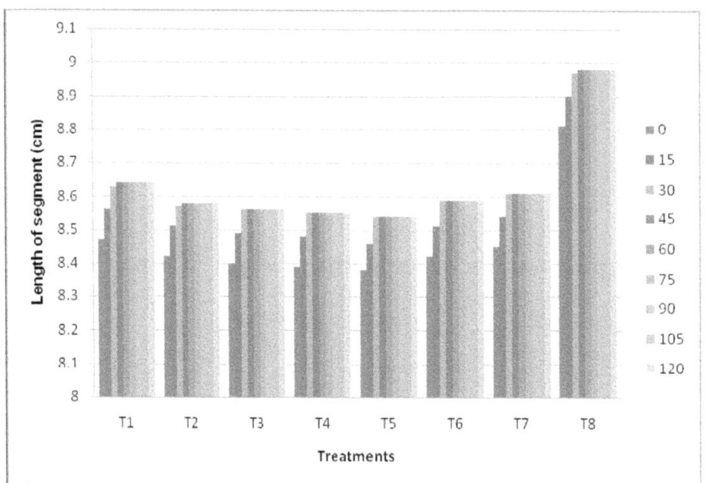

Figure 1 Effect of different sugar concentration on segments length of Nagpur mandarin segments at ambient conditions.

2. **Breath of segment**

The data in respect of segment breadth as influenced by different sugar concentrations at ambient condition was recorded upto 120 days of storage was recorded are presented depicted in Fig.2.

In general the breadth of Nagpur mandarin segments was gradually increased in all sugar concentrations up to 45 days. Then it remains unchanged up to 120 days of storage period. Maximum change in breadth from 4.18 to 4.20 cm in 50 and 45° Brix sugar solution. Minimum change in segment breadth of Nagpur mandarin 4.10 to 4.15 cm in 25° Brix sugar solution up to 45 days of storage period. After 45 days there were no change found in segment breadth of Nagpur mandarin segments up to 120 days of storage.

The increase in breadth of Nagpur mandarin segments in various sugar concentrations at ambient conditions because low concentrations of sugar syrup may get diluted and rich near saturation point, which would not help in removal of more water during osmosis process. The similar results were observed by Hope and Vitale (1972) in apple, Novakunda *et al.* (2004) in banana and Singh et al. (2008) in pineapple.

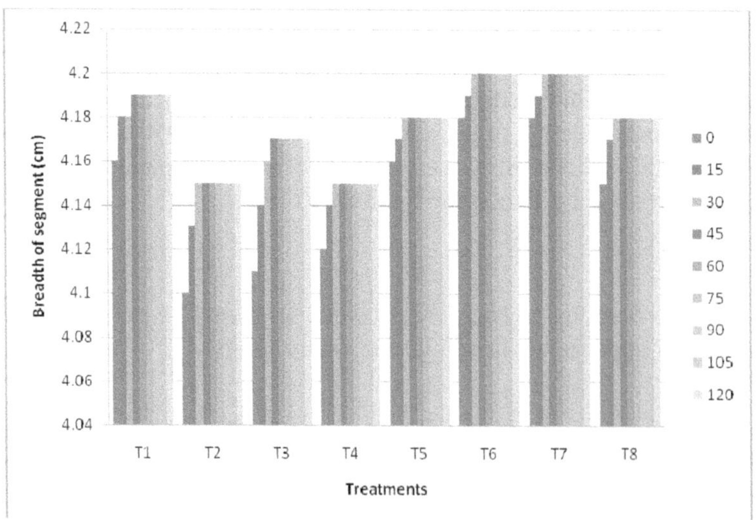

Figure 2 Effect of different sugar concentrations on segments breadth of Nagpur mandarin segments at ambient conditions.

3. Weight of segment

The data in respect of segment weight of Nagpur mandarin as influenced by different syrup concentrations at ambient condition was recorded up to 120 days of storage are presented in depicted in Fig 3. Different sugar concentrations exhibited significant difference in respect of segment weight of Nagpur mandarin segments at 15, 30, 60, 75, 90, 105, 120, days of observations.

In general the segments weight of Nagpur mandarin was increased in all sugar concentrations up to 60 days. Then it remains unchanged up to 120 days of storage period. Maximum gain in segment weight (15.45 g) was observed in 60° Brix sugar solution and minimum gain in segment weight (15.20g) in 25° Brix sugar solution. After 60 days of storage there was no change in segment weight up to 120 days of storage.

Due to higher TSS of sugar syrup, the solute gets entered inside the segment by osmosis process and weight of segment might be increased up to 60 days of storage. After 60 days of storage, migration of solute from outer side to inner side minimizes due to saturation point (Sagar and Kumar, 2009).

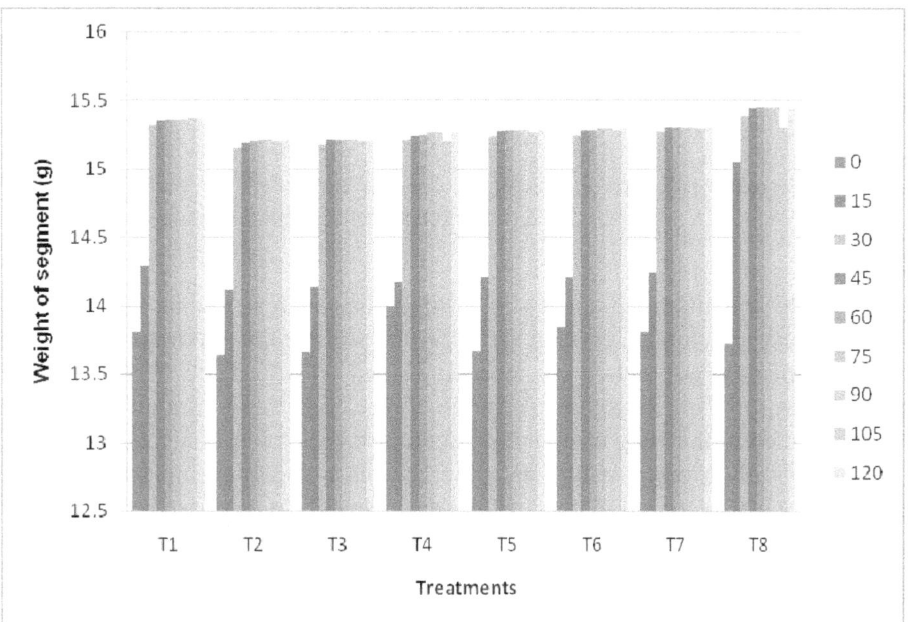

Figure 3 Effect of different sugar concentration on segments weight of Nagpur mandarin segments at ambient conditions.

4. TSS

In general, the total soluble solid content in Nagpur mandarin segments were gradually increased up to 120 days of storage. Minimum increased (1.95%) in total soluble solid at 120 days of storage was observed when the Nagpur mandarin segments preserved in 30°Brix (9.74°Brix to 9.93°Brix). However, the change in TSS was found to be more (64.57° Brix), when the Nagpur mandarin segment preserved in 60° Brix at ambient storage conditions.

The maximum TSS was recorded in Nagpur mandarin segments when preserved in 60°Brix sugar solution at 15, 30, and 45, days (12.20, 13.18, and 14.92, respectively). This treatment was closely followed by when the Nagpur mandarin segments preserved in 50° Brix sugar solution (11.08, 12.60, 13.20 respectively) at 15, 30, 45 days of storage while, minimum TSS was recorded in Nagpur mandarin segments preserved in 20° Brix (9.62, 9.65, 9.68 respectively). Increased in TSS due to increase in sugar concentrations caused by hydrolysis of polysaccharides such as starch, cellulose, pectin substance into simple substances. This indicated that during storage there was change in composition of Nagpur mandarin segments (Reddy and Chikkasubbanna, 2009) or might be due to conversion of polysaccharides into simple sugar.

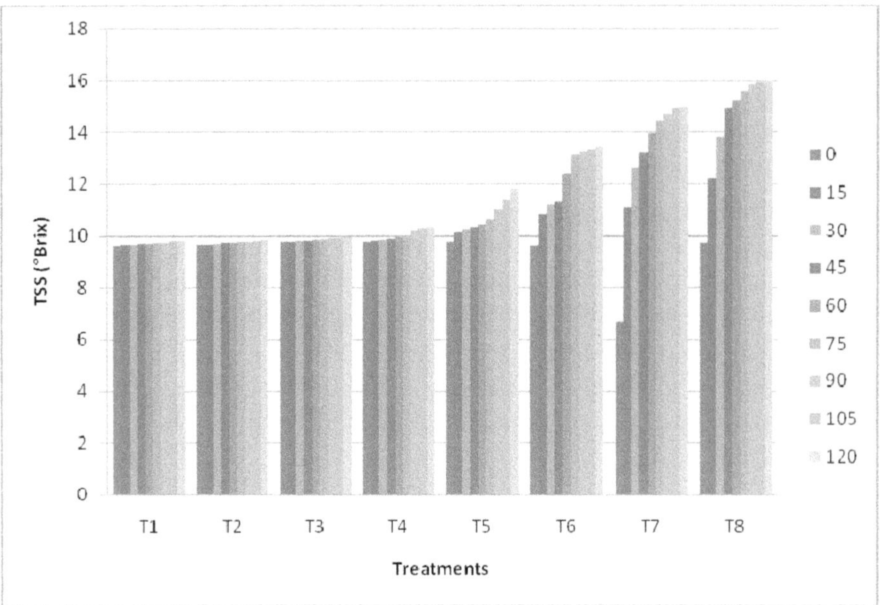

Figure 4 Effect of different sugar concentrations on TSS content in Nagpur mandarin segments at ambient conditions.

5. Titrable acidity

In general, data showed that titratable acidity of Nagpur mandarin segments was gradually decreased in all treatments during 120 days of storage. At 45 and 75 days of storage the maximum titratable acidity recorded in Nagpur mandarin segments preserved in 20° Brix (0.85 and 0.80 % respectively).This treatment was at par with 25° Brix, 30°Brix, 35°Brix sugar solution. However, minimum titratable acidity (0.73 and 0.61 % respectively) was recorded in the Nagpur mandarin segments preserved in 60° Brix. This treatment was at par with 50°Brix at 45 days of storage. At 60, 90, 105, 120 days of storage maximum acidity (0.81, 0.80, and 0.79%) titratable acidity was recorded in Nagpur mandarin segments preserved in 20°Brix sugar solution. This treatment was at par with 25° Brix sugar solution. However, minimum titratable acidity (0.61, 0.69, 0.58, 0.54 and 0.52%) was found in Nagpur mandarin segments preserved in 60°Brix syrup solution.

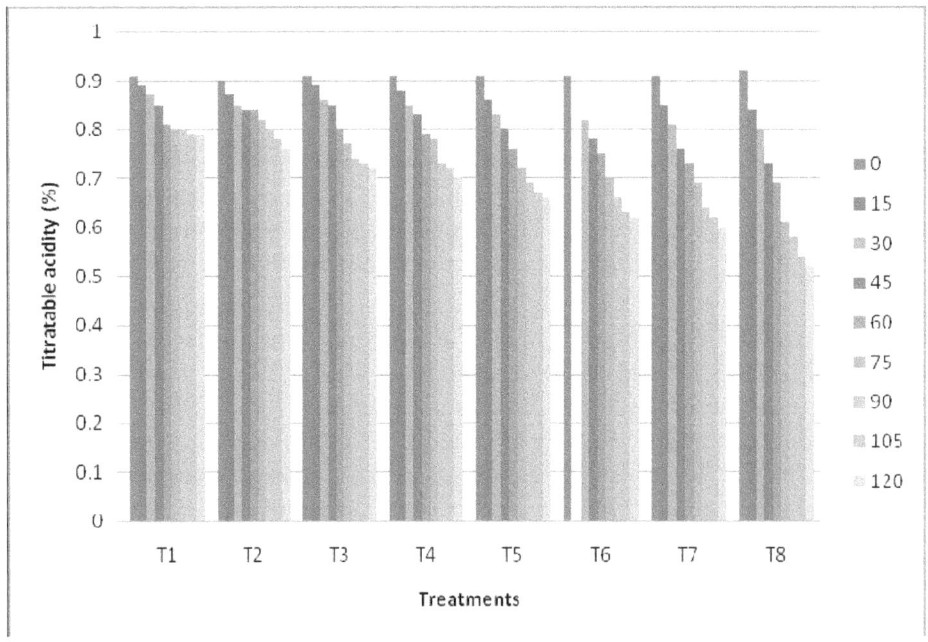

Figure 5 Effect of different sugar concentrations on acidity content in Nagpur mandarin segments at ambient conditions.

6. Ascorbic acid

In general, ascorbic acid of Nagpur mandarin segments was gradually decreased in all sugar concentrations during storage. Minimum decrease from 41.69 to 38.79 mg/100g in ascorbic acid of Nagpur mandarin segments was observed when preserved in 60° Brix sugar solution. Maximum 41.61 to 32.32 mg/100g decreased in ascorbic acid was observed in Nagpur mandarin segments preserved in 20°Brix sugar solution. Overall minimum percent decreased in ascorbic acid (6.96 %) found in 60°Brix, whereas, maximum (22.32 %) decrease in ascorbic acid was found in 20° Brix sugar solution. Leaching of ascorbic acid in hypertonic solution plays a little role in loss of ascorbic acid (Kumar and Sagar, 2009).

Ascorbic acid is a vital nutrient quality parameter and it is very sensitive due to degradation and oxidation parallel to other nutrient during food processing and storage (Veltman *et al,* 2000). The considerable reduction in ascorbic acid may be due to thermal degradation during osmosis process. Loss of ascorbic acid was less at low temperature with high concentrations of sugar syrup might be due to less degradation of ascorbic acid at low temperature at more protection by sugar of high concentrations of sugar syrup.

This results are in conformity with results of Jain et al (2006) in aonla squash, Sagar and Suresh Kumar (2009) in osmo-dehydration of mango, Naikwadi *et al* (2012) in dehydration of fig, Bhatttacherjee (2013) in aonla, Rajesh kumar and Kirad (2013) in pineapple candy, Lande (2013) in Ber preserve, Amarjeet *et al* (2016) in sapota slices and Khan et al (2016) in mango slices.

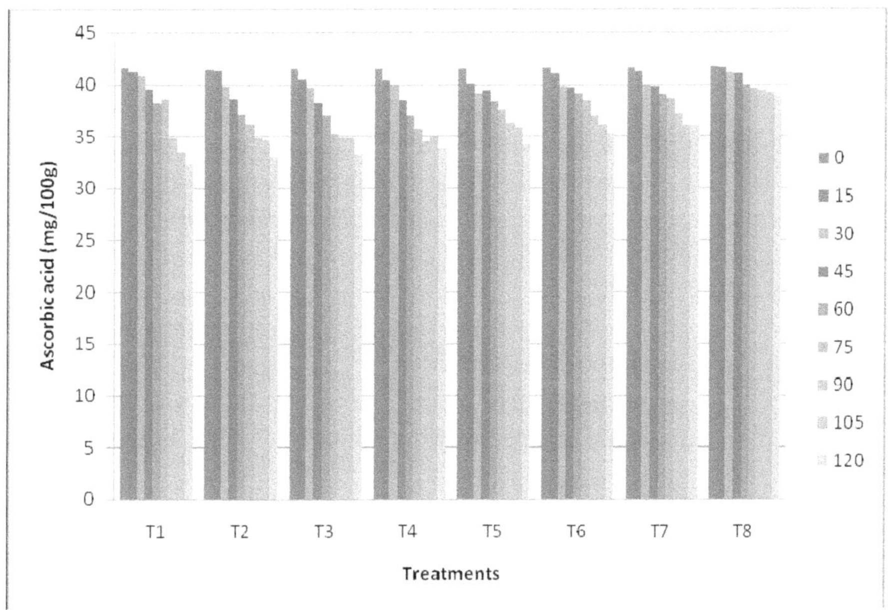

Figure 6 Effect of different sugar concentrations on ascorbic acid content in Nagpur mandarin segments at ambient conditions.

7. Overall acceptability

In general, the overall acceptability score of Nagpur mandarin segments was decreased during storage period 0 to 120 days. Maximum taste score was (8.4) was observed in Nagpur mandarin segments preserved in 60°Brix and 45 °Brix which was evaluated like very much upto 90 days of storage. However, minimum overall acceptability score (5.7) was observed in 20°Brix which was dislike slightly. After 90 days of storage up to 120 days the flavour score drastically decreased.

The overall acceptability had decreased significantly during storage period of 120 days. This may be due to the physical characteristics of the product viz., colour, taste, flavour and texture score. The decrease in overall acceptability score may be due to absorption of atmospheric moisture, dilution of sugars and changes in acidity, oxidation of ascorbic acid, as well as changes in biochemical constituents of Nagpur mandarin segments.

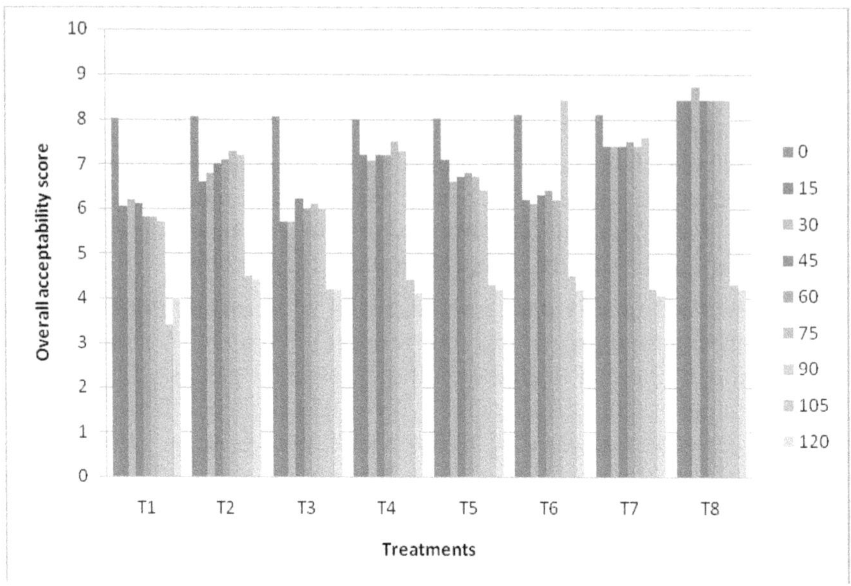

Figure 7 Effect of different sugar concentrations on overall acceptability of Nagpur mandarin segments at ambient conditions.

Conclusions

From the above investigation conducted preservation of Nagpur mandarin segments during storage preserved from different concentrations of sugar solution, following conclusions could be drawn.

(i) Nagpur mandarin segments preserved in different sugar concentrations stored under ambient conditions was found in good conditions up to 90 days of storage under ambient storage conditions.

(ii) The changes in TSS, total sugar, reducing sugar, non- reducing sugar of Nagpur mandarin segments preserved in different sugar concentrations was found increasing trend during storage period at ambient conditions. However, acidity, ascorbic acid content decreases during storage period.

(iii) Nagpur mandarin segments preserved in 60°Brix sugar solution at 120 days storage secured highest score for taste, colour, flavour, texture and overall acceptability as compared to other sugar concentrations.

References

Anonymous. (2015). Indian Horticulture Database 2014 . National Horticulture Board, Ministry of Agriculture Government of India, Gurgaon, INDIA.pp.2-3.

A.O.A.C.1990. Official method of analysis .Association of official Analytical Chemists, Washington , D.C. 16:37.

Amarjeet K, A.S. Chauhan ,K.srinivasulu, R. Ravi and V.B. kudachikar (2016). Effect of pretreatments and storage conditions on shelf life extension of sapota (Achras Zapota) Fruit and on quality of osmo-dehydrated slices.Int.J.Curr. Res. Aca.Rev.2016; 4(4) : 148-165.

Bhattacherjee, A.K., Abhay Dikshit, Sanjay Kumar , D.K. Tandon (2013). Quality of aonla candy segments in syrup prepared from steep preserved fruits in water. Indian Journal Of Natural Products and Resources . Vol. 4(1):119-112.

Etebu, E., Nwauzoma , A.B.(2014). A review on sweet orange (*Citrus sinensis* Osbeck), health, diseases, and management .American Journal of Research Communication, 2 (2).33-70.

Hope, G.W., and D.G. Vitale (1972). Osmotic dehydration : A cheap and simple method of preserving mangoes , bananas and plantains. Food Research Institute Canada Department of Agriculture Ottawa.

Singh V.P. and A.K.Singh (2006). Screening of aonla cultivars for making squash . Indian J. A RID Hort.1 (1): 44-46.

Kumar, Rajesh and K.S.Kirad (2013). Effect of different level of sugar in the preparation of pineapple (Ananascomosus L.) candy. ISSN: 2322-0600 And E-ISSN: 2322-0619, Volume 1:01-03.

Khan Shehzad, Ali Wajahat, Kamal Tariq, Khan Usman Muhammad , and Saleemgul (2016). Preservation of mango sdlices in sucrose solution with various concentrations. Department of Food Science and Technology , The University of Agriculture, Peshawar, Pakistan.

Lande A.Rekha (2013). Preparation and storage stability of ber preserve at home scale. Food science research journal, 4(2):126-129.

Naikwadi, P.M., U.D. Chavan , V.D. Pawar, R.Amarowicz (2010). Studies on dehydration of Figs using different sugar syrup treatments.J Food SciTechnol, 47 (4) : 442 – 445.

Nowakunda, K., A.Andres and Fito, P.(2004). Osmotic dehydration of Banana slices as a pre-treatment for drying processes. Proceedingsog the 14th International Drying Symposium. Sao Paulo , Brazil , Vol. C: 2077-2083.

Sagar , V.R., P.S. Kumar (2009). Involvement of some process variables in mass transfer kinetics of osmotic dehydration of mango slices and storage stability.Journal of scientific and industrial research. Vol.68,December 2009.pp.1043-1048.

Singh, C., H.K. Sharma and B.C, Sarkar (2008). Optimization of Process conditions during osmotic dehydration of fresh pineapple. J. Food Sci . Technol., 45 (4):312-316.

Reddy A.H. and V. Chikkasubbanna (2009). Studies on storage behavior of amla syrup. The Asian J. Hort 4(1) :5-9.

Application of ANSYS-CFD to Flow Problems

Suresh Kumar. N[1] and Prasanna. SVSNDL[2]

Professor[1] and Assistant Professor[2]

Department of Civil Engineering, College of Engineering (A), Osmania University,

Hyderabad Telangana State, India

nskdr@yahoo.co.in

Introduction

Water is the most essential resource for the survival of all kinds of living entities. Hence, water resources have to be properly conserved, controlled, and/or limit it's usage by appropriate means. Even the government of Telangana is planning to provide water for all needs to the extent possible by initiating different water resources engineering projects. Even the theme of *World Water Day – 2019* is *'Leaving No One Behind'*. Further, the present study also focuses on the same jargon as *'Leaving No Multi-Purpose River Valley Project without Turbulence Modelling'* for the plausible design solution. Many real life problems in hydraulic engineering involve continuous variation of flow conditions. Continuous varying flow problems are generally solved through computational hydraulics or more specifically through turbulence modeling.

Simulation models make use of mathematical equations to represent the physical interactions of given flow system. These models allow the user to assess the impact of a given input on the system without having to actually experience that event. Simulation models are easy to use and the output can be directly compared with either the experimental results or analytical solutions. Further, the huge man power, ground space and water requirements associated with physical models are completely dispensed in simulation systems. Simulation models contain physical, hydrological, operational, institutional, and legal information about the system. Hence, the modeler can examine a wide variety of potential scenarios to decide whether simulation output is reasonable or not. Moreover, simulation models help to identify the physical and institutional relationships and interactions that are esoteric and lead to needed research.

One of the best methods of simulation analysis is by adopting ANSYS-CFD software. The physical aspects of the fluid flow are basically governed by the conservation principles of mass, momentum and energy. These properties are usually expressed in the mathematical form by some assumptions that are in the form of a set of partial differential equations (PDE). This difficulty has been reduced to some extent by the advent of the digital computers for the solution of linear and non-linear PDEs in the second half of the last century with the introduction of Computational Fluid Dynamics (CFD).

The early development of CFD in 1960s and 1970s was basically for the needs of aerospace community whereas the modern CFD cuts across all the disciplines where the flow of the fluid plays a major role. CFD has not only proven to be a promising tool in the automobile field for engine applications and industrial manufacturing applications but also in the various Civil Engineering, Hydraulic Engineering, Environmental Engineering and Naval Architecture applications.

CFD is mainly adopted to elaborately interpret and clarify the characteristics of arotic flow that are not visible through physical measurements. CFD makes it possible to understand the required parameters via, Turbulence Modelling [1]. The study progresses by the use of one of the widely adopted method namely Finite Volume Method (FVM) of Fluent solver in CFD for various hydraulic engineering applications. Few of the potential areas of applications in hydraulics are detailed below:

Few of the potential areas of applications in water resources engineering are detailed below:

- Imitation of scour development around bridge piers to ascertain the ideal of shape of the bridge piers.
- Replication of peak surge height in surge tanks provided either in hydro power stations/ lift irrigation schemes for deciding the surge chamber dimensions
- Estimation of spillway design parameters viz., maximum pressure in the spillway bucket invert level, dynamic pressure on the spillway profile, trajectory angle at lip level of the bucket, distance of jet throw for flip buckets (both horizontal and vertical), assessment of cavitation, and the possible remedial measures to avert it
- Simulation of flow behaviour in air regulated siphon spillways

- Modelling of hydraulic jump for the estimation of flow surface profile on the d/s of the sluice gate, jump height and/or length.
- And many other areas where turbulence is present

Notches are the human made structures constructed for diversion of water for irrigation purpose, in adjusting the velocity in order to avoid scouring of the channel. In this paper, the experimental results obtained from the observations carried out by Husam H Alwan et al. were made use of for simulation purposes.

The objectives of the present investigations are as follows:

- To analytically estimate the coefficient of discharge from the experimental observations of Husam H Alwan et al.
- Simulate and plot the flow profile over the rectangular notch by adopting numerical modeling technique viz., ANSYS-CFD (Fluent) software

Methodology

The process of application of CFD involves a sequential procedure as detailed in Figure. 1. CFD analysis helps to a large extent in testing and examining the results, thereby reducing the effort needed for experimentation. The simulation analysis was carried out to channel flows replication of flow over a rectangular notch.

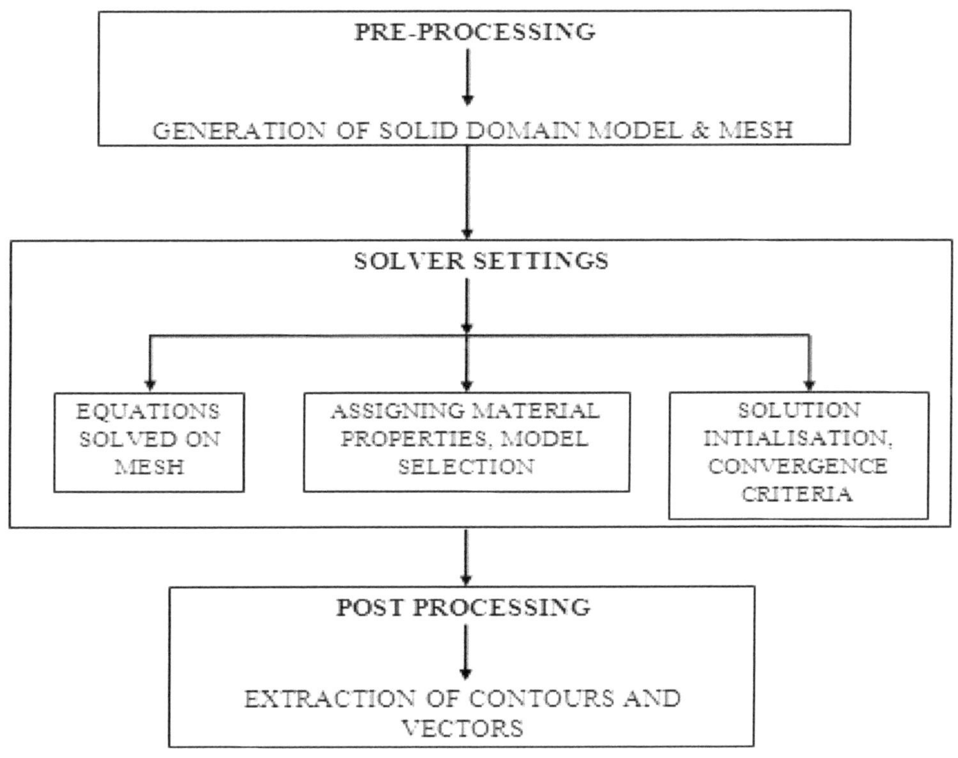

Figure 1 Modelling Approach

Experimental Observations

Husam H Alwan et al., have conducted experiments on a laboratory channel having 0.7 m length, 0.25 m wide, and 0.166 m height [2]. The sill level of the weir is at a height of 0.106 m from the bottom of the flume. The length of the notch is 0.04m. The plan and longitudinal section of the flume with notch are shown in Figure 2.

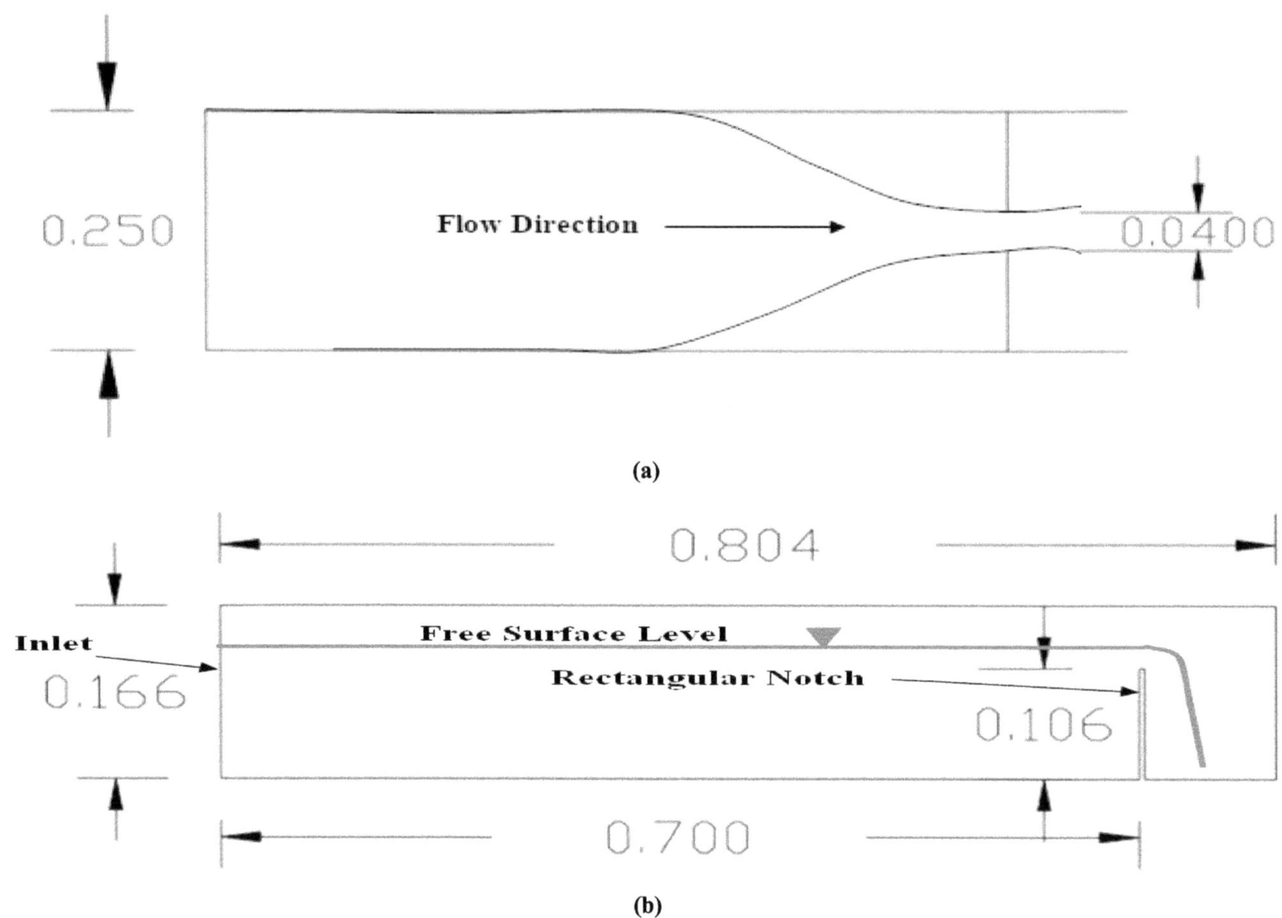

Figure 2 (a) Plan and (b) Longitudinal Section of Flume and Notch

The experimental observations of Husam H Alwan et al. were adopted for the estimation of actual discharge. The theoretical discharge was determined by the conventional eq. (1) and coefficient of discharge was also evaluated by eq. (2) and (3) [3] as detailed in Table 1. Further, the estimated velocity of approach was adopted for the simulation of flow using ANSYS-CFD.

$$Q_{th} = \frac{2}{3} * \left[\sqrt{(2 * 9.81)}\right] * L * \left\{(H + H_a)^{3/2} - H_a^{3/2}\right\} \qquad .....(1)$$

$$V_a = \frac{Q}{B * (Z + H)} \qquad .....(2)$$

$$H_a = \frac{V_a^2}{2g} \qquad .....(3)$$

$$C_d = \frac{Q_a}{Q_{th}} \qquad .....(4)$$

where, C_d = coefficient of discharge, Q_a = actual discharge in cumec, Q_{th} = theoretical discharge in cumec, L = length of the notch (0.70m), Z = height of sill level above the flume bottom (0.106m), H = height of water above the sill level of the notch in m, B = width of the flume (0.25 m), V_a = velocity of approaching flow in m/s, H_a = approaching head in m.

Table 1 Analytical Calculations

Sill level Reading	Head over the Sill (H)	Free Surface Level Reading	Time	Actual Discharge (Q_a)	Theoretical Discharge (Q_{th}) Eq. 1	Coefficient of Discharge (C_d)	Approaching velocity (V_a)
m	m	m	s	cumec	cumec	Eq. 4	m/s
0.106	0.0281	0.1341	28.67	0.000348	0.00056	0.627	0.0104
0.106	0.0329	0.1389	24.02	0.000416	0.00071	0.591	0.0120
0.106	0.0378	0.1438	18.76	0.0005330	0.00087	0.614	0.0148
0.106	0.044	0.1500	14.81	0.0006752	0.00109	0.619	0.0180
0.106	0.0487	0.1547	11.75	0.0008510	0.00127	0.670	0.0220

Simulation analysis:

The commercial numerical modeling technique ANSYS – CFD (Fluent) uses a Finite Volume approach for solving Reynolds Averaged Navier Stokes (RANS) equations. This software package is adopted in the present study to predict the flow profile over the notch. The numerical modeling analysis was carried out initially by generating the geometry based on the experimental analysis in ANSYS design modeler as shown in Figure 3. The mesh was created in ANSYS - Workbench as indicated in Figure 4. The mesh was made finer above the notch considering number of cell gaps as 25 and coarser towards the walls of the flume. Further, the processing methodology was carried out by importing the mesh assigned with boundary conditions into the Fluent solver. The solution was analyzed by means of k-ε turbulence model with the following considerations: velocity - defined at the inlet, no slip condition - applied along the walls, and pressure outlets as indicated in Figure 5.

Figure 3 Geometry of Longitudinal Section of Channel and Notch

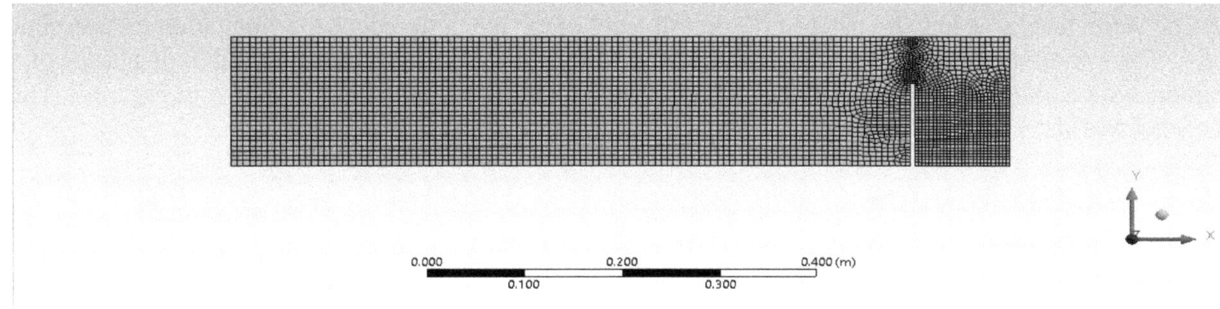

Figure 4 A View of Generated Mesh

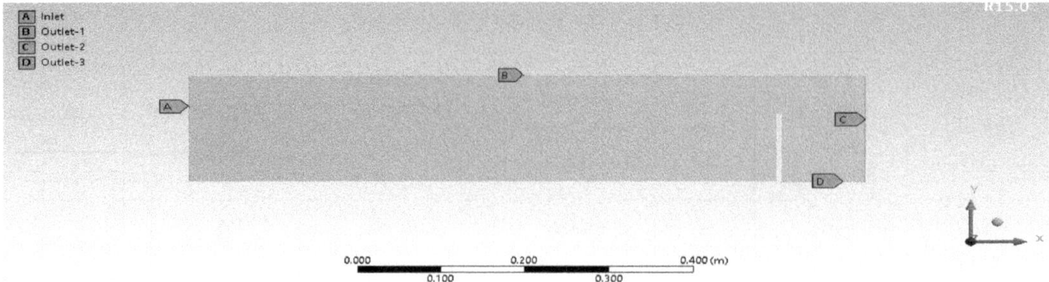

Figure 5 Boundary Conditions Adopted

The Volume of Fluid (VOF) method delivers the shape and location of constant pressure at free surface boundary. For the two dimensional steady state incompressible flows, the RANS equations are given below. The free surface elevation was evaluated by means of User Defined Function (UDF) written in C-Programming. The results were estimated by solving the following basic governing equations of two-equation model of k-ε as highlighted in eq. (4) on each control volume of the flow region [1].

$$\left.\begin{aligned} \frac{\partial k}{\partial t}+U_j\frac{\partial k}{\partial x_j}&=\tau_{ij}\frac{\partial U_i}{\partial x_j}-\varepsilon+\frac{\partial}{\partial x_j}\left[\left(\nu+\nu_T/\sigma_k\right)\frac{\partial k}{\partial x_j}\right] \\ \frac{\partial \varepsilon}{\partial t}+U_j\frac{\partial \varepsilon}{\partial x_j}&=C_{e1}\frac{\varepsilon}{k}\tau_{il}\frac{\partial U_i}{\partial x_j}-C_{e2}\frac{\varepsilon^2}{k}+\frac{\partial}{\partial x_j}\left[\left(\nu+\nu_T/\sigma_\varepsilon\right)\frac{\partial k}{\partial x_j}\right] \end{aligned}\right\} \quad \ldots..(4)$$

The first term on the LHS of eq. (4) represents the rate of change of ‘k’ and the second term explains the transport of ‘k’ by convection. While the first term on the RHS symbolizes the transport of ‘k’ by diffusion, the second term demonstrates the rate of destruction of ‘k’ and the third term illustrates the rate of production of ‘k’. The Closure coefficients and auxiliary relations are given as: $C_{e1} = 1.44$, $C_{e2} = 1.92$, $\sigma_k = 1.0$, $\sigma_\varepsilon = 1.3$.

where, k - Kinetic energy of turbulent fluctuations per unit mass, U_i - Mean velocity in tensor notation, ν_T - Kinematic eddy viscosity, ν - Kinematic molecular viscosity, τ_{ij} - Specific Reynolds stress tensor.

The Semi Implicit Method for Pressure Linked Equations (SIMPLE) coupling, was adopted which is designed exclusively for turbulence simulations. Each test run was solved for various equations such as Continuity, X - Velocity, Y - Velocity, k – equation and ε – equation leading to convergence. At each time step, a maximum of 20 iterations were taken to achieve the convergence. All the trials were observed to converge with the flow over the notch at more than 25,000 iterations. The run time of 8.0 to 10.0 h was required to establish and stabilize the flow field for each velocity over the notch. The convergence criterion of 10^{-6} was adopted in the present investigation.

The contours of multiphases viz., water and air, velocity profiles are plotted using Graphics and Animations in fluent. The water level was initially patched till the sill level of the notch to reduce the simulation time as shown in Figure 6. The variations in multiphase analysis obtained from the simulations for two different phases of water (Red color) with a value of 1.0 and air (Blue color) assigned with the zero value are shown in Figure 7. The free surface level was also extracted from the UDF programme as represented in Figure 8.

Figure 6 Patched Water Level

Figure 7 Contour of Volume Fraction by k-ε turbulence model

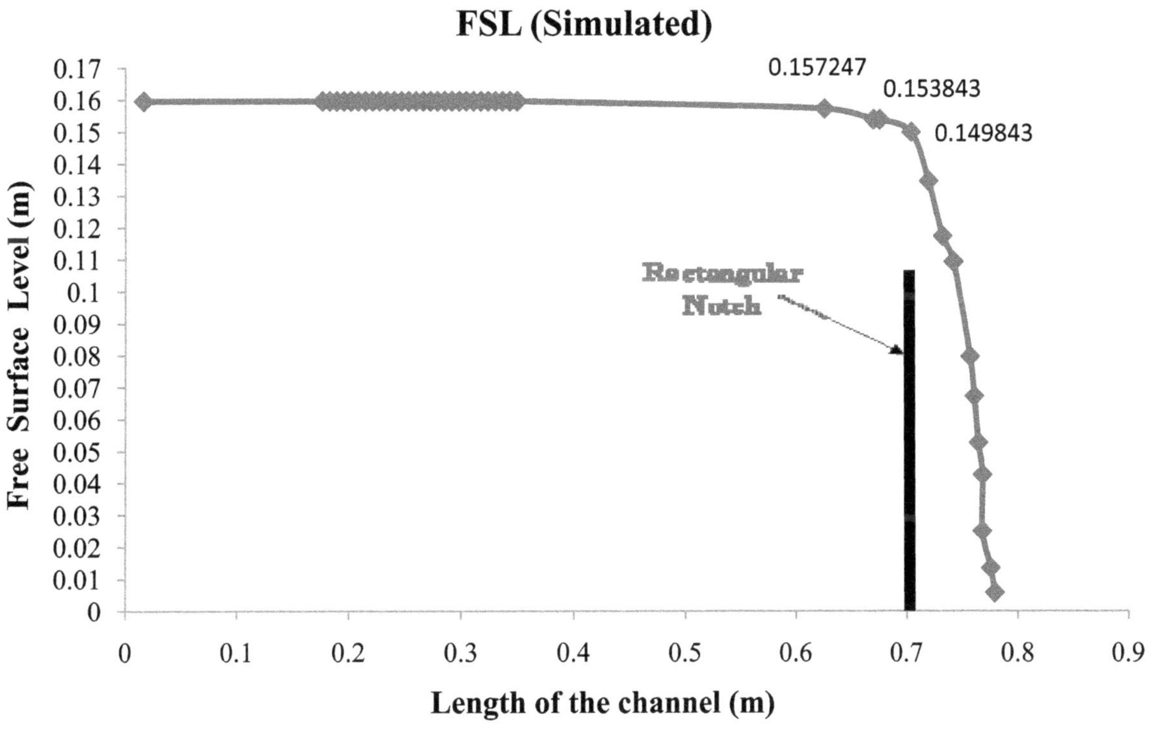

Figure 8 Simulated Free Surface Elevation over Notch

Conclusions

The following conclusions are drawn from the experimental, and present simulated analysis

1. The Free Surface Level (FSL) for various velocities as obtained from experimental results was analyzed in Fluent. The values were obtained using a User-Defined Function, by the interpretation and execution of the macro generated mainly based on the multiphase analysis.
2. The results obtained from the fluent solver simulations with multiphase turbulence model of k-ε are in consolation with the experimental results for various velocities.
3. For an approach velocity of 0.022 m/s, the head over the sill was found to be 0.0487m from the experimental observations and from the simulation analysis, the value obtained is 0.04784m.
4. The FSL from experimental observations and simulation results were found to be in fair agreement with each other for all the velocities as highlighted in the Figure 9.

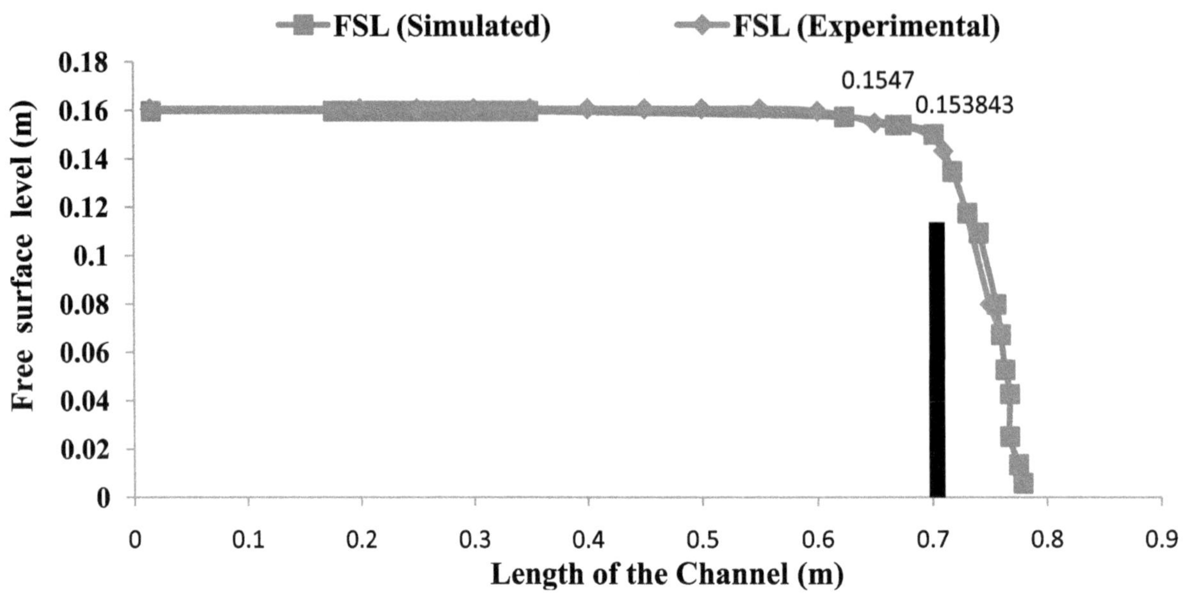

Figure 9 Comparative Free Surface Level (V = 0.022m/s)

References

Anderson John D. Jr., *Computational Fluid Dynamics – The basics with applications*, McGraw Hill Education (India) Pvt. Ltd., New Delhi, 2012, 3 – 94.

Husam H Alwan and Fadhil M Al-Mohammed, *Discharge coefficient for Rectangualr Notch Using Dimensional Analysis Technique,* IOP Conference Series Materials Science and Engineering, March-2018, DOI: 10.1088/1757-899X/433/1/012015.

Chow Ven Te (1959), *Open Channel Hydraulics*, McGraw-Hill Book Company, INC, New York.

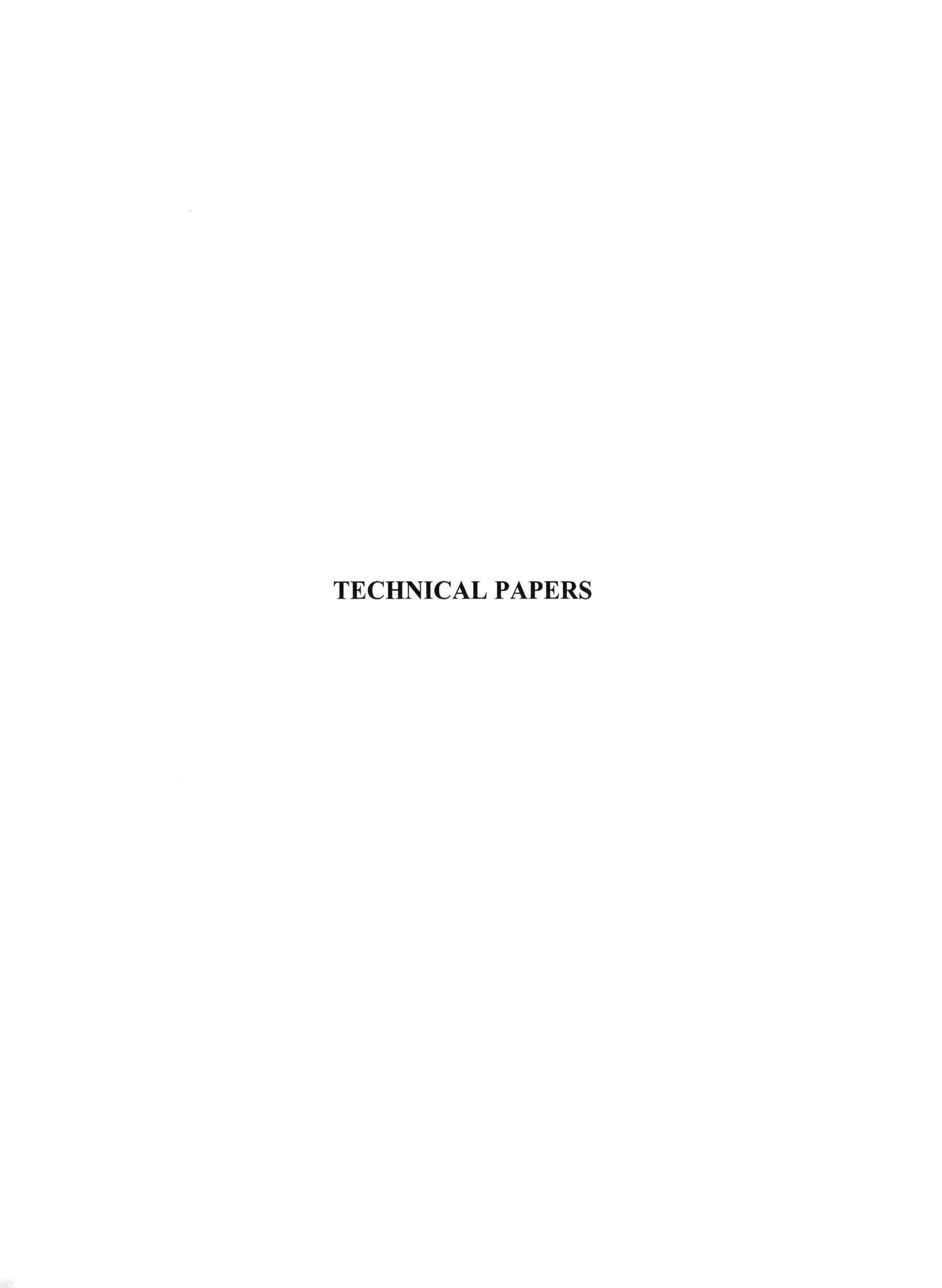

TECHNICAL PAPERS

Hypsometric Analysis of Yagachi Watershed using Remote Sensing and GIS Techniques

Praveenkumar B Salegar[1] and Shivakumar J Nyamathi [2]

[1]P.G. Student, [2]Associate Professor,

Department of Civil Engineering, UVCE, Bangalore University, Jnanabharathi, Bangalore, India.

praveenuvce93@gmail.com

ABSTRACT

Hypsometric analysis is considered an effective tool for understanding the stages of geomorphic evolution and strongly tied to flood response and for the delineation of erosional proneness of watershed. The hypsometric integral and Hypsometric curve are products of hypsometric analysis. In the present study, hypsometric analysis of Yagachi watershed with watershed atlas of India code 4B4B3 (tributary of Cauvery basin) has been carried out. Yagachi watershed lies between 75°38'31.2" to 75°54'43.2" East longitude and 13°5'52.8" to 13°23'42"North latitude with watershed area of 550.86 km^2.Yagachi catchment receives 1031mm of rainfall annually and relief from 1829m to 918m (As per ASTERDEM). DEM data with 30m resolution were used to delineate drainage system and catchment boundary subsequently. Hypsometric analysis were carried out for entire watershed with respect to stream order level (i,e. 6th,5th & 4th) , elevation- relief ratio method was employed to calculate the hypsometric integral values within GIS environment. Where, as the hypsometric curves result reveals that remarkably downward convex shapes (i,e. Monadnock Stage) as per (Strahler,1964) hypsometric curve classification hypsometric integral for entire catchment with respect to stream order level varies from 0.14 to 0.52. Hypsometric analysis is input to identify most possible soil erosion regions and suggest mechanical and vegetative measures to arrest sediment loss in the catchment.

Keywords: Hypsometric Integral, Profile Factor, ASTERDEM, Yagachi watershed, Remote sensing &GIS.

Introduction

Land degradation and topological changes within watersheds area accomplished by weathering processes, stream erosion pattern and sediment transportation by surface runoff. Moreover, the quantification and interpretation of the topological changes becomes very difficult due to complex nature of these hydrological and landform processes acting on watershed systems. In an attempt to simulate the geological stages of development and to study the influence of varying forcing factors (i.e. tectonics, climate, litho logy) on watershed topology, the hypsometry of drainage basins (area-elevation analysis) have been evaluated by the researchers. Hypsometric analysis is the relationship of horizontal cross-sectional drainage basin area to elevation (Fig 1). The hypsometric curve has been termed the drainage basin relief graph. Hypsometric curves and hypsometric integrals are important indicators of watershed conditions (Ritter et al., 2002, Kuang-Yu Cheng, et al., 2012).

Hypsometric analysis is carried out to ascertain the susceptibility of watershed to erosion and prioritize them for treatment. The slope of the hypsometric curve changes with the stage of watershed development, which has a greater bearing on the erosion characteristics of watershed and it's indicative of cycle of erosion. The hypsometric integral (HI) is also an indication of the "cycle of erosion" (Strahler, 1952, 1957; Garg, 1983). The cycle of erosion is the total time required for reduction of land area to the base level. This entire period or the "cycle of erosion" can be divided into the three stages which is, monadnock (old) (HI< 0.3), in which the watershed is fully stabilized; equilibrium or mature stage (HI 0.3 to 0.6); and in equilibrium or young stage (HI > 0.6), (Table 1) in which the watershed is highly susceptible to erosion (Strahler, 1952).

Table 1 Cycle of erosion range (Strahler, 1952)

SI. No.	Hypsometric Integral range	Remark
1.	<0.3	Monadnock(Old Stage)
2.	0.3 to 0.6	Mature Stage
3.	>0.6	Young Stage

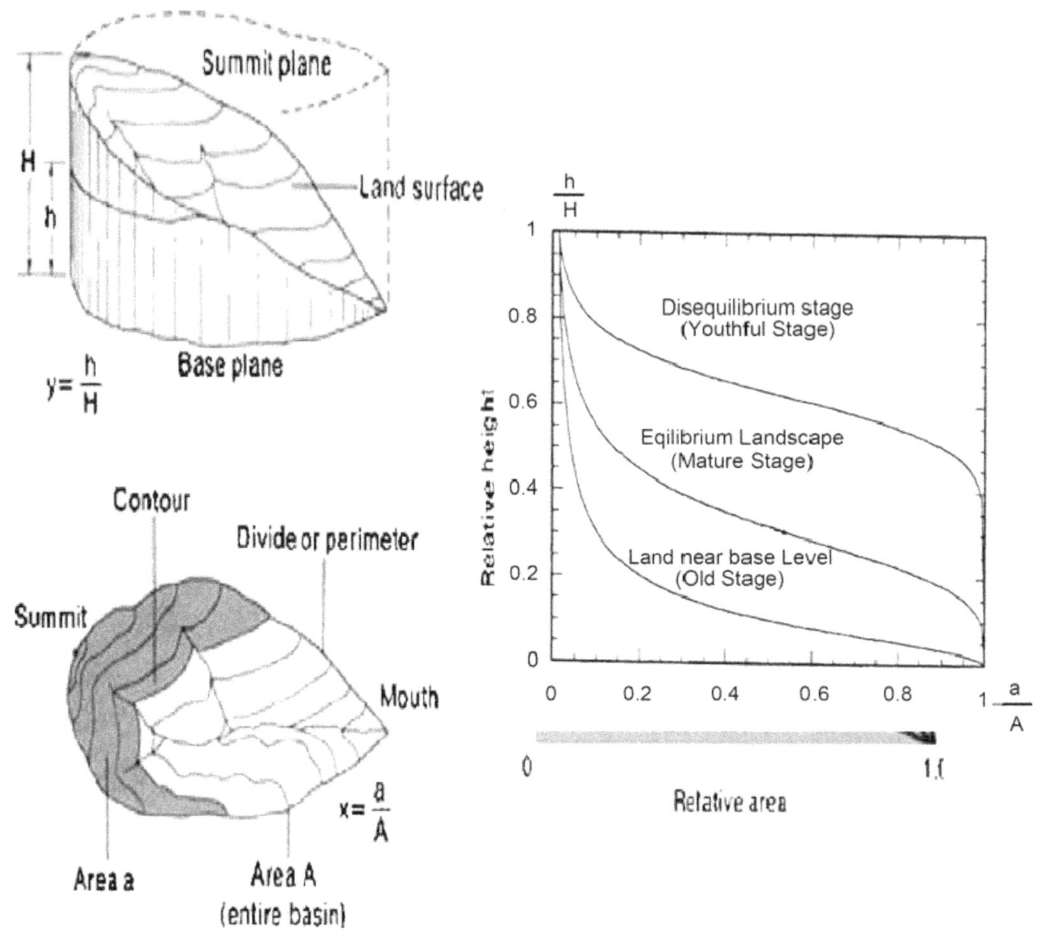

.Source: Strahler,1952.

Figure 1 Typical Hypsometric curve

Description of Study Area

Yagachi watershed lies between 75°38'31.2" to 75°54'43.2" East longitude and 13°5'52.8" to 13°23'42"North latitude with watershed area of 550.86 km^2.Yagachi catchment receives 1031mm of rainfall annually with relief ranges from 1829m to 918m (ASTERDEM Oct. 2011). The watershed is situated in between Hassan and Chikkamagaluru district of Sothern part of Karnataka (Fig. 2).

The topological information of the study area was downloaded from earth explorer and geo-referenced and extracted from DEM (representing the watershed terrain topology) using SWAT tool. Then the drainage networks for Yagachi catchment were generated and digitized using the ArcGIS10.1 software package. Stream order was assigned to each stream following the stream ordering system developed (Strahler.,1957). The entire Yagachi was found of sixth order. At stream order level hypsometric curve and Profile factor, maximum elevation, minimum elevation, were calculated using ASTERDEM in GIS platform. The attribute feature classes that accommodate these values were utilized to plot the hypsometric curves for the watershed.

Hypsometric curves and hypsometric integral is important watershed health indicator. Hypsometric analysis using GIS has been used by several researchers in India dealing with erosional topography (Pandey et al., 2004; Singh et al., 2008a; Singh et al., 2008b; Sharma and Seth 2010 and Sharma et al., 2011). The Kanhiya nala watershed under study is vulnerable to erosion looking to its undulating topography. Further, there is lack of hypsometric based studies to watershed health, which is attributable to the tedious nature of data acquisition and analysis is involved in estimation of hypsometric analysis.

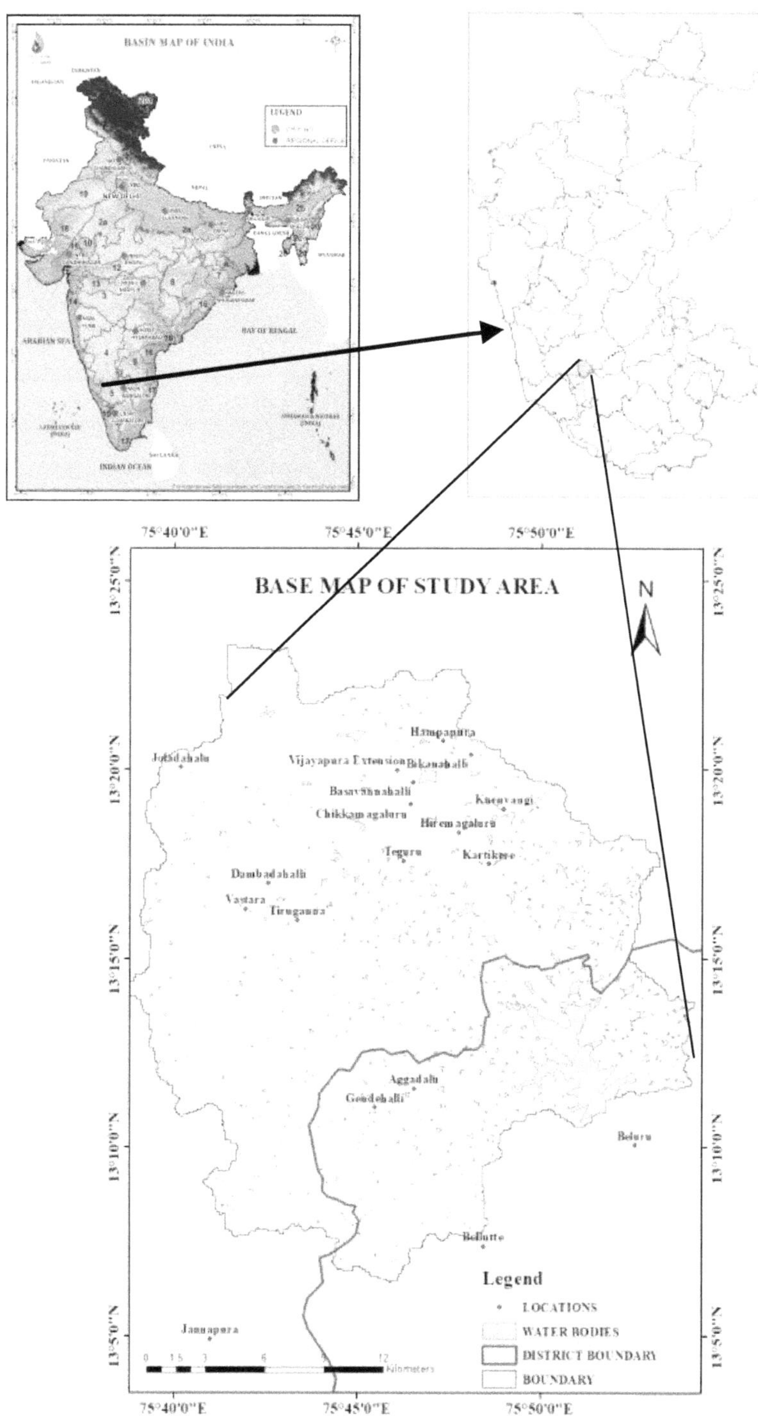

Figure 2 Base map of study area

Catchment delineation from DEM data

The DEM data was downloaded from the earth explorer with 30m resolution. Further, the DEM was processed to generate or delineate the watershed and sub-watersheds using the hydrology tool of spatial analyst module. The elevation map (Fig. 3) and drainage map (Fig. 4) of the study area respectively. The delineated Yagachi watershed with respect to steam order level 5th as North, South, Centre and North East and 4th marked as 1 to 20 are also shown in Fig. 5(a) and Fig. 5(b) respectively. The drainage network ordering was done using the Strahler stream ordering scheme (Strahler, 1964). The attribute table of the geo-referenced feature classes representing the extraction of elevation range and their enclosed area (Fig 6). The attribute feature classes containing these values were used to plot the hypsometric curve of the sub-watersheds from which the hypsometric integrals were estimated.

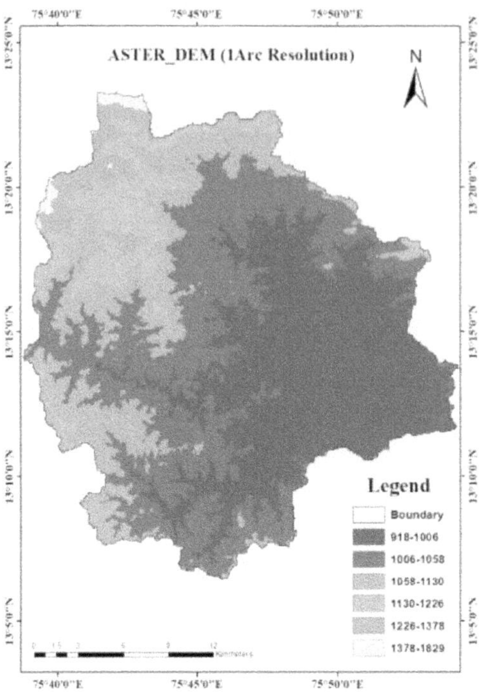

Figure 3 Digital Elevation Model of Study area.

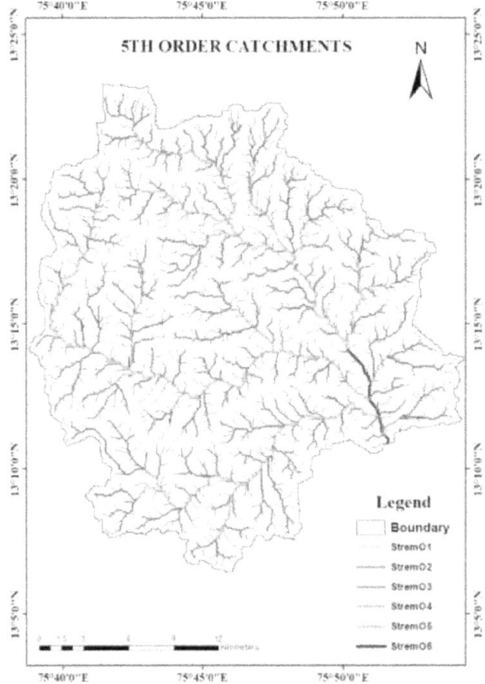

Figure 4 Drainage Network of Study area.

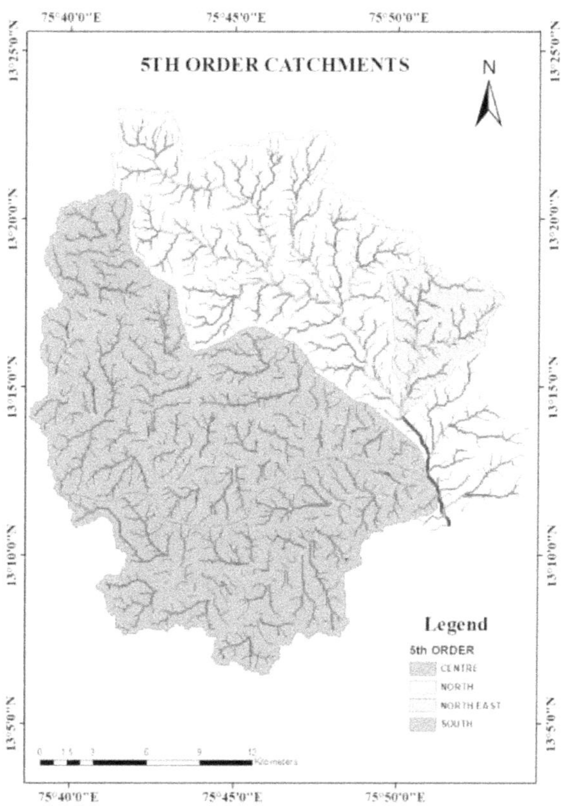

Figure 5(a) 5th Order Sub-watersheds.

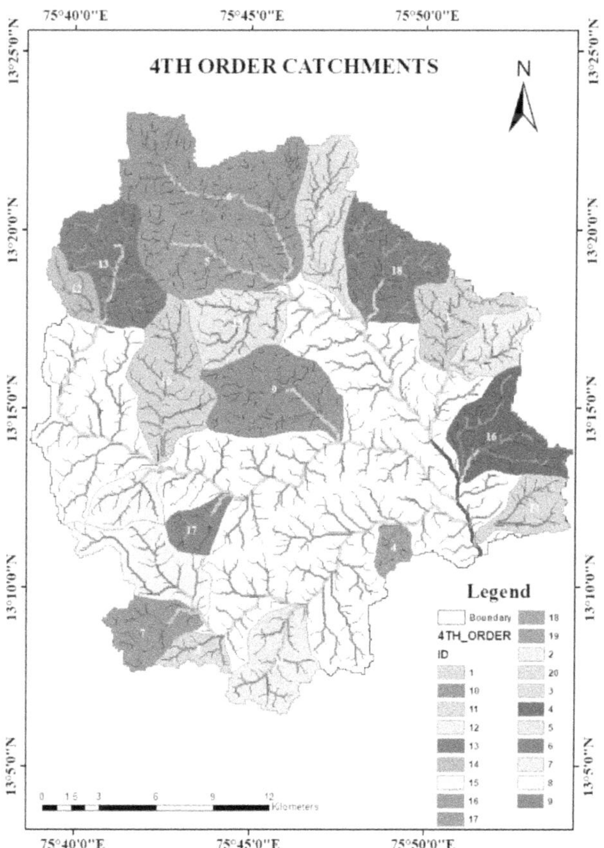

Figure 5(b) 4th Order Sub-watersheds.

Plotting of Hypsometric Curves (H_C) and Estimating HI

Hypsometric Curve

The hypsometric curve is a description of the cumulative relationship between elevation and the area within elevation intervals in a dimensionless form. The curve is plotted with the relative elevation plotted as the ordinate and the relative area within the watershed above the elevation plotted as the abscissa. The relative area is obtained as a ratio of the area above a particular elevation(a) to the total area(A) of the watershed encompassing the outlet. Similarly the relative elevation calculated as the ratio of the height of a given elevation(h) from the base plane to the maximum basin elevation(H) (Sarangi et al., 2001; Ritter et al., 2002).

There are different methods to estimate the HI values, among them elevation relief ratio method was more accurate and easy to calculate within GIS environment.(Sharma S. K. et al., 2013). The hypsometric integral was estimated (Fig. 7) using the elevation- relief ratio method proposed by Pike and Wilson(1971). The relationship is expressed as

Where, HI is the hypsometric integral;

E_{mean} is the weighted mean elevation of the watershed; E_{min} and E_{max} are the minimum and maximum elevation within the watershed.

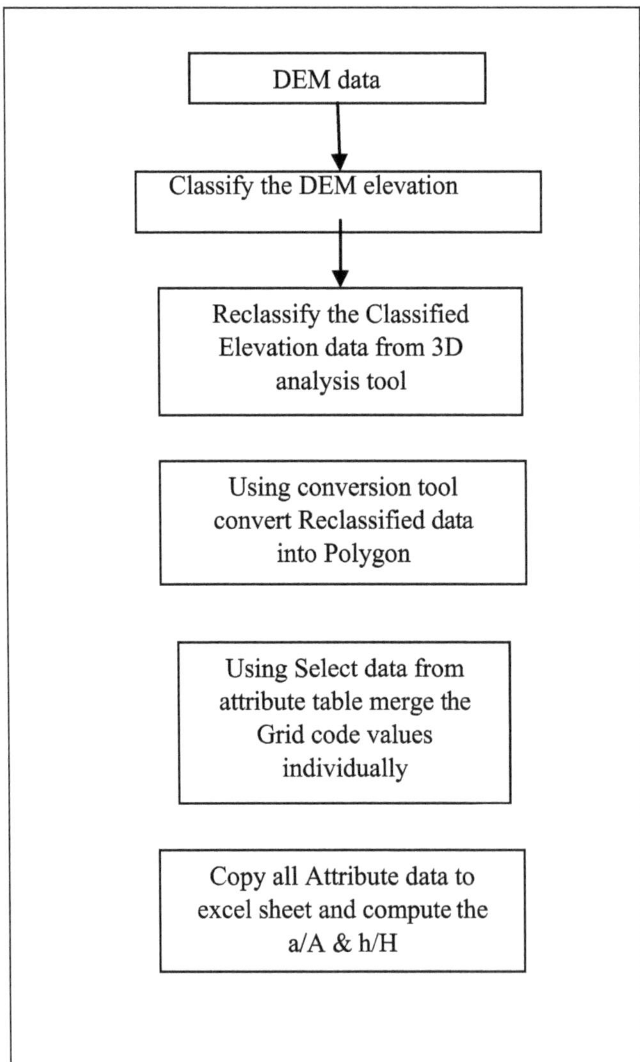

Figure 6 Extraction of a/A & h/H from DEM data

Results and Discussion

The hypsometric curves with respect to stream order level (6th, 5th, and 4th) as obtained were plotted and presented (Fig. 8). It was observed from the hypsometric curves of these stream order level that the drainage system is attaining the monadnock stage from the youth stage. The comparison between these curves (Fig. 8) indicated a marginal difference in mass removal from the sub-watersheds of study area. It was also observed that there was a combination of convex-concavo and S-shape of the hypsometric curves for the sub-watersheds under study. This could be due to the soil erosion from theses sub- watersheds resulting from the incision of channel beds, down slope movement of top-soil and bedrock materials, washout of the soil mass and cutting of streams banks.

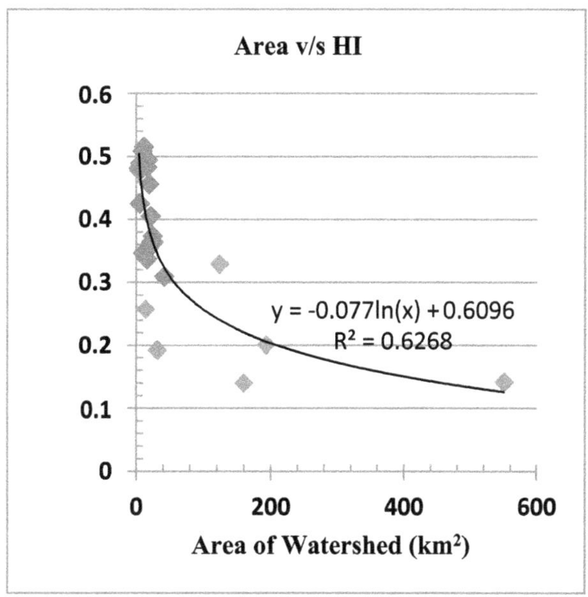

Figure 7 HI v/s Area of the watersheds.

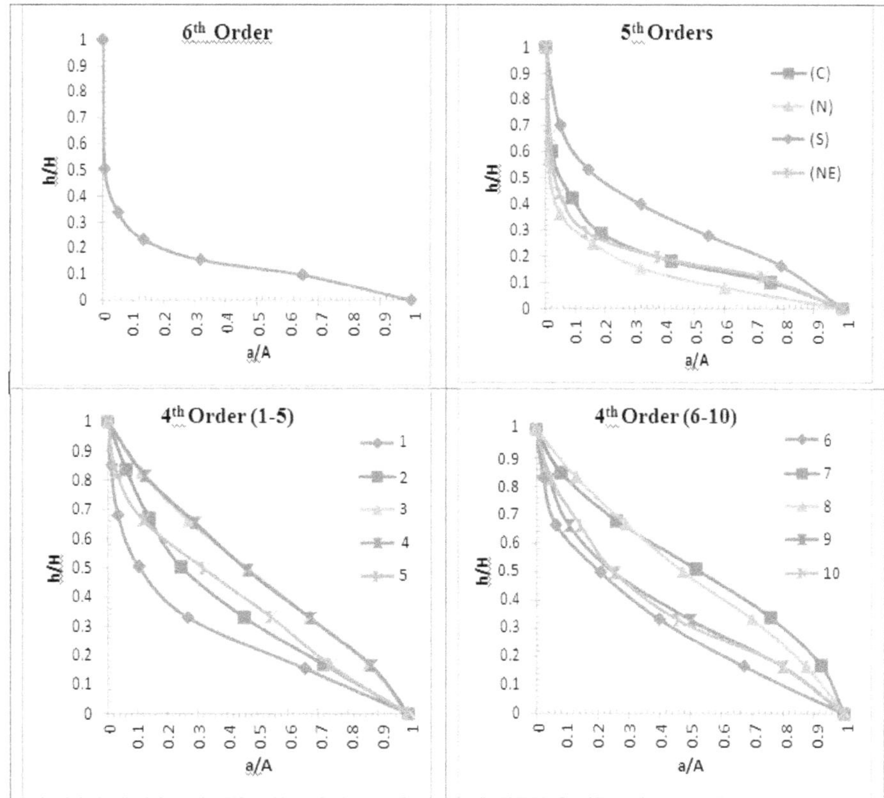

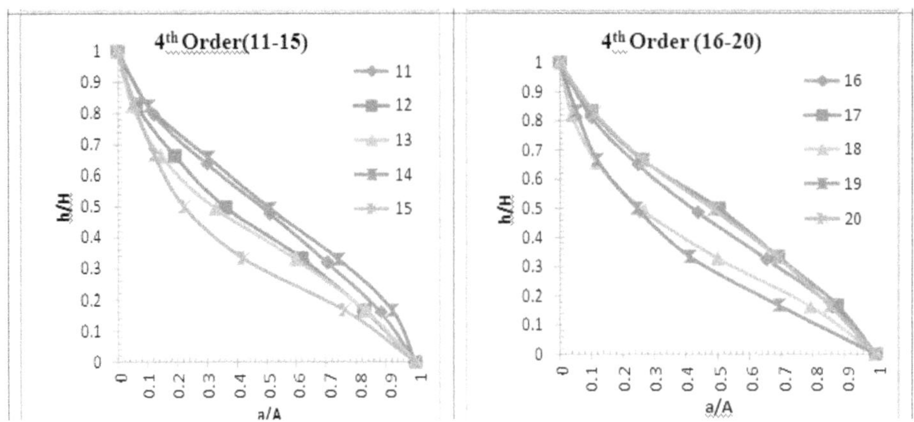

Figure 8 Hypsometric Curves of Sub-watersheds of Study area.

Table 2 Classifications of soil erosion level in the basin

Orders	Watersheds	Area (km^2)	E_{max} (m)	E_{min} (m)	EWEIG. ELE.(m)	HI	Remarks
6th Order	Entire watershed	550.86	1829	918	1,046.89	0.14	Monadnock
5th orders	NE	31.36	1242	950	1,006.25	0.19	Monadnock
	N	159.70	1829	959	1,080.96	0.14	Monadnock
	C	193.20	1548	951	1,071.04	0.20	Monadnock
	S	123.75	1164	948	1,019.10	0.33	Mature to old
4th orders	1	13.94	1242	956	1,029.68	0.26	Mondnock
	2	19.43	1392	974	1,122.06	0.35	Mature to old
	3	5.15	1103	982	1,039.88	0.48	Mature
	4	3.96	1022	930	974.32	0.48	Mature
	5	24.90	1395	1002	1,148.72	0.37	Mature to old
	6	42.07	1829	1003	1,258.48	0.31	Mature to old
	7	11.91	1164	987	1,078.17	0.52	Young to mature
	8	17.75	1153	985	1,068.02	0.49	Mature
	9	26.32	1162	962	1,034.74	0.36	Mature to old
	10	23.45	1305	996	1,108.63	0.36	Mature to old
	11	16.96	1080	967	1,021.61	0.48	Mature
	12	5.84	1548	1060	1,267.50	0.43	Mature
	13	22.06	1544	1053	1,251.74	0.40	Late Mature
	14	9.78	1020	935	978.22	0.51	Young to mature
	15	10.30	1175	965	1,037.73	0.35	Mature to old
	16	19.86	1019	945	978.76	0.46	Mature
	17	6.47	1143	981	1,060.25	0.49	Mature
	18	22.95	1160	977	1,043.56	0.36	Mature to old
	19	16.50	1266	996	1,086.99	0.34	Mature to old
	20	9.15	1155	998	1,073.66	0.48	Mature

Conclusion

Hypsometric analysis of watershed expresses the complexity of denudation processes and the rate of morphological changes. Therefore, it is useful to comprehend the erosion status of watersheds and prioritize them for undertaking soil and water conservation measures. In this study, efforts were made to estimate the hypsometric integral values of the Yagachi watershed with respect to different stream order levels by elevation-relief ratio method and HI ranges from 0.14 to 0.52. Moreover, The results of hypsometric integrals revealed that the fourth order sub-watersheds 7th and 14th is more prone to erosion in comparison to other sub-watersheds, which would necessitate construction of soil and water conservation structures like check dams, Percolation ponds, Contour trenches etc., at appropriate locations for the stream orders to arrest soil loss and conserve water. Further, the stream orders which are having hypsometric integral values less than 0.5 (i.e., approaching monadnock/Old stage)needs minimum mechanical and vegetative measures to arrest sediment loss but may require more water harvesting type structures to conserve water at appropriate locations in the watershed for conjunctive use of water.

References

Garg, S.K., (1983) "Geology- the Science of the earth". *Khanna Publishers*, New Delhi.

Harward,(1990)."Role of hypsometry and platform in basin hydrologic response." *Hydrological Processes* 4, 373-385.

Kuang-Yu Cheng, Jih-Hao Hung, Hung-Cheng Chang, Heng Tsai, Quo-Cheng Sung (2012). "Scale independence of basin hypsometry and steady state topography." *Geomorphology* 11(1), 171-172.

Markose, V., and Jayappa, K. (2011). "Hypsometric Analysis of Kali River Basin, Karnataka, India, Using Geographic Information System." *Geocarto International*, 26, 553-568.

Pandey, A., Chowdary, Y.M. and Mal, B.e. (2004). "Hypsometric analysis of watershed using Geographic Information System." *J. Soil & Water Cons. India,* 3, 123-127.

Pike, R.J. and Wilson, S.E. (1971). "Elevation-relief ratio, hypsometric integral and geomorphic area- altitude analysis. Geological Society of America Bulletin 82: 1079-1084.

Ritter, D.F., Kochel, R.C. and Miller, J.R. (2002). Process Geomorphology. McGraw Hill, Boston.

Shailesh Kumar Sharma, Sanjay Tignath etc., (2013) "Use of geographical information system in hypsometric analysis of Kanhiya Nala watershed" *International Journal of Remote Sensing & Geoscience (IJRSG)* 2(3), 30-35.

Sharma, S.K. and Seth, N.K. (2010). "Use of Geographical Information System (GIS) in assessing the erosion status of watersheds." *Sci-fronts A journal of multiple science.* 4 (4), 77-82.

Sharma, S.K., Seth N.K., Tignath, S. and Pandey, R.P. (2011). "Use of Geographical Information System in hypsometric analysis of watershed." *Journal of Indian Water Resources Society.* 31(3-4): 28-32.

Singh, O. and Sarangi, A. (2008). "Hypsometric Analysis of Lesser Himalayan Watersheds Using Geographical Information System." *Indian Journal of Soil Conservation*, 36, 148-154.

Strahler, A.N. (1952). "Hypsometric (Area-Altitude) Analysis of Erosional Topography." *Geological Society of America Bulletin*, 63, 1117-1141.

Strahler, A.N. (1957) Quantitative Analysis of Watershed Geomorphology. *Transactions*, *American Geophysical Union*,38, 913-920.

Strahler, A.N. (1964). "Quantitative Geomorphology of Drainage Basins and Channel Networks." 4(2). In: Chow V.T., Ed., *Handbook of Applied Hydrology*, McGraw Hill, New York, 439-476.

Groundwater Quality and Hydrogeochemistry for the Sustainable Development of Sub-Urban Area: A Case Study from Visakhapatnam, Andhra Pradesh

N. Subba Rao[1] and Maya Chaudhary[2]
[1]*Department of Geology, Andhra University, Visakhapatnam*
[2]*Department of Geology, Mohanlal Sukhadia University, Udaipur*
Corresponding author: maya@mlsu.ac.in

ABSTRACT

In recent time, the surrounding areas of Visakhapatnam, Andhra Pradesh are rapidly growing up. A comprehensive understanding of groundwater quality and hydrogeochemistry is essential for their sustainable development. In view of this, one sub-urban area of Visakhapatnam is selected for the present study. The area is geologically underlain by the khondalites of the Precambrian age with association of charnockites, pegmatites and quartzites as intrusive bodies. Drinking water quality index categorized the area into the excellent water quality zone (23%), good water quality zone (50%) and poor water quality zone (27%). Whereas the industrial water quality index classified the area into excellent water quality zone (17%), good water quality zone (60%) and poor water quality zone (23%). Trilnear diagram suggested that 73% of the area comes under the fresh water type and the rest (27%) under the transition water type. Gibbs diagrams demonstrated that most area (70%) is controlled by soil-water-rock interactions and the rest (30%) by evaporation due to gradual increase of sodium and chloride ions. The bivariate diagrams stated that the reverse ion exchange is the dominant process, following ion exchange and mineral dissolution. The principal component analysis (PCA) yielded PC1, PC2 and PC3, which represent the pollution, hardness and alkalinity processes, respectively. The spatial distribution of positive PC scores delineated the influences of processes on the groundwater specific area. The study suggests the treatment of water before using it for drinking and industrial purposes.

Keywords: Groundwater quality, principal component analysis, Visakhapatnam, hydrogeochemistry

Introduction

The primary source of water for more than 1.5 billion people in the world is groundwater (Veils *et al.* 2017). However, its quality is also an important issue of concern to human health, especially around the urban areas (Wang *et al.* 2013; Li *et al.* 2016; Adimalla and Li 2018). From the last two decades, the assessment of groundwater quality and groundwater geochemistry have been studied extensively such as Kumar *et al.* (2006) evaluated the hydrogeochemical processes (water-rock interactions, dissolution and weathering of mineral and ion exchange) governing the groundwater environment in Delhi. Hydrogeochemical characteristics of groundwater in the Alleppey district, Kerala were reported by Shaji *et al.* (2009) and said that the aquifer mineralogy is the main source of variation in the hydrogeochemistry of groundwater. Marghade *et al.* (2015) elucidated the geochemical characterization of groundwater, using principal component analysis, from the northeastern part of Nagpur urban area, Maharashtra and found that the quality of groundwater is governed by geogenic (rock weathering and reverse ion exchange) and anthropogenic sources (mainly urban wastes). Jassas and Merkel (2015) conducted a study on geochemical assessment of groundwater quality for drinking and irrigation purposes in Northern Iraq and found that the groundwater quality is mainly controlled by dissolution of carbonate minerals, alumina-silicate weathering and cations exchange. Kim and Park (2016) explained the hydrogeochemical characteristics of groundwater in an agricultural area of Hongseong, Korea, using principal component analysis, and stated that the nitrate contamination is the main factor for the variation of groundwater chemistry. Subba Rao (2018) explained the causes of changes in the groundwater quality with respect to water-soil-rock interactions and anthropogenic activities in parts of Andhra Pradesh.

Niranjan Babu *et al.* (1997) emphasized the importance of testing the chemical quality of groundwater, before any developmental activities in the surrounding urban areas. In view of this, an attempt has been made in one of the sub-urban areas of Visakhapatnam, Andhra Pradesh (Figure 1) to assess the groundwater quality and hydrogeochemistry for the sustainable development.

Study Area

The study area (17°40′–17°45′N and 83°5′–83°10′E) is situated in the southwestern part of Visakhapatnam, Andhra Pradesh, India (Figure 1). It covers an area of about 80 km^2. The climate is semi-arid with a temperature of 17°C in winter (December to February) to 39°C in summer (March to May). Annual rainfall is about 1,000 mm.

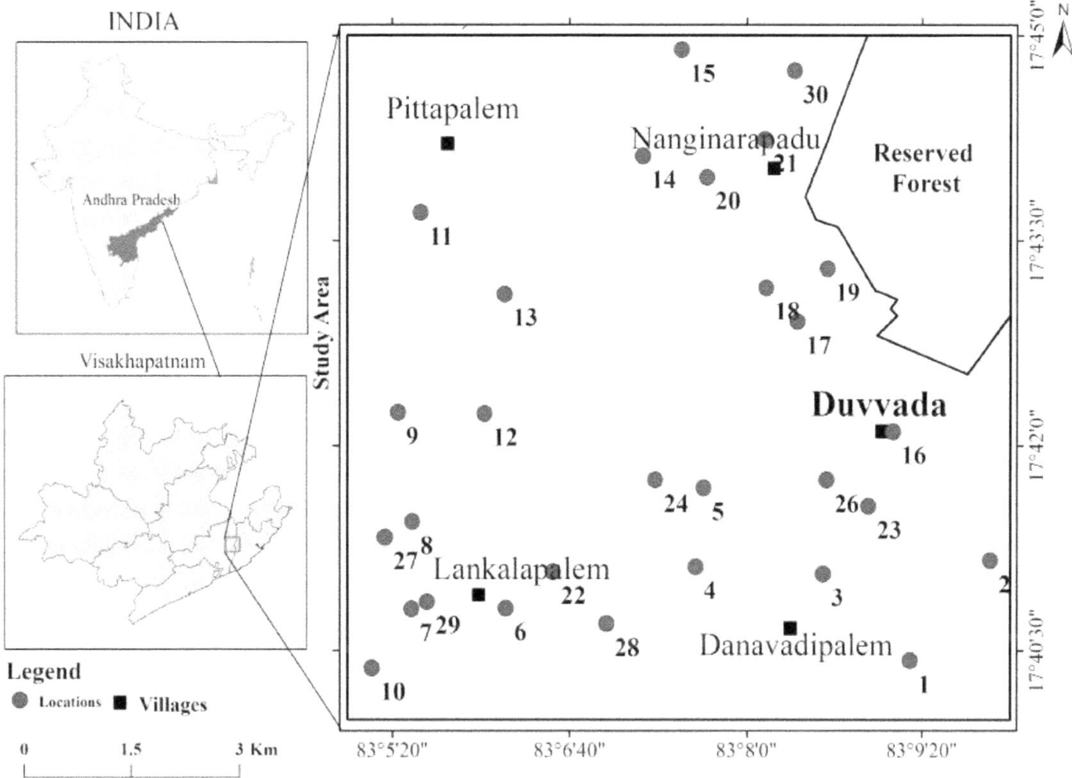

Figure 1 Location of the study area

It is characterized by undulating topography, sloping towards the southeastern part. The rock type of the area is mainly khondalites (garnet-ferrous-sillimanite-gneisses) of the Precambrian age and charnockites (pyroxene-granulites), pegmatites and quartzites occur as intrusive bodies, geologically. The essential minerals in the rock are quartz, feldspar, pyroxene, sillimanite, biotite, apatite and opaques of medium to coarse grained texture. The general foliation trend of the rocks is NE–SW with a dip of 70–80° SE.

Groundwater occurs mainly in two zones, one in the weathered zone (1–16 m) under unconfined conditions and second in the fractured rocks (12 – 60 m) under semi- confined conditions. The depth of water table ranges from less than 4 to more than 12 m below ground level, following the topography. Most of the lands, before converting into urbanization, belong to the agricultural fields. Usage of unlimited amounts of chemical fertilizers for higher crop yields is not uncommon in the area. Disposal of sewage wastes and open defection are very common phenomena.

Methodology

A total 30 groundwater samples were collected in 500 ml polyethylene bottles from the present study area during February 2013 (Figure 1). They were analyzed for determination of pH, electrical conductivity (EC), total dissolved solids (TDS), total hardness (TH as CaCO3), calcium (Ca^{2+}), magnesium (Mg^{2+}), sodium (Na^+), potassium (K^+), bicarbonate (HCO_3^-), carbonate (CO_3^{2-}), chloride (Cl^-), sulphate (SO_4^{2-}) and nitrate (NO_3^-), following the standard water quality procedures (APHA 2012). Drinking water quality index was computed to assess the suitability of groundwater quality for drinking purpose, following Ramakrishnaiah *et al.* (2009) and Ramyapriya and Elango, 2018). Similarly, the industrial water quality index (IWQI) was also computed.

Results and Discussion

The pH ranges from 6.5 to 8.7, with an average of 7.91 (Table 1), indicating alkaline character. The EC is observed in between 310 and 2,100 µS/cm and its average is 1,223.67 µS/cm. The concentration of TDS is from 200 to 1,365 mg/L and its average is 796.33 mg/L. In 76.67% of the groundwater samples, the TDS is more than its safe limit of 500 mg/L, which causes gastrointestinal irritation. The TH varies from 115 to 805 mg/L with an average of 403.33 mg/L. The TH exceeds its standard limit of 300 mg/L prescribed for drinking water in 80% of the total groundwater samples, which develop scale on well structures (BIS 2012). The Ca^{2+} varies from 32 to 168 mg/L, being an average of 91.20 mg/L, which is more than its threshold limit of 75 mg/L recommended for drinking water (BIS 2012) in 56.67% of the total groundwater samples. The Mg^{2+} is from 8 to 128 mg/L with an average of 40.53 mg/l, which exceeds the desirable limit of 30 mg/L in 66.67% of the total groundwater samples. The Na^+ is in the range of 5.9 to 271 mg/L and its average is 100.50 mg/L, which is greater than the highest desirable limit of 200 mg/L recommended for drinking water in 16.67% of the total water samples. It causes hypertension (BIS 2012). The K^+ is observed in between 0.2 and 26 mg/L with an average of 5.25 mg/L. The safe limit of K^+ (12 mg/L) exceeds in 13.33% of the groundwater samples (BIS 2012).

The HCO^-_3 varies from 100 to 720 mg/L and its average is 337.33 mg/L (Table 1). In 50% of the groundwater samples, the concentration of HCO^-_3 is more than its desirable limit of 300 mg/L. The concentration of Cl^- varies from 30 to 290 mg/L, being an average of 125 mg//L. In 3.33% of the total groundwater samples, the Cl^- content is higher than the recommended limit of 250 mg/L for drinking water, which causes salty taste. The content of SO_4^{2-} is observed from 2 to 91 mg/L and its average is 37.86 mg/L, which is below the safe limit of 200 mg/L prescribed for drinking water in all the groundwater samples. The value of NO^-_3 in the groundwater is in between 1.3 and 84.7 mg/L, with an average of 16.38 mg/L, which is more than the desirable limit of 45 mg/L in about 13.33% of the groundwater samples, which causes blue baby syndrome (BIS 2012).

Table 1. Summary of chemical composition of groundwater and drinking water quality standards (BIS 2012)

Chemical parameter	Min	Max	Average	Standard deviation	Coefficient of variation (%)	Drinking water quality standards
pH	6.5	8.7	7.91	0.72	9.11	6.5 to 8.5
EC ((µS/cm)	310	2,100	1,223.67	529.78	43.29	-
TDS (mg/L)	200	1,365	796.33	344.24	43.22	500
TH (mg/L)	115	805	403.33	152.59	37.83	300
Ca^{2+} (mg/L)	32	168	91.20	35.62	39.06	75
Mg^{2+} (mg/L)	8	128	40.53	28.64	70.67	30
Na^+(mg/L)	5.9	271	100.50	88.75	8.83	200
K^+ (mg/L)	0.2	26	5.25	6.11	116.38	12
HCO_3^- (mg/L)	100	720	337.33	167.55	49.59	300
CO_3^{2-} (mg/L)	10	70	26.84	17.01	63.38	-
Cl^-(mg/L)	30	290	125	70.64	56.51	250
SO_4^{2-} (mg/L)	2	91	37.86	28.97	76.52	200
NO_3^-(mg/L)	1.3	84.7	16.38	20.25	123.63	45

Table 2 Statistical summary of chemical composition of groundwater

Chemical parameter	Min	Max	Average	Standard deviation	Coefficien t of variation (%)	Contribution of ions to the total concentratio n
pH	6.5	8.7	7.91	0.72	9.11	-
EC (μS/cm)	310	2,100	1,223.67	529.78	43.29	-
TDS (mg/L)	200	1,365	796.33	344.24	43.22	-
TH (mg/L)	115	805	403.33	152.59	37.83	-
Ca^{2+} (mg/L)	32	168	91.20	35.62	39.06	36.73
Mg^{2+} (mg/L)	8	128	40.53	28.64	70.67	26.89
Na^{+}(mg/L)	5.9	271	100.50	88.75	8.83	35.29
K^{+} (mg/L)	**0.2**	6	5.25	6.11	116.38	1.09
HCO^{-}_{3} (mg/L)	100	720	337.33	167.55	49.59	50.25
CO^{2-}_{3} (mg/L)	10	70	26.84	17.01	63.38	8.13
Cl– (mg/L)	30	290	125	70.64	56.51	32.05
SO_4^{2-} (mg/L)	2	91	37.86	28.97	76.52	7.16
NO^{-}_{3} (mg/L)	1.3	84.7	16.38	20.25	123.63	2.41

Drinking water quality index

The calculated value of drinking water quality index (DWQI) varies from 29.07 to 134.96 with an average of 78.71. The DWQI is classified into five types: excellent water quality, when it is <50; good water quality, varies from > 50 to 100<; poor water quality, in between>100 and 200<; very poor water quality, from >200 to 300<; and unsuitable water quality, when it is more than >300 for drinking purpose. According to the classification of DWQI, 23.33% of the groundwater samples come under the excellent water quality category, 50% under the good water quality category and the rest 26.67% come under the poor water quality category for drinking purpose (Figure 2). Therefore, water treatment is essential in the groundwater samples collected from 26.67% of the wells, before using for drinking purposes. Otherwise, the public suffer a lot from water-borne diseases.

Table 3 Assigned weights of each drinking and industrial water quality parameter

Chemical parameter	Drinking purpose			Industrial purpose		
	Desirable limit	Unit weight	Relative weight	Desirable limit	Unit weight	Relative weight
pH	8.5	4	0.100	7	4	0.160
EC(μS/cm)		5	0.125	-	-	-
TDS(mg/L)	500	5	0.125	1,000	5	0.200
TH (mg/L)	-	-	-	300	4	0.160
Ca^{2+} (mg/L)	75	2	0.050	-	-	-
Mg^{2+} (mg/L)	30	2	0.050	-	-	-
Na^{+} (mg/L)	200	4	0.100	-	-	-
K^{+} (mg/L)	12	1	0.025	-	-	-

HCO_3^- (mg/L)	300	3	0.075	400	3	0.120
Cl^- (mg/L)	250	4	0.100	500	4	0.160
SO_4^{2-}(mg/L)	200	5	0.125	100	5	0.200
NO_3^- (mg/L)	45	5	0.125	-	--	-
Total		∑ 40	∑1.000		∑25	∑1.000

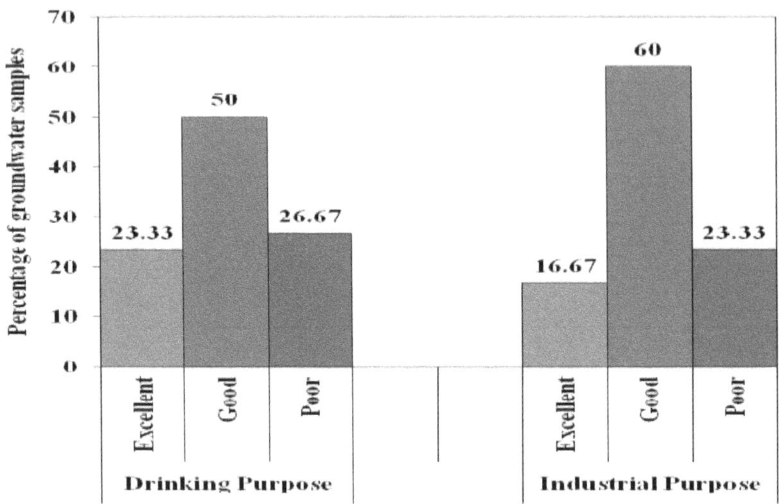

Figure 2 Classification of groundwater quality for drinking and irrigation purposes

Industrial water quality index

The values for industrial water quality index (IWQI) are from 33.12 to 118.32 with an average of 77.20. The IWQI is also classified into five types and values are varies same like DWQI. As per the classification of IWQI, 16.67% of the groundwater samples come under the excellent water quality category, 60% under the good water quality category and the rest 23.33% come under the poor water quality category for industrial purpose (Fig. 2). Thus, the water quality treatment is essential in the groundwater samples collected from the 23.33% of the wells, before using for industrial purpose. Otherwise, the industrial sector can also face problems related to the incrustation and corrosion properties, which will give impact on the industrial output and thereby the industrial income can come down.

Geochemical types of groundwater

To identify the variation occurring in the chemical composition of groundwater, the percentages of ions were plotted in trilinear diagram (Piper 1944), which help for characterization of geochemical evolution of groundwater. About 73% of groundwater samples fall in zone 5 (Fig 3a), indicating the carbonate hardness water type. The remaining 27% of groundwater samples fall in the zone 9, which shows a mixed or transition water type, where the groundwater cannot be identified as neither cations nor anions dominated by 50% of their total ionic concentrations. The distribution trend of the groundwater samples from carbonate hardness (fresh water) water type to mixed (transition) water type suggests that the original chemical quality of groundwater is mainly controlled by soil-rock-water interactions (zone 5) and is subsequently modified by mixed water type (zone 9) due to gradual influences of anthropogenic activities on the aquifer system, following the flow-path.

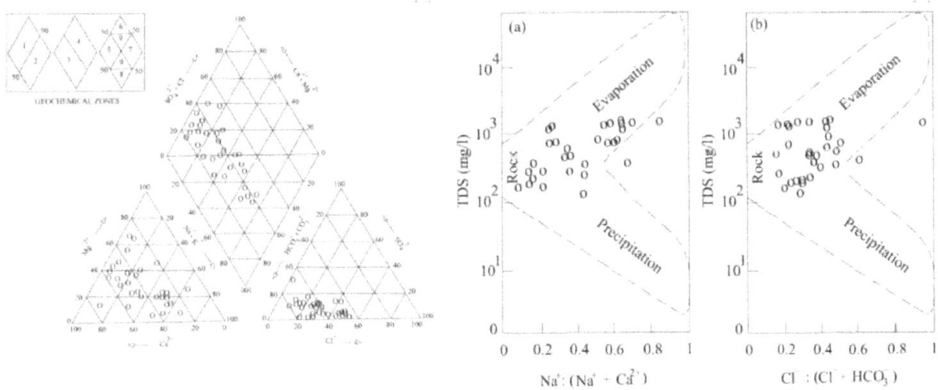

Figure 3(a) Characterization of geochemical groundwater types and controlling factors of groundwater chemistry

Controlling factors of groundwater chemistry

To identify the controlling factors of groundwater chemistry, Gibbs diagrams (Gibbs 1970) are adopted here. Seventy percent of groundwater samples are observed from the zone of rock domain (Figure 3b), where the groundwater quality belongs to fresh water type (TDS < 1,000 mg/L), which is caused by weathering and dissolution of soil salts and/or rock-forming minerals. The rest of groundwater samples (30% fall in the zone of evaporation domain, where the quality of groundwater belongs to brackish water type (TDS > 1,000 mg/L). This is a result of higher concentrations of Na^+ and Cl^- ions, causing the higher TDS, due to gradual influences of domestic effluents and irrigation-return-flows on the groundwater system. Consequently, the groundwater samples move towards evaporation domain from the rock domain.

The interactions of soil-rock-water mainly tend to contribute different chemical constituents to the groundwater system. Venugopal *et al.* (2009) used a diagram $Ca^{2+} + Mg^{2+}$: $HCO^-_3 + SO_4^{2-}$ to explain the role of weathering and dissolution processes on the groundwater body. In the present study area, 86.67% of the groundwater samples fall below the equiline of $Ca^{2+} + Mg^{2+}$: $HCO_3^- + SO^{2-}$ (Fig. 4a), indicating the reverse ion exchange as the dominating process, which controls the aquifer chemistry. Whereas the rest of the groundwater samples (13.33%) are found above the 1:1 line of $Ca^{2+} + Mg^{2+}$: $HCO_3^- + SO^{2-}$, which suggests the involvement of ion exchange process as the second factor in the groundwater system. The plotting of chemical data in $Ca^{2+} + Mg^{2+}$ *vs* Na^+ diagram (Figure 4b) also supports both the reverse ion exchange and ion exchange processes in 76.67% and 23.33% of the groundwater samples, respectively.

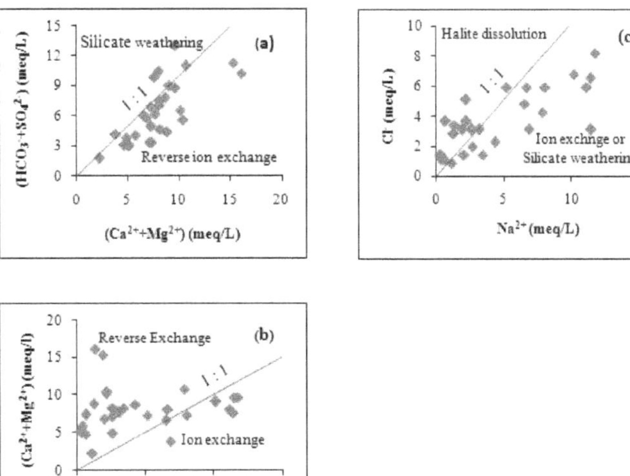

Figure 4 Relationships (a) $Ca^{2+} + Mg^{2+}$ *vs* $HCO_3^- + SO_4^{2-}$, (b) $Ca^{2+} + Mg^{2+}$ *vs* Na^+, (b) Na^+ *vs* Cl^-

In order to evaluate the effect of halite dissolution and ion exchange or silicate weathering, Meyback (1987) suggested a ratio Na^+/Cl^- . If the ratio is more than one, it reflects the ion exchange or silicate weathering process, while that is less than one, it indicates the halite dissolution. In the present study area, the ratio Na^+/Cl^- varies from 0.18 to 3.71, with an average of 1.16. In 53.33% of the groundwater samples, the ratio Na^+/Cl^- is more than unity (Fig. 4c), which indicates the ion exchange or silicate weathering, as the dominating factor controlling the chemistry of groundwater, while the rest of the groundwater samples (46.67%) show the ratio Na^+/Cl^- below unity, signifying the halite dissolution, as the second governing process of the groundwater chemistry.

Principal component analysis

In principal component analysis (PCA), the thirteen chemical variables (pH, EC, TDS, TH, Ca^{2+}, Mg^{2+}, Na^+, K^+, $HC\text{-}O_3^-$, CO_3^{2-}, Cl^-, SO_4^{2-} and ${}^3NO_3^-$) were transformed into three orthogonal principal components (PC1, PC2 and PC3), which account for about 77% of the total variance of the data matrix with eigenvalues more than one. The PC1 has strong positive loadings (< 0.600) on Na^+ (0.944), EC (0.822), TDS (0.822), Cl^- (0.796), K^+ (0.766), NO_3^- (0.708) and HCO_3^- (0.623), the PC2 on Mg^{2+} (0.936) and TH (0.886) and HCO^-(0.708), and the PC3 on CO^{2-} (0.907) and pH (0.902), indicating the influences of soil CO2, weathering and dissolution of plagioclase feldspars, and man-made pollution (domestic wastes, irrigation-return- flows, and potassium and nitrogen fertilizers) in PC1, the dissolution of CO2, weathering and dissolution of ferromagnesium minerals (biotite and garnet) in PC2 and the dissolution of soil CO2 in PC3. They represent the pollution, hardness and alkalinity controlled processes, respectively.

Spatial distribution of controlling processes of groundwater chemistry

To ascertain the influence of processes of groundwater chemistry on the groundwater specific area, the highest positive principal component scores (PCS), which are more than 1.0, are taken into consideration. The values of PCS1, PCS2 and PCS3 of pollution, hardness and alkalinity processes vary from 1.397 to 2.272, 2.88 to 3.156 and 1.013 to 1.232, respectively. Accordingly, the groundwater samples (4, 5, 11, 21, 23 and 29) of pollution process is observed from the northeastern (21), northwestern (11), southwestern (29), central (4 and 5) and eastern (23) parts, while those (2 and 20) of hardness process are found in the northeastern (20) and southeastern (2) parts. The groundwater samples (14, 17, 24 and 26) of alkalinity process are mainly spread northern (14), eastern (17 and 26) and central (24) parts (Figure 1). Since the pollution process appears to be dominant one in the present study area, which is followed by alkalinity and hardness processes, the necessary treatment is essential before taking the water for both drinking and industrial purposes.

Conclusions

Groundwater quality is of alkaline nature. Water quality index classifies the 27% and 20% of the study area comes poor water quality type for drinking and industrial purposes, respectively. Pipers diagram suggests that the geochemical evaluation of groundwater from carbonate hardness (73%) to mixed type (37%) due to gradual increase of Na^+ and Cl^- ions. Gibbs diagrams indicate that most groundwater samples (70%) fall in rock domain due to influence of water-rock-interactions and the rest (30%) in zone of evaporation due to impact of anthropogenic sources. Bivariate diagrams suggest the reverse ion exchange, ion exchange and mineral dissolution as the controlling processes of groundwater chemistry. The PCI represent the influences of soil CO_2, weathering and dissolution of minerals and man-made activities, the PC2 the soil CO_2 weathering and dissolution of ferromagnesium minerals and the PC3 the dissolution of soil CO_2, which measure pollution, hardness and alkalinity controlled processes, respectively. The influences of these processes on the groundwater specific area are identified. Necessary treatment is essential before taking the water for drinking and industrial purposes.

References

Adimalla N. and Li, 2018. Groundwater quality for drinking and irrigation purposes and potential health risks assessment: A case study from semi-arid region of South India. Expo Health doi:10.1007/s12403-018-0288-8.

APHA, 2012. Standard Methods for the Examination of Water and Wastewater.

Amer. Public Health Assoc. Washington DC.

BIS, 2012. Indian standard specifications for drinking water. IS:10500, Bureau of Indian Standards, New Delhi.

Gibbs RJ. 1970. Mechanism controlling world's water chemistry. Science 170: 1088- 1090.

Jassas H, and Merkel B. 2015. Assessment of hydrochemical evolution of groundwater and its suitability for drinking and irrigation purposes in Al-Khazir Gomal Basin, Northern Iraq. Environ Earth Sci 74: 6647–6663. Doi: 10.1007/s12665-015-4664-4.

Kim HS, and Park S. 2016. Hydrogeochemical Characteristics of Groundwater Highly Polluted with Nitrate in an Agricultural Area of Hongseong, Korea. Water 8(8): 345. doi:10.3390/w8080345.

Kumar M, Ramanathan AL, Rao MS, and Kumar B. 2006. Identification and evaluation of hydrogeochemical processes in the groundwater environment of Delhi, India. Environ Geol 50:1025-1039.

Li P, Li X, Meng X, Li M, and Zhang Y. 2016. Appraising groundwater quality and health risks from contamination in a semiarid region of northwest China. Expo Health 8(3):361-379. doi: 10.1007/s12403-016-0205-y.

Marghade D, Malpe DB, and Subba Rao N. 2015. Identification of controlling processes of groundwater quality in a developing urban area using principal component analysis. Environ Earth Sci. 74(7):5919-5933. doi: 10.1007/s12665- 015-4616-z.

Marghade D, Malpe DB, and Zade AB. 2011. Geochemical characterization of groundwater from northeastern part of Nagpur urban, Central India. Environ Earth Sci 62:1419-1430.

Marghade D, Malpe DB, and Zade AB. 2012. Major ion chemistry of shallow groundwater of a fast growing city of Central India. Environ Monit Assess 184:2405-2418. (Both are not in text)

Meybeck M. 1987. Global chemical weathering of surficial rocks estimated from river dissolved leads. Am J Sci 287:401-428.

Niranjan Babu P, Subba Rao N, Chandra Rao P, and Prakasa Rao J. 1997. Groundwater quality and its importance in the land developmental programmes. Ind J Geol 69: 305–312.

Piper AM. 1944. A graphical procedure in the geochemical interpretation of water analysis. Trans Am Geophys Union 25:914–923.

Ramakrishnalah CR, Sadas Hivalah C, and Ranganna G. 2009. Assessment of water quality index for the groundwater in Tumkur Taluk, Karnataka state, India. E-J Chem 6(2):523-530.

Ramyapriya R, and Elango L. 2018. Evaluation of geogenic and anthropogenic impacts on spatio-temporal variation in quality of surface water and groundwater along Cauvery river, India. Environ Earth Sci 77: 2. doi: 10.1007/s12665-017-7176.

Shaji E, Vinayachandran N, and Thambi DS. 2009. Hydrogeochemical characteristics of groundwater in coastal phreatic aquifers of Alleppy district, Kerala. J Geol Soc India 74: 585-590.

Subba Rao, N. 2018. Groundwater quality from a part of Prakasam district, Andhra Pradesh, India. Appl Water Sci 80:30 doi: 10.1007/s13201-018-0665-2.

Veils M, Conti KI, and Biermann F. 2017. Groundwater and human development: Synergies and trade-offs within the context of the sustainable development goals. Sustain Sci doi:10.1007/s11625-017-0490-9.

Venugopal T, Giridharan L, Jayaprakash M, and Periakali P. 2009. Environmental impact assessment and seasonal variation study of the groundwater in the vicinity of river Adyar, Chennai, India. Environ Monit Assess 149:81-97.

Wang Y, Wang P, Bai Y, Tian Z, Li J, Shao X, Mustavich LF, and Li BL. 2013. Assessment of surface water quality vai multivariate statistical techniques: a case study of the Songhua River Harbin region, china. J Hydrol Environ Res 7:30-40.

Re-inventing Water Infrastructure: An Exploration of Accumulation by Dispossession and Creative Destruction in Kanpur, UP

Pawan Kumar
Centre for Gandhian Thought and Peace Studies, SSS Central University of Gujarat, Gandhinagar
pawan.k0507@gmail.com

ABSTRACT

There is an intrinsic geographical relationship between water, infrastructure, and political rule (Wittfogel, August, 1957). This paper explores the relationship between water, infrastructure, and democracy in the context of changing property relations in the era of neo-liberalism. David Harvey's theory of accumulation by dispossession (AD), and Joseph Schumpeter's concept of creative destruction (CD) provides the political-economic method which enables a discussion on water conflicts around new- infrastructure (for instance, concept of water-ways is linked to the politics of new- infrastructure) that facilitates an understanding of neoliberalization strategies for new- infrastructure beyond existing infrastructure. Discourses on relationship between water and infrastructure, and a case study of water resources of Kanpur (selected in the proposed smart city project of India and the proposed third Water Way on Ganga River water basin) will be taken as evidences of this paper to analyse the dynamics of property relations to water infrastructure. Through content and critical discourse analysis the paper concludes that the relationship between water, infrastructure, and democracy are shaped by changing property relations. Further this change simultaneously undermines people's rights over natural resources through invocation of accumulation by dispossession and creative destruction process. This paper is an outcome of my on-going Ph.D. research.

Keywords: Water, Infrastructure, Accumulation by Dispossession, Creative Destruction, Democracy.

Summary

Approach	Political economy & ecology
Water in Kanpur	As Commons
Infrastructure	Economic-technical, and commons in nature
Democracy (political rule)	State, corporate and elite's supremacy through political, and economic (material) & social power
Concepts & Results	Hegemony, economic production, urbanisation and industrialisation, ideology produces creative destruction and accumulation by dispossession

Introduction

This paper from ecological and political economy perspective seeking to highlight on creative destruction and accumulation by dispossession process causing a rift in hydro-social cycle, and a centralised control of market over water resources of Kanpur dispossessing people's right to water.

Theoretical outline and existing literature

The nexus between modes of water governance and forms of rule is debated extensively in the social sciences (Steward, 1955a; Butzer, 1976; Worster, 1985). Wittfogel's (1957) exploration of the causal linkages between large-scale irrigation systems and the emergence of centralised bureaucracies, and conceivably authoritarian rule, led major debates. Despite the fact, Wittfogel's hypothesis of 'hydraulic society' have been criticised as technological determinism, selective, ideologically motivated by anti-communism and so on (Millon et al., 1962; Obertreis et al. 2016).

At the same time, this paper takes consideration of different concepts and perspective which do not draw on Wittfogel's theorisation to explain this relationship[1]. In oriental despotism, there is no implicit definition or the category of infrastructure. Though, it was highlighting on large scale irrigation infrastructure. But the recent critiques of Wittfogel works and water discourse highlights on possibilities of other kinds of infrastructures related to water. For instance, in the 1960s and 70s, the emergence of the environmental, and anti-globalisation movements in the western world, and in the southern world most notably in India and Brazil, these relationships have been arguably conceived as modernity's hydraulic mission' (Allan, 2006) and modern water' (Linton, 2010) and challenged by invoking the concepts such as ecological rift', water justice', right to water', right to city', water as an element of biological life', environmental flow', echo-hydrology' etc. in the era of neo-liberalism (Poff and Zimmermann, 2010; Olden et al., 2012).

Recent scholarship in economics mostly deriving from neo-liberal project depicts water, water services and water infrastructure as commodity. According to these perspectives, if water is being treated as a commodity, then the intervention of capital in various forms would be a necessary outcome, for instance payment for ecosystem services approaches (Wendland et al., 2010), 'willingness to pay' for water services (Whittington et al., 1991), 'benefit sharing' at (transboundary) basin level (cf. Crow and Singh, 2000; Turton, 2008). This can be interpreted as capitalisation ofwater'.

Other critical intervention has been made to challenge the commodity status of water. They argued that water resources, and its governing aspects are difficult to commodify or marketise. For instance Bakker (2003) has called water as an 'uncooperative commodity' in the context of the privatisation of urban water supply in England and Wales. The privatisation of large-scale canal irrigation was designated a 'neo-liberal fallacy' in the late 1980s (Moore, 1989)[2]. It is important here, to look at capitalism in the era of neo-liberalism in the light of evolutionary perspective of Schumpeter to locate developments and innovations in accumulation process in the context of the natural resources, the only market left for the domination.

Creative Destruction (CD)

This process of environmental destruction is inherent in the capitalist mode of production. In this process, the crises in accumulation engine keeps emerging occasionally, and hence, the enforced destruction of a mass of productive forces' (Marx and Engels [1848] 1998, p. 42). This is what Schumpeter conceived as a process of _creative destruction'.

According to Karl Marx (1981) in the capitalist mode of production, contradictions (class struggle) and crisis (production and accumulation) are inherent. It maintains its accumulation process through violent destruction of capital which further linked with new forms of accumulation mechanism. That is how destruction keeps the process alive.[3]

> The opening up of new markets, foreign or domestic, and the organizational development from the craft shop and factory to such concerns as U.S. Steel illustrate the same process of industrial mutation—if I may use that biological term—that incessantly revolutionizes the economic structure from within, incessantly destroying the old one, incessantly creating a new one. This process of Creative Destruction is the essential fact about capitalism (Schumpeter, 1976: 83).

---

[1]However, this paper draws from his observation and further advancement in this debate of a possible relationship between political rule and water controlling infrastructure (Boomgaard, 2007; Palerm- Viqueira, 2009; Stride et al., 2009; Akhter and Ormerod, 2014; Banister, 2014; Linton and Budds, 2014).

[2]Although, Molle and Berkoff (2007) have depicted the history of the idea of 'water pricing' and argued that water pricing was not being successful in the context of the allocation, and efficiency enhancing work. This was even before the age of tradeable water rights' which emerged in the 1990s, what we call 'market triumphalism' (Rosegrant and Binswanger, 1994; Peet and Watts, 1993).

[3]The concept of creative destruction is advanced by both Schumpeterian and Marxists. The former, emphasises on the –creative□ side of the concept, and the later on the –destructive□ side of the concept.

The process of CD refers to the technological innovation and free market processes embody capitalist dynamism of the destruction of previous forms of capital accumulation and natural resources.

In the context of natural resources, the scholars have attempted to study creative destruction in the context of corporate response to climate change, biodiversity loss, changes to the nitrogen cycle, ozone depletion, and ocean acidification. They have depicted corporate response to environmental crisis as political myth (Wright, & Nyberg, 2014).[4]

To hide these catastrophic effects of capitalism, the narratives of _green' or _natural' capitalism, _ecological modernism', and _organisational sustainability' are created (Chiapello 2013). Further, in the backdrop of environmental crisis in the name of _rational' and _efficient' corporate responses, a corporate driven discourse reinforces the existing logic of corporate capitalism and hence, there is naturalisation of neo- liberalism (Wright, & Nyberg, 2014).

In this sense, this paper demonstrates that the logic of capitalism incorporates criticism in the form of unending self-destructive tendency (Boltanski and Chiapello 2005). These narratives and political myths, appropriate on one hand critiques of capitalism (in the name of _green', _sustainable', and _environmentally friendly' products) into its re-appropriative fold of creative destruction process resulting further into new forms of profit strategy such as accumulation by dispossession, and hence self-destruction.

David Harvey defined neoliberalism as creative destruction and accumulation by dispossession. He argued:

> By this I mean the continuation and proliferation of accumulation practices that Marx had treated as 'primitive' or 'original' during the rise of capitalism. These include the commodification and privatization of land and the forceful expulsion of peasant populations (as in Mexico and India in recent times); conversion of various forms of property rights (e.g. common, collective, state) into exclusive private property rights; suppression of rights to the commons; commodification of labour power and the suppression of alternative (indigenous) forms of production and consumption; colonial, neocolonial and imperial processes of appropriation of assets (including natural resources); monetization of exchange and taxation, particularly of land; the slave trade (which continues particularly in the sex industry); and usury, the national debt and, most devastating of all, the use of the credit system as radical means of primitive accumulation (Harvey, 2006: 153).

Appropriation of rights and global commons (land, air, and water) is a key of accumulation by dispossession[5]. It (accumulation by dispossession)[5] takes place in social-spatial context. In others words, this is a process outside of developed capitalism. The social-spatial have two features: 1) the social is related to dispossession of people's rights over means of subsistence, for instance global commons and 2) invention of new sites of exploitation. The exploitation sites are geographically linked with third world countries raw materials and natural resources. Harvey argued:

---

[4] Similarly, the 1960s and 70s were marked by the arrival of land mark reports and books relating to water politics, scarcity, and conflicts around natural resources, and finally the environmental crisis (Carson, 1962; Hardin, 1968; Ehrlich, 1968; Meadows, 1972; Goldsmith; 1972; Heilbroner, 1974; Simon and Kahn, 1984) etc. These scholars highlighted that these environmental catastrophes are resulting from the inherent contradiction of a capitalist imaginary of endless economic growth and unending exploitation of nature (Foster, 2000).

[5]Other features of accumulation by dispossession suggested by David Harvey includes the commodification and privatization of land and the forceful expulsion of peasant populations; conversion of various forms of property rights – common, collective, state, etc. – into exclusive private property rights; suppression of rights to the commons; commodification of labour power and the suppression of alternative, indigenous, forms of production and consumption; colonial, neo-colonial and imperial processes of appropriation of assets, including natural resources; monetization of exchange and taxation, particularly of land; slave trade; and usury, the national debt and ultimately the credit system (Harvey, 2004) in –The _new' imperialism: accumulation by dispossession, Socialist register, 40(40).

> –Neoliberalism has in effect swept across the world like a vast tidal wave of institutional reform and discursive adjustment, and while there is plenty of evidence of its uneven geographical development, no place can claim total immunity (with the exception of a few states such as North Korea)□ (Harvey 2006, p. 145).

Harvey (2006) talks about four strategies of accumulation by dispossession: 1) privatization of commons such as water, energy and telecommunications; public institutions such as universities; culture and natural resources); 2) financialization referring to a process of bringing activities and resources into global financial system, 3) state redistribution strategies, refers to mechanism of the state to relocation of wealth and assets from poor, to the rich in society. It may be defined in terms of re- visioning tax codes into corporate fold, and 4) management and manipulation of crisis. These refer to new sites of exploitation.

Harvey, also suggest that liberation is possibly can lead within capitalism through people's resistance against accumulation by dispossession in the form of re- appropriation of public spaces and means of production (natural resources). This was also a central argument of Harvey's classical exposition of the right to city (Harvey, 2003). In this sense Harvey's theory is useful to take into account of neo- liberalisation of water and other natural resources. It helps to study the dialectical relations between the process and struggle for appropriation and re-appropriation of commons in Kanpur.

Highlighting Water Conflicts in Kanpur

Kanpur has remained one of the most conflicted zones on water in India. It is facing three level conflicts over water. The first proxy war is on the drinking water which is between the people and the corporations over water scarcity and the use of groundwater (Times of India, Jun 17, 2016; May 8, 2014; May 31, 2013, Jan 13, 2017). The conflict over the Groundwater waged against illegal use and expropriation by corporates and multinationals like Coca Coal and Pepsi in the area. The other conflict is on the construction of dams and power projects, and displacement of people due to dams. The second conflict is over the environment degradation, water pollution and loss of biodiversity of the river like Ganga. Finally, there are territorial disputes at domestic, national and global level, and diversion of the Ganga River's water (Raman, 2016).

Exploring the nexus: abundance-crisis-scarcity in Kanpur

Kanpur has the biggest water supply system in Uttar Pradesh with water in abundance due to the presence of river Ganga, the river Pandu, (surface water sources), rainwater and groundwater etc[6]. The State Government and the Municipal Government are primarily responsible for water resources, and water supply and sewerage system. Kanpur Jal Sansthan (KJS) deals with water supply and sewerage system. In 1975, Kanpur Jal Sansthan (KJS) was constituted as a specialized and decentralised body under the U.P. Water Supply & Sewerage Act. The Kanpur Nagar Nigam (KNN) deals with the social infrastructure such as education, health and medical services. The capital works are carried out by U.P. Jal Nigam (UPJN) (JNNURM, August, 2006).

In the context of water resources of Kanpur, the studies have mainly been conducted to explicate the inefficiency of urban water sewage system leading to pollution in general. Further, they have also argued that availability of water resource is not a problem in major cities situated on the bank of the Ganga river of India including Kanpur (for detail please see Sharma & Rajput, 2017; Singh, 2017: 467; Gupta & Naik, 2017; Shamsad, 2017; Viswanathan, 2008: 203).

Kanpur is the biggest city of Uttar Pradesh and eighth biggest city in India. According to the 2001 census, among the big towns of Uttar Pradesh, the growth of Kanpur has been phenomenal in the context of demography, and urbanization. In earlier times it was known as Manchester of India. Now it is the biggest commercial capital of Uttar Pradesh.

---

[6]The Pandu River is a tributary of Ganga which passes through Kanpur. The source of Pandu River is a lake in Farrukhabad. The river runs for around 120 kilometers and meets the Ganga in Fatehpur (Eco- Friends, Dec. 31, 2002). Retrieved from http://www.ecofriends.org/reports/008pandu.htm.

It is also a major city connected to the National Waterway No 1 (NW-1, The Ganga North India). The NW No-1 is followed via Ganga-Bhagirathi-Hooghly River System connecting Haldia-Kolkatta (Calcutta) - Farakka - Munger - Patna - Varanasi - Allahabad[7].

Although, the NW-1 is debated extensively by several scholars and professionals such as media personals, Scientist, Environmentalists, Politicians, political activists and so forth. Some of them argue that despite of the Ganga Action Plan (1985), the BJP launched the Ganga Rejuvenation programmer to clean river. Further, in the name of Rejuvenation of Ganga, the National Waterway has been proposed for trade, tourism and transportation with 11 major terminals including 1 at Kanpur. The project will impose huge environmental, cultural, social political or event inter-state tension threat to river‘s ecosystem by lock of free flowing Ganga into a series of reservoirs of 100 km length each. Further, the self-purifying and preserving ecological capacity of the Ganga would destroy as explored by National Environmental Engineering Research Institute (NEERI), an apex research organization under the Indian Council for Scientific Research with headquarters at Nagpur and a century ago, scientist Hankin (1896) reported that the waters of the Ganges and Jamuna rivers in India marked by antibacterial action. Edward Twort (1915) and Felix d‘Herelle (1917) autonomously described isolating filterable entities capable of destroying bacterial cultures and producing small cleared areas on bacterial lawns seemingly implying that discrete particles were involved. There are arguments that the NW-1 is a commercialization, privatization and neo-liberalization of the Ganga River‘s Water (Arise India Forum, 2012; Times of India, Jun 7, 2014; Jhunjhunwala, B. 2014a; 2014b; 2009; 2015; SANDRP, 2016).

Resistance against accumulation by dispossession

The major movement against depletion and extraction of groundwater by Coca Cola and Pepsi was led by Jal Adhikar Yatra (JAY) (Water Rights March (WRM)) in Kanpur. The rally was started from the Coca-Cola plant in Mehndiganj, Varanasi on 10th September 2006. The rally went through fifty districts in Uttar Pradesh including Kanpur and also neighboring districts within the state of Madhya Pradesh, Haryana, New Delhi and Rajasthan. The activists met with people from diverse backgrounds including farmers, labourers, Students from schools and colleges, teachers, lawyers, Political parties, social struggles in villages, towns and cities.

The rally debated these concerns with people and united those using protests, sit-ins, meetings, songs, street theater, Hasthakshar abhiyan, Parcha vithran, documentary and debates. The Jan Chetna Kalamanch team from Lok Samiti Varanashi made street theaters and songs on the issue to raise awareness. The major concern of these debates were the questions on the relationship between water and humanity, water scarcity, the role of Pepsi and Coca-Cola in expropriation, corporatization, and commodification of groundwater from the community. The impact of Coca-Cola plant in the lives of the average people in Kanpur and Mehndiganj and so forth, and the impact of water privatization, water pollution, and Groundwater depletion on people and communities, and water conservation and management as if people mattered (Jal Adhikar Yatra Report, 10th September 2006 – 05th October 2006; Drew, 2008).

A one day Area Water Partnership (AWP) Conference was held on 22 July, 2012 by Shramik Bharti, Kanpur; Society for Action in Community Health (SACH) and Society for Promotion of Wastelands Development (SPWD), New Delhi at Jajmau, Kanpur, an industrial city on the banks of the Ganga which also happens to be a major centre for leather tanneries, which let out toxic effluents. The Ganga at Kanpur is unclean and synonymous with pollution and is a proof of the government‘s failure to clean up the river. The Conference outcome was that the real agenda of the state is in commercialization of Ganga water, all in the name of cleaning the river.

As regards Ganga, 550 new dams are being proposed on 17 streams of the river in Uttarakhand. The Ganga water from Uttarakhand Himalayas is not reaching in Kanpur at all and gets diverted for other uses like thermal power plants, atomic power plants, irrigation and domestic water supply, on the way through barrages, dams and canals. Kanpur is getting polluted. Water of Ramganga and Kali River is polluted by domestic sewage, the paper mills and sugar mill sewage.

---

[7] The information is available at the website of Ganga Waterway Information Centre. http://www.gangawaterway.in/activity/90-comments-on-tor-for-nw-1-studies-submitted-to-inland- waterways-authority-of-india.html.

Various transnational waste water treatment companies preferred the centralized solutions to these problems. The World Trade Organization (WTO), World Bank (WB) and International Monetary Fund (IMF) have always preferred neoliberal solutions since 1980s, thereby the present conflict and crisis are the result of such solutions. The water supply and sewage is also managed by Jal Board and Jal Sansthan which are practically under the control of state government and Kanpur Nagar Nigam is powerless. In the spirit of the 74th Constitutional Amendment Act, the powers for these functions should be transferred to Kanpur Nagar Nigam, and the planning of solutions to these problems should be worked out by Kanpur Nagar Nigam in a participatory planning process.

A strong case was made against the privatization of water. It was suggested that in the name of cleaning the Ganga River, the commercialization of Ganga water is the real agenda. The draft water policy 2012 suggests that water is an economic good and therefore by implication cannot be provided free. Some are proposing that holy water of Ganga should be bottled even before Hardwar and sold in the market. The barrage was constructed in Kanpur to provide supply of 200 mild of drinking water, while people are not interested in the piped water and using only ground water for domestic use. Now it is being suggested that people cannot use hand pumps without permission of the government (Sengupata, 2014; Bharti, 2012).

India‘s 80% irrigation and drinking water source is groundwater. It is also alarming that the groundwater is depleting day by day. The State government formed 70 paani- panchayat and 10 bhujal-sena consist of villagers is an attempt to check groundwater depletion, and to sensitization of local people and communities. It intended to aware people about the utilization of groundwater judiciously while remaining alert to prevent its pollution. At least 30% of the development blocks in UP have pull out groundwater without recharging the resource ever. The state government has executed State Groundwater Conservation Mission (SGCM) in 'stressed' blocks. The rural mission would be implemented in 271 blocks, including Bundelkhand and Vindhyan region, while the urban part of the campaign will cover 22 cities, including Lucknow, Kanpur, Allahabad, Varanasi, Agra, Aligarh, Meerut, Ghaziabad and GB Nagar (Times of India, 17 July, 2017).

There are a numbers of petitions have been filed before the National Green Tribunal (NGT) principal bench and the Supreme Court (SC) New Delhi against the industries and other water managing bodies causing water scarcity and water pollution[8].

Re-inventing Water Infrastructures in Kanpur

Since the mid-twentieth century, global urban metabolisms have faced environmental, social, economic, and demographic challenges (Yigitcanlar and Lee 2014). It is important here to invoke concept of political myths (rational and efficient) of corporate capitalism in the context of responding to urban challenges. The concepts such as intelligent cities (Komninos 2002), smart cities[9] (Yigitcanlar 2015), information city, digital city, and sustainable city (Yigitcanlar 2006) are invoked in policy discussion in the backdrop of many urban problems including drinking water supply (issues around water resources as a whole are not being taken into consideration). In short, the smart city symbolises a new kind of technology-led urban utopia (Townsend, 2013).

Hollands, (2008), argued that the concept of smart city is highly ideological concept, hiding fundamental challenges and problems. It has been assumed that IT can deal with economy, environment and in various aspect in a more rational and efficient way. Critiques of urbanisation have argued that these cities are being proposed and promoted along the line of neo-liberal strategies to capture urban commons and governing infrastructure of the same (Harvey, 1989)

---

[8]For instance in the matter of Supreme Court of India case M.C. Mehta vs. Union Of India & Ors on 12 January, 1988, there are a number of cases held in NGT Judgement of the National Green Tribunal regarding Ganga river pollution, 13/07/2017, Order of the National Green Tribunal regarding

[9]The smart city concept started from e-governance movements, IT companies, and governments in Europe and the United States in the late 1990s and early 2000s (Coe et al., 2001). Smart cities primarily focus on ICT integration in modern urban infrastructures (Alawadhi et al., 2012).

This is because the rhetoric of the corporate smart city limited space for participation and democratic decision-making, due to the profit motive of global IT, software, engineering, construction and utilities companies (Haque, 2012; Hill, 2013). This is reflected in the business-friendly, pro-urban policies of the government, most notably the current Smart Cities Mission (SCM) (Government of India (GoI), 2014a)[10]. The prevention and control of pollution and rejuvenation of river Ganga, 17/04/2017, NGT notice to MoEF, UP on plea against Signature City Project, Order of the National Green Tribunal regarding Ganga pollution, 07/03/2017, Order of the National Green Tribunal regarding pollution of river Ganga in Segment _B‘ of Phase-I, 19/10/2016 (1988 AIR 1115, 1988 SCR (2) 530; India Environment Portal, 2016; 2017). The detailed order of these cases held before the NGT and the Supreme Court of India are available at the website of India Environment Portal http://www.indiaenvironmentportal.org.in/. language of smart city documents is highly questionable in the context of inclusion of dispossessed people. It is demarcated in various projects based on public-private partnership where state has a very limited role to play[11].

Although, various documents and articles on smart city, including the Ministry of Urban Development‘s Smart Cities Concept Notes of September and December 2014 talks about more greater participation in decision making process, inclusivity, constraints over land acquisition, financing and inclusiveness, but none of the article and document have invoked the human right to water and sanitation, existing water availability, water crisis, water scarcity in various cities selected in smart city project (GoI, 2014a, 2014b)[12]. Scholars have further criticised these kind technology driven projects as _corporate urbanism‘ (Sood, 2015).

Further, critiques have argued that smart cities the concentrates on elite power through private interests in urban development (Datta, 2015; Economic and Political Weekly 2010). Rahul Mehotra, argued that they are _founded on capital and investment, but don‘t consider the human being as part of this equation‘ (LiveMint, 2015). From commons point of view, a link between democracy (at the municipal level) and natural resources such as water, land, knowledge commons, space commons, people‘s collective rights over these commons have intricately tied up with question of appropriation through the processes of dispossession and elite accumulation, state-citizen power relations and rural livelihoods and capacities in India (Baka, 2013; Levien, 2012; Narain, 2009).

Conclusion

The water, infrastructure, and democracy are linked through the concept of commons in Kanpur. The urbanisation process after invocation of neo-liberal policies in the context of management of urban resources inherently linked with the process of creative destruction of old infrastructure of urban resource management institutions thereby facilitating accumulation by dispossession through new infrastructures such as smart cities and water ways. Therefore, in order to counter these linkages and processes, scholars invokes the concept of _the right to city‘ and _cities as commons‘ where people‘s resistance is inevitable. These processes can be interpreted as _appropriation by direct states and the markets strategies of accumulation by dispossession and re-appropriation of urban commons by peoples struggle. The scholars have pointed out that re-invention/reorganisation of infrastructures is mediated through property regimes in liberal democratic societies. The property regime produces new-

[10]This is to be done through the promotion of instruments including clean technology use, widespread information and communication technology (ICT) reliance, financing via public private partnerships (PPPs) and private sector investment, improved citizen consultation, and _smart‘ or e-governance initiatives rolled out for urban local bodies (ULBs) (GoI, 2014). These cities would offer: [g]ood quality but affordable housing, cost efficient physical, social and institutional infrastructure such as adequate and quality water supply, sanitation, 24 x 7 electric supply, clean air, quality education, cost efficient health care, dependable security, entertainment, sports, robust and high speed interconnectivity, fast & efficient urban mobility. (GoI, 2014).

[11]This is reflected in several underwhelming urban policies in India in recent years. For instance, the infrastructure-focused Mega City Scheme, rolled out in 1993 in Bangalore, Chennai, Hyderabad, Kolkata and Mumbai, (Chakravorty, 1996; GoI, 2004), subsequently launched in 2005, the Jawaharlal Nehru National Urban Renewal Mission (JNNURM), these urban policies were criticized for funding delays, incomplete devolution of power and ineffective urban land acquisition processes (GoI, 2012a; Weinstein et al., 2014). Despite the urban governance as provisioned under the 74th Constitutional Amendment Act (Aijaz, 2008; Weinstein, 2009).

[12]A document of United State India Business Council (USIBC) on smart city represent over 300 companies in 13 different sectors consist of members from both India and the U.S. would have a greater control of corporates.

infrastructures to facilitate or shift physical access, through material and billing regimes for drinking water in the era of neo-liberalism.

References

Aijaz, R. (2008). Form of urban local government in India. *Journal of Asian and African Studies,* 43(2), 131– 154.

Aijaz, R. and Hoelscher, K. (2015). *India's Smart Cities Mission: An Assessment. ORF Issue Brief 124.* Observer Researcher Foundation: New Delhi.

Akhter, M. and Ormerod, K.J. (2014). The irrigation technozone: State power, expertise, and agrarian development in the U.S. West and British Punjab, 1880-1920. *Geoforum,* 60, 123-132.

Alawadhi. S. Aldama-Nalda, A. Chourabiet, H. et al. (2012) Building understanding of smart city initiatives. *Lecture Notes in Computer Science,* 7443, 40–53.

Albino, V. Berardi, U. and Dangelico, R. (2015). Smart cities: Definitions, dimensions, performance, and initiatives. *Journal of Urban Technology,* 22(1), 3–21.

Allan, J. A. (2006). IWRM: The new sanctioned discourse. Integrated water resources management: Global theory, emerging practice and local needs, 38-63.

Arise India Forum, (2012). Self-preservation and self-purification differentiate Ganga from all other rivers on the planet. Retrieved from http://www.ariseindiaforum.org/self-preservation-self-purification-differentiate- ganga-rivers-planet.

Baka, J. (2013). The political construction of wasteland: Governmentality, land acquisition and social inequality in South India. *Development and Change*, 44(2): 409–428.

Banister, J.M. (2014). Are you Wittfogel or against him? Geophilosophy, hydro- sociality, and the state. *Geoforum,* 57, 205-214.

Bharti, Shramik (2012, 31 July). Strengthen participatory urban governance to prevent pollution in Ganga at Kanpur and recognize the need to look for decentralized solutions. Retrieved from http://www.indiawaterportal.org/articles/strengthen-participatory-urban-governance-prevent-pollution-ganga-kanpur-and-recognise-need.

Boltanski, L. and Chiapello, E., (2005). *The new spirit of capitalism*. London: Verso.

Boomgaard, P. (Ed). (2007). *A world of water. Rain, rivers and seas in South-East Asian histories.* Leiden: KITLV Press.

Bottici, C. (2007). *A philosophy of political myth*. Cambridge: Cambridge University Press.

Bulkeley, H. (2013). *Cities and climate change*. Routledge.

Butzer, K.W. (1976). *Early hydraulic civilisation in Egypt. A study in cultural ecology.* Chicago: Chicago University Press.

Chakravorty, S. (1996). Too little, in the wrong places? Mega city programme and efficiency and equity in Indian urbanisation. *Economic and Political Weekly*, 31(35– 37), 2565–2572.

Chiapello, E. (2013). Capitalism and its criticisms. In: P. Du Gay and G. Morgan, eds. *New spirits of capitalism? Crises, justifications, and dynamics*. Oxford: Oxford Universiy Press, 60–81.

Coe, A. Paquet, G. and Roy, J. (2001). E-governance and smart communities a social learning challenge. *Social Science Computer Review*, 19(1), 80–93.

Crow, B. and Singh, N. (2000). Impediments and innovation in international rivers: The waters of South Asia. *World Development,* 28(11), 1907-1925.

Datta, A. (2015). New urban utopias of postcolonial India: _Entrepreneurial urbanization' in the smart city. *Dialogues in Human Geography*, 5(1), 3–22.

Davies, M.I.J. (2009). Wittfogel's dilemma: Heterarchy and ethnographic approaches to irrigation management in eastern Africa and Mesopotamia. *World Archaeology,* 41(1), 16-35.

Economic and Political Weekly (2010) Private cities. *EPW,* 45(12), 7–8.

Foster, J.B. (2000). *Marx's ecology: materialism and nature.* New York: Monthly Review Press.

Gabrys, J. (2014). Programming environments: Environmentality and citizen sensing in the smart city. *Environment and Planning D: Society and Space,* 32(1), 30–48.

GoI (2004). Mid-Term Appraisal of the Tenth Five Year Plan 2002–07. *Planning Commission*: New Delhi.

GoI (2012a). Performance Audit of Jawaharlal Nehru National Urban Renewal Mission. *Comptroller and Auditor General of India*: New Delhi.

GoI (2014a). Draft Concept Note on Smart City Scheme (September). *Ministry of Urban Development*: New Delhi.

GoI (2014b). Draft Concept Note on Smart City Scheme (December). *Ministry of Urban Development*: New Delhi.

Gupta, K. & Naik, A. B. (2017). *Water Supply and Sewerage system in Kanpur.* Lap Lambert: Academic Publishing.

Harvey, D. (2003). The right to the city. *International journal of urban and regional research,* 27(4), 939-941.

Harvey, D. (2006). Neo□Liberalism as creative destruction. Geografiska Annaler: Series B, *Human Geography*, 88(2), 145-158.

Hollands, R (2008) Will the real smart city please stand up? City 12(3): 303–320. Idiculla Mat.hew (2015). *Crafting City Spaces: New Spatial-Legal Regimes in India.*

Paper presented at the RC21 International Conference on _The ideal city: Between myth and reality'. Urbino (Italy) 27–29 August 2015.

Jal Adhikar Yatra, (10th September 2006 – 05th October 2006). A report on the rally from Mehndiganj, Varanasi to Parliament (New Delhi). Retrieved from https://ashadocserver.s3.amazonaws.com/707_Report_Jal_Adhikaar_Yatra.pdf.

Jhunjhunwala, B. (2009). Challenges before River Authorities. Retrieved from http://gangawaterway.in/blogs/94-challenges-before-river-authorities.html.

Jhunjhunwala, B. (2014). Additional Representation to IWAI regarding NW-1. Retrieved from http://www.gangawaterway.in/activity/91-representation-to-world- bank-on-tors-for-nw-1-studies.html.

Jhunjhunwala, B. (2014). Danger Ahead! Ganga will become series of Stagnant Pools. Retrieved from http://gangawaterway.in/blogs/93-danger-ahead-ganga-will- become-series-of-stagnant-pools.html.

Jhunjhunwala, B. (2015). Comments on TOR for NW-1 Studies Submitted to Inland Waterways Authority of India. http://www.gangawaterway.in/activity/90-comments- on-tor-for-nw-1-studies-submitted-to-inland-waterways-authority-of-india.html.

JNNURM, (Agust, 2006). Final Report of the Kanpur City Development Plan. Retrieved from http://jnnurm.nic.in/wp-content/uploads/2010/12/CDP_Kanpur.pdf.

Jong, M., Joss, S., Schraven, D., Zhan, C., & Weijnen, M. (2015). Sustainable–smart– resilient–low carbon–eco–knowledge cities; making sense of a multitude of concepts promoting sustainable urbanization. *Journal of Cleaner Production*, 109, 25 –38.

Joss, S., & Molella, A. P. (2013). The eco-city as urban technology: Perspectives on Caofeidian international eco-city (China). *Journal of Urban Technology*, 20(1), 115- 137.

Kayastha, S. L. & Kumra, V. K. (1990). An appraisal of Urban Environment and Suggestions for Improvement in Kanpur region. pp. 3-25. In Nandan, R. & Sinha, P. (eds.). *Environment and Human Response: Selected Essays in Geography*. New Delhi: Concept Publishing Company.

Komninos, N. (2002). *Intelligent cities: Innovation, knowledge systems, and digital spaces.* New York: Taylor & Francis.

Lee, S. H., Han, J. H., Leem, Y. T., & Yigitcanlar, T. (2008). Towards Ubiquitous City: Concept, Planning, and Experiences. *Igi Global,* 2, 148–169.

Lees, S.H. (1989). On irrigation and the conflict myth. *Current Anthropology* 30(3): 343-344.

Levien, M. (2012). The land question: Special economic zones and the political economy of dispossession in India. *Journal of Peasant Studies,* 39(3–4), 933–969.

Linton, J. (2010). *What is water?: The history of a modern abstraction*. UBC Press.

Linton, J. and Budds, J. (2014). The hydrosocial cycle: Defining and mobilizing a relational-dialectical approach to water. *Geoforum*, 57, 170-180.

LiveMint (2015) I have no idea what a smart city means: Rahul Mehrotra. Available at: http://www.livemint.com/Politics/VFwRPtdqCtg6lyBeacEfZI/I-have-no-idea- what-a-smart-city-means-Rahul-Mehrotra. html (accessed 5 December 2015).

Marx, K. and Engels, F. [1848] (1998). *The communist manifesto*. London: Verso.

McMichael, A. J. (2000). The urban environment and health in a world of increasing globalization: issues for developing countries. *Bulletin of the World Health Organization* 78(9), 1117-1126.

Millon, R.; Hall, C. and Diaz, M. (1962). Conflict in the modern Teotihuacan irrigation system. *Comparative Studies in Society and History*, 4(4), 494-524.

Mitchell, W.P. (1973). The hydraulic hypothesis: A reappraisal. *Current Anthropology*, 14(5), 532-534.

Nandan, R. & Sinha, P. (eds.). (1990). *Environment and Human Response: Selected Essays in Geography.* New Delhi: Concept Publishing Company.

Narain, V. (2009). Growing city, shrinking hinterland: Land acquisition, transition and conflict in peri-urban Gurgaon, India. *Environment and Urbanization* 21(2), 501– 512.

Obertreis, J., Moss, T., Mollinga, P., & Bichsel, C. (2016). Water, infrastructure and political rule: Introduction to the special issue. *Water Alternatives*, 9(2).

Olden, J.D.; Kennard, M.J. and Pusey, B.J. (2012). A framework for hydrologic classification with a review of methodologies and applications in ecohydrology. *Ecohydrology,* 5(4), 503-518.

Palerm-Viqueira, J. (2009). Governance and management of irrigation systems. *Water Policy* 11(3), 330-347.

Peet, R. (1985a). Introduction to the life and thought of Karl Wittfogel. *Antipode,* 17(1), 3-21.

Poff, N.L. and Zimmerman, J.K. (2010). Ecological responses to altered flow regimes: A literature review to inform the science and management of environmental flows. *Freshwater Biology*, 55(1), 194-205.

Raman, Shri Papri (2016, March 05). India's Water War: City Workers Fight Corporate Privatization Efforts. Retrieve from http://www.truth- out.org/news/item/35029-india-s-water-war-city-workers-fight-corporate- privatization-efforts.

SANDRP, (2014). A river map for rejuvenatingGanga. Retrieve from https://sandrp.wordpress.com/2014/10/21/a-river-map-for-rejuvenating-ganga/.

SANDRP, (2016). Digging Our Rivers 'Graves?. Retrieved from https://sandrp.wordpress.com/2016/02/19/digging-our-rivers-graves/.

Satterthwaite, D. (1999). *The Earthscan reader in sustainable cities*. Earthscan/James & James.

Schnaiberg, A. and Gould, K.A. (1994). *Environment and society: the enduring conflict*. New York: St. Martin's.

Schumpeter, J. A. (1976). *Capitalism, socialism and democracy*. Routledge: London and New York.

Sengupta, Sushmita (2014, 18 July). Ganga's burden of pollutants from Kanpur to Varanasi. Down to Earth. Retrieved from http://www.downtoearth.org.in/blog/gangas-burden-of-pollutants-from-kanpur-to- varanasi-45382.

Shaban, A. & Sharma, R. N. (2007). Water Consumption Patterns in Domestic Households in Major Cities. *EPW*, 2190-2197.

Shaban, A. (2008). Water Poverty in Urban India: A Study of Major Cities. Seminar Paper, UGC – Summer Programme, submitted to UGC- Academic Staff College Jamia Millia Islamia, New Delhi, June 30 – July 19, 2008. Retrieved from http://jmi.ac.in/upload/publication/Water_Poverty_in_urban_India.pdf.

Shafi, M. & Raza, M. (eds.). (1989). *Spectrum Of Modern Geography: Essays In Memory Of Prof. Mohammad Anas*. New Delhi: Concept Publishing Company.

Shamsad, (2017). Accessibility of Water for houseless population: A case for Kanpur city Uttar Pradesh. In Sharma, P. & Rajput, S. (eds.). *Sustainable Smart Cities in India: Challenges and Future Perspectives*. pp. 515-532, Switzerland: Springer International Publishing.

Sharma, P. & Rajput, S. (2017). *Sustainable Smart Cities in India: Challenges and Future Perspectives*. Switzerland: Springer International Publishing.

Singh, A. K. (2017). An Issue of Governance in urban water supply: A study of Varanasi, Uttar Pradesh. In (eds.). *Sustainable Smart Cities in India: Challenges and Future Perspectives*. p. 467, Switzerland: Springer International Publishing.

Singh, H. H. et. al. (eds.). (1985). *Geography and Environment: Issues and Challenges*. New Delhi: Concept Publishing Company.

Sivaramakrishnan, K. (1995). Colonialism and forestry in India. *Comparative studies in society and history*, 37(*1*), pp. 3-40.

Sood, A. (2015). Industrial townships and the policy facilitation of corporate urbanisation in India. *Urban Studies*, 52(8), 1359–1378.

Srivathsan, A. (2015). A trip down the rabbit hole of Modi's smart cities wonderland. *The Wire*, 5 June.

Stride, S.; Rondelli, B. and Mantellini, S. (2009). Canals versus horses: Political power in the oasis of Samarkand. *World Archaeology,* 41(1), 73-87.

Swyngedouw, E. (2004). *Social power and the urbanization of water: Flows of power*. Oxford: Oxford University Press.

Swyngedouw, E. (2009). The political economy and political ecology of the hydro- social cycle. *Journal of Contemporary Water Research & Education* 142(1), 56-60.

Swyngedouw, E. (2014). 'Not a drop of water': State, modernity and the production of nature in Spain, 1898-2010. *Environment and History*, 20(1), 67-92.

Times of India, (2017, July17). Govt. wages war against depleting groundwater. Retrieved from http://timesofindia.indiatimes.com/city/lucknow/govt-wages-war-against-depleting-roundwater/articleshow/59624488.cms.

Times of India, (Jun 13, 2017). Kanpur administration acts after fish die in lake. Retrieved from http://timesofindia.indiatimes.com/city/kanpur/kanpur-administration-acts-after-fish-die-in-lake/articleshowprint/59119441.cms.

Times of India, (Jun 17, 2016). Ganga runs dry, drifts miles away from Ghats in Kanpur. Retrieved from http://timesofindia.indiatimes.com/city/kanpur/Ganga-runs-dry-drifts-miles-away-from-ghats-in-Kanpur/articleshowprint/52788701.cms.

Times of India, (Jun 7, 2014). A waterway from Varanasi to Kolkata?. Retrieved from http://timesofindia.indiatimes.com/india/A-waterway-from-Varanasi-to- Kolkata/articleshowprint/36165685.cms.

Times of India, (May 31, 2013). Water crisis deepens sparks protests in Kanpur. Retrieved from http://timesofindia.indiatimes.com/city/kanpur/Water-crisis-deepens-sparks-protests-in Kanpur/articleshowprint/20358796.cms.

Times of India, (May 8, 2014). Women protest against water crisis in Kanpur South. Retrieved from http://timesofindia.indiatimes.com/city/kanpur/Women-protest-against-water-crisis-in-Kanpur South/articleshowprint/34815630.cms.

Townsend, A. M. (2013). *Smart cities: Big data, civic hackers, and the quest for a new utopia.* New York: WW Norton & Company.

Turton, A. (2008). A South African perspective on a possible benefit-sharing approach for transboundary waters in the SADC region. *Water Alternatives,* 1(2), 180-200.

Viswanathan, V. N. (eds.). (2008). *Human Rights: Twenty First Century Challenges.*

New Delhi: Kalpaz Publication.

Weinstein, L. (2009). Democracy in the globalizing Indian city: Engagements of political society and the state in globalizing Mumbai. *Politics & Society*, 37(3), 397– 427.

Weinstein, L. Sami, N. and Shatkin, G. (2014). Contested developments. In Shatkin G (ed.). *Contesting the Indian City: Global Visions and the Politics of the Local.* Wiley Blackwell: Chichester, UK.

Wendland, K.J. Honzák, M. Portela, R. Vitale, B. Rubinoff, S. and Randrianarisoa, J. (2010). Targeting and implementing payments for ecosystem services: Opportunities for bundling biodiversity conservation with carbon and water services in Madagascar. *Ecological Economics*, 69(11), 2093-2107.

Whittington, D. Lauria, D.T. and Mu, X. (1991). A study of water vending and willingness to pay for water in Onitsha, Nigeria. *World Development* 19(2), 179-198.

Wittfogel, K.A. (1957). *Oriental despotism. A comparative study of total power.* New Haven, CT: Yale University Press.

Worster, D. (1985). *Rivers of empire. Water, aridity, and the growth of the American west.* Oxford: Oxford University Press.

Wright, C., & Nyberg, D. (2014). Creative self-destruction: corporate responses to climate change as political myths. *Environmental Politics*, 23(2), 205-223.

Yigitcanlar, T. (2006). Australian local governments' practice and prospects with online planning. *URISA Journal*, 18(2), 7–17.

Yigitcanlar, T. (2015). Smart cities: an effective urban development and management model? *Australian Planner,* 52(1), 27–34.

Yigitcanlar, T., & Kamruzzaman, M. (2014). Investigating the interplay between transport, land use and the environment: A review of the literature. *International journal of Environmental Science and Technology,* 11(8), 2121–2132.

York, R. (2004). The treadmill of (diversifying) production. *Organization & Environment,* 17 (3), 355–362.

Design of Water Distribution Network for Continuous Supply: A Case Study

Rane A. R.[1], Joshi R. A.[2] and Regulwar D.G.[3]

[1]Post Graduate Student, [2]Assistant Professor Department of Civil Engineering, Vishwakarma Institute of Information Technology, Pune, Maharashtra State, India.

[3]Professor, Department of Civil Engineering, Government College of Engineering, Aurangabad, Maharashtra State, India.

amol.17p057@viit.ac.in, rahul.joshi@viit.ac.in, regulwar@gmail.com

ABSTRACT

The rapidly increasing water demands are exhibiting tremendous amount of stress on water supply system. Continuous water supply is more reliable than intermittent water supply. As time and money required to store water by consumer is less, also in the aspect of water demand and operational head loss, continuous water supply system comes out to be superior to intermittent water supply system. In this paper, an attempt is made to design continuous water supply network with raised water demand. A case study of K-east ward Mumbai, Maharashtra State, India is presented in this paper. For simulation, model is created using EPANET 2.0 and analyzed with two different daily water demand patterns with different peak factors to design water distribution network for the area. Extended period simulation is carried out for 24hrs with 1-hour time step. Simulation of result showed change in pressure with respect to time, containing positive and negative pressures in certain part of distribution network.

Keywords: Water Distribution Network, Continuous Water Supply, EPANET

Introduction

The increasing demand of water have started exhibiting greater amount of stress on urban water supply systems. The metropolitan cities of India are facing challenges in water supply field. Intermittent type of supply is, in general, provided in many of the cities of India. To tackle this inconvenience in use of water for the whole day, consumers stores required water and on next supply hours the un utilized storage is flushed to drain. While proposing continuous supply of water it is considered that consumer will not store water as adequate supply and pressure will be there, resulting less wastage of water. Continuous water supply is more reliable than intermittent water supply. As time and money required to store water by consumer is less, also in the aspect of water demand and operational head loss, continuous water supply system comes out to be superior to intermittent water supply. Master planning of areas significantly affect the water distribution pipe size. Factors like location of Elevated Service Reservoir (ESR), Node/Junction Elevation, and Demand controls the pipe diameter. Walski (1995). In intermittent supply pipes remain vacant in non-supply hours and can be subjected to the risk of injection of contaminations. Rao *et al.* (2015). In valve operation when flow is restricted, back flow in pipes occurs causing pressure drop at nodes. Kumar *et al.* (2016). In intermittent supply, pipes remain empty the majority of available head of ESR contributes to fill the empty pipes of supply network. Municipal Corporation of greater Mumbai (MCGM) is planning for round the clock water supply with raised demand to 268 lpcd, under the water distribution improvement program for K-East ward of Mumbai. In this study an attempt is made to design water distribution network for the study area using EPANET 2.0

Study Area

K- East ward of Mumbai is located in latitude – 19.0884° to 19.1420° and longitude 72.8502° to 72.8815°. Water supply to the area is done through four Ground Service Reservoir (GSR) located in hilly part in north region of K-east ward. The entire supply to area is by gravity. Intermittent type of supply is carried out using valve operations, Projected population for the year 2019 for K – East ward is 690990. Demand and service junction are 1220 which are serving total demand of 282 MLD.

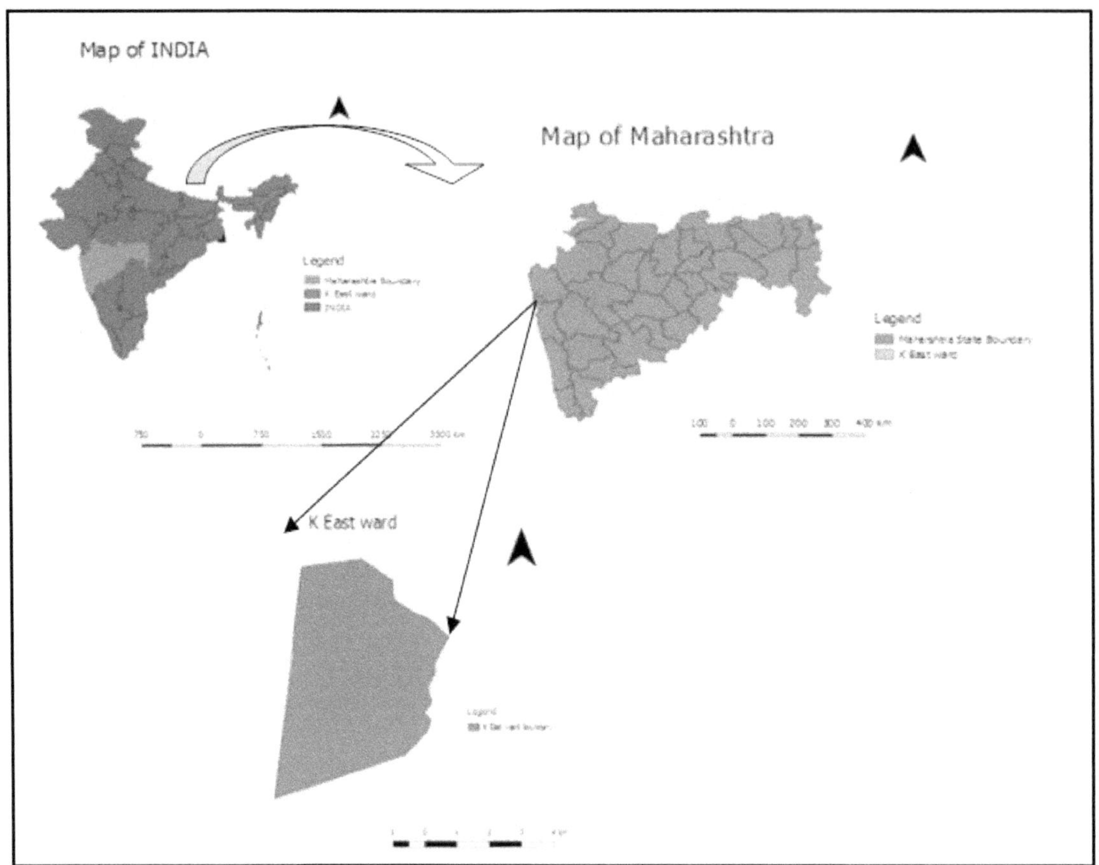

Fig. 1 K – East Ward Mumbai, Maharashtra, India.

Methodology

For simulation, model is created using EPANET 2.0 with the data obtained from municipal corporation of greater Mumbai. Population is forecasted for intermediate stage year 2034 and final stage year 2049 as per the practices adopted by the Maharashtra Jeevan Pradhikaran. The average of Arithmetical Increase Method, Incremental Increase Method and Geometric Progression Method is considered. Junctions are classified into service junction and demand junctions depending upon connection to size of pipe diameter and demand is allocated only to demand junctions.

Table 1 Table of Projected Population.

Year	Arithmetical Increase Method	Incremental Increase Method	Geometric Progression Method	Average
2019	678002	670420	724548	690990
2034	805507	765542	1000460	857170
2049	933011	836968	1381441	1050473

Based on population, demand is calculated considering the supply of 268 lpcd. Which is far more than general Indian standard. Simulation and design is done for demand driven analysis. Extended period simulation is carried out for 24 hrs with one-hour time step using two different hourly demand patterns.

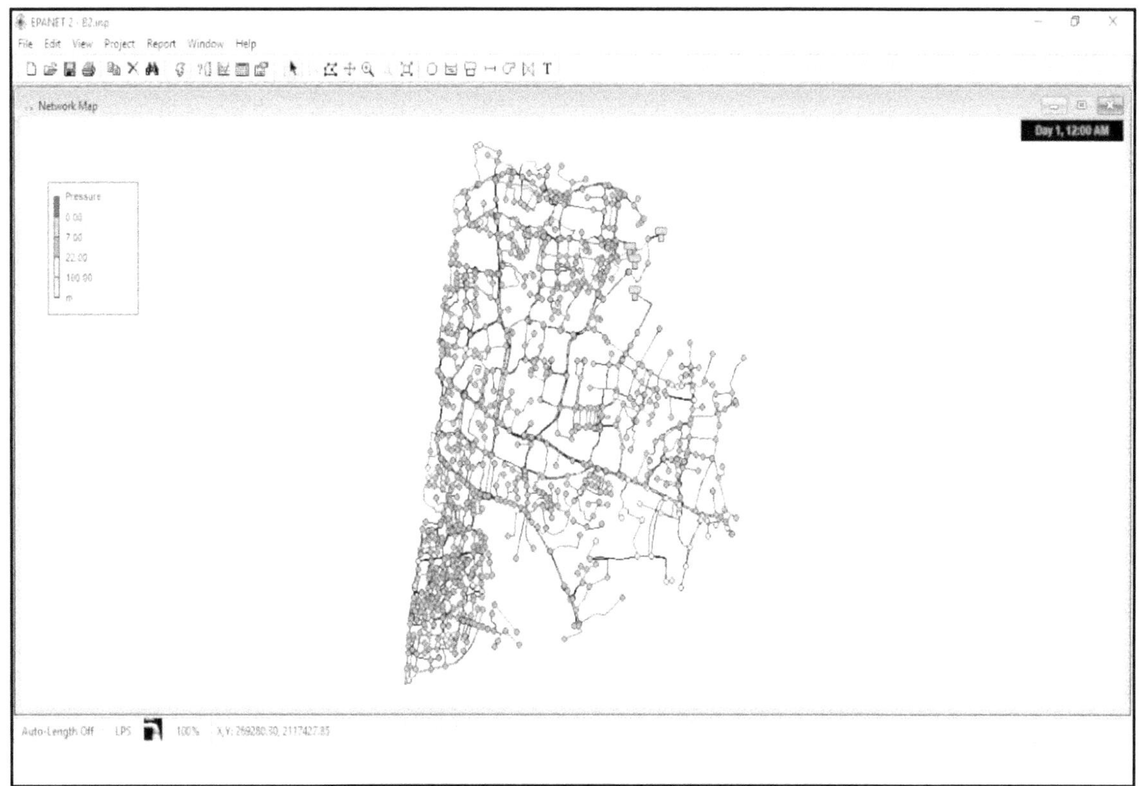

Fig. 2 Water distribution network modelling using Epanet 2.0

First hourly demand pattern is adopted from manual of CPHEEO 1991. Which has the peak factor 3.

Table 2 Table of Hourly Demand Pattern given by CPHEEO.

Time from Start (Hrs)	1	2	3	4	5	6	7	8	9	10	11	12
Multiplayer	0.1	0.1	0.1	0.5	1	1.5	2.5	3	2.5	2	1	1
Time from Start (Hrs)	13	14	15	16	17	18	19	20	21	22	23	24
Multiplayer	0.5	0.5	0.5	1	1	1	1.5	1.5	0.5	0.3	0.3	0.1

Second hourly demand pattern is used by Gholizadeh *et al* (2016). Which has the peak factor 1.75.

Table 3 Table of Hourly Demand Pattern used by Gholizadeh *et al* (2016).

Time from Start (Hrs)	1	2	3	4	5	6	7	8	9	10	11	12
Multiplayer	0.8	0.6	0.3	0.6	0.4	1	1.2	1.25	1.3	1.4	1.6	1.75
Time from Start (Hrs)	13	14	15	16	17	18	19	20	21	22	23	24
Multiplayer	1.6	1.5	1.4	1.4	1.25	1.2	1.1	1	0.9	0.85	0.8	0.8

Result and Discussion

Formulation of District metered areas (DMA) with respect to GSR present in network was difficult as all GSR are located in single location. Simulation of model with two different hourly demand pattern showed large variation in number of junctions having negative pressure. This number varies with respect to time. It is observed that most of the junctions have negative pressure in morning hours between 6 am to 11 am which resembles higher demand of water. The graph 1 & 2 shows variation in pressures at junctions for two different hourly demand patterns with respect to time. Only 12 junctions from different areas which are being served under the GSR - 4 were selected for graphical representation and result of 35 junctions for minimum pressure for both peak factors is given in appendix table 4 due to space limitation. Similar pattern of pressure is observed in the rest of the network.

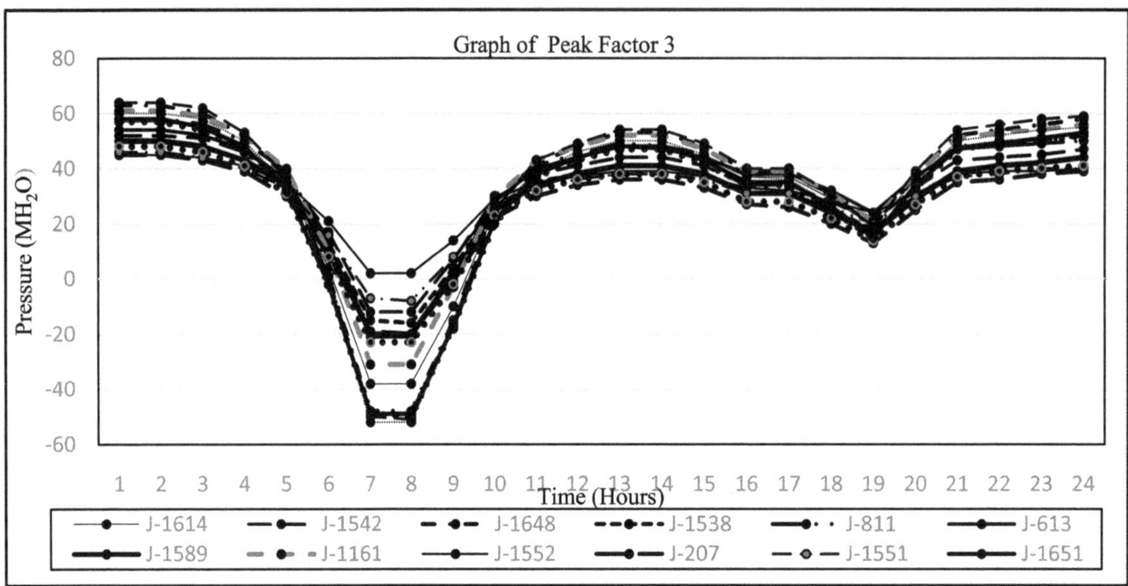

Fig. 3 Graph of Pressure Vs Time.

Simulation with the peak factor 1.75 shows there is fall in pressure as well but, pressure does not fall below 7 mH_2o which is minimum pressure criteria given by CPHEEO.

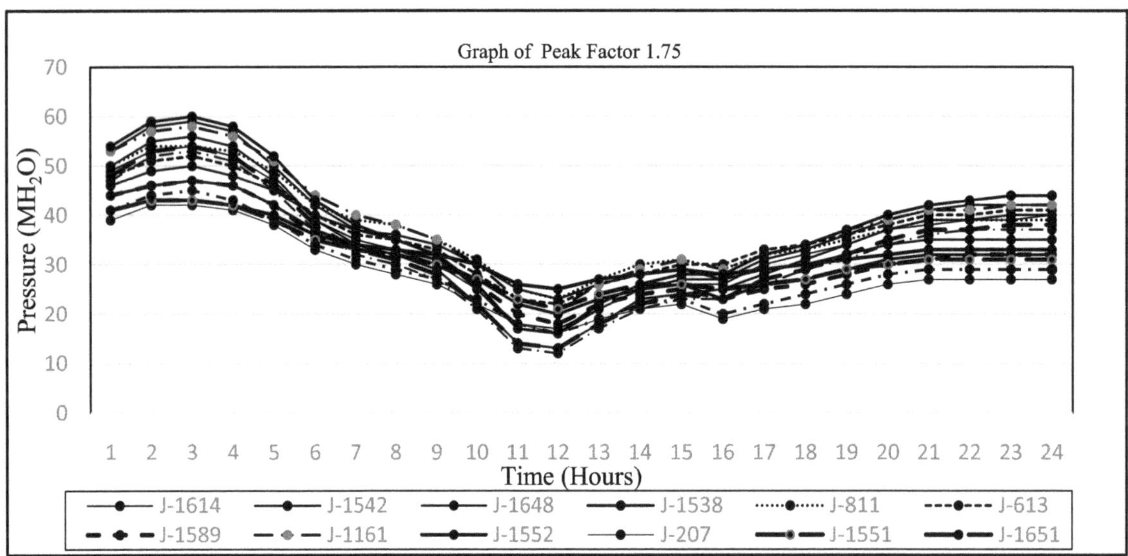

Fig. 4 Graph of Pressure Vs Time.

Conclusion

The simulation of network showed large variation in pressure on peak hours including negative pressure at several junctions. This highlights the necessity to derive accurate hourly demand pattern for higher water demand which not only will resemble urban water consumption pattern but also have minimum peak factor. As to tackle the negative pressures at junction planners and designers tends towards increasing the diameter of supplying pipe to increase the pressure at junction which results in uneconomic design.

Future Scope

To minimize the cost of pipe use of linear programming, fuzzy linear programming, and genetic algorithm can be done to obtain the optimal size of diameter of pipe which must satisfy the constraints of customer demand and minimum nodal pressure criteria.

References

Walski T M, "*Optimization and Pipe-Sizing Decisions.*" ASCE. J. Water Resource. Planning and Management 1995, 121(4): 340-343.

Kumar, G. P., Geetha, P., & Shanmugasundaram, G. A. (2015). *Real Time Water Utility Model Using GIS: A Case Study in Coimbatore District.* Intelligent Systems Technologies and Applications, 115– 123. doi:10.1007/978-3-319-23258-4_11

Rao, Rameshwar & Naik, Mg & Naresh, Aadhi. (2015). "*Simulation of Continuous Water Supply Distribution System: A Case Study Part of Hyderabad Metropolitan City.*" HYDRO 2015 INTERNATIONAL, 20th International Conference on Hydraulics, Water Resources and River Engineering.

Gholizadeh M H, Hamedi A, Niazi A, Eazi M, "*Optimal Design of Water-Supply Pipe Systems Using Economic and Technical Analysis.*" ASCE. World Environmental and Water Resources Congress 2016. doi:10.1061/9780784479889.030.

Manual on Water Supply and Treatment. CPHEEO 1991. Ministry of Urban Development, Govt. of India.

Appendix

Table 4 Table of minimum pressure for peak factor 3 & 1.75

Node Id.	Elevation (m)	Demand (m^3/day)	Pressure (Minimum) (m H_2O) For Peak Factor 3	Pressure (Minimum) (m H_2O) For Peak Factor 1.75
J-568	32.11	44	-20	16
J-595	20.28	45	-52	12
J-613	25.46	43	-21	22
J-705	21.57	77	-50	12
J-811	20.28	71	-20	23
J-1032	18.49	76	-54	12
J-1133	17.11	44	-50	14
J-1161	16.35	46	-31	21
J-1162	23.54	23	34	46
J-1288	11.98	38	-48	18
J-1538	27.86	48	-16	20
J-1542	25.57	87	-12	23
J-1545	30	85	-23	16
J-1551	32.92	67	-8	21
J-1552	33.91	27	2	25
J-1556	28.84	44	-20	36
J-1560	23.47	53	-3	30
J-1562	22	25	16	39
J-1576	31.95	21	-24	17
J-1583	27.53	24	-15	20
J-1589	27.48	92	-20	18
J-1601	20.12	29	-30	46
J-1603	25.94	26	-1	30

Node Id.	Elevation (m)	Demand (m^3/day)	Pressure (Minimum) (m H_2O) For Peak Factor 3	Pressure (Minimum) (m H_2O) For Peak Factor 1.75
J-1606	20	26	2	35
J-1614	21.65	19	-38	17
J-1623	17.02	63	-42	18
J-1627	20.59	55	-41	16
J-1634	18.06	25	-52	13
J-1647	19.59	17	-52	12
J-1648	15.58	39	-48	16
J-1651	20.32	31	-49	13
J-1654	13.99	42	-51	16
J-1997	33.17	47	-10	30
J-2002	20.19	63	-32	18
J-2078	22.37	38	-13	25

Study of Rainfall Pattern using Extreme Value Distributions

Shreenivas Londhe[1] and Nikhilesh Gandhre[2]
[1]*Professor in Civil Engineering,* [2]*Post Graduate Student,*
Water Resources and Environmental Engineering, Vishwakarma Institute of Information Technology, Pune
shreenivas.londhe@viit.ac.in, nikhilesh.218M0005@viit.ac.in

ABSTRACT

The study aims to analyze the rainfall pattern in various districts of Maharashtra using extreme value distributions. The data used for analysis is from the year 1901-2000. The data is divided in two sets with first set from 1901 to 1950 and second set from 1951 to 2000.Rainfall for return periods 50 & 100 years is calculated using Gumbel & Log Person distributions. There is a considerable difference between the rainfall values of each return period between both the sets for all the three distributions. An increase in rainfall is seen in the Raigad, Sindhudurg, Pune, Kolhapur, Sangli, Nanded, Gadchiroli, Chandrapur districts while a decrease is seen in Ratnagiri, Nashik, Ahemadnagar, Jalgaon, Aurangabad, Beed, , Amravati, Buldhana, Yavatmaal, Nagpur,. This may also be attributed to effect of climate change over last 100 years.

Keywords: Rainfall Pattern, Extreme Value Distribution, Gumbel, Log Pearson

Introduction

Rainfall is the most important hydrological parameter and is largely responsible for deposition of fresh water on the globe. It provides a suitable condition for many types of ecosystems as well as water for hydroelectric projects and crop irrigation. Rainfall has a unique characteristic of diversifying with respect to space and time thus resulting in variations from place to place and time to time. Analysis of rainfall distribution over an area helps in planning strategies for irrigation and cropping pattern. Probability and frequency analysis of rainfall enables us to reach at conclusions which can be used to prevent floods, droughts and can be applied to the development of water resources like dams, canals, etc. In recent years floods as well as drought like conditions are being observed in various areas of Maharashtra. Changes in rainfall pattern in any phase might affect an agro based economy which largely depends upon the monsoon. Extreme Value Distributions are used in this analysis to study the changes in rainfall pattern over a period of 100 years.

Study Area & Materials

The state of Maharashtra is considered as a study area. It is situated at 19.7515° N latitude and 75.7139° E longitude having a total area of 307,713 km². Out of 36 districts, the study is being carried out on 18 of them. Annual precipitation data of these 18 districts for a period of 100 years i.e from 1901 to 2000 is collected from Indian Meteorological Department (IMD).

Methodology

The 100 years precipitation in mm data of each district is divided into two sets i.e 1900-1950 and 1951-2000.Rainfall for returns periods of 50 and 100 is calculated using extreme value distributions for each data set. Details about the extreme value distributions used are given below. The fitness of a particular distribution to a particular data is judged using Chi-squared test details of which are given in subsequent sections.

Extreme Value Distributions

Extreme Value Distributions arise as limiting distributions for maximums or minimums (extreme values) of a sample of independent, identically distributed random variables as the sample size increases. In hydrology these distributions can be used to predict the extreme events such as maximum rainfall, floods and droughts. In this analysis Gumbel & Log Pearson Distributions are used for analysis of rainfall pattern.

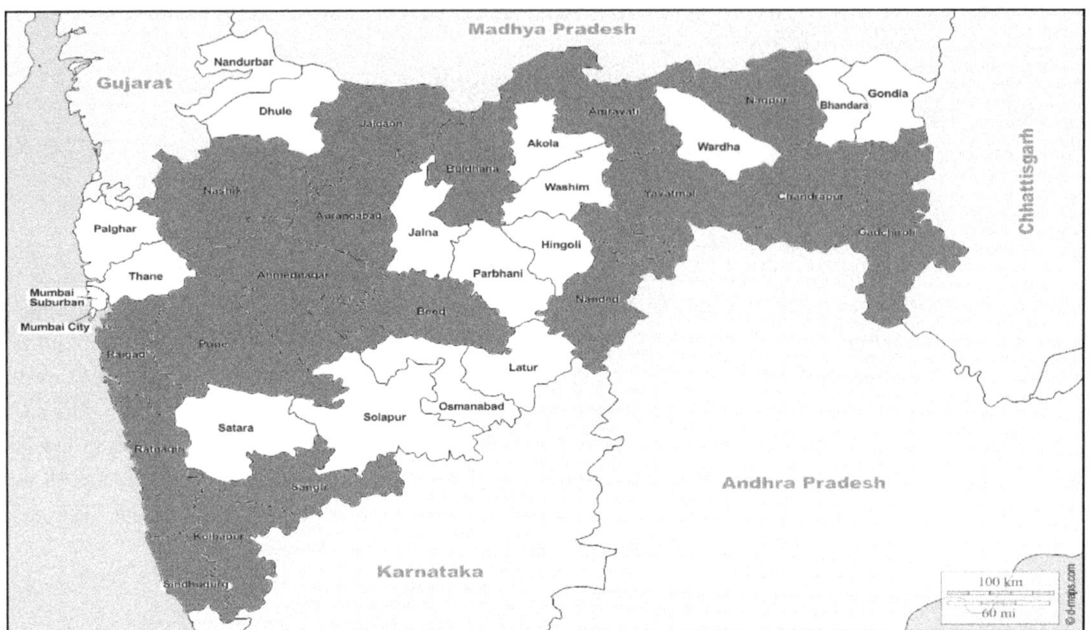

Figure 1 Map of Study Area

Gumbel Distribution (1941) (Subramanya, 2008)

The following steps are used to analyze the annual maximum rainfall probability distribution.

(i) Means of annual rainfall ($\bar{X}$)

Standard deviation (sn)

(ii) Reduced variate is calculated using the formula.

$$Yt = -\ln(\ln(T/(T-1))) \quad \text{.....(1)}$$

Where

Yt = Gumbel reduced variate, a function of 'T'

T = return period in years

(iii) Frequency factor K for Gumbel Distribution is calculated using the formula

$$K = (Yt-\overline{Yn})/Sn \quad \text{......(2)}$$

Both $\overline{Yn}$ (reduced mean) & Sn (reduced standard deviation) of the Gumbel Variate and are functions of sample size N and their values are available in tabular form for several of N (Subramanya, 2008)

(iv) Magnitude of rainfall for the given return period is calculated by

$$Xt = \bar{X} + (K * sn) \quad \text{.....(3)}$$

Xt = magnitude of rainfall for given return period 'T'

K = Frequency factor

$\bar{X}$ = Mean of annual

sn = standard deviation

Log Pearson Distribution (Subramanya, 2008)

In this method, the sample (i.e. Z in this case) is first transformed into logarithmic form before analyzing. For Log Pearson, Kz is a function of which is calculated by both the return period and the coefficient of skew (Cs). The values of Kz are given in the book (Subrmanya, 2008) Following steps were followed for implementing Log Pearson distribution.

(i) Log X = Z of all rainfall data

(ii) Mean ($\bar{Z}$) of the log values

Standard deviation (Sn) of the log values

Coefficient of skew of the log values (Cs)

(iii) The value of frequency factor (Kz) was taken from the statistical table in Engineering Hydrology by K Subramanya,2008 corresponding to Cs to "T" (Return period).

(iv) Magnitude of rainfall Xt for a given return period 'T' is calculated by formula

$$Zt = \bar{Z} + (Kz * Sn) \quad .....(4)$$

Where Xt = Antilog (Zt)

Test for good fit

Chi Square Test

The test intends to evaluate the difference between the sample data and the probability distribution. The formula for Chi Square test is given by

$$X^2 = \sum_{i=1}^{N} \frac{(Oi-Ei)^2}{Ei} \quad .....(5)$$

Where

χ^2 = calculated chi-square

Ei = frequency that is hoped regarding to the class division

Oi= frequency on the same class

N =number of class intervals

The formula of Ei is as follow

$$Ei = \frac{n}{N} \quad .....(6)$$

Where:

n = number of data

N= number of class intervals

The criterion for goodness is value of $\chi 2$ calculated should be less than value of χ^2 from the table (Reddy, 1997). The χ^2 value is read out from the table by knowing two parameters

(a) Degree of freedom=k-h-1

Where, k = number of class intervals

h = number of estimated parameters

(b) Level of Significance

Results and Discussions

Table 1 shows consolidated results of Gumbel and Log Pearson distributions at all the districts

Table 1 Rainfall for 50 yr & 100 yr Return periods (mm)

RAIGAD	GUMBEL			LOG PEARSON		
XT	1900-1950	1951-2000	DIFF	1900-1950	1951-2000	DIFF
50	4636.34	4995.36	359.02	3992.03	4566.57	574.55
100	4972.11	5339.27	367.16	4065.81	4734.09	668.28
SINDHUDURG	GUMBEL			LOG PEARSON		
XT	1900-1950	1951-2000	DIFF	1900-1950	1951-2000	DIFF
50	4172.98	4540.79	367.81	3592.80	4010.97	418.16
100	4457.11	4828.09	370.98	3648.55	4082.13	433.58
RATNAGIRI	GUMBEL			LOG PEARSON		
XT	1900-1950	1951-2000	DIFF	1900-1950	1951-2000	DIFF
50	5243.53	5203.57	-39.96	5026.40	4970.20	-56.20
100	5623.24	5569.24	-53.99	5290.51	5217.42	-73.08
PUNE	GUMBEL			LOG PEARSON		
XT	1900-1950	1951-2000	DIFF	1900-1950	1951-2000	DIFF
50	1661.21	1790.72	129.52	1482.79	1565.53	82.73
100	1802.52	1933.99	131.47	1550.06	1614.70	64.63
SANGLI	GUMBEL			LOG PEARSON		
XT	1900-1950	1951-2000	DIFF	1900-1950	1951-2000	DIFF
50	1079.56	1287.72	208.17	1016.94	1195.41	178.48
100	1169.68	1407.30	237.62	1084.06	1289.57	205.51
KOLHAPUR	GUMBEL			LOG PEARSON		
XT	1900-1950	1951-2000	DIFF	1900-1950	1951-2000	DIFF
50	2560.60	2787.23	226.63	2451.26	2561.96	110.70
100	2799.48	2971.24	171.76	2668.16	2655.69	-12.47
NASHIK	GUMBEL			LOG PEARSON		
XT	1900-1950	1951-2000	DIFF	1900-1950	1951-2000	DIFF
50	1717.90	1524.25	-193.65	1536.07	1369.28	-166.79
100	1865.52	1643.82	-221.71	1604.99	1421.13	-183.86
AHMEDNAGAR	GUMBEL			LOG PEARSON		
XT	1900-1950	1951-2000	DIFF	1900-1950	1951-2000	DIFF
50	1049.83	950.22	-99.62	974.89	805.61	-169.29
100	1148.89	1029.47	-119.42	1043.35	826.43	-216.92
JALGAON	GUMBEL			LOG PEARSON		
XT	1900-1950	1951-2000	DIFF	1900-1950	1951-2000	DIFF
50	1244.64	1162.61	-82.03	1153.45	1030.70	-122.74
100	1349.37	1253.32	-96.05	1220.58	1061.79	-158.78

AURANGABAD	GUMBEL			LOG PEARSON		
XT	1900-1950	1951-2000	DIFF	1900-1950	1951-2000	DIFF
50	1331.04	1166.38	-164.66	1135.03	1091.99	-43.03
100	1456.37	1259.75	-196.62	1172.79	1237.29	64.50
BEED	GUMBEL			LOG PEARSON		
XT	1900-1950	1951-2000	DIFF	1900-1950	1951-2000	DIFF
50	1352.68	1253.12	-99.56	1088.92	1253.12	164.20
100	1493.33	1365.66	-127.68	1126.93	1365.66	238.72

NANDED	GUMBEL			LOG PEARSON		
XT	1900-1950	1951-2000	DIFF	1900-1950	1951-2000	DIFF
50	1636.78	1783.59	146.81	1571.53	1684.40	112.87
100	1792.94	1949.86	156.92	1722.97	1824.45	101.48
AMRAVATI	GUMBEL			LOG PEARSON		
XT	1900-1950	1951-2000	DIFF	1900-1950	1951-2000	DIFF
50	1616.14	1598.98	-17.16	1465.93	1388.21	-77.71
100	1756.56	1739.40	-17.16	1543.41	1461.96	-81.45
BULDHANA	GUMBEL			LOG PEARSON		
XT	1900-1950	1951-2000	DIFF	1900-1950	1951-2000	DIFF
50	1325.30	1220.46	-104.84	1174.54	1102.64	-71.89
100	1435.06	1317.54	-117.52	1219.40	1149.26	-70.14
YAVATMAL	GUMBEL			LOG PEARSON		
XT	1900-1950	1951-2000	DIFF	1900-1950	1951-2000	DIFF
50	1619.92	1618.32	-1.61	1423.55	1514.94	91.40
100	1750.35	1749.50	-0.85	1470.66	1608.22	137.57
NAGPUR	GUMBEL			LOG PEARSON		
XT	1900-1950	1951-2000	DIFF	1900-1950	1951-2000	DIFF
50	1878.94	1695.57	-183.36	1806.23	1645.92	-160.31
100	2026.30	1820.74	-205.56	1916.58	1740.18	-176.40
CHANDRAPUR	GUMBEL			LOG PEARSON		
XT	1900-1950	1951-2000	DIFF	1900-1950	1951-2000	DIFF
50	2069.81	2147.16	77.35	1815.09	1996.65	181.56
100	2227.95	2326.03	98.08	1866.40	2119.83	253.43
GADCHIROLI	GUMBEL			LOG PEARSON		
XT	1900-1950	1951-2000	DIFF	1900-1950	1951-2000	DIFF
50	2234.04	2316.35	82.31	2013.70	2229.86	216.16
100	2402.11	2501.43	99.31	2081.31	2389.28	307.98

Raigad, & Sindhudurg districts show an increase rainfall by more than 300mm.Pune, Kolhapur, Sangli, Nanded, Buldhana districts show an increase in rainfall by more than 100mm.More than 100mm decrease is seen in the Nashik, Ahemdnagar, Aurangabad, Jalgaon, Beed districts. Nagpur district shows an increase in rainfall by 150mm.Chandrapur, Gadchiroli districts show an increase by 75mm.Amravati & Yavatmaal districts also show a decrease less than 20mm.These changes in rainfall can be attributed to the effect of climate change over rainfall pattern at each of the location discussed above.

Chi square test

The test is carried at 5% significance level (Suhartanto et al. (2018))

(i) Gumbel

Degree of Freedom = k-h-1 = 10-2-1 = 7

k = number of class intervals = 10

h = number of derived parameters = 2 ($sn, \overline{X}$)

Value of χ^2 from the table = 14.067

(ii) Log Pearson

Degree of Freedom = k-h-1 = 10-3-1 = 6

k = number of class intervals = 10

h = number of derived parameters = 3 (Z,Sn,Cs)

Value of χ^2 from the table = 12.59

Table 2 shows consolidated results of all the locations while table 3 gives results of χ^2 test which indicate whether the distribution is suitable for a particular data set or not.

Table 2 χ^2 calculated

DISTRICT	GUMBEL		LOG PEARSON	
	1900-1950	**1951-2000**	**1900-1950**	**1951-2000**
RAIGAD	13.6	7	66.4	47.8
SINDHUDURG	15	10	72	72
RATNAGIRI	6	7.2	58.4	63.6
PUNE	6.8	5.6	65.6	65.6
SANGLI	5.6	10.8	49.6	52
KOLHAPUR	8	6.8	81.2	62
NASHIK	6	4.4	56.4	64.8
AHMEDNAGAR	7.2	6.8	64.4	69.6
JALGAON	8.8	14.8	56.4	93.6
AURANGABAD	21.2	9.6	52.7	68.8
BEED	2.8	5.2	86.6	60.8
NANDED	8	8.8	61.6	71.6
AMRAVATI	2.4	4.4	67.6	84.4
BULDHANA	8.8	11.2	69.2	54.6
YAVATMAL	8.4	5.2	82	76.8
NAGPUR	8.4	12.8	66.4	57.2
CHANDRAPUR	8	6	71.2	62
GADCHIROLI	6	16	35.6	45.6

Table 3 Chi Square Test Results

GUMBEL	DISTRIBUTION SUITABLE	LOG PEARSON	DISTRIBUTION NOT SUITABLE
χ^2 calculated < 14.067		χ^2 calculated > 12.59	

Conclusions

Districts considered for analysis follow Gumbel Distribution. Maximum increase is seen in the district of Nanded (9%). Minimum increase is seen in the districts of Chandrapur and Gadchiroli (3% to 4%). Maximum decrease is seen in the district of Aurangabaad (13%) while minimum decrease is seen in Yavatmaal (0.1%). Raigad, Sindhudurg, Sangli, Pune, Kolhapur districts show an increase in rainfall (7% to 8%).In Nashik, Nagpur, Ahmendnagar districts the decrease is by (10% to 11%) while Jalgaon, Beed, Buldhana districts show a decrease of (7% to 9%.) Amravati & Ratnagiri district shows a decrease of (1%). The rainfall trend has increased in the exterior parts of Maharashtra while it has decreased in the interior parts.

References

Reddy J (1997), Stochastic Hydrology, Laxmi Publlications (New Delhi) Pvt.Ltd,

Subramanya K (2008), Engineering Hydrology, Tata McGraw Hill Education (India) Pvt.Ltd,

Suhartanto. E. et al. (2018) "Frequency Analysis for Estimating Design Rainfall - Case Study on the Upstream Lesti Sub-watershed" *Asian Journal of Applied Science and Technology (AJAST)*, 2,(1), 87-100.

Approaches to Mining Waste Land Development: A Review

Munnoli P M[1], Udagatti N[2], Chavan A[3], Harshitha K[4] and Joshi I[5]

[1]Professor, [2,3,4,5]Final Year students, Department of Civil Engineering

S D M College of Engineering and Technology Dharwad.

[1]Corresponding author: munnolipm@gmail.com

ABSTRACT

Mining exploitation in India and the World is a continues and age old activity for fulfilling natural resource demand of minerals. This has created an imbalance in the ecosystem and a huge quantity of mining dumps and waste lands, which have adverse effect on humans, animals and plants.

The paper presents the review on various studies on improvement of mining soil properties and waste land development along with mining extraction laws. The review revels that the industrial waste like rise husk ash, saw dust, lime, Fly ash, press mud, vermi-compost etc are being utilised to improve the properties of the mining soil.

Keywords: Mining, natural resource, industrial waste

Abbreviations

UCS :Unconfined compressive strength	C:Carbon
CBR:California bearing ratio	RHA :Rice husk ash
SOC:Soil organic carbon	SWD:Sawdust
N:Nitrogen	TGA:Total geographical area
CBA: Cement bottom ash	SOM: Soil organic matter
MCDR: Mineral conservation and development rules	ACC: Associated cement companies

Introduction

The mining industry whose products are metals and minerals on which national economy of any country depends and there has been a progressive increase in the production of metals and minerals over the past 50-60 years has made the progress on one side, ecological imbalance on plants, animals, humans, water and atmosphere on the other side are inevitable (Cooke and Johnson 2002; Anderson 1992; Gowdy and Mc Dainiel1999; Sengupta 1993).

India is essentially known as tropical country. After independence govt of India took measures to develop in the fields of agriculture , industries, water resource , mining as a natural resource management, which led to anthroprogenic activities leading to alarming pollution in a hapazard way. The mining industry growth on economic fronts India contributes around 10.5% to the GDPA (MMII 2010) on one side created mining waste lands unattented on the other side.

Wastelands are lands which are economically unproductive, ecologically unsuitable andare subjected to environmental deterioration (Dhir, 1991). Wasteland encompasses ecological, economic, and technological aspects of land in its entirety. In earlier days, the wasteland was considered to be the barren land which was not used for agriculture purpose (Waste land classification and genesis).Stamp (1948) defined the wasteland as that piece of land which had been previously used but has been abandoned and for which no further use has been found, like old quarries, mine spoils etc.

Degraded land without vegetation is waste land (Balasubramanian, 2015). Land is a non-renewable resource, degradation of land is caused by deforestation, Excessive use of Fertilizers and Pesticides, Over grazing, Salination, Water-logging, Desertification, Soil erosion, Landslides, (TENTH FIVE YEAR PLAN 2002-07). Approximately 68.35million hectare area of land is lying as wastelands in India at present which is more than 20% of the country's geographical area. India's Geographical area of 3.2 m.sq.km out of which 0.63 million sq.km are existing as wastelands (Balasubramanian, A2015).

Residual of soil and aggregate obtained from mine during the process of the extraction of the ore is mine waste. Problems created by mine waste are classified as ecological, environmental and problems affecting human interest. Mine wastes are classified as:

1. Unprocessed waste (overburden waste). 2. Processed waste

Unprocessed wastes are classified as: a) Lateritic soils and aggregates b) Lumpy ores. c) Phyllitic clays d) Limonitic clays. e) Magni-ferous clays. (D'souza, J.,*et.al* 1995).

Fly ash was successfully used for stabilizing exapansive clays. The strength characteristics of fly ash stabilized clays are measured by means of unconfined compressive strength or from CBR values. Depending upon the soil type, the effective fly ash content for improving the engineering properties of the soil varies between 15 to 30%. Rice husk ash (RHA) is obtained from burning of rice husk. The husk is the by product of the rice milling industry. 10% of rice grain is rice husk. On burning the rice husk, about 20% becomes RHA (Brooks, 2009).

Stabilization is the process of fundamental change in the chemical properties of soils by adding stabilizers either in, wet or dry conditions to increase the strength and stiffness of originally weak soils. The geotechnical engineers design the foundations and other structure on the soil after investigation of the type of soil, its characteristics and extent. It was found that utilization of industrial waste like saw dust ash is an alternative to stabilize the soil for various construction purposes. When activators like lime is added to sawdust ash the results are very encouraging. SDA, an industrial waste can be efficiently used in soil stabilization. This can reduce construction cost of roads particularly in rural areas of developing countries like India. The existing soil at a particular location may not be suitable for the construction due to poor bearing capacity and high compressibility or even sometimes excessive swelling in case of expansive soil.The improvement of soil at the site is indispensible due to rising cost of the land, and there is a huge demand for high rise buildings. There is a need to concentrate on improvingthe properties of soils using cost effective practices like treating with industrial waste those having cementitious value. In this study, the materials like saw dust ash (SDA) and lime are used to improve the geotechnical properties of soil. (Shawl, z.z., et, al.2017).

Mine dumps: Are those lands where waste debris is accumulated after extraction of minerals. Included are the mine/quarry areas subject to removal of different earth material (bothsurface and sub-surface) by manual and mechanized operations. Large scale quarrying and mechanical operations result in creation of mine dump. It includes surface rocks and stone quarries, sand and gravel pits, soil excavation for brick kilns, etc. (Boori, 2010 wasteland atlas of India).In regard the state wise distribution of wasteland of India in percentage is represented in Table 1.

Table 1 Percentage Distribution of land use pattern effects in India.

Type of land	**States**	**Percentage**
Land with or without scrub	Madhya Pradesh, Gujarat Maharashtra	19 11 19
Alkalinity or salinity	Gujarat Uttar Pradesh &Tamil Nadu	0.020sq.km
Shifting cultivation	Manipur ,Assam, Nagaland, Mizoram	0.03514m.sq.km
Underutilized degraded notified forest land	Andhra Pradesh and Madhya Pradesh	16
Sandy areas	Rajasthan, Assam.	0.05sq.km
Waste land due to mining industries	Bihar, MadhyaPradesh , Rajasthan, Tamil Nadu.	0.0012sq.km
Barren lands	Jammu kashmir, Andhra Pradesh, Rajasthan	0.065sq.km

Type of land	States	Percentage
Steep slopes	Jammu kashmir, Himachal Pradesh, Maharashtra, Uttar Pradesh	0.008sq.km
Wasteland area	Karnataka	7.06%
Source:Boori, M.S 2010 wasteland atlas of India		

Karnataka is one of the southern states and the topography is varied with low lying coastal plains the rugged Western Ghats. The total geographical area of Karnataka is 191791sq.km, out of which 14438.12sq.km is under wasteland which accounts for 7.53%of its geographical area (table no 1). Under-utilised/degraded notified forest scrubland dominated is the major wasteland category, which covers an area of 5245.32sq.km. in terms of the change in the wastelands as compared to 2003, there has been an increase to the tune of 901.54sq.km. The district with higher percentage of wasteland is Ballari with 17.37% of its area under wastelands, while Dakshin Kannada with 2.13% of its total geographic area, and has the least wastelands. (Boori, M.S2010 wasteland atlas of India).

Karnataka has 266 iron ore mines, out of which 134 are located in forest areas. Bellary district in Karnataka alone has 148 mines spread over an area of 10598 hectares of land

National Environment Engineering Research Institute (2004) report contained how to minimize dust from mining, storage and transportation, safe disposal of mining waste, restoration and stabilisation of abandoned mines and using scientific mining techniques.

The total geographical area of India is 191791.00 sq.km. and total wasteland area is 13536.58 sq.km. and total land area in percentage is 7.06%. District wise distribution of wastelands in Karnataka state is represented in Table 2.

Table 2 Distribution of wastelands in Karnataka state

District	%of TGA	District	%of TGA
Bangalore(r)	4.98	Haveri	6.01
Bangalore(u)	2.98	Hassan	4.64
Belgaum	10.66	Kodagu	2.04
Bellary	12.96	Kolar	7.24
Bidar	6.63	Mandya	7.88
Bijapur	3.14	Mangalore	4.31
Bagalkote	10.64	Mysore	3.26
Chamrajnagar	3.02	Raichur	10.92
Chikmagalur	8.43	Koppal	6.08
Chitradurga	14.11	Shimoga	4.63
Davangere	8.44	Tumkur	9.41
Dharwad	3.43	Uttar kannada	5.60
Gadag	8.00	Udupi	7.15
Gulbarga	4.83		
Source: Karnatka state remote sensing applns center, Bangalore (2003).			

The social and legislative context of mining in many parts of the world today means that some form of land use goals will be set, prior to the granting of mining permission for new mine. Reclamation considerations are in corporate into the mine planning such that it becomes a major governing factor in the mining operations, waste disposal, and site closure (Cooke and Johnson, 2002).

The approaches to restore a mining waste land are divided in to three categories namely Social approaches, Engineering & Technological approaches and Legislative approaches.

The social approaches will cover involvement of the stake holders like public, mine owners, and government examples are social agro forestry, providing and developing grazing lands, Hill management society, providing marketing facilities etc.

Social approaches: It is the reclaimed land that remains indefinitely and is required to meet the major goal of sustainability, means maintenance of land use patterns for future generations (Haigh 1993.) The ecological restoration is the term applied to integrated development of mining waste land guided by ecological principles SER 1996;(Hobbs 1999; Bradshaw 1990; Pastorok .et al.,1997). The minerals and sustainable development project (MMSD) has been set up for collaborating in a global strategic initiative which will guide the future mining industry (World Economic Forum; MMSD 2001).

Kumar, M., Kumar, S (2017) have provided a case study to analyze the wasteland development plan in Jind district .The wasteland included salt affected land, sand dunes ,mining dumps ,industrial wasteland etc .These lands were reclaimed according to the characteristics of particular type of wasteland.Kivinen.s (2017)concluded that sustainable reuse of post mining areas requires efficient evaluation of land use potentials, limitations and the overall suitability of sites for local land use needs. Increasing sizes of new mining projects, calls for paying more attention on the characteristics and functions of the future post mining landscapes in order to avoid negative land use development leading to degraded and underused area from the environmental, economic and social perspective.

On the social front there are works reported to reclaim barren lands and mining waste lands. In sukho major project near Chandigarh Mittal (1986) has reported the success of integrated water shed management of barren lands due to overgrazing and subsequent filling of silt in Sukhana Lake. The key to success is the people's participation at all levels of interventions by providing separate grazing land, Hill management society, use of organic manures etc. Mittal (1990).

Development of wastelands and degraded lands (2002).

Table 3 Causes of land degradation

Causes of degradation	Area(million hectares)	Percentage of total area
Water erosion	107.12	61.7
Wind erosion	17.79	10.24
Ravines	3.97	2.28
Salt-affected	7.61	4.38
Water logging	8.52	4.9
Mines & quarry waste	-	-
Degraded land due to shifting cultivation	4.91	2.82
Degraded forest lands	19.49	11.22
Special problems	2.73	1.57
Coastal sandy areas	1.46	0.84
TOTAL	173.64	100.0
Source: Development of wastelands and degraded lands (2002).		

Technological approaches

Bharathan *et.al.,* (2018) have studied soil stabilization using silica fumes and cement .The proportion of silica fume: cement when mix in ratio of 75:10:15 should increase strength and permeability and also suggested further experimental work to be carried out to pinpoint effect of silica fumes on soil characteristics.

Dodor.D.E *et.al.,* (2018) concluded that the co-application of bio char and cattle manure can potentially stabilize C in manure amended sandy soils, albeit with a temporary mineral N limitations to plants. Amadi & Okeiyi (2017) the study states that use of quick and hydrated lime improves the UCS&CBR values and stabilizes the lateritic soil and improves the engineering properties. The results were obtained by the comparison of different percentages of quick and hydrated lime.

Gupta.A *et.al.,* (2017) concluded that the geotechnical and chemical properties of derdged soil couldn't meet the criteria for highway sub grade materials and thus using cement bottom ash mix (1:1) has a significant role in this study. From the results of CBR,wetting and drying and concentration of heavy metals in leachate ,it is found that the dredged soil stabilization with 10C and 20CBA(10%cement +10%bottom ash) both optimally fulfilled the acceptable criteria required for using it as a highway sub grade materials and 20CBA is the most suitable mixing proportion.

Nnochiri,E.S *et.al.,* (2017) shown that addition of saw dust ash and lime increases the optimum moisture content and decreases the maximum dry density of the soil, UCS is increased and saw dust ash lime mixture acts as cheap stabilizing agent for lateritic soil.

Shawl, Z.Z *et.al.,* (2017) concluded that the addition of saw dust ash (SDA) improved the properties of the virgin soil, making it good for sub-base material whereas on addition of lime, the plasticity index reduced drastically when compared with the soil stabilized with sawdust ash alone.

Chattopadhyay .S.(2006) he concluded that there have been effects of sustainable regeneration of derelict mine sites and the reclamation of derelict mines and he discussed about issues of bio-physical reclamation and current trends in planning by some best practices of case studies .

Jha, J.N and Gill, K.S (2006) have studied the influence of different mix proportions of lime & RHA on compaction, strength, CBR values and durability characteristics of soil which have shown increased strength & durability.

Bhuvaneshwari.S *et.al.,* (2005) have studied that stabilization of expansive soil using fly ash, as fly ash has a cementious value when it is mixed with water forms the cementious compound and improves the strength and compressibility characteristics of soil thus improving the engineering properties and stabilizing the soil.

Cooke.J.A and Johnson.M.S (2002) provides a restoration planning model is presented where the presence and absence of top soil conserved on the site has been given the status of primary practical issues for consideration in ecological restoration in mining.

Buss, J *et.al.,* (2018) have concluded that the growth of plants was seen in the mining soil by the application of fertilizer treatment in greenhouse experiment.

Rowley, C.M *et.al.,* (2018) have concluded that the addition of ca2+ generally improves the microbial conditions for the decomposition by increasing the PH and reducing stress from H+, it can conterintuitively reduce respiration rates through the stabilization of SOC.

Wang, D *et.al.,* (2017) results suggested that bio char can enhance the physical protection of soil organic matter (SOM) in yolo soil by increasing the proportion of C stored within macro aggregates and thus offer a novel mechanism by which bio char may contribute to soil aggregation and C sequestration. This mechanism is dependent on soil texture as bio char had minimal impacts on aggregation and microbial communities in a coarser textured soil.

Sheron, A *et. al.,* (2010) concludes that the top soil is essential component for land reclamation in mining areas. Productivity of soil can also be increased by adding various substitutes such as hay, saw dust, bark mulch, wood chips, wood residue sewage sludge, animal manure as they stimulate the microbial activity which provides the nutrients (N, P) and organic carbon to soil.

The soil aggregation Munnoli (2009) and water holding capacities of soil Munnoli (2011) characteristics have been reported.

A associate of ACC cements has recovered the cement mining land and converted the large open excavated depression into a water lake with recreation fountain's and vegetation around.On the industrial front the land with high pH and with no vegetation has been reported to be reclaimed using vermi-compost by Singh, 1997. Similarly Munnoli, P M (1998) has reported the vegetation recovery for Jagajit Industry in Punjab. Sesa Goa mining industry has reported use of vermicompost of canteen waste for its mining waste land (Singh 1997;).

Legislative approaches

Mineral conservation and development RULES, 1998, As per NATIONAL MINERAL POLICY, 2008(Annual report 2017-18) rules:

Mining operations under mining lease:

1. No holder of mining lease shall commence mining operations in any area except in accordance with the mining plan approved.
2. If the mining operations are not carried out in accordance with the approved mining plan, it is found to be incorrect. With applicable laws including these rules, the competent authority may by order, suspend all mining operations. Such operations are required to restore under the approved mining plan.
3. Every certified mining plan shall be subjected to review and updation in every 5yrs, as per provision of sub-rule.
   A. Of rule 17 of the minerals (other than automic and hydro carbons energy minerals) concession rules, 2016.
   B. Mining lease holder shall submit the mining plan to the competent authority for review at least 180 days before the expiry of 5 yr period for which it is approved on the last occasion, for mining operations for a period of 5 subsequent years.

Prospecting and mining operations:

1. It shall be carried out in such a way to ensure systematic development of mineral deposits conservation of minerals and protection of the environment.
2. The prospecting licence holder shall carry out exploration as per the norms provided in the minerals (evidence of mineral contents) rules, 2015.
3. The mine lease holder shall carry out detailed exploration (G1 level) over the entire potentially mineralised area under the mining lease, within a period of 5 yrs from the date of opening of mine after execution of such mining lease.
4. In case of existing mining leases detailed exploration (G1 level) over the entire potentially mineralised area under the mining lease, within a period of 5 yrs from the date of commencement of these rules.
5. Mineral resources shall be periodically estimated based on exploration carried out at the resourcesand shall be estimated up to the threshold value of the mineral, as may be notified by Indian Bureau of Mines ,from time to time and the updated resources shall be furnished in the review of the mining plan of 5 yr interval.

Reclamation and rehabilitation as per MCDR, 1988:

As per section 23-

1. A part of mine during the subsistence of the lease except with prior permission in writing of the controller general or the authorised officer.
2. The owner of every mine shall send to the controller general, a note in the form-D of his intention to abandon a mine or a part of mine so as to reach them at least 90 days before the intended date of such abandonment.
3. Such notice shall be accompanied by plans and sections on a scale not less than1cm=10m setting 4th accurately the work done in the mine up to the time of submission of the notice including the measures envisaged for the protection of abandoned mine or part thereof, the approaches to the environment
4. The authorised officer may buy an order in writing made before the proposed date of abandonment, prohibit abandonment or allow it to be done with such conditions as he may specify in the order.
5. The leaseholder shall not abandon a mine or part thereof unless a final mine closure plan duly approved by the regional controller of mines by the state govt. In this behalf, as the case may be implemented. For this purpose, the lease shall be required to obtain a certificate from the regional controller of mines by the state govt.
6. As per section 34, every holder of prospecting licence or mine lease shall undertake the phased restoration, reclamation and rehabilitation of lands affected by prospecting or mining operations and shall complete this work before the conclusion of such operations and the abandonment of prospector mine.

The review suggested that the various authors have worked in the direction of engineering manures using waste materials and integrated approach to adopt biotechnological measures in alliance with social measures can bring in overall changes and sustainable restoration of waste lands which has a greater scope in India and as well as in the world to apply and save the mother earths ecological balance between natural resources and human demands.

Therefore the present project work is taken up with a focus to conduct investigations on mining waste soil mixed with definite proportion of lime as suggested by other authors and using RHAand SWD in different proportions.

Investigations will also be conducted on mining soil to demonstrate soil erosion control measures using vermi-compost of cow dung.

Further based on the study of vegetation previously existed through ethno-botanical study isolation of micro organisms especially bacteria have to be isolated and inoculated in to the mining waste soil for investigating soil aggregation, water holding capacity, and plant growth nutrient characteristics.

Acknowledgement

We sincerely thank Dr. Prakash M Munnoli for his encouragement and support during the preparation of the draft paper and editing process till the completion. We also thank Dr. S B Vanakudre principal SDMCET for his continuous support and encouragement throughout our study. We sincerely thank Prof V S Hegde for his support during the initial ideas for the preparation of project and draft paper. We also thank Mr. Mruthunjay Reddy for his support during the visit to mining site at Hospete.

References

Amadi, A.A., Okeiyi, A (2017) "Use of Quick and Hydrated Lime in Stabilization of Lateritic Soil: Comparative Analysis of Laboratory Data", International Journal of Geo-Engineering, vol.8, No-3, 1-13.

Anonymous (2010): Mineral and Mining Industry in India 2010.

Anonymous: Development of Wasteland and Degraded Lands.581-593.

Anonymous: Wasteland development ch-5, 49-51.

Balasubramanian, A (2015) "The Wastelands in India", 1-14

Barathan, R., Giridharan, A., Saranya, P (2017) "Soil Stabilization Using Silica Fume and Cement" International Journal of Civil Engineering, 78-82.

Bhuvaneshwari, S., Robison, R.G., Gandhi, R (2005) "Stabilization of Expansive Soils Using Flyash"Flyash India, vol.8, No-5, 1-10.

Boori, M.S (2010) "Wasteland Atlas of India-2010.

Bradshaw, A.D. (1990).The reclamation of derelict land and the ecology of ecosystems.InRestoration ecology: A synthetic approach to ecological research. Edited by W.R.Jordan, M.H.Wong, J.W.C.Wong, and A.J.M.Baker.Lewis, Boca Raton, Fla.pp.153-162.

Brooks, R.M., (2009) "Soil Stabilization with Fly ash And Rice husk Ash" International Journal of Research and Reviews in Applied Sciences, vol.1, No. 3,209-217.

Buss, J., Stratechuck, K., Pinno, B.D (2018) "Growth and Competition among Understory Plants Varies with Reclamation Soil and Fertilization".1-8.

Chattopadhyay, S., (2006) "Sustainable Regeneration of Derelict Mine Sites" ITTPI Journal 3:4, 42-48.

Cooke, J.A., and Johnson, M.S. (2002) "Ecological Restoration of Land with Particular Reference to the Mining of Metals and Industrial Minerals: A review Of Theory and Practice"Enivronmental Revolution, Vol.10, 41-71.

Dodor, D.E., Amanor, Y.J., Attor, F.T., Adjadeh, T.A., Neina, D., Miyittah, M (2018) "Co-application of Bio char and Cattle Manure Counteract Positive Priming of Carbon Mineralisation in Sandy Soil" Environmental Systems Research.vol.7, No-5, 1-9.

Dhir, B.M., (1991), "Mining, blasting and deforestation and their effects" Paper presented at ISTE Winter school on environmental engineering, held from 16-27 Dec.1991.

D'souza, J., Lobo, B., M.Souza (1995) "Effect of mining on microbial ecology of Bicholim Taluka, Goa. Possible restoration processes in the region. Technical report submitted to ministry of environment and forests, New Delhi by the department of microbiology, Goa university.

Gupta, A., Arora, V.K., Biswas, S (2017) "Contaminated Dredged Soil Stabilization Using Cement And Bottom Ash For Use As Highway Subgrade Fill" International Journal Of Geo-Engineering 8:20,1-13.

Haigh, M.J. (1993).Surface mining and the environment in Europe.Int.J.Surf.Min.Reclaim. 7:91-104.

Hobbs, R.J. (1999). Restoration of disturbed ecosystems. In Ecosystems of disturbed ground. Edited by L.Walker. Ecosystems of the World, Elsevier, and Amsterdam.pp.691-705.

Jha,J.N and Gill,K.S(2006) "Effect Of Rice Husk Ash On Lime Stabilization Of Soil"IE(I)Journal-CV,vol.87,33-39.

Kivinen, S (2017) "Sustainable Post-Mining Land Use: Are Closed Metal Mines Abandoned or Re-Used Space?" vol.9, No-1705, 1-18.

Kumar, M., Kumar, S (2017) "A Case study of Development Plan in Jind District (Haryana)", International Journal of Academic Research and Development, vol.2, No-5,831-836.

Mittal, S.P. and Singh, G. (1990) "Resource conservation and management in watersheds of Shivalik foothills", International Symposium on Water Erosion, Sedimentation and Resource Conservation. Oct. 9-13, Dehradun, 330-339.

Mittal, S.P., Mishra, P.R., Grewal, S.S. and Agnihotri, Y. (1986) "Success story of Sukhomajri watershed management project", Indian Journal of Soil Conservation, vol. 14 (3), 1-8.

Munnoli P M (1998)"A study on management of organic solid waste of agro based industries through vermiculture biotechnology", ME Thesis, TIET Patiala, India, 11-30.

Munnoli P M (2011) Water holding capacity of earthworms vemicompost made of sugar industry waste (press mud) in a mono-and polyculture vermireactors, The Environmentalist. vol.31 (4), 394-400.

Munnoli, P.M. and Bhosale, S (2008) "Soil aggregation by vermicompost of press mud", Current Science, vol.95 (11), 1533-1535.

Nnochiri,E.S., Emeka, H.O., Tanimola, M (2017) "Geotechnical Characteristics Of Lateritic Soil Stabilized With Sawdust Ash-Lime Mixtures" The Civil Engineering Journal 1",Vol. 1,No- 7,66-76.

Pastorok, R.A. et. al., (1997).An ecological decision framework for environmental restoration projects. Ecol. Eng. 9:89-107.

Rowley, C.M., Grand, S., Verrecchia, E.P (2018) "Calcium-mediated Stabilization of Soil Organic Carbon" Biogeochemistry, vol.137, 27-49.

Shawl, Z.Z., Parkash, V., Kumar, V (2017) "Use Of Lime And Sawdust Ash In Soil Stabilization" International Of Innovative Research In Science,Engineering And Techonology,vol.6,No-2,1682-1689.

Sheron, A., Sheron, A.S., Poonia, P (2010) "Soil Reclamation of Abandoded Mine Land by Revegetation: A Review" International Journal of Soil, Sediment and Water, vol.3, No-2.

Singh, N.B. (1997)"Dissertation on Development of process package for organic solid waste management through vermiculture Biotechnology, in organic waste generating industries in Punjab" 91-93.

Wang, D., Fonte, S., Parikh, S.J., Six, J., Scow, K.M (2017) "Bio char Additions Can Enhance Soil Structure and the Physical Stabilization of C in Aggregates" vol-303,110-117.

Xiao, W., Hu, Z., Fu, Y (2014) "Zoning of Land Reclamation In coal Mining Area and New Progresses for The Past 10 years ", International Journal Of Coal Science Technology, vol.1, No-2,177-183.

Integrated River Flow Modeling Using Regression

Shreenivas Londhe[1] and Sayali Shirgave[2]

[1]Professor in civil Engineering, [2]Post graduate student,

Water Resources and Environmental Engineering, Vishwakarma Institute of Information Technology, Pune

shreenivas.londhe@viit.ac.in, shirgave.sayali@gmail.com

ABSTRACT

Accurate estimation of flow in a river having one or more tributaries is though important a challenging task. When two or more rivers meet the increased flow in the combined river may prove to be dangerous for downstream locations particularly in the monsoon season. The present work is an attempt to predict flow at downstream station using flow at 2 upstream stations (Irwin Bridge and Warna dam) one of which is on a different river using linear and non-linear regression techniques. Discharge at two gauging stations, one on Warna River and Krishna River are available from 2005 to 2016. After confluence the discharge data is available on the combined river (Krishna) on downstream side (Ankli Bridge, Sangli, Maharashtra). The discharge at Ankli Bridge is predicted using discharges at two upstream stations mentioned earlier using Linear and non-linear regressions.

Introduction

Flood being a natural disaster, mostlyduring monsoon season, cannot be avoided but the severity of flood can be reduced by providing proper flood management. Giving an alert or prior warning to people can save damage to lives, properties etc.

The flood situations becomes dangerous at the confluence of rivers particularly in monsoon season. Confluence of rivers can be defined as the junction where two or more rivers or streams join together to form a single channel. The discharges from both the rivers are added together at confluence and it is carried to the downstream side of the combined river. To predict the flow at the downstream station in the river having one or more tributaries becomes a necessary task as it may become dangerousfor downstream locations along the river.

This can be achieved by using physics based models which solve Saint Venant's equation to yield water levels and discharges on the downstream side. However such models require exogenous data such making their use a specialized task. In the absence of such a data which is a very common case in countries like India predicting floods on downstream of any confluence becomes a highly difficult though most required task. Data driven techniques, both hard and soft work on the data rather than physics of the underlying process. Regression is the most primary methods used in data driven techniques which is used in the present study to predict discharge at a downstream station using discharge at two upstream stations on two different rivers. The two upstream stations selected are on the river Krishna and Warna (tributary of Krishna River) and downstream station at Ankli Bridge on combined Krishna River In the state of Maharashtra The flood discharge at Ankli Bridge is predicted using both linear and non-linear regression.

Literature review

As mentioned in the 'Introduction' various methods are being used for predicting discharge at a locations using discharge at upstream locations(s) which follow physics of the process. However considering use of regression in the present work earlier works which used data driven techniques are only discussed in subsequent paragraphs.

Dixit et al. (2018) compared the traditional as well as data driven technique of Genetic Programming for routing of flow. It is observed that predicted discharge obtained by using GP nearly matches the observed discharge. On the other hand, there is poor prediction done by Muskingum method.

In previous work by Daud s.et al (2011), the study findings show that ANN was successfully applied to the available data sets for SwatRiver and the resulting models showed encouraging results with respect to the measures of performance such as R2, AAE, MAE, and RMSE. The study proved that long-term data available at one station can be used to obtain missing stream flow data for stations where long-term data is not available.

Study area

The present study is carried out for a reach from Irwin Bridge (upstream station) located on Krishna River (A) and Warna dam(B) (another upstream station) located on Warna River to the Ankli Bridge (downstream station) located on KrishnaRiver situated after their confluence at Haripur. The daily data of flow through river (cusecs) from 27/7/2005 to 16/10/2016 was collected from Sangli irrigation division, Sangli (govt. of Maharashtra). The distance between upstream station Irwin bridge (A) and downstream station (C) is approximately 12 kilometers. The bed slope of the reach is found to be 0.00025. The details of upstream and downstream stations are provided in the table 1

Table 1 Locations of gauge stations.

GD Sites	District	Taluka	Latitude	Longitude
Irwin Bridge	Sangli	Miraj	$16^0$51'37.08"	$74^0$33'25.56"
Warna Dam	Sangli	Shirala	$17^0$08'10"	$73^0$51'50"
Krishna-Warna confluence (Haripur)	Sangli	Miraj	$16^0$49'30.83"	$74^0$37'10.33"
Ankli Bridge	Sangli	Miraj	$16^0$48'19"	$74^0$33'58"

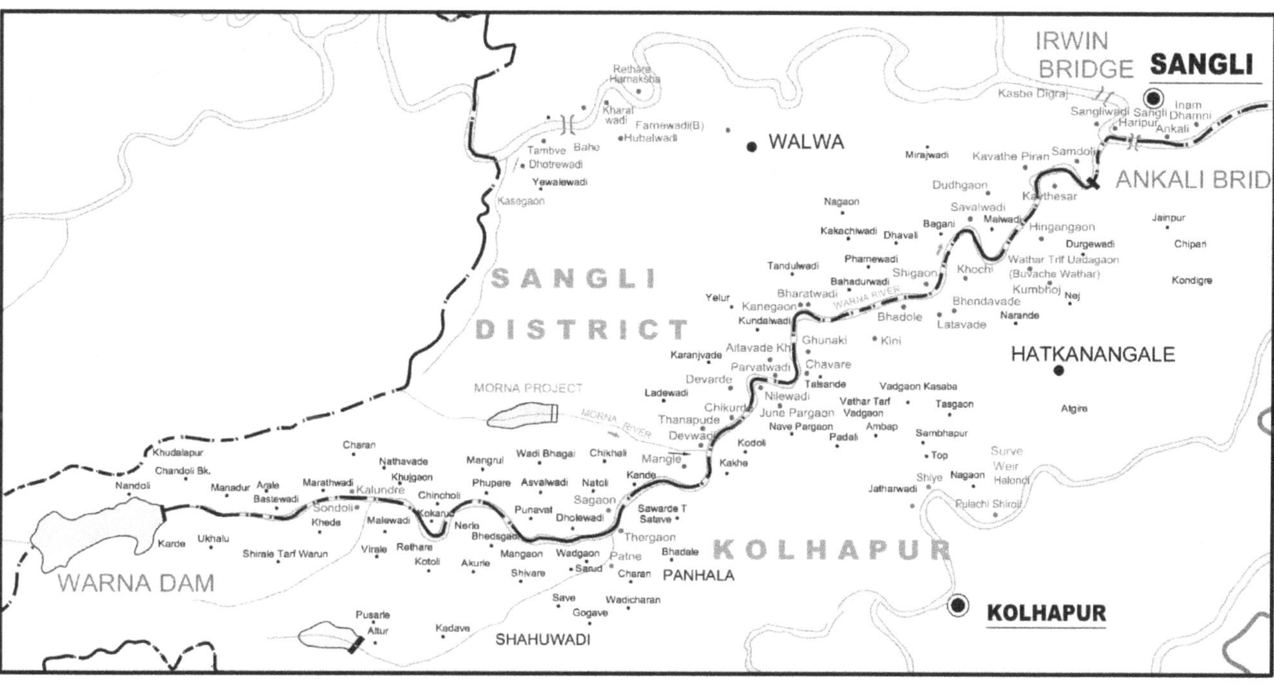

Figure 1 The pictorial view of the gauge stations is provided below.

Methodology

Regression

In many problems there are two or more variables that are inherently related and it may be necessary to explore the nature of this relationship. Regression analysis is a statistical technique for modelling and investigating the relation between two or more variables. Suppose there is a *dependent variable y* that is related to *k independent variables* say $x_1,x_2,.........x_k$.The relationship between these variables is characterized by mathematical model called a *regression equation.* This regression equation is fitted to a set of observed data. (Reddy, 1997)

A. Simple Linear Regression

Let X and Y be the independent and dependent variables respectively. Let (x_1,y_1), (x_2,y_2),.......... (x_n,y_n) be the pairs of concurrent observations taken on X and Y. Then Y= a + bX can be possible regression equation wherea and b are constants which can be obtained from the regression analysis.(Reddy,1997)

B. Polynomial Fitting

Curvilinear data which do not indicate a linear scatter diagram even under any of the transformations can be fitted by a polynomial of the form $Y=\alpha_0 +\alpha_1 x +\alpha_2 x^2+\ldots\ldots\ldots\ldots\ldots+\alpha_p x^p$ where $\alpha_0,\alpha_1,\ldots\ldots\ldots,\alpha_p$ are the regression coefficients to be determined and p is called the degree of polynomial.(Reddy, 1997)

Results and Discussion

The table 2gives the regression equations obtained for Ankli Bridge. The equations are derived using the different methods of regression and their results are compared using correlation coefficient (R) between the observed and predicted values. It cab seen that results of Logarithmic fitting are less accurate compared to Linear and polynomial fitting

Table 2 Equations obtained and coefficient of correlation for Ankli Bridge

Regression	Equation	R
Linear fitting	y=1.0506x+4591.7	0.9625
Polynomial fitting	$y = -(4E^{-12})x^3+(4E^{-7})x^2+(1.1231)x+3160.4$	0.9632
Logarithmic fitting	y = 36825ln(x) – 305491	0.8403

Additionally scatter diagrams are also plotted to judge the accuracy of prediction which confirm the same observation. The scatter diagram shows a balanced scatter though for few discharges the values are underpredicted.

All The scatter plots are plotted with combined discharges of upstream stations on X axis and the discharge at downstream station on Y axis. The values are found to be reasonably accurate for all the methods of regressions.

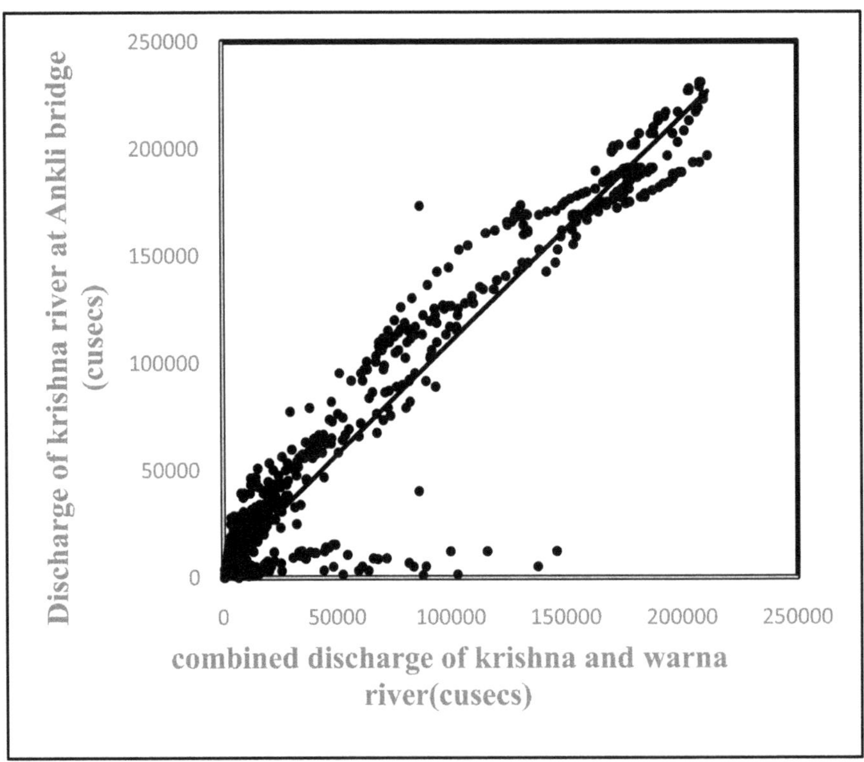

Figure 2 Scatter plot for linear regression.

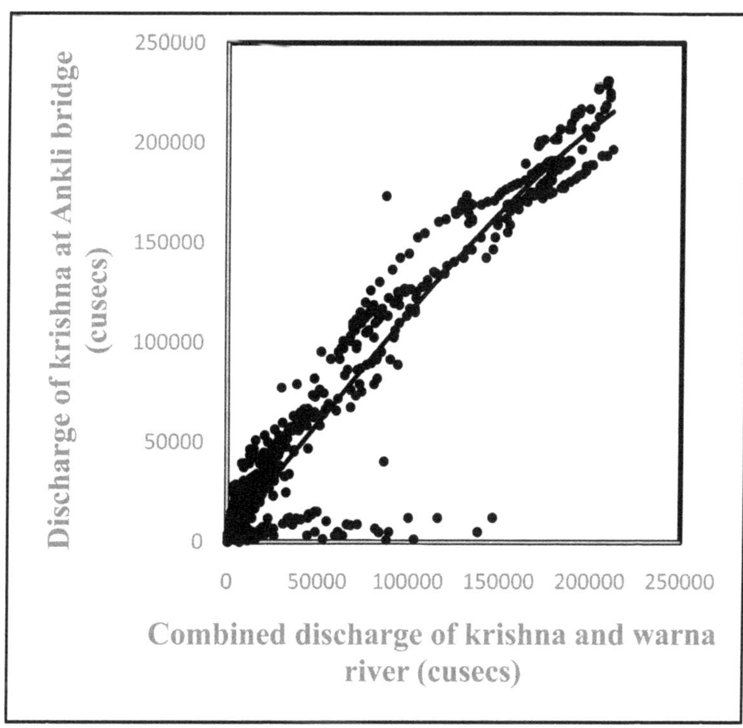

Figure 3 Scatter plot for polynomial fitting

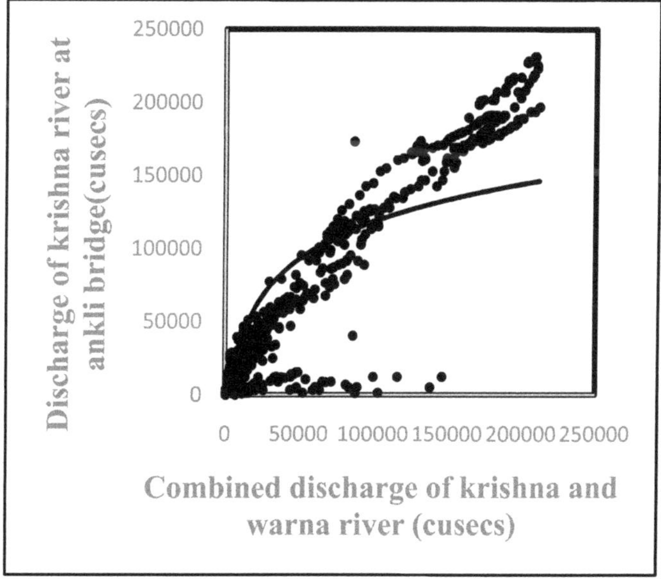

Figure 4 Scatter plot for logarithmic fitting

Model Validation

The data from 2005-2016 were used to calibrate the models which were then tested for unseen data measured in the year 2018. The accuracy of results was judged by calculating the correlation coefficient as well root mean squared error between the observed and predicted values. Table No. 3Shows 'R' and RMSE values which indicate a reasonable performance of all the models.

Table 3 Details of errors measured along with coefficient of correlation

Regression	RMSE (cm^3/sec)	Correlation
Linear fitting	1252.14	0.9077
Polynomial fitting	13238.84	0.9077
Logarithmic fitting	2463.05	0.9212

The scatter plot for the data 2018 is shown below in figure Nos.5, 6, 7 which confirm the above results However in this case the results of logarithmic regression are also par with other methods, in fact a shed better.

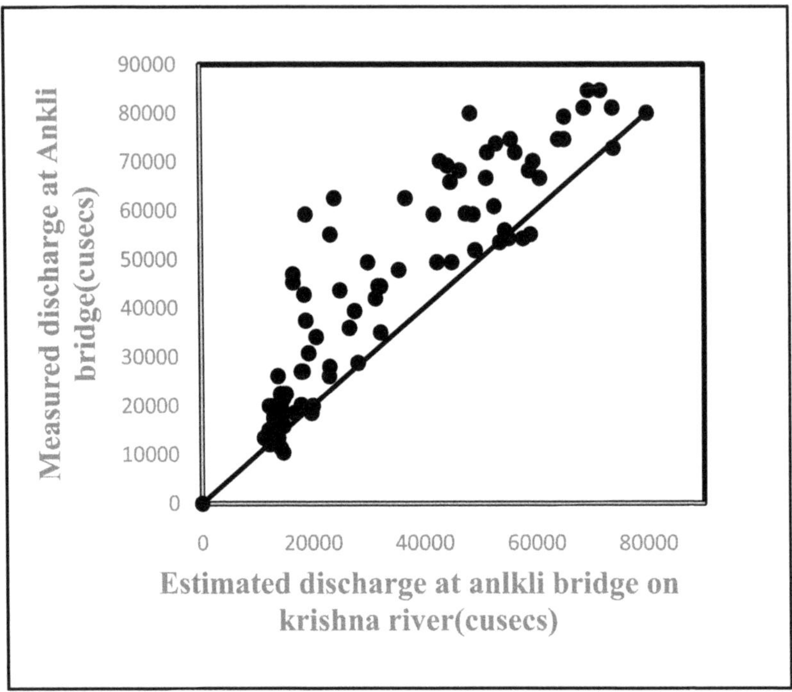

Figure 5 Scatter plot for estimated values using linear fitting.

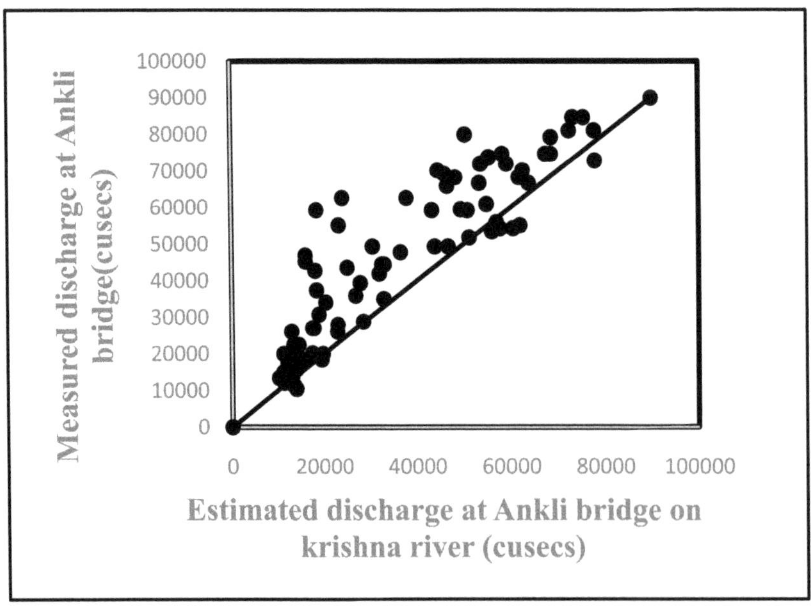

Figure 6 Scatter plot for estimated values using polynomial fitting.

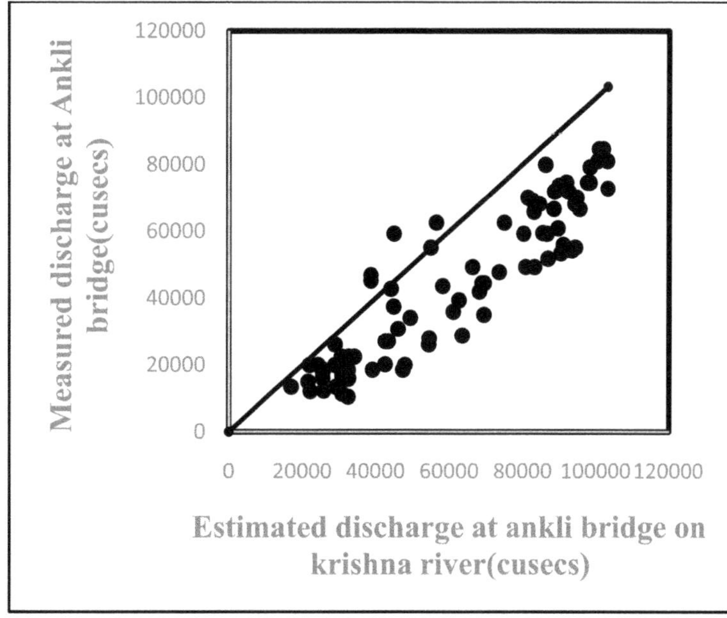

Figure 7 Scatter plot for estimated values using logarithmic fitting

Conclusion

In the present work prediction of discharge at a downstream station on river Krishna is done using discharges at two upstream stations, one on river Krishna and one on river Warna before the confluence. It is observed that all the modes give reasonable results both in calibration and validation phase. Despite of not considering the parameters like temperature, topography etc. this method gives the estimation of water levels with a reasonable accuracy. Other data driven techniques like Artificial Neural Networks, Genetic Programming can also explored for better accuracy.

References

Asati, S., & Rathore, S. (2012)."Comparative study of stream flow prediction model,"*International Journal of sciencesBiotechnology and Pharma Research, 1*, 139-151

Dixit P. et al. (2018) "Flood routing: a comparison of theoretical and data driven technique," National conference on water, environmental& society, 195-200

P. Jayarami Reddy, Stochastic Hydrology, Laxmi publications (p) Ltd, New Delhi, 1997 Sangli irrigation division, Sangli (govt. of Maharashtra).

Suleman Daud et al. (2011) "Stream Flow Modeling of River Swat Using Regression and Artificial Neural Networks (ANNs) Techniques,"*Advanced Materials Research* Vols. 255-260 (2011) pp 679-683.

Veiga V. B.et al. (2015) "Development of flow forecasting models in the Bow River at Calgary, Alberta, Canada," Water 2015, 7, 99-115.

Exploring Environmental Migration
Exploring Indigenous Knowledge as a Means to Sustainable Development

Anjali Dalal
PhD Research Scholar, Jawaharlal Nehru University, New Delhi, India
anjalidtt@yahoo.co.in

ABSTRACT

A widespread view that is gaining ground in the current trends of human migration is related to environmental change induced migration that has a potential to evolve into a global crisis by displacing a large number of people from their homes and forcing them to flee. International reports postulates that a billion people could be permanently displaced by climate change related phenomenon such as droughts, floods and hurricanes. The economic development induced migration, due to environmental degradation, is again a new phenomena introduced in post - colonial societies. Moreover, there is no internationally agreed definition or law, for the term "climate refugee" and the extent to which these displaced persons constitute a separate identifiable group. Simultaneously, it is also seen that, indigenous people at many places have immensely contributed to restoration of biodiversity in the depleted landscapes. Some indigenous groups manipulate the local landscape to augment its heterogeneity, and some have been found to be motivated to restore biodiversity in degraded landscapes. The focus of this study is environmentally induced displacement as a result of anthropogenic factors. The study would try to answer the issues of environmental migration from the perspective of the feasibility of indigenous knowledge in environmental conservation so as to work in direction of sustainable development with economic growth and expansion. The methodology is based on secondary literature review as well as analysis of data on indigenous knowledge in various research studies. This has been proved in the end, with an example of a Hiware Bazar village in Mahararastra, that the indigenous knowledge leading to sustainable development could have a possible implication of reverse human migration. It was thus, realised that few local communities possess immense knowledge practices to impede the process of environmental migration by introducing sustainable development techniques.

Keywords: Environmental migration, Environmental refugee, Climate change, Indigenous knowledge, Hiware Bazar village

Introduction

A widespread view that is gaining ground in the current trends of human migration is related to environmental degradation induced migration that has a potential to evolve into a global crisis by displacing a large number of people from their homes. A direct impact of ecological deterioration is visible in form of climate change and its impact in form of global warming, increasing intensity of floods and droughts, etc. In fact, the total number of people at risk of sea-level rise alone in Bangladesh could be 26 million, and 20 million in India (Myers 1996; Watson et al. 1998).In 2017, displacement has been caused primarily by extreme weather events, especially flooding (8.6 million) and storms(7.5 million) in South and East Asia. (IDMC report, 2018) However, focus of my study is yet another crucial and less explored area of enquiry into the anthropogenic reasons acting as a push factor to environmental change and subsequent migration. In fact, events leading to extreme weather conditions could be the result of certain human induced developmental projects (post-industrialisation age) that proved detrimental to the environment in the long run. They are commonly associated with unsustainable modern techniques of growth leading to environmental degradation in form of rising air and water pollution, loosening of soil fertility, depletion of natural resources, poor planning, industrial accidents and the loss of traditional agricultural practices with overuse of fertilisers and cutting of forest resource. The ecological de-stability has become an important reason for the loss of economic opportunities and sustainability of life, that has led to human migration. Many areas in the Western part of India has been victimised to these developmental models, under the garb of economic progress (for e.g, Green Revolution). For instance, in Punjab, modern agricultural technologies improvised production and worked as a 'pull factor' to workforce from within and across the states. However, the after-effects of these technologies proved unsustainable with the overuse of fertiliser and off-seasonal practices that diminished the production and income in the longer run, which became a 'push factor' to out-migration for an alternative occupation. In Rajasthan, 67% of area is affected by desertification andsoil degradation as a result of unsustained agricultural production that led to huge out-migration. With a breakdown of traditional water management practices and in adequate scientific water-management systems, irrigation is insufficient and inefficient thus directly hurting livestock rearing, that is the major economic activity of some villages. Similarly, man-made water crisis due to mismanagement of water resource is quite alarming in Maharastra. With 50% villages likely under 'drought like situation' and the farmer's suicide rate at 7 cases per day, displacement of people from the state is enormously increasing due to declining economic opportunities as a result of deleterious living environment[1]. There is, thus, an urgent need of mainstreaming adaptation strategies for human migration in order to protect this vulnerable group mainly in the developing countries.

Conversely, human migration is not a recent phenomena. Since ages, people have been dealing with deteriorating environment by developing sustainable techniques. These communities have developed their unique sources of passing down the knowledge acquired through their experiences of ecological preservation which include story telling arts, cultural values, healing techniques, ballads, songs and verses. Over a period of time, such indigenous knowledge system has assisted in preserving different cultures, languages and traditions along with the biodiversity with which we share this earth. Overall, this article will focus replacing the modern with indigenous techniques of development so as to promote sustainable development. It further postulate the idea of restoration of such techniques that could address the problem of environmental migration and bring back the already displaced people.

Framing Environmental Migration: Ravenstein (1995) was the major proponent to conceptualise migration theory based on push- pull factor, that is, unfavourable conditions push people out of place, while favourable conditions pull them to another place. Though the model was build on employment factors, it can was easily adapted to explain few climate change phenomenas as well (for e.g., the migration of people from Bangladesh to India as a result of depleting fishery sector and degradation of Sundarbans forests). The theorisation of environmental migration started in 1997 with the "Alarmist School" (Norman Myers, 1998), that predicted a huge data on climate change induced migration, with "at least 25million environmental refugees ... and by 2025 over 200 millionpeople would be displaced worldwide as a result of theimpacts of a changing climate". Critical to this school of thought, rose the "skeptics", who found these figures exaggerated with no empirical evidence and poor modelling (Brown 2008, Koser 1996, Black 2001).Such "skeptics"have urged the adoption of more critical perspectives so asto avoid allowing such misperceptions to spread.Although my focus is not based on the long term climate change related migration, yet the given statistics helps in establishing the seriousness of the migration issue related to environmental change.

Another set of discourse includes displacement or forced migration due to developmental factors or improper methods of economic growth. Inmany cases the drivers of such displacements are related to 'resource wars', that is depletion or expropriation of natural resources that belong to the local community, in the name of economic development (Phillippe le Billon, 2007). Whatever the reasons for dislocation, the outcomes arenonetheless often quite similar: homelessness, landlessness,the loss of livelihoods and connection to important culturaland/or religious spaces, and in many cases physical andmental harm (Michael Cernea, 1997). It has, therefore, given rise to a new category, called "internally displacedpersons" (IDPs), that is, human migration within borders, especially as a result of development-induced displacement as a result of improper construction of dams, roads, parks (Bogumil Terminski, 2013). Akey challenge is the fact that displacement due to developmentis justified along nationalist or economiclines without analysing proper causal elements. An important question is; how do people who have been displaced not be some "natural disaster" but due to long-term processes of environmental degradation led weakening of their socio-economic capacities justify their need for analternative mode of development? How do communities who haveadapted to certain temporary forms of arrangement due to ecological change be made sure of possibility of sustainable future or reverse migration of their own villages?

Indigenous knowledge techniques- Responses to Sustainable Development: The focus of this article is to approach the crisis of human migration using the framework of sustainable development as a means to prevent environmental degradation. In the contemporary period, environmental displacement is conceptualised, negotiated, and governed at various levels. The World Summit on Sustainable Development has introduced the meaning and importance of this term in a broad framework and as a means to economic growth. Many governmental

---

[1]TNN | Jul 12, 2018, 04:5
Retrieved at:http://timesofindia.indiatimes.com/articleshow/64954221

organisations and social activists have now focused on different models of sustainable development to counter the faulty processes of development and reduce environmental migration.

I would rather exclaim that indigenous knowledge (IK)must be considered as a means to sustainable development that could address the problem of recent migration as well as promote reverse human migration. This is so because humans have lived in communities of hunters, gatherers, etc. since the dawn of civilisation and have developed their unique techniques to cope up with natural surroundings. Such knowledge techniques have been acquired, preserved and transferred from generation to generation.Moreover, indigenous knowledge provides the basis for problem-solving strategies for local communities, especially the poor. It represents an important component of global knowledge on development issues. Yet, it is an under-utilised resource in the development process. Learning from IK, by investigating first what local communities know and have, can improve understanding of local conditions and to provide a productive context for activities designed to help the communities. Utilising its importance in environmental migration theories could open new dimension in the research of environmental change induced displacement as well as understanding and correcting the developmental models in post-industrial societies.Some of these techniques are worth mentioning, that have been part and parcel of "lived experiences" of the communities, as well as kept the environmental sustainability intact over generations.

1. **Animal Protection and Women**: Women in Gujarat have developed their ways to combat the drought conditions and still provide nutrient rich food to the humans and animals dependent on them. Women have saved their animals from drought using desi babool fruits (Acacia Nilotica Delile, fig. 3), Vilayati babool (Prosopis juliflora DC), Mithi Imli leaves, maintained their health using Pilu (Salvadora Persica Linn, fig. 2), Kahdo and Umro (Ficus Hirta Vahl); and fed their camels soft buds and leaves of S. Persica Linn and S. Oleoides Decne[2]. Doob (Cynodon dactylon Pers.), Kajaro, Char and Ghjnathiyo grasses are also used. These women have always indicated that a portion of balanced diet isimportant for both animals and human beings and gradually developed their own strategies for maintaining it. Many other plant leaves have been used by women at the time of famines. For instance, Kanjara, tandarda, kharkhori (Leptadenia pyrotecnnica Decne), vaseti and ekad leaves and buds of phag (Pueraria tuberosa DC) are used to make food at extreme famine situations. Women, or 'the custodians of indigenous agro-biodiversity,' have improvised their systems through local groups to preserve and modify these plants and grasses to maintain proper and nutrient rich supply. If these women pass on their rich knowledge, much advancement could be made in agricultural research.(Ranjay K. 2003)

Fig. 1 Thor (Euphorbia neriifolia)

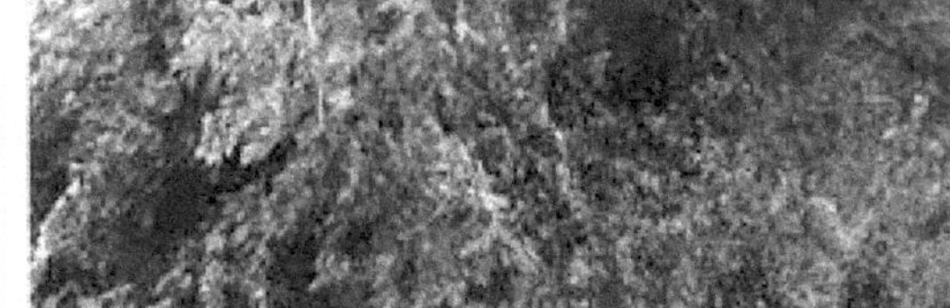

Fig. 2 Pilu (Salvadora persica)

Figure 3 Babool leaves (on the left) Source: google, wikipedia images.

2. **The Sacredness of Preservation:** Another techniques of establishing traditional knowledge is preservation through 'sacredness' such as few trees or practices are considered unique in itself and meant to be religiously followed or protected. For instance, Meetei communityof Manipur and Assam follows some traditional practices and live in harmony with the nature. Umang Lais (sacred groves), teak, lemon, ginger, eucalyptus, bamboo, fishes, water fowls, snails etc. are an important part of Meetei food, religion, tradition and culture, and protected with full security. With the idea of conservation many of these are not consumed at certain periods(L. Jeetendra 1998). Similarly, Kumaon Himalayas are known for Pani Dhara— marriage solemnized in the presence of a water body,and Kumbh Vivah— practice of marrying a bride to a pitcher of water

[2] Ambasta SP, The Useful Plants of India, (Publications & Information Directorate, CSIR, New Delhi), 2000.

representing the groom, that is locally recognized. (Panna, 1920)This shows the honouring of water and its consideration as sacred. Some of these practices still continue and people understand the importance of conservation of natural resources as they are linked to the existence of their unique identities.

3. **Traditional Agricultural Practices:** In study of indigenous practiccs in Bangladesh, it was found that farmers had unique techniques of saving crops from insects, rather than switching to fertilisers and pesticides. For instance, farmers of Sunamganj (North East Bangladesh), Gajipur, Rajshahi (near Indo-Bangladesh border) use a variety of ingredients to secure their plants in place of insecticides and pesticides promoted by government in Green Revolution. They mix ash with kerosene oil and spray it to control aphid infestation, similarly powder of Neem leaves is mixed with water and spread to repel insects in rice fields. By digging deep circular trenches around the plants of brinjal, cauliflower and cabbage they have developed a method to control caterpillar attacks. Potatoes are covered with dry sand and rice husk, they are protected from tuber moths (Chowdhary et al. 1996). These techniques are still prevalent there, and continued since ages, thus making their farming techniques organic and pure in nature.
4. **Merging of Traditional and Scientific Knowledge:** An integrated nutrient management system was developed by National Agricultural Research Systems in association with NGOs for resource poor farmers of Maharashtra. This system is a meeting ground for indigenous practices and technical knowledge which focuses on sustainable growth of crops. This was specially beneficial to those farmers who cannot afford expensive agrochemical products to manage productivity. In another illustration to this idea is the Green Belt Movement (GBM) of Kenya, that depicts a possible reconciliation to a scientific and traditional approach of environmental preservation. It was visible in their women's opposition to Nairobi government for introducing Shamba System, wherein, communities were allowed to grow crops along with exotic tree species. The Shamba system resulted in the indigenous forests degradation, associated with the invasion of foreign species. In response to this, GBM pilots a Clean Development Mechanism (CDM)[3] project, wherein, individual farmers and groups, plant a mixture of exotic and native species of trees only on farms and public lands, while they only plant 'native tree'[4] species in forests. This illustration is a reflection of the idea that CDM is a combination of both traditional and modern knowledge system by using the exotic and native species at appropriate planting grounds. More such examples have to be deliberated so as to conciliate both the techniques into sustainable living.

   Traditional medicinal values have been recognized as alternate healing methods and maintained sustainable ecological development. The merging of smaller tradition into greater tradition has posed a threat to these communities and their vast source of unexplored knowledge. Privatization and commercialization of native lands and resources along with continued marginalization and alienation in this era of globalization has put them at risk. To safeguard these people and their resources and knowledge the state must recognize them and work with them in policy making to ensure their survival and environmental conservation.

Conclusion

An incredible example of reverse migration could be seen in a village called Hiware Bazar, in Ahmednagar district, Maharashtra. During 1970's, low rainfall, drought, extreme poverty, unemployment and naked hills were the features that shocked the village. Most people starting to migrate to near urban cities, Mumbai and Pune for better wages. Yet their linkages with their own land kept them nostalgic. Years later, few youth of the village, gathered together and tried to solve the problem by reviving traditional knowledge of water resource management. This was done with the support of Panchayat and employment guarantee schemes initiated by the government. However, the youth clubbed these schemes with their own traditional practices, such as rain water harvesting by constructing Jal Pandal (water made such that rain water flows directly into the irrigation field). Many other experiments were conducted in farming system related to appropriate plantation that suits the trough prone area and becomes

[3]The experience highlights the need for governments' policies and programmes to be aligned with specific requirements for carbon projects. CDM calls for projects to certify there will be no 'leakage'.

[4]The latter species include: Prunus Africana, Olea African, Cordia Africana, Juniperus Procera, and Hagenia among others, depending on the ecological zones.

productive for the villagers. This gradual led to income generation and helped in social transformation as well. Mostly women were part of all these activities and emerged as great managers and supports of both household and agricultural field. Almost 40 families returned to the village between 1992 to 2002, according to Panchayat records. Today, a fourth of the village's 216 families are millionaires. Hiware Bazar's sarpanch, Popat Rao Pawar, says "just over 50 families have an annual income over Rs 10 lakh. The per capita income of the village is twice the average of the top 10 per cent in rural areas nationwide (Rs 890 per month)". "The number of wells has increased from 97 to 217. Irrigated land has gone up from 120 ha in 1999 to 260 ha in 2006. Grass production went up from 100 tonnes in 2000 to 6,000 tonnes in 2004", as per Panchayats official report. Sakhubai Thange, a 70-year-old villager who has been cutting grass for the last 25 years, recalls the time when overgrazing had made grass scarce. "The efforts put in by the people of the village for soil and water conservation have created a surplus," she says. In fact, the village has now become an ideal for youth as well as older generation to revive their own indigenous knowledge systems and stop unsustainable development and forced environmental migration.

References

Ambasta SP, "The Useful Plants of India", Publications & Information Directorate, CSIR, New Delhi, 2000.

Anthony Richmond, "Reactive Migration: Sociological Perspectives on Refugee Movements," Journal of Refugee Studies 6, no. 1 (1993): 7–24.

Astri Suhrke and Annemarie Visentin, "The Environmental Refugee: A New Approach," Eco decision, September 1991, 73.

Bogumil Terminski, "Development-Induced Displacementand Resettlement: Theoretical Frameworks and CurrentChallenges," Development 10 (2013): 101.

Elizabeth Lunstrum, "Reconstructing History, Grounding Claims to Space: History, Memory, and Displacement in the Great Limpopo Transfrontier Park," South African Geographical Journal 92 no. 2 (2010): 129–43.

Essam El-Hinawi, "Environmental Refugees", Nairobi: UNEP, 1985: 4.

Gregory White, "Climate Change and Migration", Oxford: Oxford University Press, 2011.

JoAnn McGregor, "Refugees and the Environment," in Geography of Refugees: Patterns and Processes of Change, ed. R. Black and V. Robinson, 157–71 (London: Bellhaven, 1993).

Khalid Koser, "Changing Agendas in the Study of Forced Migration: A Report on the Fifth International Research and Advisory Panel Meeting, April 1996," Journal of Refugee Studies 9, no. 4 (1996): 53–66

Koko Warner, "Global Environmental Change and Migration: Governance Challenges," Global Environmental Change 20, no. 3 (2010): 402–13.

Lal Panna,"Kumaon Local Customs", Superintendent Government Printing Press, United Provinces, Allahabad, 1920, 7.

Mark Dowie, "Conservation Refugees: The Hundred Year Conflict between Global Conservation and Native Peoples", Cambridge: MIT Press, 2009.

Michael Barnett, "Humanitarianism with a Sovereign Face: UNHCR in the Global Undertow," International Migration Review 35, no. 1 (2001): 244–77.

Michael Cernea, "The Risks and Reconstruction Model forResettling Displaced Populations."World Development 25,no. 10 (1997): 1569–87.

Norman Myers, "Environmental Refugees," Population and Environment: A Journal of Interdisciplinary Studies 19, no. 2 ,1999: 167–82.

Olli Brown, "Migration and Climate Change", Geneva: International Organization for Migration, 2008.

Peter Vandergeest, Pablo Idahosa, and Pablo S. Bose, eds., "Development's Displacements: Economies, Ecologies and Cultures at Risk" , Vancouver: UBC Press, 2007.

Phillippe le Billon, "Geographies of War: Perspectiveson 'Resource Wars,'" Geography Compass 1, no. 2 (2007):163–82.

Ravenstein, E. 1885. The laws of migration. Journal of theStatistical Society 28:167–235

Richard Black, "Environmental Refugees: Myth or Reality?," New Issues in Refugee Research Working No. 34 (Sussex: Centre for Development and Environment, University of Sussex, 2001), 1–20.

Richard Black, Neil Adger, Nigel Arnell, Stefan Dercon, Andrew Geddes, and David Thomas, "The Effect of Environmental Change on Human Migration," Global Environmental Change 21 (2011): S3–S11.

Singh Ranjay K, "Agro-biodiversity and food security in risk- prone agro-ecosystems: An appraisal of tribals wisdom", Int Symp Alternative Approach to Enhancing Small-Scale Livelihoods and Natural Resources Management in Marginal Areas-Experience in Manson Asia, (United Nations University, UN House, Tokyo, Japan), (October 29-30), 2003, 10.

Singh, L. Jeetendro et al.(1998), "Environmental Ethics in the Culture of Meeteis from North East India" in Song Sang-yong et al.(eds), Bioethics in Asia in the 21st Century, Christchurch: Eubios Ethics Institution.

UNHCR, "Climate Change, Natural Disasters and Human Displacement: A UNHCR Perspective," in Climate Change and Displacement Reader, ed. S. Leckie, E. Simperingham, and J. Bakker (New York: Earthscan, 2012), 149.

Ground Water Quality Assessment in and Around Medchal Town Telangana State, India

Putta Praveen[1], T. Saideekshitha, G. Swathi, K.J. Anusha, K.Kavya, and M. Anjaneyulu[2]

[1]CMR Institute Of Technology, Hyderabad [2]Mahathma Gandhi University, Nalgonda

praveenreddy.putta@gmail.com

ABSTRACT

The water quality assessment was conducted in and around Medchal town, Telangana State. Study area covers around 13 sq. km in the survey of India toposheet No 56k/6 and lying between the N17^{0}36'15"to17^{0}38'45"and E78^{0}28'17" to78^{0}30'00".The ground water quality assessment is an important to make sure sustainable safe use of water. However the overall water quality in the Medchal area is difficult due to the special variability of industrial contaminations and wide range of indicators in analysis data set. Attempted ground water quality based on the integrated analysis of Physcio and chemical parameters. Using the GIS contouring methods pH, Turbidity, electrical conductivity (EC), Chlorides (Cl), Carbonates (CO3) and Bicarbonates (HCO3) have been prepared. From the water quality standards (WHO) the quality of water is predominantly moderate to good.

Keywords: Ground Water Quality, Physico- chemical, Medchal and Telangana State

Introduction

Groundwater resources are dynamic in nature and affected by such factors as the expansion of irrigation activities, industrialization and urbanization, hence monitoring and conserving this important resource is essential.

Medchal, a major industrial area of Telangana for last decade, is situated in the NE of the Hyderabad city. The geology varies in age from Archean to Recent and includes schists, granites, pegmatite, feld- spathic gneisses with associated bands of quartzites and amphibolites. In the study area, the occurrence and movement of groundwater is restricted to semi-weathered/ weathered and fractured rocks. The thickness, size, extent and openness of the weathered zone and the interconnec- tions of fractures govern the groundwater movement in and around medchal town. The resistivity of the water-bearing

weathered/fractured rocks varies from 120 to 150 ohm; the depth to ground water generally ranges from 150 to 200 m below ground level.

Urbanization and the unregulated growth of the population of the Medchal district have

altered the terrain and slope morphology of the area. As a consequence, changes have taken place in the surface drainage and industrial effluents system which indirectly affects the hydrogeology while the water infiltration ratio has resulted in a lowering of the groundwater. Inadequate environmental protection measures in the industries as well as waste dumps, drainage system have resulted in significant water pollution.

For any city, a ground water quality map is important for drinking and irrigation purposes and as precautionary indication of potential environmental health problems. Singh and Lawrence (2007) prepared a groundwater quality map in GIS for Chennai city, Tamilnadu, but a ground- water quality assessment in the Medchal district is much more difficult due to the spatial variability of multiple contaminants and wide range of indicators that could be measured. There were many publications related to identification of groundwater potential zones from resistivity surveys as well as from remote sensing data (Singh et al. 2007) but no systematic updated publication/database was available for a water quality index map for this study region.

The main objective of the present work was to make a ground water quality assessment using GIS, based on the available physico-chemical data from 18 locations (Fig. 1)

a b

Fig. 1 (a) Indicates location map of the study area, (b) indicates locations of collected water samples marked on Toposheet

For medchal, rain fall is the main source of ground water, irrigation and industry in medchal district. Disposal of Sewage, Industrial wastes and other human activities effects of medchal ground water were highly polluted (Venkatachalapathy and Karthikeyan, 2013).

The condition of ground water may be polluted with pathogen, toxic metal, chemical compounds such as pesticides, herbicides and other industrial waste becomes waterborne outbreaks (Begum, and Harikrishna, 2008). The quality of the shallow aquifers in and around the textile, bleaching and dyeing units, which use a wide variety of chemicals and dyes at medchal, Bhavani and Chennimalai and their environs are highly polluted due to the indiscriminate discharge of untreated effluents... As per the survey conducted, about 40 dyeing units and 2 tanneries are in operation in catchment area and expected to discharge the trade effluent (both treated and untreated) either directly or indirectly through drain (Sivakumar et al., 2010).

The present study was aimed at analyzing some important characteristics of wastewater considered herein for the medchal area of Telangana. Physic-chemical parameters such as pH, EC ,TDS , chloride content, carbonates, bicarbonates, turbidity etc. were carried out.

Methodology

The present study has been carried out in following steps:

Geographical and morphological features of study area: Medchal-Malkajgiri is the one of district of 33 districts of Telangana State,and it is located in the NE of Hyderabad.Medchal is newly industrial developing town from past decade.The water quality assessment was conducted in and around Medchal town, Telangana State.

Sampling points: Water samples collected from different bore wells. From various sources of water at different depth and covered maximum populated area, industrial area, commercial area and residential colonies of medchal were analyzed.

Physico-chemical analysis of water quality: In this research we have done the these parameters like Electrical Conductivity, PH Chlorides, Carbonate , Bicarbonate, Total Dissolved Solids and Turbidity

GIS analysis

The various thematic layers on hardness, pH have been generated using GIS contouring methods with Arcview 3.2a. The spline contouring method has been used for generating the contours needed for the creation of a triangulated network for each thematic layer. Spatial distribution maps for, pH, TDS, HCO3,CO3, EC, Chlorides and Turbidity. The flow chart in Fig. 2 was followed to develop a ground water quality classification map from thematic layers (Fig. 3a–j) based on the WHO (1984) and Indian (ISI 1983) Standards for drinking water. The classification of water quality is essential for an assessment of suitability for domestic, agriculture or industrial uses.

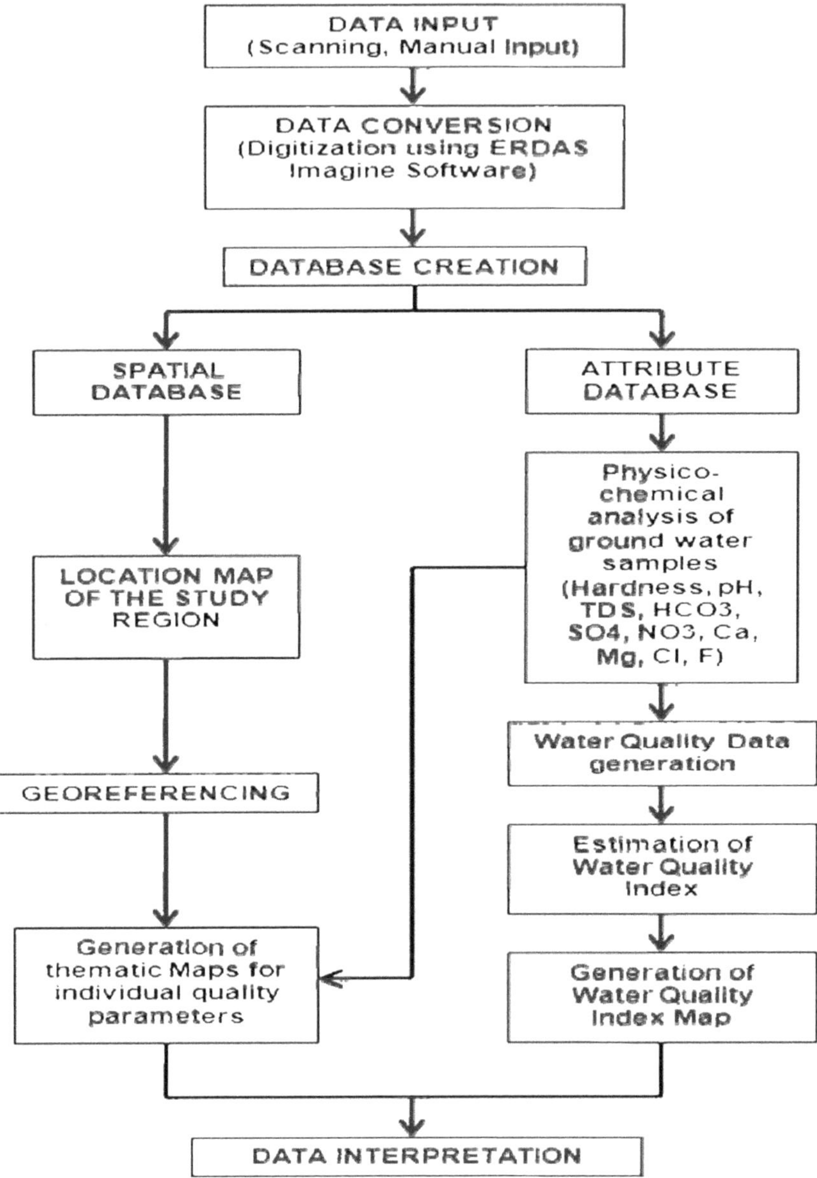

Fig. 2 Flow chart of the method adopted

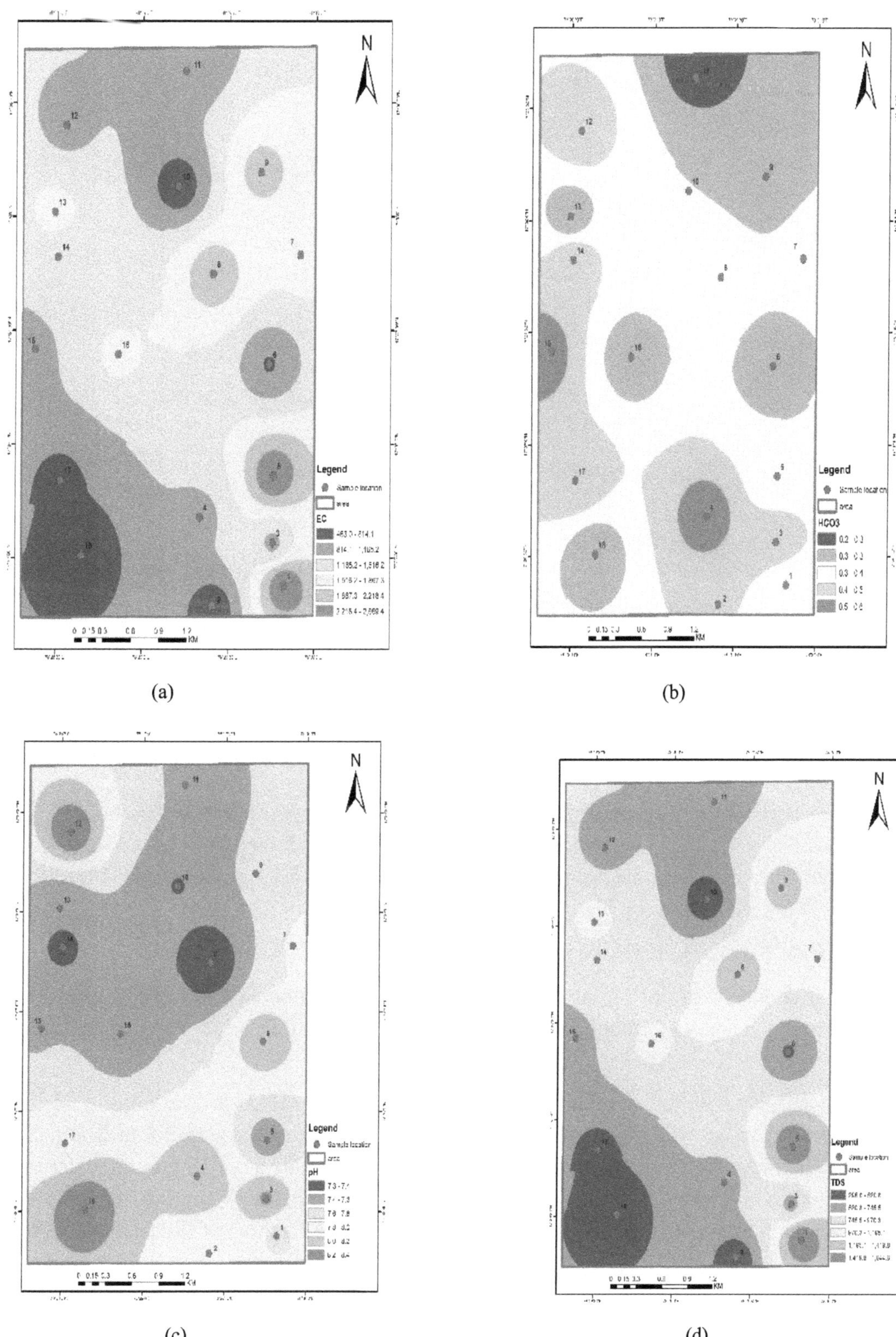

(a)

(b)

(c)

(d)

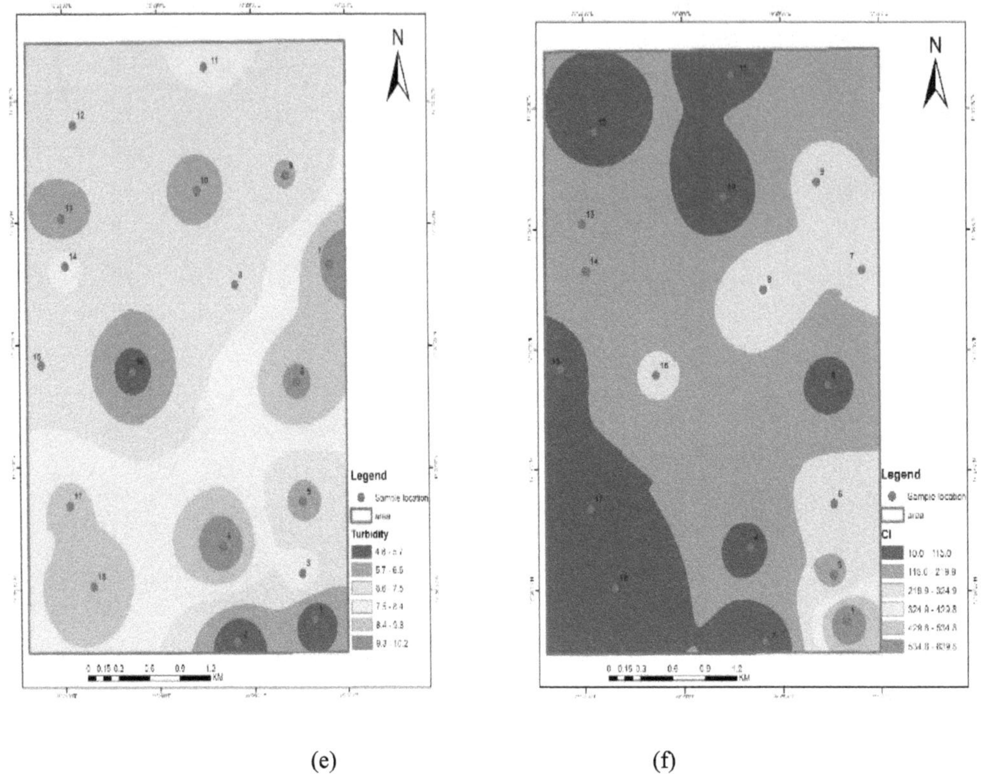

(e) (f)

Fig. 3 (a) Distribution of EC levels in Medchal area (b) Distribution of HCO3 values in Medchal area (c) Distribution of P^H in Medchal area (d) Distribution of TDS in Medchal area (e) Distribution of Turtbidity in Medchal area (f) Distribution of Cl in Medchal area.

EC signifies the total concentration of ionized constituents of water pH value range was between 0.91to and 5.89 to with an average of in both areas. All samples were found with in the maximum permissible limit sample. After analyzing 18 ground watersample the value of electrical conductivity ranged between 463μS/cm to 2570 μS/cm of an average 1299.2 μS/cm *(Fig. 3 a)*

Carbonate + Bicarbonate (CO32- + HCO3-) conc. was calculated Indirectly from alkalinity.

In the present study after analyzing of 18 ground water sample, chloride concentration was observed between 43 mg/l to 640mg/l . 18 sample for in and around medchal town , the value of chloride was ranged from 0 ppm to 0.89*(Fig. 3b)*

P^H value indicates whether a solution is acidic or alkaline. In principle, pH up to 7 of any liquid is said to be acidic and 7 indicate neutral water while 7 to 14 is said to be alkaline. The pH value 6.5 to 8.5 is essential for drinking water; beyond the range water will be harm to health. During the analysis of 18 ground water sample from different areas of Medchal(industrial area) , it observed that the value of pH was ranging from 6.8 to 8.48 with an average of 8.29 . The maximum pH value 8.48 was obtained in the industrial area of Medchal*(Fig. 3c*).

The conductivity of a solution of water is highly dependent on its concentration of dissolved salts, because of this reason we are conducting the TDS test the laboratory test is carried out all 18 samples by using filter paper at a temp of 103-105 c that can be observed in values in between 296-1645*(Fig. 3d*).

Turbidity levels remains between 4.8 to 10.21 NTU this water can be suitable for drinking purpose.in the study area 2 samples exceeds the permissible limits*(Fig. 3e*).

All chloride salts are highly soluble, removed from water by participation except under the influence of freezing or evaporation, chloride also free from effects of exchanges, adsorption and biological activity , thus if water one takes chloride into solution, it is difficult to remove the chloride through natural process.

In the present study after analyzing of 18 ground water sample, chloride concentration was observed between 30 ppm to 2500. 18 sample for in and around medchal town , the value of chloride was ranged from 10 ppm to 640(*Fig. 3f*).

Conclusion

The above study indicates that groundwater of variable quality exists in the Medchal area. However, despite the heavy industry, an analysis of the chemistry of 18 bore wells indicates it is generally suitable for drinking purposes, except where TDS exceeds the permissible of 500 mg/l, in the study area nearly 9 samples are exceeds the permissible limits . Two samples are exceeds the permissible limits of turbidity. The water is also suitable for industrial purposes and is acceptable for irrigation, with a low alkalinity and moderate to high salinity risk.

Remedies and control measures

Prevention is better than cure. Since. Purification of contaminated groundwater is an expensive and time taking process, it is better to try and collectively work together to avoid groundwater contamination at source stage itself. Since this study has identified contamination from sewage as the main cause for groundwater contamination in the study area - all steps to curtail it and prevent further deterioration should be taken up immediately.

References

Grimwood, M.J. & Dixon, E. 1997. Assessment of risks posed by List II metals to Sensitive Marine Areas (SMAs) and adequacy of existing environmental quality standards (EQSs) for SMA protection. Report to English Nature.

Parr, W., Clarke, S.J., Van Kijk, P, & Morgan, N. 1997. Turbidity in English and Welsh Tidal Waters. Report to English Nature.

Scott, C.R., Hemingway, K.L., Elliott, M., De Jonge, V.N., Pettwick, J.S., Malcolm,

S. & Wilkinson, M. 1999. Impact of nutrients in estuaries - Phase 2. Report to English Nature and the Environment Agency.

DETR. 1997. The Urban Waste Water Treatment (England and Wales) Regulations 1994. Working document for dischargers and regulators. A guidance note issued by the Department of the Environment, Transport and the Regions and the Welsh Office.

SEPA. 1998. Scottish Bathing Waters 1998 Bathing water quality results.

Central Ground Water Board, GOI. Ground Water Quality in Shallow Aquifers of India. 6.2010.

WHO Guidelines for Drinking Water Quality, Recommendations, WorldHealth Organization, Geneva, Vol.1, 1984, 130.

Impact of Climatic Parameters on Agriculture Productivity Forecasting through Machine Learning Algorithm

Chitty Venkata Abhiram[1] and Laxmi Manasa Gorugantu[2]
[1]*Department of Mathematics, Statistics and Computer Science*
Pragati Mahavidyalaya Degree College, Osmania University, Hyderabad, India
[2]*Department of Computer Science and Engineering,*
ACE Engineering College Jawaharlal Nehru Technological University, Hyderabad, India
manasa.gorugantu@gmail.com, cvabhiram@outlook.com

ABSTRACT

The rise in earth's average temperature is because of increased volumes of CO_2 followed by greenhouse gases released by burning of fossil fuels. The consequences are glaciers melting, droughts, heat waves, ocean acidification, species extinction and so many. The rise in temperature, erratic monsoon, rainfall, humidity, limited access to water sources adversely affects agriculture production affecting food security. Machine Learning (ML) has been proven to forecast out of experience and guide the policy makers to take appropriate decisions at right time. In agriculture sector, ML algorithms provide a prediction for crop, soil and water management, crop production, disease detection, crop quality. In this work, we use ML algorithm (Linear Regression) to predict the future climatic parameters in different parts of India and its influence on crop yield.

Keywords: climate change, crop yield, food security, machine learning algorithm, climate smart agriculture

Introduction

Sustainable Development

Sustainable development and economic growth are complementary to each other. The World Commission on Environment and Development defined Sustainable Development as the one which meets the needs of the existing generation without causing any damage to resources meant for future generations [10]. New technological developments and innovations are always welcomed without which alone development can't be achieved. Technology always provide remedial measures to the societal needs duly taking care of sustainable development. Eco friendly technological development aims for more productivity, recycling, alternate materials, improvised processes, pollution control and limited usage of resources. While making the policies, the above factors need to be considered. The recycling hub for spinning yarns and weaving of textiles at Panipat, India is an example of sustainable development and economic growth with the ST interventions on the concept of recycling [1].

Need for Green Technologies

The change in the society is purely dependent on the technological advancement that is taking place around the world. The green technologies are focused towards mitigation and adaption initiatives with increase in efficiency and innovation over the existing system. While developing these technologies, care is required to meet the local challenges and taking the community into future impacting lives at large. The companies are focusing on developing new and innovative projects like installation of solar panels on the roof top at New Delhi [6], generation of electricity from both and wind and water under The Netherlands' livable wind turbine project-EWICON [7], generation of energy through people's footsteps in collaboration with Samsung at South Africa [8] which simultaneously provide sustainability as an added product.

Climate Change

The earth's average temperature has risen because of increased volumes of CO_2 followed by greenhouse gases released by burning of fossil fuels and other human activities. The consequences are melting of the glaciers, droughts, heat waves, ocean acidification, species extinction and so many. Drought and subsequent acute water crisis in Cape Town was mainly because of below average rainfall due to high atmospheric temperatures [5]. There were so many restrictions on usage of water less than 50 liters per day. The present drought is the result in gradual decrease in rainfall over period of several years. The lesson is that the City Urban Planners have no appropriate technological plans to predict future risks. The impact of Climate change is in the form of water may be excess or less. Scarcity of water may lead to food security, political uncertainty. During the past, the wars were for land and so on. In the days to come, the wars would be for natural resources. Though there is no direct relationship between climate change and wars, certain conditions would influence the factors leading to conflicts. Various researchers in their studies mentioned that for want of basis amenities like water, food safety, there would be conflicts hampering sustainable development. Drought led to Syrian war is the classical example [2].

UN Efforts

In order to resolve global conflicts, the UN climate conferences have great significance to save the planet earth to maintain sustainable. At Conference of Parties (COP+23) climate engineering was introduced aiming to conserve CO_2 and produce liquid solution which will be used for other industrial processes [9]. For this there are around 300 projects underway in around 30 countries across the world. Climate change also leads to some of the business opportunities. It leads to new innovations and development of technology for the protection of climate. At COP+23 Bonn, UNWHO and UNFCCC signed a MoU. The two important components of the MoU are to handle public health issues arising due to Climate change and to mitigate through capacity building activities.

Effect of Climate Change on Agriculture Sector

As per the Food and Agriculture Organization of the United Nation, out of the 40 countries around the globe, 31 countries alone in Africa are in need of food requiring external assistance [4]. After formation of People's Republic of China, a new initiative was announced to kill rats, flies, mosquitoes and sparrow. The result was that locusts damaged the crop and due to lack of grains, more than 50 million people died due to starvation of food. The disturbance in biodiversity has impact on development. In 1992 Rio Earth Summit much emphasis was on biodiversity [11]. Several research studies revealed the direct relationship between biodiversity and climate change. Among the other sectors, agriculture sector play an important role in Indian economy growth. The contribution of the agriculture sector alone to our country's GDP is about 17-18% and more than 50% of our country's population is engaged in agriculture activities [3]. Agriculture and its subsidiary sectors like animal husbandry, fisheries and forests are highly prone to climate change. The rise in temperature, erratic monsoon, rainfall, humidity, limited access to water sources adversely affects agriculture production affecting food security. For higher yields fertilizers and pesticides will be used extensively and will have direct impact on human health. The policies relating to economic growth of our country should focus to develop invest in low carbon technologies, promote sustainable resilient food systems through sustainable agricultural practices knowledge sharing.

Artificial Intelligence and Machine Learning

To know about the changes in climate first we need a technology for handling data, which is very large also known as big data. This data generally collected from local governance or organizations. Climate change is a complex action, requires a well based technology also known as Artificial Intelligence (AI). Machine Learning (ML), a subset of Artificial Intelligence is a growing field which can help in identifying patterns in the given data set. ML in today's world is formally defined as acquiring knowledge and expertise from the previous experience, occurrence, event without programming to do a well-defined set of instructions. So, training data (data from which knowledge is acquired) is the main constituent on which these ML techniques are tuned. As the volume of data is rapidly growing day-by-day, ML algorithms continue to give more and more accurate results. The last few decades have been phenomenal for ML as it has been applied in many diverse fields such as finance, health care, finance, cybersecurity etc. We choose to apply one of the ML algorithm on rainfall data and predict it's trend in future. Based on these prediction results, we propose mitigation strategies to have sustainability in future.

Experiment Setup

Regression Analysis

Regression is one of the most widely used statistical machine learning technique which is proven to be a powerful algorithm of predicting a value of an output variable from the given input variable [14]. The algorithm is trained on the data set in such a way that it establishes a static relation between any input and output. Consider the following input-output pairs: (x1, y1),(x2, y2), . . . ,(xm, ym). The Linear Regression approach is to approximate the values of ϑ_1 and ϑ_2 to form a straight line with equation being $y = \vartheta_1 * x + \vartheta_2$ such that overall loss function (mean-square: sum of squares of difference between actual and predicted "y" value) is minimum with respect to all the data points. This process is done in several iterations such that loss function values decreases in every step using an optimization process. This optimization algorithm updates the values of ϑ_1 and ϑ_2 such that loss function approaches it's minimal value. There are several such techniques available but one of the widely used is Gradient Descent algorithm. Figure 1 shows the variation of loss function after every iteration.

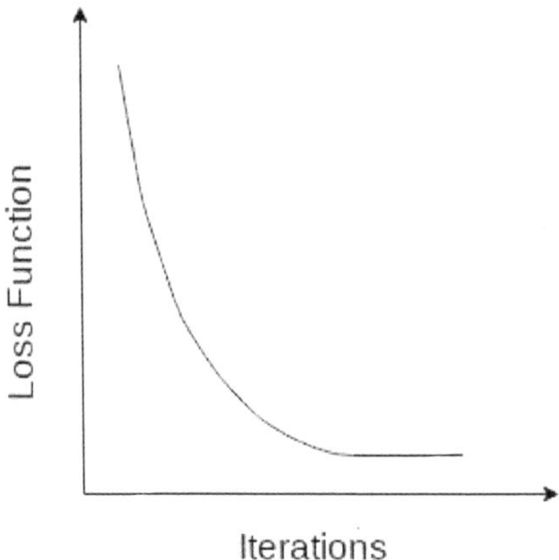

Figure 1 Optimization of Loss Function

Data Set

Rainfall data which is present on kaggle [13] is used as training data for our experiments. The data consists of amount of rainfall occurred from the year in various parts of India. It has recorded data from the year 1900 to 2015 based on which the future predictions are made by the algorithm. As a sample data set, we have considered data of Arunachal Pradesh, East Rajasthan, Coastal Karnataka which represent different regions (North-East, West, South respectively) of India. The pictorial representation of these data points is shown in Figures 2,3,4 (year on x-axis and amount of rain on y-axis).

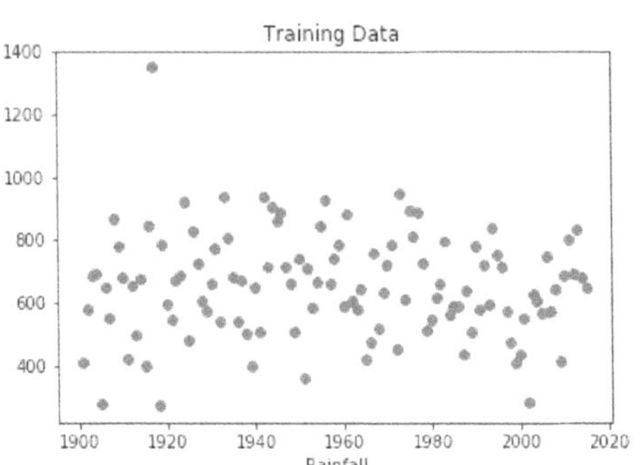

Figure 2 Arunachal Pradesh Prediction

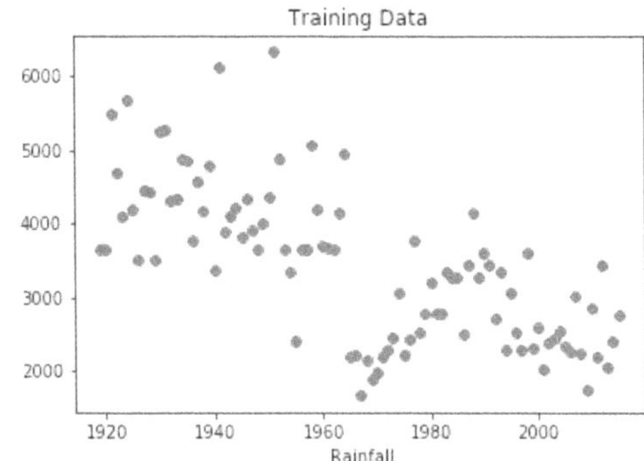

Figure 3 East Rajasthan Prediction

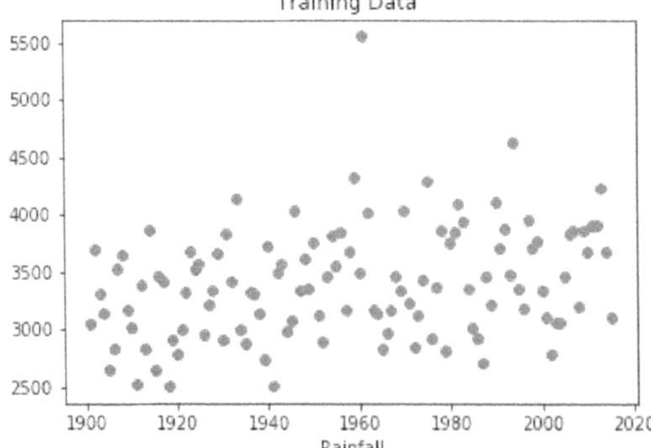

Figure 4 Coastal Karnataka Prediction

Simulation methodology

As a part of running our experiment, we used python programming language to read the data sets from the appropriate files (csv, xlsx) and Google's Tensorflow [15] to simulate Linear Regression Algorithm. Tensorflow is an open source library which was developed for running several Machine Learning (regression, SVM, decision tree) and Deep Learning algorithms (Neural Networks).

Results and Discussions

We applied the Linear Regression Algorithm on the three data sets and plotted a straight line whose loss function is as minimal as possible. The plot for Arunachal Pradesh, East Rajasthan, Coastal Karnataka are given in Figures 5,6,7 respectively. The analysis for three different regions are given below:

1. **Arunachal Pradesh:** As per the data pattern and the fitted line (going downwards) from the Figure 5, it is quite clear that rainfall is decreasing over the years. If the prediction is made along the line, then rainfall in north east part of India continues decline in the coming years to come.
2. **East Rajasthan:** It is vivid from the fitted line to the data points that rainfall is being constant and limited within a certain bound. So, we can conclude that rainfall is steady and will remain uniform over the years in the western part of India.

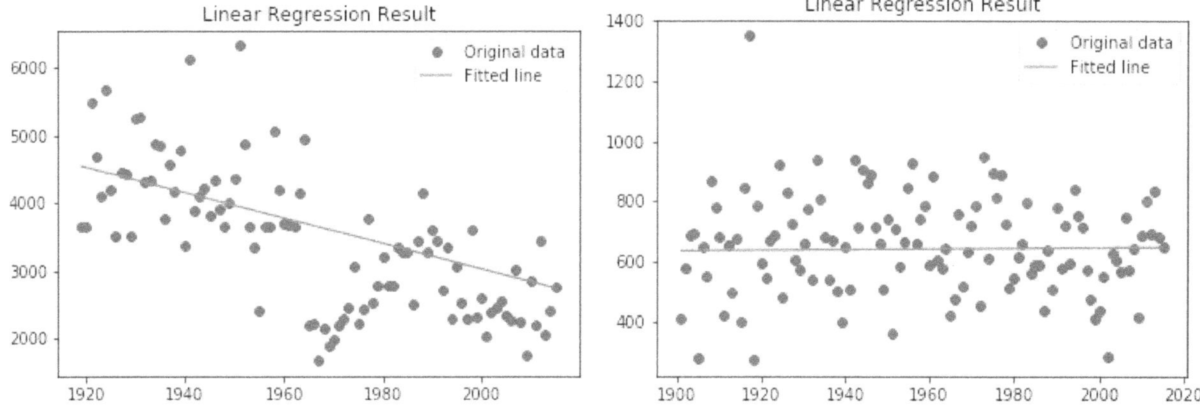

Figure 5 Arunachal Pradesh Prediction

Figure 6 East Rajasthan Prediction

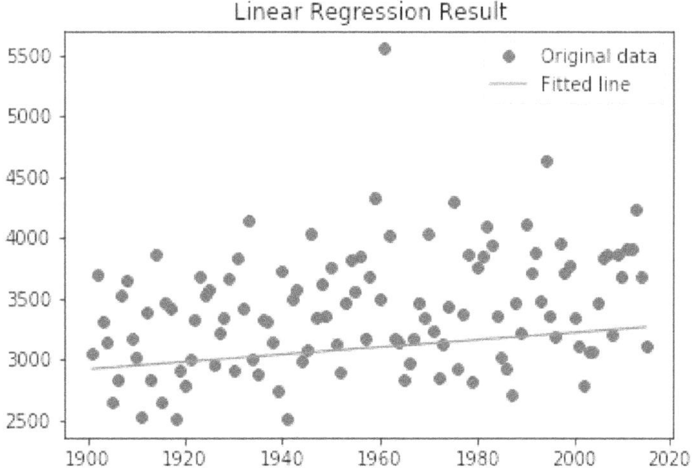

Figure 7 Coastal Karnataka Prediction

3. **Coastal Karnataka:** It can be seen from Figure 7 that rainfall is increasing slowly over the years. So, the south Indian regions may experience more rainfall than the previous year if the preciction is followed as per the fitted line.

Mitigation and Adoption Strategies

Some of the mitigation and adoption measures for better crop yield are listed below:

Agriculture

The climate change is showing negative impact on agriculture sector globally and has direct adverse effects on soil fertility, microbial activity preventing plant growth causing food security. The erratic climatic parameters and unexpected drought or floods have greater impact on crop pro- duction. Through innovative approaches, the policies should focus on developing and promoting climate resilient crops through sustainable agricultural practices and innovative approaches. In the context of limited water resources, the other option to enhance higher crop yield by adopting suitable crop rotation, multiple cropping patterns and cropping intensity through organic farming methods and to take the support of local agriculture university duly establishing proper market linkage.

Water

After rejuvenation and restoration of minor irrigation sources, the groundwater level will be increased, quality gets improved and fluoride content levels will be reduced considerably. The fertilizers usage could be minimized as the silt can be used as manure which is highly nutrient. Re- vival of tanks also helps increased aquaculture activity through improved feeding and harvesting techniques. Involving the local community and with the active participation of intergovernmental coordination, the activities like afforestation, nursery raising, plantation around the tanks and lakes could be taken which addresses and tackle the problem of rural employment and poverty reduction and to some extent urban migration can be reduced. With the new inventions and technological advancement, there is a need to bring reforms in agriculture sector to attain food security duly protecting the earth's ecosystem by balancing economic growth as well social equality otherwise people will be at risk for hunger and poverty globally in the days to come.

Conclusion

Many agencies in India are actively involved in predicting various parameters using artificial intelligence. In this paper, machine learning algorithm has been proposed and future climatic parameters in certain parts of India are evaluated. It may be seen that the crop yield is directly proportional to the climatic parameters and therefore, need for developing climate resilient crops is inevitable. In future applications, few more locations and data sets will be selected [12].

Acknowledgement

We are thankful to our faculty members and mentors of Pragati Mahavidyalaya Degree College and ACE Engineering College, Hyderabad for their encouragement and guidance from time to time. We are also equally thankful to NCWES–2019 organizers for their encouragement and support.

References

Pant, Suman and Nagpal, Neet Responses of public towards shoddy industry at Panipat. In *Asian J. Home Sci.,*, 11 (1) : 117-123, DOI: 10.15740/HAS/AJHS/11.1/117- 123.

Kelley, C. P., Mohtadi, S., Cane, M. A., Seager, R., Kushnir, Y. Climate change in the Fertile Crescent and implications of the recent Syrian Drought *Proceedings of the National Academy of Sciences, 112(11), 3241-3246.*, 2015.

Madhusudhan L Agriculture Role on Indian Economy. *Bus Eco J 6: 176. doi:10.4172/2151- 6219.1000176*, 2015.

FAO, Crop Prospects and Food Situation. Quarterly Global Report, *December. http://www.fao.org/documents/card/en/c/CA2726EN.*, 2018

WATER OUTLOOK 2018 REPORT Revision 25 - updated 20 May 201 *Produced by Department of Water and Sanitation City of Cape Town*, 2018.

https://www.teriin.org/blog/solar-rooftops-delhi-neighbourhood-takes-leap,

https://www.dezeen.com/2015/04/20/designs-unveiled-for-giant-rotterdam-wind-turbine- you-could-live-inside/

http://www.pavegen.com/samsung

https://news.un.org/en/story/2017/11/636342-bonn-climate-engineering-risky-should-be- explored-experts-say-un-conference

Report of the World Commission on Environment and Development: Our Common Future, 1987

United Nations Conference on Environment Development Rio de Janerio, Brazil, 3 to 14 June 1992

http://academic.udayton.edu/kissock/http/Weather/citylistWorld.htm

https://www.kaggle.com/rajanand/rainfall-in-indiarainfall%20in%20india%201901- 2015.csv

Nguyen et al. Author age prediction from text using linear regression Proceedings of the 5th ACL-HLT Workshop on Language Technology for Cultural Heritage, Social Sciences, and Humanities, 2011

Girija, Sanjay Surendranath Tensorflow: Large-scale machine learning on heterogeneous distributed systems Software available from tensorflow. org

Significance of Strange Table and SCS-CN Methods in Estimating Runoff in Various Catchments of Telangana State, India

M. Vamshi Raj

M.Tech (HWRE), IIT Kanpur, Assistant Executive Engineer, Office of Chief Engineer, Hydrology & Investigation, Irrigation & CAD Department, Government of Telangana, 2nd Floor, Jalasoudha, Errummanzil, Hyderabad.

vamshiraj.mtechcivil.iitk@gmail.com, iamvamshi.raj@gmail.com

ABSTRACT

Hydrology has undergone an almost complete transformation over the past century, from its empirical origins in the early 20th century, to become a fully-fledged and key component of Earth System science by the early 21st century. Many researchers have developed different methodologies for assessing water availability at ungauged locations. The accurate estimation of runoff is necessary for managing our limited water resources, in order to cater to the needs of all involved stakeholders for the sustainable growth of our ecosystem. However, it remains a challenge for hydrologists to estimate runoff accurately as it is influenced by physiographic parameters of catchment, hydro meteorological processes like rainfall, interception, evapo-transpiration, infiltration as well as changes in anthropogenic activities like in land use/land cover, construction of cross drainage structures, etc in the catchment. As a result, it becomes imperative to ignore few hydrological processes of lesser importance and by taking into account only crucial parameters influencing the flow processes in the catchment to simplify the methodology in computation of runoff. Researchers like Strange (1892)have followed the philosophy in arriving runoff based on experimental results by considering only rainfall and type of catchment as the only influencing parameters. However, with the advancements in the field of hydrology like ease of computation, availability of data, Geographic Information System (GIS), the methodologies for computation of yield have changed drastically to methods like SCS-CN method(1969) which depends primarily upon Curve Number (accounting more parameters like soil type, antecedent moisture condition, landuse/landcover). The objective of the present study is to identify key parameters contributing to the difference in runoffs obtained by two popular methods like Stranges table and SCS method in catchments with different rainfall and physiographic characteristics like Himayatsagar project, Kotpally project, Tammileru projectwhich can be used in selection ofsuitable method for estimating runoff.

Introduction

The assessment of runoff in an ungauged catchment is one of the classic problems faced by the hydrologists as it is essential to predict the water available in a catchment with reasonable accuracy. Water tends to move differently on the earth surface as it moves down in varied landscapes encountered in the catchment till it reaches the outlet point. This movement depends on many parameters like land use, land cover, type of soil, slope etc., leading to the complexity in estimation of runoff in catchments. Though, there are many approaches available for estimation of runoff at the outlet of the catchment, very little research is done on the scopeof these different approaches catchments in relation with keyparameters influencing runoff. In this study, an attempt is being carried out to explore the applicability of the two popular used for runoff estimation –Strange Table & SCS-Curve Number method with changing landscape scenarios in catchment areas of three different projects namely Himayatsagar project, Kotpally project of Krishna Basin and Tammileru project. Although, both the methods are empirical in nature, the SCS-Curve Number method takes into account the effect of parameters like soil type, antecedent moisture condition, and landuse/landcover in comparison to the Strange Table method which considers the only rainfall and type of catchment.

Study Area

Three projects were selected for the study – Himayatsagar, Kotpally and Tammileru. Himayatsagar project is located on Musi River, near Rajendranagar in Rangareddy District. Kotpally project is located on Kotpallyvagu, a tributary of Bheema River in K-6 Sub-basin near Dharur in Vikarabad District. Tammileru project is located near Chintalapudi, West Godavari District of Andhra Pradesh.

Materials and Method

Data

The daily rainfall dataof the monsoon period of 28 years from 1988 to 2015 of all the influencing rain gauge stations of each of three projects considered in the study is as follows.

Table 1 Details of the Rainfall Data and Projects catchment areas.

S. No.	Name of Project	Catchment Area (Sq.km)	Influencing Rain Guage Stations considered	Period of Rainfall Data taken
01.	Himayatsagar	1054	Rajendranagar, Moinabad, Chevella, Pargi, Pudur and Shabad	Monsoon (June 1st – Oct 31st) rainfall data from 1988-2015
02.	Kotpally	330	Bantaram	Monsoon (June 1st – Oct 31st) rainfall data from 1988-2015
03.	Tammileru	597.26	Sattupally	Monsoon (June 1st – Oct 31st) rainfall data from 1988-2015

The arithmetic mean method is used to arrive at the weighted rainfall of the catchment areas of the three projects.In this study, Arc-GIS 10.4 for the catchment area delineation and land-use/land-cover classification is done using based on National Remote Sensing Centre (NRSC), Hyderabad dataand soil classification is done using National Bureau of Soil Survey and Land use planning (NBSS & LUP) data for the catchment areas of the three projects – Himayatsagar, Kotpally&Tammileru.

Methodology

The following two methods used in the present study:

1) ***Strange Table Method:*** Strange (1892) studied the available rainfall and runoff in the border areas of present day Maharashtra and Karnataka and has obtained a linear relationship between them based on catchment characteristics. He classified the catchments into three different types – good, average and bad as per the resultant runoff from the catchment. This approach is preferred to many other recently developed ones, due to its simplicity and easy adaptability (Latha et al., 2012).
2) ***SCS-CN Method:*** This method is developed by soil conservation services (SCS) of USA in 1969, is a simple, predictable and stable conceptual method for estimation of direct runoff depth based on storm rainfall depth. It depends on only one parameter, curve number (CN). In this approach, an attempt has been made to represent the spatial heterogeneity in the catchment involving parameters like soil type, land-use, land-cover and antecedent moisture condition in a single dimensionlesscurve number. Generally, the SCS-CN method equation is given below;

$$Q = \frac{(P-\lambda S)^2}{P+(1-\lambda)S}, \quad \text{For } P > \lambda S,$$

$$Q = 0, \quad \text{For } P \leq \lambda S,$$

Here l varies from 0.1 to 0.3 depending upon antecedent moisture condition

(AMC) of soil and soil type (Ministry of Agriculture, Handbook, 1972),P is rainfall in mm, Q is runoff in mm, S is the maximum retention capacity of soil which in turn depends on Land-use and Land-cover, given by

$$S = \frac{25400}{CN} - 254$$

Where CN is curve number taken from the tables, Chapter-7, SCS Handbook, Section-1972.

It is evident that the SCS-CN method takes into account the spatial heterogeneity considering the landuse/landcover and soil types in evaluating curve number for each catchment and also antecedent moisture condition (AMC) of the catchment whereas the Strange Table follows a linear approach expressing runoff only as fraction of rainfall.

Results and Discussion

The average monsoon rainfall for the period of data used is found to be higher in catchment area of Thammileru project in comparison to the Himayatsagar &Kotpally (shown in fig 1). The land-use/land-cover analysis indicate that the khariff only is predominant type in comparison to the other types in all the three projects (shown in fig 2, fig 4 & fig 6).On the other hand soil data analysis indicate clay percentages in Thammileru and Kotpally (79.43% & 60.53%) are significantly higher in comparison to the Himayatsagar (14.01%) (shown in fig 3, fig 5 & fig 7).The annual monsoon runoff obtained from SCS-CN method is compared with annual monsoon runoff from Strange Table method from 1988 to 2015 in all the three projects. It is seen that themostly annual runoff obtained from Strange Table is more in comparison with the SCS-CN method in all the three projects (except two years-1991 & 2008 in Himayatsagar, except one year-1994 in Thammileru & except five years-1988, 1989. 2002, 2004 & 2014 in Kotpally) (shown in fig 8 to 10).Though, there is a difference in annual runoff obtained SCS-CN method and Strange Table, the annual runoff in all the projects following relatively similar trends in both the methods (shown in fig 8 to 10).Although, the absolute percent difference between the annual monsoon runoffs obtained from the two methods is not showing any significant pattern in the three projects, the average percentage difference of the 28 years (1988-2015) tend to be higher in case of Thammileru project (51.30%) in comparison to the other two projects-Kotpally (42.27%) and Himayatsagar (26.96%) (shown in fig 11). Also, it is interesting to note that the clay fraction in the catchments of these projects is directly proportional to the average percentage difference between annual monsoon runoffs of the twomethods (R^2=0.9917) (shown in fig 13).It is also found that the area used for agriculture for khariff crops only is inversely proportional to the difference between the runoffs of the two methods(R^2=0.9227)(shown in fig 12).In contrast, it is found out that parameters like average monsoon rainfall and catchment area play a lesser role in contributing to the difference between the runoffs of the two methods (R^2=0.6571 & R^2=0.5217) (shown in fig 14 & fig 15). It is also observed that the forest cover present in the three projects is in proportional to the annual monsoon rainfall (R^2=0.9953) provingthe significant role forests play in attracting monsoons (shown in fig 16).

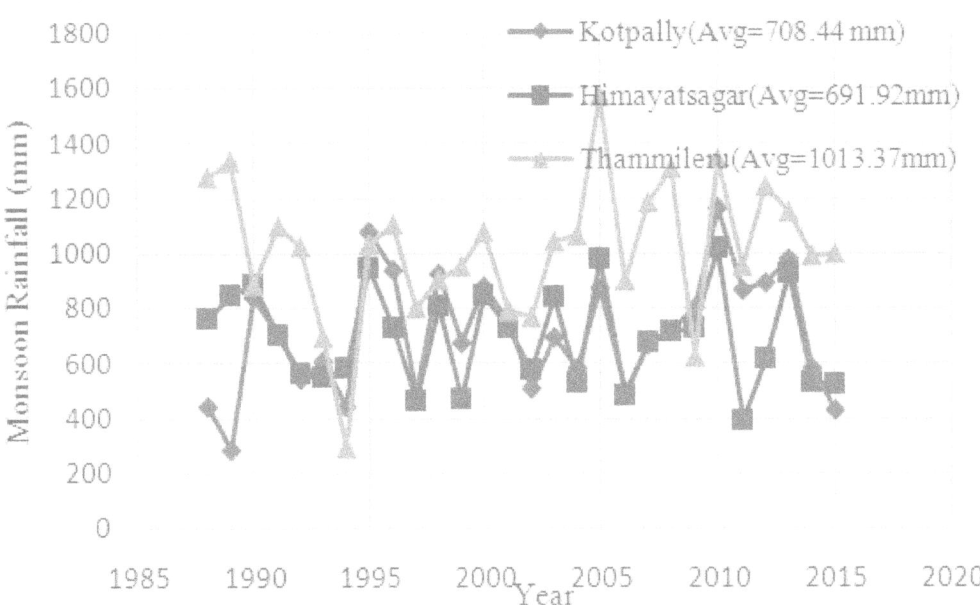

Figure 1 Time Series of Annual Monsoon Rainfall from 1988 to 2015

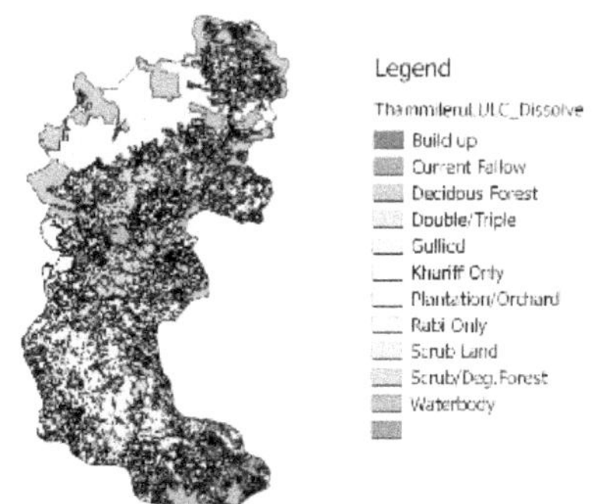

Figure 2 LULC Map of Thammileru Project

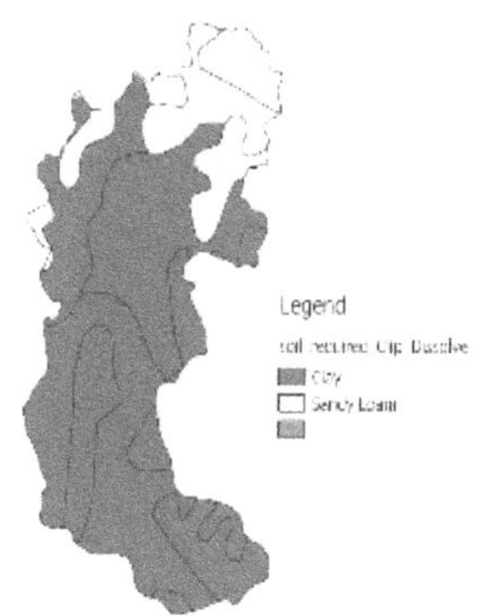

Figure 3 Soil Map of Thammileru Project

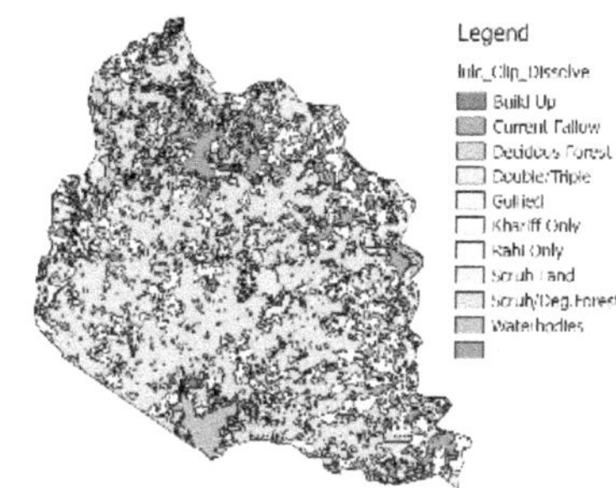

Figure 4 Soil Map of Kotpally Project

Figure 5 LULC Map of Kotpally Project

Figure 6 LULC Map of Himayatsagar Project

Figure 7 Soil Map of Himayatsagar Project

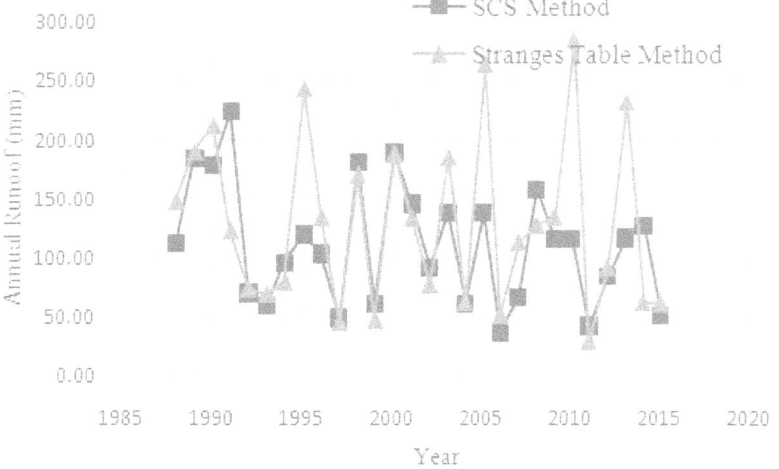

Figure 8 SCS Vs Strange Table Method (Himayatsagar)

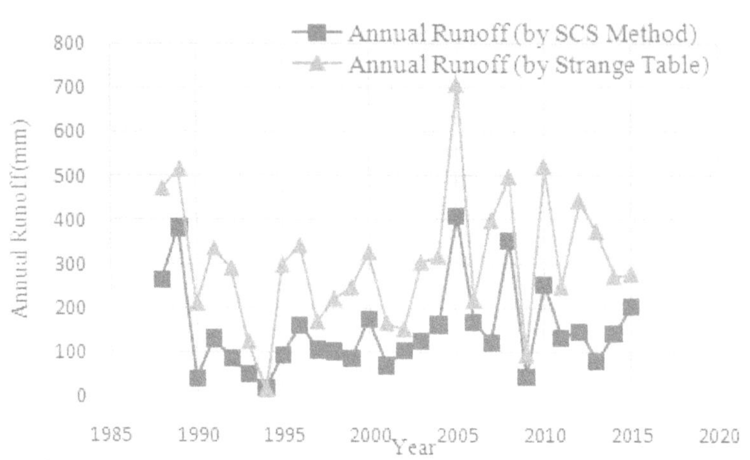

Figure 9 SCS Vs Strange Table Method (Thammileru)

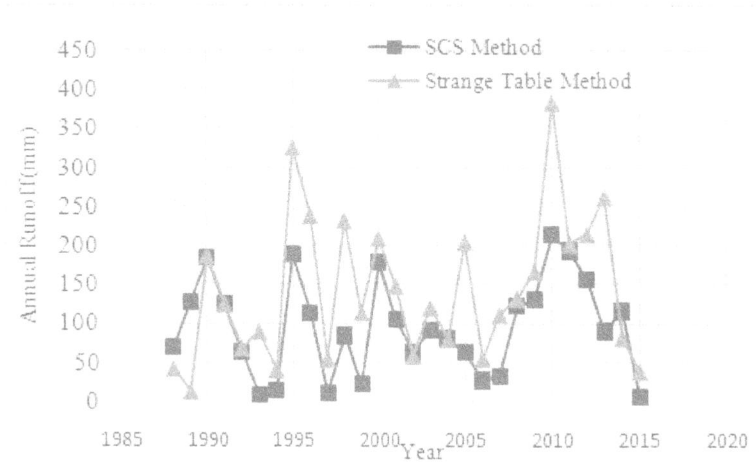

Figure 10 SCS Vs Strange Table Method (Kotpally)

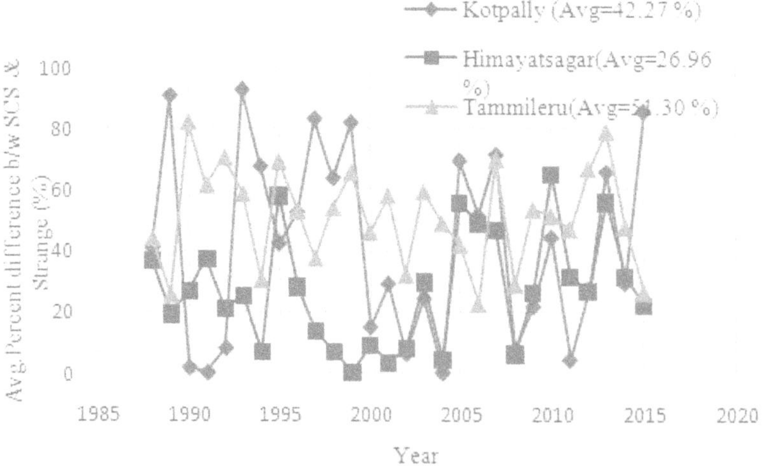

Figure 11 Time series of Percent Difference between Annual Monsoon Runoff(%) obtained from SCS & Strange Table over the period(1988-2015)

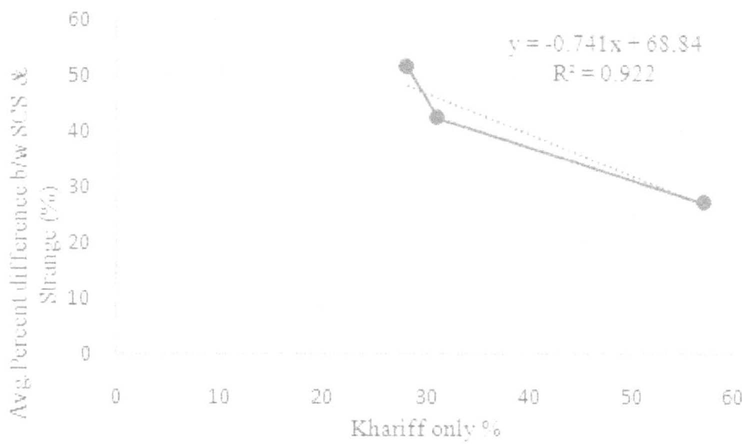

Figure 12 Khariff only (%) vs Average Percentage Difference b/w SCS & Strange Runoff (%)

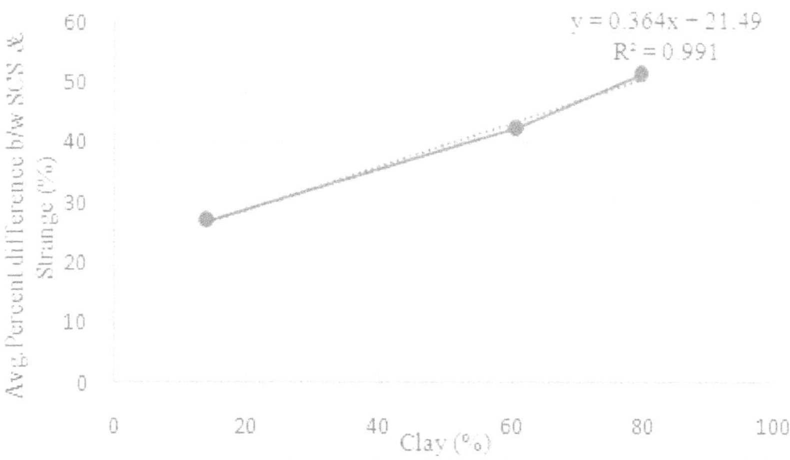

Figure 13 Clay (%) vs Average Percentage Diffrence b/w SCS & Strange Runoff (%)

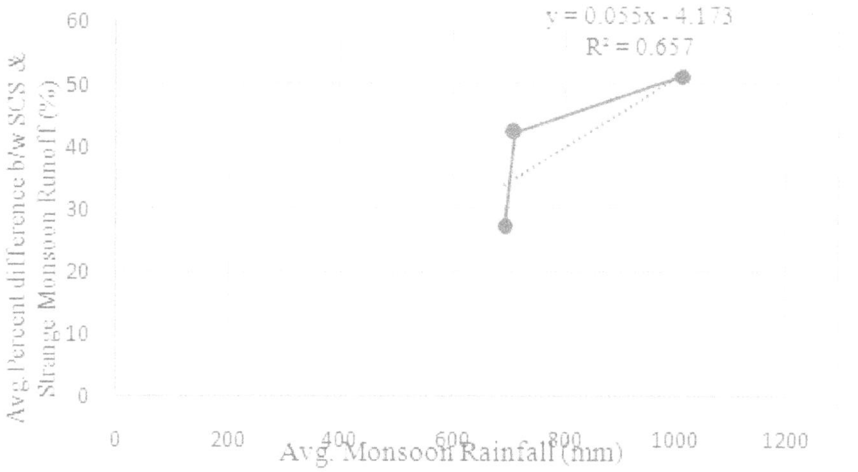

Figure 14 Average Monsoon Rainfall (mm) vs Average Percentage Difference b/w SCS & Strange (%)

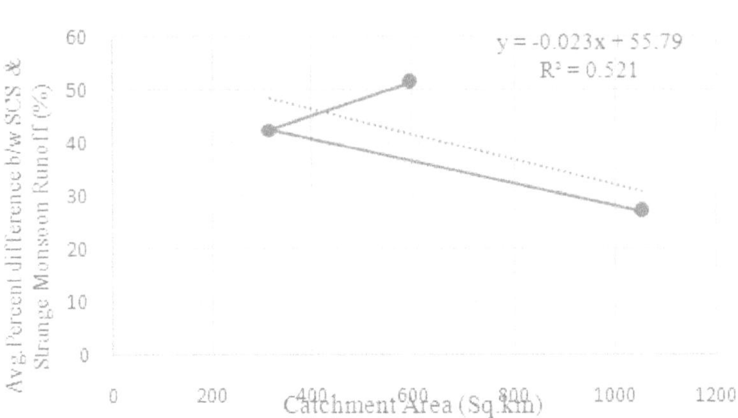

Figure 15 Catchment Area (Sq.km) vs Average Percentage Difference b/w SCS & Strange(%)

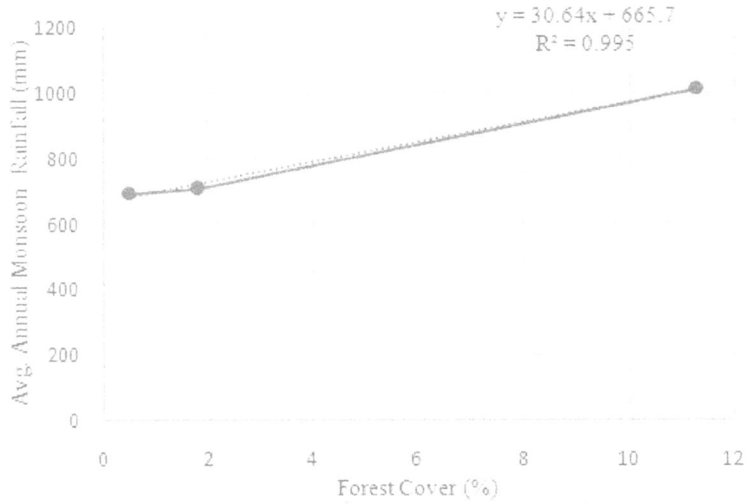

Figure 16 Forest Cover (%) vs Average Annual Monsoon Rainfall (mm)

Conclusions

1. It can be inferred from the study of three catchment areas that soil parameters like clay content and land-use/land-cover parameters like khariff only are the key parameters causing the difference in computed runoff results between the strange table and SCS-CN method.
2. It can be concluded that catchments like Himayatsagar with lesser area with clay content and higher areaused for agriculture type of khariff only, are relatively better for simple linear rainfall-runoff methods like Strange Table than SCS-CN method giving lesser variation in the end results.
3. Subsequently, it can also be concluded that following a generalized approach may not be appropriate for all the catchments without studying the critical parametersinfluencing runoff in the catchment.

References

Latha M., Rajendran M., Murugappan., 2012, Comparison of GIS based SCS-CN and Strange table Method of Rainfall-Runoff Models forVeeranam Tank, Tamilnadu, India.

Adsorptive Removal of Monocrotophos, An Organophosphorus Insecticide using Different Adsorbents from Surface Water

Jalindre S.S.[1], Bhatkhande D.S.[2], Khamparia S.[3] and Joshi R.A.[4]

[1]Post Graduate Student, [2]Professor, [4]Assistant Professor, [3]Assistant Professor

[1,2,&4]Department of Civil Engineering, [3]Faculty of Science and Technology,

Water Quality Centre of Excellence, Vishwakarma University, Pune, Maharashtra State, India.

sourabh.17p067@viit.ac.in, dhananjay.bhatkhande@vit.edu. shraddha.khamparia@vupune.ac.in, rahul.joshi@viit.ac.in

ABSTRACT

Monocrotophos, an organophosphorus insecticide is extensively utilized in India for protection of various crops. Continuous usage of monocrotophos has severely affected both surface as well as ground water to a large extent. In the present study, four different absorbents namely activated carbon, saw dust, fly ash and maize cobs (activated with $CaCl_2$, HCl, NaOH) were investigated for effective removal of monocrotophos from its aqueous solution. Based on adsorption studies carried out, followed by comparative analysis, activated carbon was found to be best suitable for complete removal of monocrotophos. This higher rate of adsorption of monocrotophos onto activated carbon suggested its onsite applicability for inhibiting this toxic insecticide from contaminating the water resources.

Keywords: Pesticide, Monocrotophos, Insecticide, Adsorbent, Water pollution.

Introduction

India is known to be an agriculture-based country, and it is evidenced that about 30% of the agricultural produce is lost due to attack of pests. In India there has been extensive usage of various pesticides with an aim to protect the crops from different types of pests, weeds and insects. Hence agrochemicals like insecticides, fungicides and herbicides have been in increasing demand for protection of crops in one way or the other [1]. Among all of these, Monocrotophos (MCP) is known to be one of the widely used insecticide for protection of important crops such as wheat, cotton, sugarcane, groundnut, tobacco, maize and paddy.

But, the demerit lying during application of such insecticides is that it pollutes and affects the ground as well as river water pollution severely. As per the news given in Hindustan Times, a total number of 442 deaths of farmers and farm workers were reported who died after inhaling such poisonous pesticides in India. Among all of these, 94% of deaths were from Punjab (223), and Maharashtra (183). The deaths were reported in between the financial year 2013-14 to 2017-18.

Recently in August 2018, 3 farmers died in Maharashtra, because of the use of MCP. After these deaths, MCP was banned for 60 days. But it was lifted due to defined clause in insecticide rule according to which, States have no right to ban but can only control the licensing. The World Health Organization (WHO) has placed MCP in Class 1B, a category reserved for highly hazardous pesticides [2].

Monocrotophos ($C_7H_{14}NO_5P$) is an organophosphorus compound and used as a systemic insecticide and its action is mainly on organs such as skin, eyes and central nervous system. MCP can cause various diseases such as the damage of respiratory system, trigger runny nose, tears, pain, chest discomfort, coughing, localized sweating and involuntary muscle contractions, blurred vision and pupil constriction [3]. It is highly toxic to bees, birds, mammals and aquatic invertebrates. The acute oral lethal dosage of MCP (LD_{50}) for rats was found to be 21 mg kg^{-1} [4]. According to WHO, the ingestion of MCP (120 mg) may be fatal to human beings [3].

MCP being cheap is widely utilized by farmers to protect their crops from the attack of various insects. Even after the life threatening hazard farmers could not stop using such type of insecticides. So the need of the hour is to bring onsite solution wherein both the farmers and crops could be benefitted equally. The main objective of the current research is to evaluate the potential of different adsorbents capable of adsorbing MCP sprayed on crops.

Fig. 1 Structure of MCP

Method and Materials

Information was collected regarding the use of insecticides and pesticides in Maharashtra State [3]. The most hazardous insecticide which have taken lives of 3 farmers in Yavatmal District, Phoskill was identified as most dangerous from all of them. This insecticide contains MCP which was 01B category i.e. extremely poisonous drug according to WHO [2]. Literature review carried out to find the treatment, technology/ method for removal of MCP from surface water directed towards adsorption technique to be the efficient amongst all others [4,5,6,7,8]. This technique can be applied on site and is also cheap. So different absorbents such as Activated Carbon, Saw Dust, Fly Ash and Maize cobs (Activated with $CaCl_2$, NaOH, HCl) were experimented to remove the monocrotophos from laboratory prepared sample. After treatment of different adsorbents the samples were analyzed using Thermo Scientific Genesis 10S UV VIS Spectrophotometer.

All the chemicals ($CaCl_2$, NaOH, HCl) utilized were of AR Grade. MCP was acquired from agrochemical agency.

Experimental

Sample Preparation

Batch studies were performed using 10 ppm of MCP solution from insecticide named Phoskill containing 68% Monocrotophos Tech. Serial dilutions were simultaneously prepared for calibration. The λ_{max} value of MCP was found to be 212 nm using UV VIS Spectrophotometer.

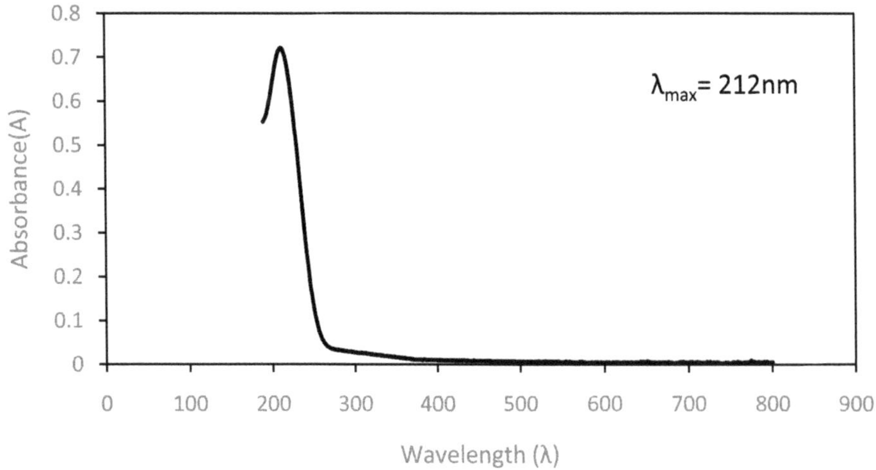

Fig. 2 λ_{max} of Monocrotophos

Preparation of Adsorbents

Activated Carbon (AC) and fly ash (FA) were used without any treatment during the adsorption studies. Sieve analysis of saw dust (SD) was carried out by passing through 600 microns and retained on 75 micron IS Standard

Sieves. Four different sizes of saw dust were obtained. Dirt and dust particles were removed by multiple washings. Then these samples were dried using hot air oven and were used as adsorbent during studies.

Small pieces of dried maize cob (MC) were activated with $CaCl_2$ (1N), HCl (5N) and NaOH (5N). Initially MC was soaked for 24hrs for activation and then dried using hot air oven for 12hrs.

Adsorption Studies

10 ppm solution of MCP was prepared using standard method. 1gm of prepared adsorbents were taken and was mixed with 50ml of 10ppm MCP solution. The mixture was shaken periodically for 60min at 400-500 rpm on a magnetic stirrer. The supernatant solution was filtered through Whatman filter Paper No.41 and checked for its absorbance using UV VIS Spectrophotometer.

Result Analysis

Batch studies were performed with different adsorbents and percentage removal of MCP was calculated using equation No. 1. Results obtained after adsorption studies are depicted in Table No. 1.

$$\%\,Removal\ of\ MCP = \frac{Initial\ Concentration - Final\ Concentration}{Initial\ Concentration} \times 100 \quad .....(1)$$

Table 1 Comparative analysis of uptake of MCP by AC, SD, FA, MC

Sr. No.	Adsorbents	Amount of monocrotophos After Adsorption (mg)	Removal of MCP (%)
1.	AC 1100 IV	1.08	96.12
2.	AC 900 IV	27.8	0
3.	AC 600 IV	3.74	86.55
4.	AC 600 IV(Coal Base)	27.8	0
5.	AC 1200 IV(Coconut Shell)	22.40	19.41
6.	SD (600 Micron)	27.8	0
7.	SD (300 Micron)	27.8	0
8.	SD (212 Micron)	27.8	0
9.	SD (150 Micron)	27.8	0
10.	Fly Ash	27.8	0
11.	MC (without activating)	27.8	0
12.	MC Activated with $CaCl_2$(1N)	27.8	0
13.	MC Activated with HCl (5N)	27.8	0
14.	MC Activated with NaOH (5N)	27.8	0

On comparing the values of percentage removal of MCP by different adsorbents, it is clear that AC of 600 IV(86.55%), 1100 IV(96.12%), 1200 IV(19.41%) were adsorbing MCP with different efficiencies. Since AC 1100 IV performed best amongst all the tested adsorbents, further investigation was carried out by varying different dosage. Batch studies were performed to check percentage removal of MCP at different variations of activated carbon which is shown in Table 2. 10 ppm solution of MCP was prepared using standard method as mentioned in 3.3.

Table 2 Percentage removal of MCP at different AC 1100 IV dosage

Sr. No.	AC 1100 IV (g)	Amount of MCP Remained After Adsorption (mg)	Removal of MCP (%)	Adsorbate(mg)/ Adsorbent(g) Qe	Equillibrium Concentration (mg/L) Ce
1	0.3	9.03	67.55	3.25	1.3
2	0.5	7.95	71.44	2.86	0.71
3	0.7	6.44	76.08	2.32	0.55
4	0.8	5.82	79.10	2.09	0.52
5	1.0	1.08	96.12	0.4	0.48

The adsorption phase concentration after equilibrium was calculated using Eq. 2 as described below;

$$Q_e = \frac{(Co - Ce)V}{m} \quad .....(2)$$

where,

Q_e = Adsorbent (i.e., solid) phase concentration after equilibrium, mg adsorbate/g adsorbent

C_o = Initial concentration of adsorbate, mg/L

C_e = Final equilibrium concentration of adsorbate after absorption has occurred, mg/L

V = Volume of liquid in the reactor, L

m = mass of the adsorbent, g

A graph was plotted between Q_e versus C_e to gain knowledge about the mechanism of the adsorption taking place between AC 1100 IV and MCP. It is evident from the Fig. 3. that the adsorption followed Freundlich isotherm. Fruendlich isotherm is defined as follows,

$$Q_e = \frac{x}{m} = K_f \, C_e^{\,1/n} \quad .....(3)$$

Where,

Q_e = x/m = mass of adsorbate adsorbed per unit mass of adsorbent, mg adsorbate/g adsorbent

K_f= Fruendlich capacity factor, (mg adsorbate/ g activated carbon) * (L water/ mg adsorbate) $^{1/n}$

C_e = equilibrium concentration of adsorbate in solution after absorption, mg/L

1/n = Fruendlich Intensity Parameter

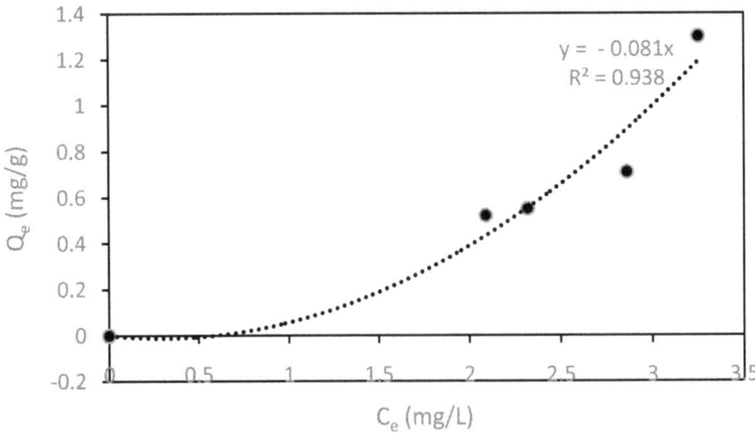

Fig. 3 Graph of Equilibrium Concentration vs Adsorbate/Adsorbent

R^2 close to unity shows favorable and multilayer adsorption. It also directed towards physical adsorption.

Conclusion

Amongst all the adsorbents i.e. 1100IV, 900IV, 600IV (Coal Based), 900IV (Coconut Shell), 1200IV Activated Carbon, maximum adsorption was exhibited by 1100IV i.e. 96.12% and the lowest by 1200IV i.e. 19.41%. It was observed that in case of Saw dust, Fly Ash, and Maize Cob (with and without activation) negligible uptake of took place directing towards inefficacy of adsorbing MCP.

Results indicated that amongst all the tested adsorbents, Activated Carbon having 1100IV showed maximum adsorption. Also activated carbon seems to be good and cheapest option available for the onsite removal/treatment of MCP. Based on the experimental studies carried out, application of Activated carbon over the surface of soil could be highly recommended to farmers using MCP, as Activated carbon is highly efficient in adsorbing it, and thus hindering it to percolate to ground as well as surface water. Again at high temperature when exposed to sunlight, adsorbed MCP from Activated carbon again comes in air and again act as an insecticide. So there is dual use of insecticide to the farmers in single spray, which saves money as well as time. As per the observations and literature review, it can be concluded that MCP cannot be banned at this moment. So awareness among farmers, farm workers and insecticides/pesticides producers and sellers should be created in order to minimize the harmful effects of MCP on humans as well as ecology by using activated carbon in the mentioned way.

Acknowledgements

Mr. Saurabh Jalindre is thankful to Dr. D.S.Bhatkhande and Dr. S.Khamparia, the Director and Deputy Director of Water Quality and Excellence Centre, Vishwakarma University, Pune for their support and guidance.

Abbreviations

MCP - Monocrotophos

WHO - World Health Organisation

UV VIS - Ultraviolet Visible

LD50 - Lethal Dose 50

$CaCl_2$ - Calcium Chloride

HCl - Hydro Chloric Acid

NaOH - Sodium Hydroxide

AC- Activated Carbon

SD - Saw Dust

FA - Fly Ash

MC- Maize Cob

References

Anonymous (2000) Year Book in Pesticide Information. January–March Issue, pp. 6–7. New Delhi: Food and Agricultural Organization (FAO).

World Health Organization (2004) The WHO recommended classification of pesticides by hazard and guidelines to classification. Geneva, Switzerland.

World Health Organization (WHO 1993) International Programme on Chemical Safety: Monocrotophos: health and safety guide. Geneva, Switzerland.

Janghel EK, Rai JK, Rai MK, Gupta VK "A new and highly sensitive spectrophotometric determination of monocrotophos in environmental, agricultureal and biological samples", *Journal of Chinese Chemical Society*.

K. Sahithya, Devlina Das, Nilanjana Das, (2016), "Adsorptive removal of monocrotophos from aqueous solution using biopolymer modified montmorillonite - CuO composites: Equilibrium, kinetic and thermodynamic studies", *Journal of Process Safety and Environment Protection.*

Tejomyee S. Bhalerao, Pravin R. Puranik, (2009). "Microbial degradation of monocrotophos by Aspergillus oryzae", Jalgaon, India. *Journal of International Biodeterioration and Biodegradation.*

Ismail M, Murtaza Sayed, Hasan M Khan and William J Cooper (2014), "Analysis of Pesticides in Water Samples and Removal of Monocrotophos by γ-Irradiation", Peshawar , Pakistan. *Journal of Analytical and Bioanalytical Techniques.*

Anju Agrawal, Ravi S. Pandey, Bechan Sharma (2010), "Water Pollution with Special Reference to Pesticide Contamination in India", Kanpur, India. *Journal of Water Resources and Protection.*

Fuzzy Logic Based Reservoir Operation: A Case Study

Nilesh Sathe[1], Shreenivas Londhe[2] and Preeti Kulkarni[3]
[1]PG student, [2]Professor, [3]Associate Professor,
Vishwakarma Institute of Information Technology,
nsathe81@gmail.com, shreenivas.londhe@viit.ac.in, preeti.kulkarni@viit.ac.in

Abstract

Reservoir operation is an important task to be carried out throughout the entire life span of the reservoir. The conventional methods used for reservoir operation are time consuming and complex for interpretation. To reduce this complexity of interpretation, a fuzzy rule-based model is developed for reservoir operation. Fuzzy model works on 'if-then' principle. This principle includes construction of membership function, development of fuzzy rules, implication and defuzzification. Stochastic dynamic programming is used for construction of membership function instead of experts opinions which in turn has helped to increase the system performance measure. Centroidal method is used for defuzzification. The programming for fuzzy logic is done in MATLAB environment. The methodology is illustrated through the case study of Khadakwasla reservoir built on Mutha river in Pune (Maharashtra), India. Inflow, storage, rainfall and time period of the year are considered as inputs and release as output. Results derived from the method above are satisfactory and implementable.

Keywords: Fuzzy model, Stochastic dynamic programming, MATLAB, Mutha river

Introduction

Reservoir operation is an important process which is to be carried out throughout entire life of the reservoir. Reservoir operation is generally operation of the sates of the reservoir, states of the reservoir is generally considered as initial storage, inflow in the reservoir during the time period and many more according to the user like soil moisture, rainfall etc.as vedula and Mujumdar (1992) There are number of operating policies available for operation of the reservoir but the most common policy implemented in India in practice is standard operating policy as in the work by Panigrahi and Mujumdar (1992). There are again various techniques of operation but are limited due to some complications. Russel and Campbell (1996) also emphasised that due to the 'high degree of abstraction' necessary for efficient application of optimization techniques, the applicability of most reservoir operation models is limited. The managers and reservoir operators are often uncomfortable with the sophisticated optimisation techniques used in the models, which are made much more complex by the inclusion of stochasticity of hydrologic variables as Panigrahi and Mujumdar (1992). The fuzzy logic approach may provide a promising alternative to the methods used for reservoir operation modelling, because, as Russell and Campbell (1996) mention, the approach is more flexible and allows incorporation of expert opinions, which could make it more acceptable to operators. As Shrestha et al. (1996) confirm that fuzzy logic is an appropriate tool to consider the impreciseness of variables like inflows, in reservoir operation modelling.

The fuzzy logic based Modeling of a reservoir operation is a simple approach, which operates on an 'if-then' principle as Loucks aadn van Beek(2007), where 'if' is a vector of fuzzy explanatory variables or premises such as the present reservoir storage, the inflow, the demand, and time of the year. The 'then' is a fuzzy consequence such as release from the reservoir.

In this article, a fuzzy rule-based reservoir operation model is developed for a single purpose reservoir. with the difference that stochastic dynamic programming (SDP) is used for construction of membership function, and generation of rule base for the reservoir operation. As Stochastic dynamic programming is a widely accepted optimization method for deriving efficient reservoir operating policies as Tilmant et.al (2002). Thus, fuzzified version of SDP is a attractive version of reservoir operation. SPD model and fuzzified SDP model with the same discretization scheme and same hydraulic information gives similar system performances as shown by Tilmant et.al (2002).

Methodology

This study is implemented for operation of khadakwasala dam. It is situated on mutha river in pune, maharastra. Which is operational since 1869. Standard operating policy is used for operation of the reservoir. Though a constant supply is provided as drinking water supply through the pipes so there isn't any requirement for considering it as separate output variable, as it has no significance on operation. The methodology used here is to operate the release from the reservoir for irrigation through RBC by SDP and fuzzy logic. As rainfall and inflow as random variable. Storage also becomes random variable as it depends on inflow.

Fuzzy logic operation is carried out in MATLAB as follows. For construction of membership function, as triangular membership functions are used, each membership function is divided into four parts part 1,2,3 and 4, i.e. A-B, B-C, C-D, and D-E respectively as shown in "figure- 1". Part 1 is assigned with all zeros, part two is defined by equation of straight line i.e. y = m1. x + c1, where y is membership value of the variable, m1 is positive slope. X is value on x axis and c1 is negative intercept. Point c is assigned with y=1. Part 3 is defined by equation of straight line i.e. y=m2.x + c2, where y is membership value of the variable, m2 is negative slope. X is value on x axis and c2 is positive intercept. And part four is assigned with all zeros. Value of Y for each membership function for a given X (crisp input) is obtained, is fuzzification of input. Minimum value from the obtained y values are considered as alpha cut for the output variables membership function. If alpha cut occurs twice for same membership function then take maximum value of alpha cut. As shown in "figure 2".

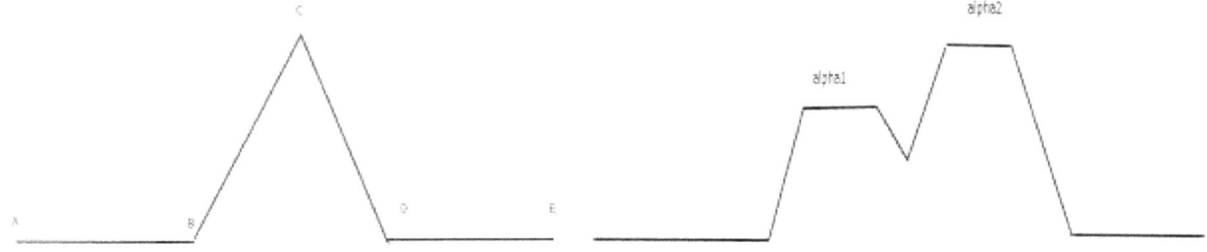

Figure 1 **Figure 2**

Then CG of the area covered is calculated by the algebraic equation (1) to get the crisp value of the output. The equation is implemented in discreet domain in MATLAB thus integral is replaced by summation. Numerator consists summation of all alpha values multiplied by their indices (x values). And denominator is summation of alpha values.

Input for the fuzzy system are inflow in the reservoir, initial storage, time period of the year, rainfall downstream of the reservoir and output is release from the reservoir. The significance of the parameters is as shown in "figure-3". Fuzzy logic works on 'if-then' principle and centroidal method is used for defuzzification to get the crisp output of the release.

Figure 3 Significance of input and output parameters.

Fuzzification

Fuzzification is converting the crisp value of input into a fuzzy input as shown by Ross(1995). The input value is always a crisp value and limited to universe and the result of fuzzification is fuzzy degree of membership, that the value lies between interval 0 to 1. It can be explained as the degree to which a particular value belongs to a class interval is on how we define the limits of the membership function. Here for construction of membership function SDP is used instead of experts opinion and thus ranges of membership function is derived from SDP discretization. Discretization of the data is done on trial and error basis, such that none of the discretised value should lie in a trapping state. Discretization is distributing span of variable into n number of equal spaces and assigning each with a representative value. For example, storage has min value 34.1 and maximum value 85.91 and it is divided into 5 equal number of spaces. Like from 34.1-44.46, 44.62-54.82 and so on. Membership functions of various parameters are as shown in "figure 4". As the number of membership function increases error decreases and vice versa by Jain and Singh (2003). With increase in number of membership function number of rules increases so it becomes linguistically complicated, so it is essential to keep optimal number of membership function. All the parameters i.e. inflow, storage, rainfall, release is assigned triangular membership function as there is linearity in the data and triangular membership function are convenient for operation.

(a) Storage (b) Inflow

(c) Rainfall (d) Release

Figure. 4 Membership functions

Formulation of fuzzy rule base

There are various methods used for formulation of rule base, fuzzy rules can be formed from the actual historical data available as shown by Abrahart (2008). Here in this study SDP discretization is used for formulation of fuzzy rule base using 12-month inflow transition probability. Four inflow class intervals, five storage class intervals, and five rainfall class intervals are used. The data required for SDP is storage, inflow, and rainfall discretization scheme and making sure that it should not stack in trapping state, transition probability matrix and system performance measure as shown in the work by Vedula and Mujumdar (1992). With 12 time period a year (i.e. from July to June) and Transition probability matrix is defined for all the parameters. To formulate the fuzzy rule Reservoir operation is simulated for 11 years, these values are compared with the membership function constructed. The data thus generated will have the values matching the membership function. And this database is used to formulate the fuzzy rules for individual time period. For example, if *storage* is low, and inflow is low medium, and *rainfall* is medium and *time period* 8 then *release* is medium high.

Defuzzification

As Mamdani inference system used, it is necessary to carry defuzzification of the output to get the crisp output value. There are various methods for defuzzification, in this study centroidal method is used for defuzzification to get the crisp value of release. As triangular membership functions are used so centre of gravity gives most precise result compared to other methods. Centroidal method is also called as centre of gravity or centre of area. By Ross(1995). it is given by algebraic equation.

$$z* = \frac{\int \mu c(z).\, zdz}{\int \mu c(z)dz} \quad .....(1)$$

Where,

z is range of variables z^* is defuzzified value, μc. membership value

□ μc(z)dz denotes area of the region bounded by the curve μc.

Result and discussion

Aim of this study is to provide implementable operating policy for reservoir operation. Storage is classified as 'low', 'low-medium', 'medium', 'med-high', and 'high' in the range of [1-90]. Inflow is classified as 'low', 'medium', 'high', and 'very-high' in the range of [1-500]. Rainfall is classified as 'low', 'low-medium', 'medium', 'med-high', and 'high' in the range of [1- 2300]. Release is classified as 'low', 'low-medium', 'medium', 'med-high', 'high' and 'very- high' in the range of [1-120]. Rules are generated for each month as from the data it is observed that release also depends on time period of the year i.e. months. So, there are distinct rules according to time period. Which leads to appropriate result from the inference.

Figure 5 Values of actual release and fuzzy inference output.

For simulation, rule generated are used as knowing the reservoir storage, inflow in the reservoir and rainfall in time period 't', and appropriate rule is applied. Then to obtain the crisp value of the release centroid method for defuzzification is implemented by Which we obtain the crisp of the release for particular time period and input parameters. "Figure 5" shows the values of actual release and fuzzy inference output.

Conclusion

In this article a fuzzy logic-based system is constructed for reservoir operation. The advantage of fuzzy logic-based reservoir operation is that complex operation policies are avoided so operators can feel comfortable for using this model. As A. Tilmant et.al. (2002) point out, the fuzzy logic approach by itself, is not an alternative to the more conventional optimisation techniques. Rather, it provides an opportunity for the reservoir operators to participate in formulating the rule base, and hence may be more acceptable to them than the policies derived using complex optimization models. In this study SDP model is used for construction of membership function and formation of fuzzy rules instead of involvement of experts opinion or group of experts. As Stochastic dynamic programming is a widely accepted optimization method for deriving efficient reservoir operating policies by A Tilmant et.al (2002). Fuzzified version of SDP is an attractive version of reservoir operation, it gives better system performance measure than fuzzy operation alone. As the advantage of fuzzy rule-based model is that it is easy to develop and operate. The model is transparent and easy to understand due to its rule-based structure, which mimics the human way of thinking, even when preset release rules are not applied. Thus, makes convenient for users to operate.

References

Abrahart, Robert J (2008), “Practical Hydroinformatics”, Springer

A. Carvllo et.al. (2004) “Optimal fuzzy decision strategies for reservoir management”, ASCE.

A. Tilmant et.al. (2002) “Comparison of Fuzzy and Nonfuzzy Optimal Reservoir Operating Policies”, ASCE.

Bijaya P. Shrestha et.al, (1996) “Fuzzy Rule-Based Modeling Of Reservoir Operation”, ASCE

D. P. Panigrahi, P. P. Mujumdar, (2000) “Reservoir Operation Modelling with Fuzzy Logic” Water Resources Management 14: 89–109.

Loucks,Daniel,P; , (2017) “water resources system planning and management”. Springer

S. k. jain and v. p. singh (2003), “Water resources system planning and management”, book ELSEVIER

Timothy J. Ross (2010), “Fuzzy Logic with Engineering Applications”, book Wiley publication.

Vedula and Mujumdar, (1992) “water resources system planning and management”.

Open Dumped Municipal Solid Waste – Effects on Peripheral Water Resources – A Case Study

Senthilkumar. G[1] and Murugappan. A[2]
Assistant Professor, Professor, Department of Civil Engineering
Annamalai University, Tamil Nadu
Corresponding Author: cdm.gsk@gmail.com

ABSTRACT

The quality of ground water is largely influenced by the geochemistry, geological formations of the region and by the land use pattern. Presently due to exponential growth of population, urbanization, industrialization, irrigation, livestock watering complimented by growth of economy and the change in life style, has caused enormous stress on all the water bodies. Specifically in the recent decade due to the non availability of the surface water, the ground water is being continuously explored and it reaches an alarming situation that in many locations the water table level has gone down drastically. The available ground water in many placesgets contaminated due to anthropogenic activity.Hence assessing quality of ground-water and developing strategies to protect aquifers from contamination are essential for its sustainable use. In the present study ground water quality analysis were carried out in and around the municipal solid waste open dump site located at Anandathadavapuram road, Tiruviladur,Mayiladuthurai, in Nagapattinam District of Tamilnadu. The physico-chemical parameters pH, EC, TDS, Hardness, Chlorides, Sulphate, Na, K, Ca, Mg, and Alkalinity were tested and the test for the presence of heavy metals Cu, Cr, Ni, Cd, Zn and Pb were carried out using Atomic Absorption Spectroscopy for four successive seasons. The results of the study reveals the fact that the ground water and the surface water body located in the vicinity of the open dump site is contaminated due to sustained dumping / leaching of municipal solid waste and due to other anthropogenic activity and needs treatment before its use for human consumption. The paper also discusses the water resources planning measures for conserving and preserving the ground water for its sustainable use.

Keywords: Municipalsolid waste, ground water quality, water resource planning, heavymetals, sustainable use of water resources.

Introduction

The accelerated growth of industries in India has led to migration of people from villages to the urban centers for employment, education, for improved living environment and in turn generates thousands of tones of Municipal Solid Waste (MSW) daily. With urban population growing at a rate of 3 to 3.5% per annum, the annual increase in overall quantity of solid waste is assessed at about 5% and it translates to 260 million tonnes per year by 2047 (MOF, 2009). The unscientific disposal of municipal solid waste causes adverse impact on all attributes of the environment and human health (Sharholy et al., 2005, Kansal, 2002, Gupta et al., 1998). The refuse after disposal on to the land undergoes a series of biological and chemical reactions as the waste biodegradation progresses.

In general the pollutants leached from the landfills migrates through soil, have every possibility to reach the water table and may leads to deterioration of ground water quality in neighborhood areas of landfill site. The prevention of migration of leachate from the open dumped municipal solid waste can be achieved by providing a single or double liner barrier system on the ground with a facility for the collection of excess leachate and to treat it in the effluent treatment plant prior to its disposal.

The volume of leachate is influenced by infiltration and percolation of rainwater through the various layers in a landfill. The characteristics of MSW leachate varies significantly among landfills depending on the composition of waste, climatic conditions, age of landfill and the technology adopted for land filling (Kjeldsen et al., 2002; Renou et al., 2008; Ziyang et al., 2009). The pollutants present in the MSW landfill leachate includes dissolved organic matter, inorganic macro components, heavy metals and xenobiotic organic compounds.

A clear understanding of flow regime, transport and fate of contaminants from open dumping site is very essential for thorough understanding of the possible impact of open dumping on the ground water as well as to evolve a suitable mechanism for mitigation or prevention of pollution due to leachate migration.

Study Area

The study area is located in Nagapattinam district which lies on the East coast of Tamilnadu.Mayiladuthurai is a first grade Municipal Town situated in Nagapattinam District at 11°7' North latitude and 79°39' East longitude at a distance of 281km from Chennai, the capital city of the state of Tamilnadu, India.

The annual mean maximum and mean minimum temperature recorded are 39.4°C and 33.8°C respectively. The summer prevails during the period April to June, with the maximum temperature observed during the months of May and June often exceeding 40°C. The town receives rainfall generally in two spells, South West monsoon extending from June to August and North East monsoon from October to December. The average rainfall of the town is around 1150mm, out of which 70% is realized during North-East monsoon.

Though the Mayiladuthurai block does not possess any mineral reserves, high fertile alluvial soil is predominant along the stretches of river Cauvery. The entire Delta region shows wide variations in the positions of sand, clay and silt both laterally and depth-wise, resulting in wide variation in permeability values and consequently the yields of wells from depth to depth in the same location. The aquifer in this region occurs at about 2 to 3m below ground level in major part of Cauvery basin (CGWB, 2008). Based on the experimental analysis the top 0.25m depth of soil contains organic content varying between 4 – 6%. Fine alluvium from 0.5 to 2.0m depth and fine sandy soil with a capacity to bear water is present from 2.4m to 3.1m.

Ground water occurs in both confined and semi confined state. The aquifer system is quaternary shallow aquifer and lower Miocene deeper aquifer, both hydraulically interconnected. The Pliocene shallow aquifers are represented by sands, gravels, sandy clays and variegated clays. Heterogeneity in lithology is also noticed. Ground water occurs under water table as well as in confined conditions and is developed by means of shallow tube wells and dug well cum bore wells. The depth of tube well ranges between 30 to 100m.The specific capacity values ranges between 13.43 to 870 lpm/ m and transmissivity 1202m^2/day (CGWB 2008).

Status of Municipal Solid waste

The Municipal Solid Waste generation in Mayiladuthurai town is around 36 tonnes / day, at a per capita waste generation rate varying between 300 – 400 gms. The solid wastes generated are disposed off in the open dumping site located in Anandathandavapuram road at Tiruvilandur.

The open dumping site is being operated from the year 1980 to till date. The dumping site is located at a distance of 1.2km in the Northern part of the town. The waste from all the 36 wards are being dumped, which spreads over a planned area of 4.62 acres and now spreads and overflowing to an extent of 6.5 acres with a height of open dumped waste varying between 4 to 5m. Since there is no liner system at base for prevention of leachate migration, the waste which is open at the top as well as at the bottom, simply receives the precipitation and allows it to pass through the various layer of the unconsolidated waste resulting in increased moisture in the cell. The rain water percolated into the soil beneath and eventually pollutes the ground water in the surrounding areas. In addition, the surface runoff from the heaped solid waste contaminates the small water bodies located in the periphery of the open dump site. The residents residing in and around the open dump site are not using the ground water for drinking due to its objectionable quality.

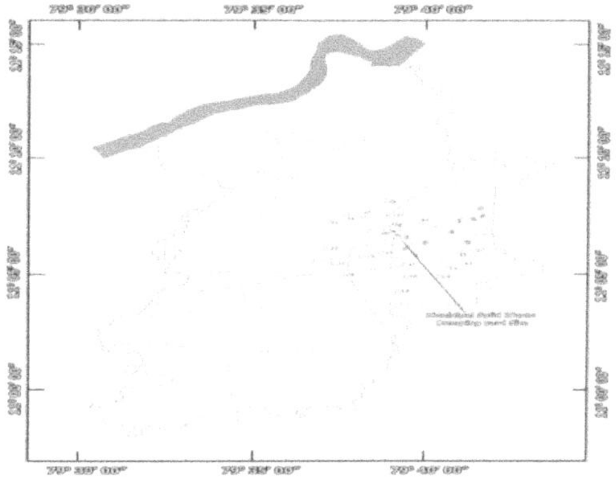

Figure 1 Location Map of the Study Area

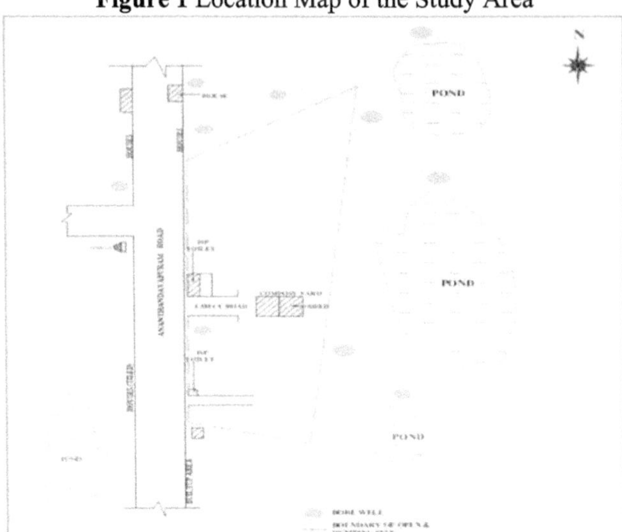

Figure 2 Location of Bore Wells in and around Dumping Site

The Tiruvilandur open dump site is surrounded by human dwellings, along the stretch of Anandhathandavpuram road in the East, barren land followed by two water bodies in the West, few houses followed by agricultural land in the North and an urban location with houses, shops, followed by National highway at a distance of 0.8km connecting Chennai and Kumbakonam in the South. The location map and the layout of the of the open dump site are given in Figure 1 and 2.

Solid Waste Sampling and Characterization

In order to have more reliable and representative sampling, two truckloads of waste originating from residential and commercial wards were analyzed at the point of disposal, just inside the dumping site. In the quartered quarter, samples were segregated into different categories and measured for individual components like food waste, paper, plastics, wood; glass, tin cans, textile, ceramics, rubber, vegetable waste etc, and weighed individually.

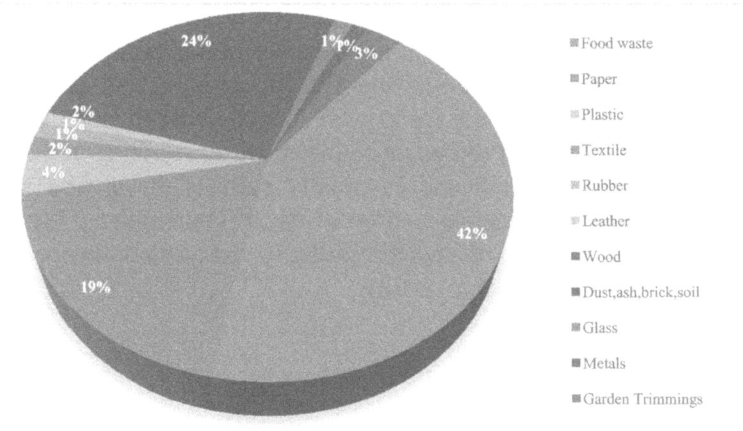

Figure 1 Composition of MSW - Mayiladuthurai Town

The Figure 1 show the percentage distribution of various constituents of solid waste. The percentage distribution of food waste, paper, plastic, textile, rubber, leather, wood, dust / ash / brick / soil, glass, metals and garden trimmings are varied in the range of 37 – 42%, 19 – 21%, 3.25 – 3.68%, 1.64 – 3.0%, 1.4 – 2.4%, 1, 2 – 2.04%, 24 – 27%, 1.16 - 2.30%, 1.23 – 1.62%, 2.86% respectively. It is observed that the food waste is around 40% and 35% of the waste is trash followed by the non combustible waste dust, ash, bricks, soil, glass and metals. Based on the composition analysis it is observed that the presence of various component are in the order of, food waste > Dust ash brick soil > paper > Plastics > Textile > wood > rubber > glass > metals > leather > Garden trimmings.

Looking at the composition of the waste generated in developing countries, one can conclude that the biodegradable portion dominates the bulk of MSW.

Sampling of Leachate

The composition of leachate varies largely both in space as well as in downstream direction (Peter Kjeldsen, *et.al.,*1995). For European Union condition, for landfills without any protective cover on top it is estimated that approximately 15 to 60% of rainfall may enter into the waste and turns out as leachate (Ehrig, 1983). Also if the waste is not compacted like in Indian condition, the rate of generation of leachate will be more. Since there is no existence of leachate collection mechanism or leachate streams for collection of leachate, pit is excavated to a depth of 4 to 5 foot after a day's time the oozed out leachate are collected in a laboratory cleansed plastic cans after rinsing the cans by leachate. The collected sample of leachate were shifted to the laboratory, stored at 4°C and analysed within 48hours. The analysis on the leachate sample was carried out using the guidelines given in Standard methods for the Examination of Water and Waste Water, (APHA, 2005).Table 1.0 gives the characteristics of leachate.

Table 1 Characteristics of Leachate – Mayiladuthurai open dump site

Sno	Parameter	Concentration
1.	pH	7.61
2.	EC	22.37
3.	TDS	14987
4.	Chlorides	2400
5.	Hardness	3100
6.	Sulphate	480
7.	COD	1510
8.	Na	483
9.	K	543
10.	Ca	388
11.	Mg	218
12.	Nitrate	124
13.	Fe	27.96
14.	COD	3920
15.	Cu	1.22
16.	Zn	2.86
17.	Pb	2.62
18.	Cr	1.5
19.	Ni	0.65
20.	Cd	BDL

(All values are in mg/l, except pH, EC in millimhos/cm)

Sampling of Ground Water

Ground water samples are obtained from 10 bore wells and from an open well located near to the Municipal Solid Waste open dumping site of Mayiladuthurai Town. Ground water samples were collected in laboratory cleansed 1litre capacity polyethylene bottles after rinsing the bottle with water to be collected and numbered as BW1, BW2,.....BW10 respectively. The bottles are then transferred to laboratory and stored at 4°C for further analysis.

The collected ground water samples are analyzed for its physico-chemical characteristicspH, EC, TDS, Hardness, Chlorides, Sulphate, Na, K, Ca, Mg, and Alkalinity and tested for the presence of heavy metal ions like Cu, Cr, Ni, Cd, Zn and Pb using AtomicAbsorption Spectroscopy. The quality characteristics of ground water samples are investigated during May 2015, January 2016, May 2016 and January 2017.

Rainfall Recharge

The recharge to the aquifer is varying with respect to soil type, land use, topography and rate of rainfall infiltration. The rainfall recharge in the Cauvery river flood plain is 45% of the surface runoff. The vast majority of the land area is used for agricultural purposes and composed of sandy formation which is recharged by 35% of rainfall (CGWB, 2009). Smaller part of the land is covered by urban dwellings which is having 25 to 35 % of rainfall as recharge to the aquifer.

Results and Discussion

In Thiruvilandur, solid waste open dumping site, the waste is disposed off on land surface directly without any pre-treatment and segregation of hazardous materials which in turn affect the livelihood of the surrounding people. Habitat near the open dumping site quite often reports the poor quality of ground water near to the municipal solid waste open dump site. Presently, groundwater around 0.8 to 1.2 km radius from open dump site is not being used for drinking purpose. It was observed from field visit that there is no surface lining to prevent the leachate entry into the ground. During monsoon seasons, the rainfall leaches the waste and the leachates generated directly infiltrates into the ground and contaminate the groundwater. Higher TDS were observed in bore wells located down gradient to the open dumping site.

Table 2 Quality Characteristics of Ground water near open dump site – May 2015

Well/ Parameter	PH	EC	TDS	Hardness	Chlorides	Sulphate	Na	K	Ca	Mg	Alkalinity	Cu	Cr	Zn	Ni	Cd	Pb
BW1	7.34	8.95	5997	3170	1263	264	1665	121	158	28	150	0.001	0.001	0.81	0.040	BDL	0.26
OW	6.53	9.17	6144	3900	1895	345	1280	142	134	40	108	0.122	0.023	1.28	0.140	0.002	0.35
BW2	7.89	8.85	5930	1200	1510	330	1550	56	543	40	103	0.140	0.013	1.13	0.080	BDL	0.22
BW3	7.64	5.65	3786	1138	260	165	280	86	124	52	146	0.013	BDL	0.05	0.017	BDL	0.08
BW4	7.92	5.85	3919	368	510	162	265	115	127	73	156	0.010	BDL	0.27	0.012	BDL	0.07
BW5	8.27	8.72	5843	1860	810	1550	820	30	180	28	120	0.011	BDL	0.03	0.022	BDL	0.017
BW6	7.46	8.23	5514	660	785	180	865	61	155	122	90	0.150	0.010	1.56	0.031	BDL	0.14
BW7	7.80	8.30	5561	640	585	195	260	38	121	78	26	0.080	0.002	0.18	0.068	0.001	0.07
BW8	7.33	5.90	3953	890	360	160	210	35	190	133	24	0.015	0.012	0.23	0.020	BDL	0.16
BW9	6.49	4.84	3242	397	410	165	90	44	76	98	48	0.011	0.001	0.24	0.010	BDL	BDL
BW10	8.22	5.15	3450	910	554	180	354	23	44	21	38	0.003	0.002	0.22	0.014	BDL	0.002

(All the values are in mg/l, except pH and EC in m mhos / cm)

Table 3 Quality Characteristics of Ground water near open dump site – Jan 2016

Well/ Parameter	PH	EC	TDS	Hardness	Chlorides	Sulphate	Na	K	Ca	Mg	Alkalinity	Cu	Cr	Zn	Ni	Cd	Pb
BW 1	7.97	5.10	3417	2320	2344	114	1583	25	182	88	137	0.121	0.021	0.16	0.003	BDL	0.054
OW	6.85	6.22	4167	2365	2575	204	1120	85	465	172	122	0.134	0.110	1.93	0.041	0.001	0.121
BW 2	7.90	5.21	3490	910	1010	120	850	45	323	145	187	0.014	0.024	0.56	0.051	BDL	0.015
BW 3	7.45	7.23	4844	1267	138	110	1260	66	145	66	170	0.033	0.012	0.054	0.003	BDL	0.065
BW 4	7.20	5.02	3363	780	670	220	550	67	74	72	156	0.043	0.011	0.013	0.002	BDL	0.247
BW 5	7.75	4.65	3115	880	800	233	530	26	189	124	187	0.211	0.002	1.110	0.021	BDL	0.136
BW 6	7.72	4.90	3283	1367	465	138	122	53	146	110	134	0.032	0.023	0.023	BDL	BDL	0.021
BW 7	7.445	5.10	3417	520	314	156	130	19	109	37	123	0.031	0.013	0.101	0.003	BDL	0.122
BW 8	7.01	6.78	4542	464	196	187	46	19	85	132	134	0.012	BDL	0.001	0.023	BDL	BDL
BW 9	7.01	2.40	1608	334	360	134	48	23	39	19	77	0.168	0.002	0.012	0.011	BDL	0.014
BW10	7.23	2.01	1346	469	424	104	134	21	54	14	44	0.001	BDL	0.001	BDL	BDL	0.012

(All the values are in mg/l, except pH and EC in m mhos / cm)

Table 4 Quality Characteristics of Ground water near open dump site – May 2016

Well/ Parameter	PH	EC	TDS	Hardness	Chlorides	Sulphate	Na	K	Ca	Mg	Alkalinity	Cu	Cr	Zn	Ni	Cd	Pb
BW 1	7.81	5.99	4013	3468	980	410	960	57	460	165	90	0.03	0.01	1.35	0.01	BDL	0.120
OW	7.67	7.76	5199	3270	2690	138	1534	82	700	343	270	0.01	0.04	1.44	0.012	BDL	0.122
BW 2	7.67	7.74	5185	1280	1389	280	1215	86	178	210	134	0.21	0.06	0.54	0.001	BDL	0.121
BW 3	8.41	4.25	2847	770	1356	590	460	63	380	230	223	0.05	BDL	0.21	0.002	BDL	0.012
BW 4	7.63	1.45	971	1010	440	432	570	53	234	189	176	0.007	0.07	0.56	0.002	BDL	0.012
BW 5	8.55	6.65	4455	1390	1280	190	225	72	156	122	111	0.011	0.08	2.07	0.102	BDL	0.011
BW 6	7.63	1.93	1293	460	510	210	243	28	132	143	132	0.023	0.09	0.30	0.010	BDL	0.120
BW 7	7.83	3.78	2532	530	720	440	184	22	187	47	81	0.043	0.016	0.21	0.030	BDL	0.121
BW 8	8.19	4.65	3115	220	556	287	210	42	27	46	76	0.002	BDL	0.20	0.001	BDL	0.021
BW 9	7.54	3.34	2237	290	630	175	220	12	39	23	110	0.012	0.08	0.10	0.031	BDL	0.011
BW10	7.65	2.33	1561	265	435	222	120	22	14	15	120	0.023	0.09	0.47	0.001	BDL	0.023

(All the values are in mg/l, except pH and EC in m mhos / cm)

Table 5 Quality Characteristics of Ground water near open dump site – Jan 2017

Well/ Parameter	PH	EC	TDS	Hardness	Chlorides	Sulphate	Na	K	Ca	Mg	Alkalinity	Cu	Cr	Zn	Ni	Cd	Pb
BW 1	7.21	5.12	3430	2686	2170	314	1654	56	295	58	58	0.011	BDL	1.64	0.001	BDL	0.134
OW	7.34	5.33	3571	2110	3451	376	1280	70	470	167	85	0.002	0.012	2.13	0.024	BDL	0.220
BW 2	6.39	5.76	3859	1580	2898	254	180	38	310	136	142	0.011	0.002	0.013	0.022	BDL	0.022
BW 3	7.86	5.23	3504	774	3421	377	890	52	254	195	145	0.020	0.002	0.234	0.053	BDL	0.142
BW 4	8.02	5.45	3651	850	890	430	710	34	265	90	122	0.020	0.012	0.12	0.032	BDL	0.021
BW 5	7.21	4.67	3128	255	706	501	152	12	380	150	34	0.001	0.011	1.12	0.021	BDL	0.014
BW 6	7.33	5.08	3403	456	1104	220	565	10	123	120	67	0.101	0.002	0.025	0.003	BDL	0.001
BW 7	6.55	4.43	2968	680	560	342	632	14	154	110	69	0.001	0.003	0.038	0.002	BDL	0.110
BW 8	6.89	3.54	2371	237	980	354	290	21	132	115	64	0.001	0.002	0.133	0.012	BDL	0.011
BW 9	7.41	2.12	1420	340	385	228	180	23	156	126	65	0.001	0.022	0.185	0.011	BDL	0.012
BW10	7.95	2.35	1574	182	321	225	250	28	122	210	24	0.002	0.002	0.045	0.001	BDL	0.110

(All the values are in mg/l, except pH and EC in m mhos / cm)

Table 6 Surface Water Quality Characteristics – May 2015

S. No	Distance from open dump site (m)	Nature of source	pH	hardness	Turbidity	EC	Chlorides	DO	Acidity	Alkalinity	Sulphate	Total Solids	TDS	BOD	COD
1	250	pond	7.98	130	35	0.838	120	3.45	BDL	10	24	1910	562	140	170
2	50	pond	6.82	210	30	0.920	165	4.10	BDL	14	55	2400	617	124	135
3	200	pond	8.10	180	13	0.990	105	3.65	BDL	BDL	12	720	663	122	116
4	100	pond	8.30	176	15	0.482	85	4.59	BDL	BDL	16	1650	323	148	120

(All values are in mg/l, except pH, EC in millimhos/cm)

It is observed from Table 2 that the pH of the ground water samples varied in the range of 6.49 to 8.27. The lower value of 6.49 observed in the Bore well (BW9). The Electrical conductivity (EC) ranges between 4.84 to 8.97mmhos / cm, a lower value is observed in well BW9, whereas higher EC is observed in samples of water obtained from OW followed by BW1 with an EC of 8.95mmhos /cm. The TDS of all the ground water samples near to the vicinity of open dump site are higher and exceeds the desirable limit of 500 mg/l and a maximum permissible limit of 2000 mg/ l prescribed by the standards for drinking water IS 10500 – 2004. A higher TDS of 6144 mg/l is observed in ground water sample obtained from OW and a lower TDS concentration of 3242 mg/l is observed in ground water sample obtained from BW9. The water samples of OW, BW1, and BW2 are having a higher TDS of 6144 mg/l, 5997 mg/l, and 5930 mg/l respectively.

The higher TDS in all the ground water samples may be due to leaching and dissolution of various contaminants from the MSW open dumping site, into the ground water system.

The assessment of ground water during the post monsoon period was carried out during January 2016 and January 2017. From Table 3 the concentration of inorganic pollutants in ground water has shown a varying trend. The TDS varied between 1346 mg/l in sample obtained from BW10 to 4844 mg/l in BW4. Similarly, the hardness, chlorides and sulphates show a declining trend in majority of wells except for samples from BW1and OW. The presence of lower concentration of inorganic parameters and the heavy metal ions may be due to dissolution, dilution of pollutants by rain and rainfall recharge. From Table 4 and 5a varied trend in pollutant concentration has been observed when compared to the previous seasons of May 2015 and January 2016. This may be due to reduction in rate of rainfall or due to variation in rate of recharge.

The hardness in ground water is due to the presence of salts of Calcium and Magnesium which is present in dissolved form. The Calcium and Magnesium in water is responsible for the formation of lather when it reacts with soap. Total hardness is expressed as calcium carbonate ($CaCO_3$) equivalent of calcium and Magnesium ions present in water. Based on the presence of hardness the ground water is classified as Soft, Moderately hard and hard. The distribution of Total hardness in ground water samples of the study area vary between a 265 mg /l to 3468 mg/l. Hardness of less than 75 mg/l will make the water tasteless and hence a desirable limit of 75 – 115 mg/l is recommended for drinking water. The samples of ground water obtained from BW1, BW2, BW4 and BW5 are having the higher concentration of Total hardness as 3468 mg/l, 1280 mg/l, 1010 mg/l, and 1390 mg/l respectively. In BW6, BW7, BW8, BW9, and BW10 the total hardness is ranging between 220 mg/l and 530 mg/l. A Hardness of less than 300 mg/l is given as desirable limit for hardness in drinking water as per IS 10500 – 2004.

It is observed that the concentration of chlorides is varying between 435 mg/l to 2690 mg/l. A higher concentration of chlorides of 1389 mg/l, 1356 mg/l, and 1280 mg/l is observed in BW2, BW3 and BW5. In the remaining wells the chloride concentration is varying between 435 mg/l to 720 mg/l. The desirable limit for concentration of chlorides in drinking water is 250 mg/l. The concentration of sulphate varied between 138 mg/l and 590 mg/l. Higher concentration of chlorides and sulphate may be due to the leaching from the open dumped municipal solid waste. The concentration of sodium and potassium varied between 130 mg/l to 1534 mg/l and 12 mg/l to 86 mg/l respectively.

The ground water samples were also analysed for the presence of heavy metal ions Cu, Cr, Zn, Ni, Cd and Pb. It can be observed from Table 1.1 that the concentration of Cu in ground water samples obtained from BW1 varied between 0.001mg/l to 0.15mg/l. A higher concentration is observed in ground water sample obtained from BW6. It is observed that the concentration of Cu in ground water samples of BW2, BW3, and BW7 are exceeding the desirable limit of 0.05 mg/l (IS 10500 - 2004). The heavy metal ion Cr varied between 0.001 mg/l to 0.023 mg/l and it is observed that the concentration of Cr in all the ground water samples is well below the desirable limit of 0.05 mg/l.

Even though Zinc is considered as an essential nutrient for a healthy body, excess concentration of Zinc is harmful to human health. The concentration of Zinc up to 5 mg/l is given as desirable limit and the maximum permissible limit for concentration of Zinc in drinking water is 10 mg/l (IS 10500 - 2004). It is observed that Zn is varying between 0.05mg/l to 1.56 mg/l which is well within the desirable limit of 5 mg/l.

The concentration of Ni varied between 0.010 mg/l to 0.140 mg/l. The heavy metal ion Ni is a ubiquitous metal, which finds increased applications in modern technologies. The heavy metal ion Cd occurs naturally in Zn, Pb, Cu and other ores which act as source to the ground and surface water. The disposal of batteries, electronic gadgets, metallic compounds, corrosion in galvanized plumbing and water mains are the reasons for the release of Cd. The desirable limit for Ni and Cd is 0.02mg/l and 0.01mg/l respectively. the Table 1.1 show that the concentration of Ni for samples BW1, OW, BW2, BW3, BW4 , BW6 are exceeding the desirable limit whereas in other samples a lesser concentration is observed. The concentration of Cd is detected as BDL for majority of the samples except OW. It is observed that the concentration of Pb in GW samples obtained from BW5, BW9, BW10 are less while in other samples the concentration of Pb is exceeding the desirable limit of 0.05 mg/l. The possible sources of lead are from used / abandoned batteries, waste from auto garages and waste from photo labs. Higher concentration of lead is toxic to all life forms. The results of the analysis on ground water samples from ten bore wells (BW) and from an open well (OW) located near to the vicinity of open dumping site shows higher inorganic and heavy metal concentration. The TDS and other inorganic pollutants are present in excess of the maximum permissible limit for drinking water prescribed by IS 10500 (2004). It was observed that the heavy metal concentration of BW water samples analyzed for the present study exceeds the desirable limit. The environmental consequences of leachate generated in Tiruvilandur, Mayiladuthurai open dump site on ground water is simulated using MT3D. It is observed that ground water is flowing downstream towards north and north east direction. Many of the contaminant will tend to follow the same gradient and direction but the magnitude of the contamination caused by the heavy metal ions depends on its respective distribution coefficient.

Surface water Quality

The samples of water obtained from the ponds located near to the vicinity of the open dump site show deterioration in quality in terms of presence of higher Total solids, TDS, BOD and COD. In spite of natural oxidation and reduction process the presence of COD in surface water show the pond water is contaminated due to migration of leachate and surface runoff originating from the open dump site. The contamination from the unplanned solid waste disposal system becomes a threat to all the environmental components (saravanan.,et.al 2010). Table 6 show the surface water quality of pond water located near the to municipal solid waste dump site.

Conclusions

The migration of leachate from the open dumped municipal solid waste contributes to soil contamination and deteriorates the quality of both soil and the ground water. Suitable measures are required for attenuation of pollutant at source and for prevention of its migration. Hence it is essential to understand the ground water flow, fate and transport of contaminants and its attenuation to evolve suitable strategies for pollution prevention. In the present study investigations were carried out to study the cause and effects of open dumping of municipal solid waste at Mayiladuthurai town. Presently the solid waste originating from the town is being dumped on ground without any measures for prevention of leachate migration. The presence of inorganic pollutant and the concentration of heavy metal ions in leachate are comparatively higher during the pre monsoon period than the post monsoon period. The lower pollutant concentration is mainly due to the dissolution of pollutant by rain, dispersion, dilution, oxidation, partial stabilization and precipitation of metals. With an average rainfall of 1150mm and the rate of recharge through the waste at 40% the leachate generation is estimated as 0.082lit/sec during south-west monsoon to 0.190lit/sec during the north – east monsoon period. Even though the quality of ground water is dynamic the characterization of the ground water samples in and around the open dumping site shows that ground water is not potable, the TDS, Hardness, Chlorides, sulphate, Na, K and heavy metal ions like Cu, Ni, Pb are found to be higher and exceeds the permissible limits prescribed by, Standards for drinking water IS 10500 – 2004. From the ground water samples collected from the bore wells located within 5 km radius from open dumping site, an improvement in quality of ground water has been observed with increase in distance from the open dump site. In order to prevent further pollution the practice opf open dumping is to be stopped immediately and the a policy of treating waste as resource is the need of the hour.

References

APHA (2005) Standard methods for the Examination of Water and Waste Water, 20th edition, AWWA, Washington D.C.

Bureau of Indian Standards for Drinking Water - Specification, IS 10500 : 2012

Central Ground Water Board (CGWB), District Ground Water Brochure, Nagapattinam

District, Tamil Nadu, (2008), Ministry of Water Resources, Government of India.

Ehrig, H.J., Sickerwasserentstehung in Altlasten und ihre Pro-blematik. Handbuch der Altlastensanierung. Hrsg:Franzius/Stegmann/Wolf/Brandt. R.v. Decker's Verlag, Heidel-berg. (1990).

Gupta,S., Krishna,M., Prasad, R.K.,Gupta, S., Kansal, A., Solid waste management in India: options and oppurtunities. Resource, Conservation and Recycling 24, (1998), pp.137-154.

Kansal, A., Solid waste management strategies for India, Indian Journal of Environmental protection, 22(4), (2002), pp.444-448.

Kjeldsen P., Barlaz M., Rooker A., Baun A., Ledin A. and Christensen T., "Present and Long Term Composition of MSW Landfill Leachate: A Review," Critical Reviews in Environmental Science and Technology, Vol. 32, No. 4, (2002), pp. 297-336.

Ministry of Finance (MOF), Department of Economic affairs, Government of India (2009), Position paper on Solid waste management sector in India.

Peter Kjeldsen., Poul L., Bjerg, PIA Winther., Kirsten Rurge., Jorn K. Pedersen, BentSkov., AnjaFoverskov and Thomas H. Christensen: Assessment of the spatial variability in leachate migration from an old landfill site, Ground Water quality: Remediation and Protection (Proceedings of the prague conference), IAHS Publications, 225, (1995), pp.365-373

Renou. S, Givaudan. J. G, Poulain. S, Dirassouyan. F and Moulin. P, "Landfill Leachate Treatment: Review and Opportunity," Journal of Hazardous Materials, Vol. 150, No. 3, (2008), pp. 468-493.

Saravanan.R., et.al, Modeling of ground water pollution due to Solid waste Dumping – A Case study, International Journal of Earth sciences and Engineering, (2010), Vol 3, No 02.

Sharholy, M., Ahmad,K., Mahmood, G., Trivedi, R.C., Analysis of municipal solid waste management systems in Delhi – a review. In: Book of proceedings for the Second International congress of Chemistry and Environment, Indore, India, (2005), pp.773-777.

Ziyang L., Youcai Z., Tao Y., Yu S., Huili C., Nanwen Z. and Renhua H., "Natural Attenuation and Characterization of Contaminants Composition in Landfill Leachate under Different Disposing Ages," Science of the Total Environment, Vol. 407, No. 10, (2009), pp.33.

Fluoride Toxicity of Ground Water in Pavagada Taluk, Tumkur District, Karnataka

Vedamurthy. B[1], Nandeesha[2], T.S. Umesha[3] and Vishal. R. Khandagale[4]
[1]UG Student, [2] Professor, [3&4]Associate Professor,
Civil Engineering Department, Siddaganga Institute of Technology, TumkurKarnataka, India.
Civil Engineering Department, Akshaya Institute of Technology, Tumkur Karnataka, India.
vedamurthy002@gmail.com, nandeeshsit@gmail.com, vishal.rk28@gmail.com

ABSTRACT

Habitat of Pavagada taluk faces large number of socio-economic and health related issues besides the usual acute shortage of potable water. The District experiences arid to semi-arid type of climate. The groundwater and a less rain for a few days in a year are the main source of water for irrigation, drinking and other domestic consumption. The quantity is low but more frightening is the quality of ground water that is being used. A survey of thirty five villages in Pavagada taluk of Tumkur district was conducted. The primary data includes the study of total dissolved solids. The Total hardness values ranged from 78 to 444 mg/l which was in most cases above the acceptable limit of 300 mg/l. The other parameters studied are pH, Electrical conductivity, TDS, Alkalinity, Calcium, Magnesium, Iron, Sulphate, Nitrate, Fluoride, Chloride and obtained results are compared with BIS 10500:2012. For the analysis of Soil some of the parameters like pH, Calcium, Sulphate, Potassium and Organic content are analyzed to understand the influence of soil contamination on quality of groundwater. The groundwater in the study area is found to be slightly acidic to basic (pH ranging from 6.63 to 9.65) and moderately hard to very hard (Total Hardness ranging from 78 mg/l and 444 mg/l) in nature. Concentration of Nitrate is above permissible limit in 21groundwater samples out of 51 samples and Fluoride concentration is above permissible limit in 37 samples locations out of 51 samples. The result suggests poor groundwater quality which is highly saline (hard water) with high Fluoride contents. These high Fluoride levels of ground water used for drinking are anticipated to be the reasons for the prevalence of dental carries and skeletal Fluorosis in Pavagada taluk.

Keywords: Fluoride, groundwater quality, habitat, hard water, Pavagada taluk.

Introduction

Water is the essence of life but what if it starts becoming the cause of morbidity and what if it engulfs the whole flora and fauna of the habitat. It needs serious attempts to know the reasons behind this continuing tragedy. Yes, water is life as it creates and preserves life but there are many places where high salt contamination in water is the main because of death. There are many places in the world where quantity and quality of water is threatening the survival of people. This problem is faced not only in the underdeveloped nations but also in developing nations where population growth rate is very high and base of the population pyramid is large. In the developing nations there are few places with lopsided development that lacks in the basic amenities to maintain the good living standard. It is not about the facilities of shelters or food: rather about the first and foremost basic need and pre requisite of life i.e. potable water. Pavagada taluk in Tumkur district is a part of semi-arid region in the district. It receives annual average rainfall of 560mm rainfall and facing the acute water shortage and poor (saline) water quality both.

In India, research on the assessment of groundwater quality especially with reference to fluoride has been carried out by various workers. Both fluoride (F-) and hydroxyl (OH-) ions have the same ionic radii and are roughly of the same size and hence they can easily replace each other in many rocks forming processes. Fluoride exists naturally in a number of different minerals.[11] The major fluoride containing minerals are: Fluorite (CaF2), Fluorapatite [Ca2F (PO4)], Cryolite (Na3AlF6) and Topaz [Al2SiO4 (OH.F) 2]. So, fluoride ions are widespread in lithosphere as compared to hydrosphere, atmosphere and the biosphere since most of the fluoride exists bound in different minerals. The quality of water in groundwater is a major concern for the mankind because human welfare is directly linked with it and so, the water quality regulating authorities such as WHO and BIS gives the fixed desirable and permissible limits of various physicochemical parameters of water. The desirable limits are safe limits and permissible limits are allowed only in the event of the absence of alternative sources.

Quality of water is an important parameter for human health. The potable water should be free from pathogenic agents and chemical constituents, pleasant to taste and usable for human and animal consumption with other domestic purposes. The ground water is characterized by multiple quality pollutants viz., coloured dyes, heavy metals, nitrates and fluoride etc. The water is a universal solvent, which contain many dissolved substances. The preliminary study carried out in the present investigation indicates that Pavagada region contains high fluoride contamination in groundwater due to its rock structure.

Table 1 Various minerals having Fluoride with their chemical composition in the particular rocks: as per who standards.

Sl. No.	Mineral	Chemical Composition	Rocks
1.	Fluorspar	[CaF2.3Ca3(PO4)2]	Pegmatite, Pneumatolitic deposits
2.	Fluorite	CaF2	Pegmatite, Metamorphosed limestone
3.	Lepidolite	K2(Li,Al)5(Si6Al2)O20(OHF)4	Gabbros, Dolerites
4.	Tremolite Actinolite	Ca2(MgFe+2)5(Si8O22)(OHF)2	Clay
5.	Rock Phosphate	NaCa2(MgFe+2)4(AlFe+3)(SiAl)8O22(OHF)2	Limestone, Fossils

Source: WHO Report: 2007.

Study Area

The study area is located in Tumkur district of Karnataka state and forms highly undulating terrain with hillocks having sporadic vegetation. It cover an area of 1362 sq.km and falls between longitude E77°30'-78° 05' and latitude N13° 55'to 14°20'.The area is covered under survey of India toposheet Nos.57 F/3,F/4,F/7,F/8 and G/8.Pavagada town is about 165 Kms north-east of Bangalore as shown in figure1. The study area forms part of hard rock terrain of Karnataka state and is drought prone (average annual rainfall 560mm). The sources of water in this area include bore well, hand pump, water tanks etc. The study area is reported to be facing a lot of problems regarding the quality of water. The residing people are facing acute problems of fluorosis which is due to deficient of excessive quality of water. Thus an effort has been made to survey the study area and analyze the quality of water by sampling and presenting the results in an interesting and attractive way so that the need for reforms is highlighted.

Methodology

Samples of groundwater are collected from bore wells & hand pumps ,during the pumping of water is taking place in a 1-liter plastic can and they were brought to the laboratory. About 51 samples were collected from various places of study area as shown in figure 2, Samples were drawn with a pre-cleaned plastic polyethylene bottle. Prior to sampling, all the sampling containers were washed and rinsed thoroughly with then groundwater. All the parameters were analyzed as per BIS procedure. Soil samples from fifteen locations were collected and analyzed for pH, EC, organic content, Sulphur content, available Phosphorous, Potassium, Calcium, Magnesium, nitrogen and micronutrients like Copper, zinc, Iron and manganese.

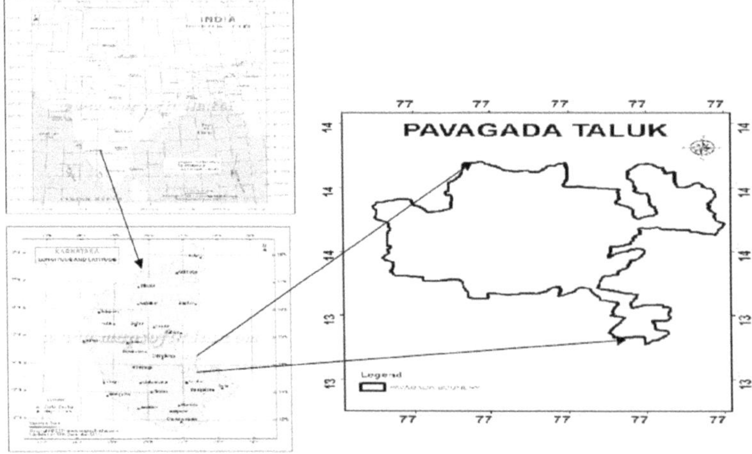

Figure 1 Study area

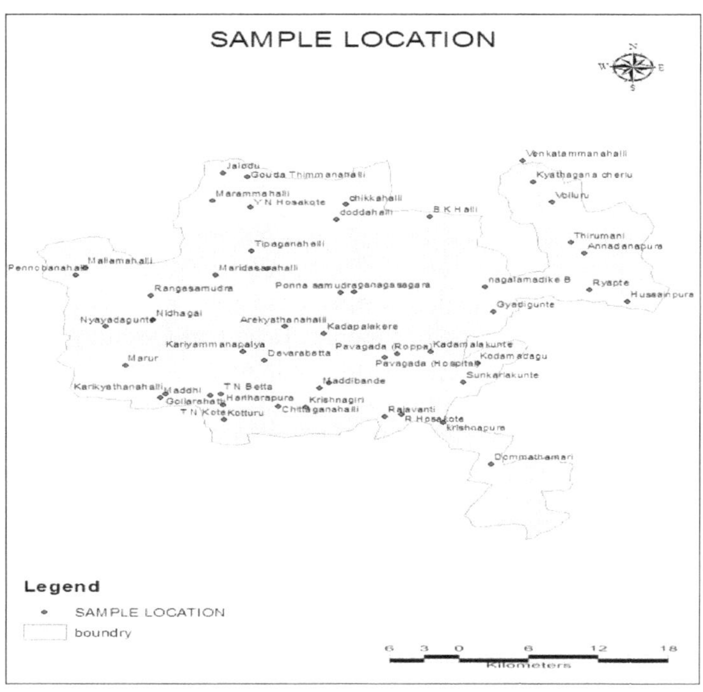

Figure 2 Sample location map

Results and Discussion

Table 2 Results of water quality analysis.

Water quality parameters	Min.	Max.	Avg.	SD
Calcium	2.14	387.02	122.69	9.654
Magnesium	0.19	65.71	8.37	1.758
Iron	0	0.6	0.03	0.093
Bicarbonate	24	172	84.42	43.88
Chloride	17.94	579.8	147.75	138.22
Nitrate	10	160	47.04	39.09
Fluoride	0.13	4.678	1.836	0.790
Sulphate	8	68	36.102	15.71
TDS	10	1020	543.89	255.06
Electrical conductivity	114	3420	1162.22	628.46
Total Hardness	78	444	256.46	82.90
pH	6.63	9.65	7.286	0.483

To assess the suitability of groundwater samples for drinking purpose, results of the physico-chemical analysis of groundwater samples were compared with Indian Standards (BIS 2012) prescribed for drinking water. The pH values of all the samples were confined within the permissible limits of BIS standards (BIS 2012), with an average value of 7.286. The electrical conductivity of water is a measure of the concentration of ionized substance in it. It may also be related to the problems like excessive hardness or mineral contamination. The values of EC varied from 114 to 3420 µS/cm. The values of total dissolved solids (TDS) in the samples varied from 10 to 1020 mg/l with an average value of 543.89 mg/l., Water with TDS more than 500 mg/l may cause gastrointestinal irritation and hence, not considered fit for drinking water supplies (BIS 2012). Bicarbonates value ranges from 24 to 172mg/l with an average of 84.42. Hardness of water is due to the presence of carbonates, sulphates and chlorides of calcium and magnesium. The concentration of Ca^{2+} in the samples varied from 2.14 to 387.02 mg/l with an average value of 122.69 mg/l. On the other hand, Mg^{2+} concentration varied from 0.19 to 65.71 mg/l with a mean value of 8.37mg/l. The total hardness varied in the range 78–444 mg/l with an average of 256.46mg/l. Based on total hardness (TH) classification, groundwater is classified into four categories, i.e. soft water (0–60 mg/l), moderately hard (61–120 mg/l), hard (121–180 mg/l) and very hard (> 180 mg/l). The majority (> 80%) of the samples falls in the category of hard water ranging from hard to very hard. Iron concentration ranges from 0 to 0.6mg/l with a mean value of 0.03mg/l. chloride content in water ranges from 17.94 to 579.8 mg/l. Nitrate value ranged from 10 to 160mg/l with an average value of 47.04mg/l and sulphate value ranged from 8 to 68 with a mean value of 36.1mg/l.

Globally, fluoride in groundwater is known to contaminate the water sources. In surface waters, fluoride can occur naturally by the deposition of atmospheric derived particles and /or by fluoride-containing soils and rocks weathering and in ground water resources by leaching of rocks and soils. The problem of high fluoride content in drinking water has become a serious environmental issue in the field of water quality management and human health. Fluoride ions have dual significance in water supplies. High concentration of F-causes dental fluorosis also called "mottled enamel" (disfigurement of the teeth). At the same time, a concentration less than 0.8 mg/l results in "dental caries". Hence, it is essential to maintain the F- concentration between 0.8 to 1.0 mg/l in drinking water. The Indian standard prescribes the limit for fluoride for water to be used for drinking as 1.5 mg/l. The concentration of Fluoride values ranges from 0.13 to 4.678mg/l with an average value of 1.836mg/l.

Table 3 Results of soil analysis.

Soil parameters	Min.	Max.	Avg.	SD
pH	7.02	8.47	7.47	0.384
CALCIUM	8.49	12.57	11.18	0.516
SULPHUR	4	41	12.71	8.48
POTASSIUM	2.23	13.54	5.98	3.48
OC (%)	0.24	1.36	0.734	0.332
IRON	0.3	2.2	0.806	0.516
COPPER	0.049	0.94	0.148	0.23
ZINC	0.053	0.349	0.095	0.077
MANGANESE	0.008	0.797	0.404	0.246
PHOSPHORUS	2.23	13.54	5.983	3.481
NITROGEN	18.94	277.7	130.91	70.26
EC	0.01	0.34	0.062	0.084

Analysis of soil is carried out to assess the role of soil contamination and its effect on altering the quality of groundwater. The pH values of the soil samples are in the range of 7.02 to 8.47 with an average value of 7.47.The electrical conductivity of water is a measure of the concentration of ionized substance in it. It may also be related to the problems like excessive hardness or mineral contamination. The values of EC varied from 0.01 to 0.34 μS/cm. The values of calcium in the samples varied from 8.49 to 12.57mg/kg with an average value of 11.18 mg/kg. Sulphur value ranges from 4 to 41mg/kg with an average of 12.71mg/kg. Organic content in the samples varied from 0.24 to 1.36 mg/kg with an average value of 0.734 mg/kg. The value of potassium varies from 2.23 to 13.54mg/kg. Available iron content in soil sample varies from 0.3 to 2.2mg/kg. The copper concentration in soil sample varies from 0.049 to 0.94 mg/kg. Zinc value varies from 0.053 to 0.349mg/kg with an average value of 0.095mg/kg. Manganese values vary from 0.008 to 0.797mg/kg. Phosphorous values vary from 2.23 to 13.54 mg/kg. Nitrogen values vary from 18.94 to 277.7mg/kg.

Table 4 Correlation matrix of water quality parameters

	F-	Ca	Mg	TH	pH	Fe	SO_4	EC	TDS	HCO_3	Cl	NO_3
F-	1											
Ca	-0.17	1.00										
Mg	0.04	-0.24	1.00									
TH	0.35	0.01	-0.04	1.00								
pH	-0.15	-0.14	0.09	-0.46	1.00							
Fe	-0.08	0.21	-0.13	-0.05	0.09	1.00						

Table 4 *Contd...*

	F-	Ca	Mg	TH	pH	Fe	SO_4	EC	TDS	HCO_3	Cl	NO_3
SO_4	0.04	0.00	0.11	0.45	-0.22	-0.13	1.00					
EC	0.11	-0.09	-0.04	0.60	-0.45	0.06	0.31	1.00				
TDS	0.19	-0.14	0.04	0.43	-0.19	0.15	0.52	0.91	1.00			
HCO_3	0.17	-0.14	-0.09	0.41	-0.20	0.13	0.21	0.25	0.33	1.00		
Cl	0.30	-0.09	-0.12	0.62	-0.40	0.06	0.21	0.74	0.18	0.33	1.00	
NO_3	0.02	0.04	-0.08	0.38	-0.20	-0.11	0.11	0.16	0.01	0.11	0.22	1.00

Correlation of fluoride concentration with other physico-chemical parameters.

In order to study relation between the fluoride concentration and other physio-chemical parameters, correlation studies were performed. The fluoride values showed positive correlation with Total hardness, TDS, EC, Bicarbonates, Chloride, Nitrate, Sulphate and Magnesium The principle ion contributing TDS are Carbonate, Bicarbonate, Chlorides, Fluorides, Sulphate, Nitrates, Sodium, Potassium, Calcium and Magnesium. Thus Fluoride value shows negative correlation with TDS (r = 0.19). In groundwater, hardness is formed mainly due to carbonate, bicarbonate, chloride, Sulphate, Nitrate of Ca+ and Mg+. Total hardness recorded positive correlation with fluoride (r = 0.35). An overall correlation analysis is almost a negative correlation between the concentration of fluoride ion with calcium, Iron and pH which shows the concentration of fluoride may reduce in the presence of high calcium content in the aquifer.

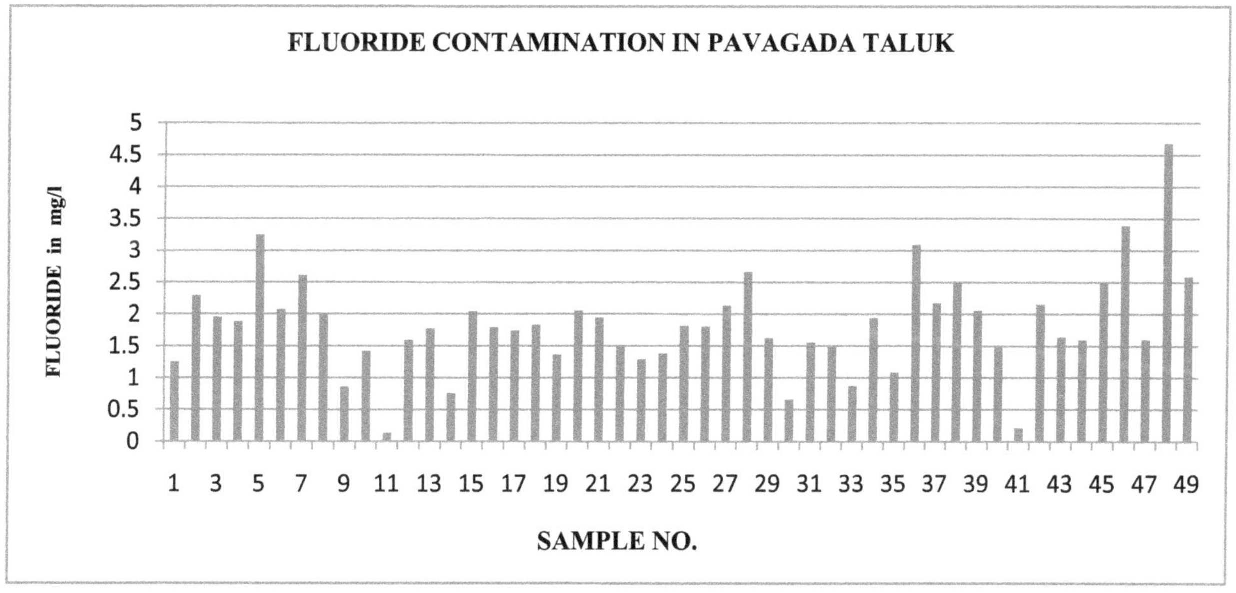

Figure 3 Concentration of Fluoride in collected water samples

All the groundwater samples collected in study area were clear without any, odour and turbidity. The data obtained after analysis of all the samples was compiled and presented in Table-2. The Maximum permissible limit of fluoride is 1.5 ppm. The fluoride concentration is minimum of 0.13 (Table-1) and maximum of 4.678 mg/l and 72.5% of the samples are exceeding the maximum permissible limit as per BIS-10500(2012).

Conclusion

The fluoride values showed positive correlation with Total hardness, TDS, EC, Bicarbonates, Chloride, Nitrate, Sulphate and Magnesium and is almost a negative correlation with calcium, Iron and pH. In the study area the fluoride concentration is exceeding the permissible limit in 37 groundwater samples out of 51 samples analyzed as per the limits prescribed by BIS. This study reveals that water from selected areas are not suitable for human consumption because of excessive limits of fluoride. Many cases of dental fluorosis among children and skeletal fluorosis in the aged person have been observed in our field visit. The reason for decreasing quality of water in this region may be due to over exploitation of groundwater for domestic and agricultural purposes and also fluoride containing minerals like Fluorite, mica, etc. present in rocks like granite and gneisses in Pavagada taluk. Hence, rapid and reliable monitoring measures are essential for keeping a close watch on water quality and health environment. The harvesting of rainwater, either directly in cisterns, tanks or by careful collection via small recharge dams, offers a potentially safe and attractive alternative solution in endemic areas and also by applying suitable artificial recharge technique to the contaminated tube wells reduces the concentration of Fluoride or else contaminated tube wells should be seal off Simultaneously, people must be provided with safe drinking water by installing reverse osmosis units.

References

Agrawal, V., A.K. Vaish and P. Vaish: Groundwater quality: Focus on fluoride and fluorosis in Rajasthan. Curr. Sci., 73, 743-746 (1997).

Arif M., Hussain J., Hussain I. and Kumar S. An Investigation of Fluoride Distribution in Ladnu Block of Nagaur District, Central Rajasthan World Applied Sciences Journal 26 (12): 1610-1616, 2013

BIS, Bureau of Indian Standards, Specifications for Drinking Water, IS: 10500, New Delhi, India, 1991.

Devi S.B. and Kamble R.K., Groundwater fluoride in East Imphal district of Manipur, IJEP, 2006,

Edmunds, W.M. and P.L. Smedley: Fluoride in natural waters occurrence, controls and health aspects. In: Medical geology. New York (Ed.: O. Selenus). Academic Press (2003).26(10), 885-891.

Gupta S., Mondal D. and Bardhan A., Geochemical provenance and spatial distribution of fluoride in Groundwater in parts of Raniganj coal field, West Bengal, India, Archives of Applied Science Research, 2012, 4 (1), 292-306.

Handa B.K., Geochemistry and genesis of fluoride containing ground water in India, Groundwater, 1975, 13(3), 275-281.

Muggler L., Fluoride in soil solution. An assessment of fluoride availability and uptake by plants in Fluoride contaminated soils, Term paper, 2009, 1-3.

Paramesha Naik D., Usha M. and Somashekar R.K., High fluoride content in groundwater-a case stud in Bagepalli, Kolar District, Karnataka State, Eco Environ & Cons., 2007, 13(2),303-305.

Pillai, K.S. and V.A. Stanley: Implications of fluoride - An endless uncertainty. J. Environ. Biol., 23, 81-87 (2002).

Trivedi, P.: Relationship between fluoride, total alkalinity, total hardness in ground water of pali district in arid and semi - Arid region of western Rajasthan. Proc. Nat. Acad. Sci. India, 58, 7-11 (1988).

UNICEF: States of the art report on the extent of fluoride in drinking water and the resulting endemicity in India. Report by Fluorosis and Rural Development Foundation for UNICEF, New Delhi (1999).

WHO, World Health Organization, Guidelines for drinking water quality, Geneva, 1993.

WHO, Fluorine and Fluorides, 1984, WHO Environmental Health Criteria. 36 Geneva WHO Guidelines for Drinking Water Quality, Vol. 1 (2nd Edition), Geneva.

Groundwater Quality and its Suitability for Irrigation and Domestic Purposes: A Case Study in Dodda Timmayana Palya, Tumakuru District, Karnataka

Kannika. R[1], Prajwal T P[2], Nandeesha[3] and Vishal. R. Khandagale[4]

[1,2]UG student, [3]Professor, [4] Research scholar,

Civil engineering dept. Siddaganga Institute of Technology, Tumkur Karnataka, India.

prajwal.t.p.10@gmail.com, nandeeshsit@gmail.com, vishal.rk28@gmail.com

ABSTRACT

In this study we evaluated some significant physio-chemical parameters of groundwater of Dodda timmayana playa to assess the quality of water for domestic and irrigation purposes. Twenty groundwater samples are collected from different drinking water sources and each water sample was analysed to determine physico-chemical parameters as per IS standards. The investigated physico-chemical parameters are: pH, Total Hardness (TH). The analysed major ions were: Sodium (Na^+), Iron(Fe^+) , Potassium (K^+), Magnesium (Mg^{2+}), Calcium (Ca^+), Fluoride(F^-), Chloride (Cl^-), Nitrate (NO_3^-), Sulphate (SO_4^{2-}). The results of all the samples are analyzed and compared with the Bureau of Indian standards (B.I.S). For the analysis of Soil some of the parameters like pH, Calcium, Sulphate, Potassium and Organic content are analyzed to understand the influence of soil contamination on quality of groundwater. Overall, the groundwater in the study area is found to be slightly acidic to basic (pH ranging from 6.93 to 8.26) and hard to very hard (Total Hardness ranging from 204 mg/l and 388 mg/l) in nature. Water quality indices are generally used as a tool to convert a large data set into a much reduced and informative form. Water quality index (WQI) by weighing arithmetic index method is used to assess the suitability for drinking and irrigation purpose. The results are analyzed in the light of USSL diagram and Wilcox diagram for better understanding. The results of physico-chemical analysis and calculated water quality parameters suggest that all water samples are suitable for irrigation purposes.

Keywords: Dodda timmayana playa, groundwater quality, Water quality indices, Piper trilinear

Introduction

Three fourth of the earth's surface is covered by water. Inspite of this apparent abundance of water, less than one percent is available for human use in the form of surface water as 97 percent is contained in oceans etc. and 2 percent is locked up in ice-caps and glaciers. However as civilization and population increases man recognizes the importance of water from a quantity view point for agriculture, transportation, drinking and domestic usage with less significance given to its chemical and biological importance. Ground water contamination is nearly always the result of human activity. In areas where population density is high and human use of the land is intensive, ground water is especially vulnerable. Virtually any activity whereby chemicals or wastes may be released to the environment, either intentionally or accidentally, has the potential to pollute ground water. When ground water becomes contaminated by the dissolved elements and gases and by presence of suspended solids, bacteria, and viruses, it is difficult and expensive to clean up. Such water is no longer fit for a specific use, such as drinking, the water is said to be contaminated. If the water becomes heavily contaminated it is said to be polluted. The source of about 90% of drinking and irrigation water is from groundwater resources in the study area and exploited through shallow wells, borehole/drilled wells, hand pump operated wells. Hence a need for routine groundwater quality assessment is required .The aim of the study was to investigate the quality of ground water by determining water quality index and classify the groundwater for drinking/domestic and irrigation purposes. Generally speaking, the presence of water with high purity in the nature is rare, even the rain water contains some dissolved gases, small particles of the soil and bacteria that are suspended in the air. In addition, after the rain water has touched the surface of the earth it will wash and carry along some organic and inorganic pollutants to groundwater. Water quality index is one of the most effective tools that helps in communicating information on the quality of water to the concerned citizens and policy makers (especially governments at all levels). It thus becomes an important parameter for the assessment and management of groundwater. WQI is defined as a rating, reflecting the composite influence of different water quality parameters. WQI is calculated from the point of view of the suitability of groundwater for human consumption. Also, important irrigation parameters such as percentage sodium (%Na), Sodium Absorption Ratio (SAR) and explained with the help of USSL and Wilcox plot so as to have comprehensive understanding of the suitability of groundwater in the study area for human consumption and irrigation purpose based on computed, groundwater characteristics, quality assessment and water quality index values.

Description of the Study Area

The study area is in Tumakuru district, Karnataka, India. Location map of study area is shown in fig.1. The study area is having an area of 150 acres consisting of 95 houses and 350 people. The main crops are Ragi, Paddy, Jower, Tordal and Coconut. The fertilizers used are organic fertilizers, urea, 20-20 and DPA. An effort has been made to survey the study area and analyze the quality of water by sampling and presenting the results in an interesting so that the need for reforms is highlighted.

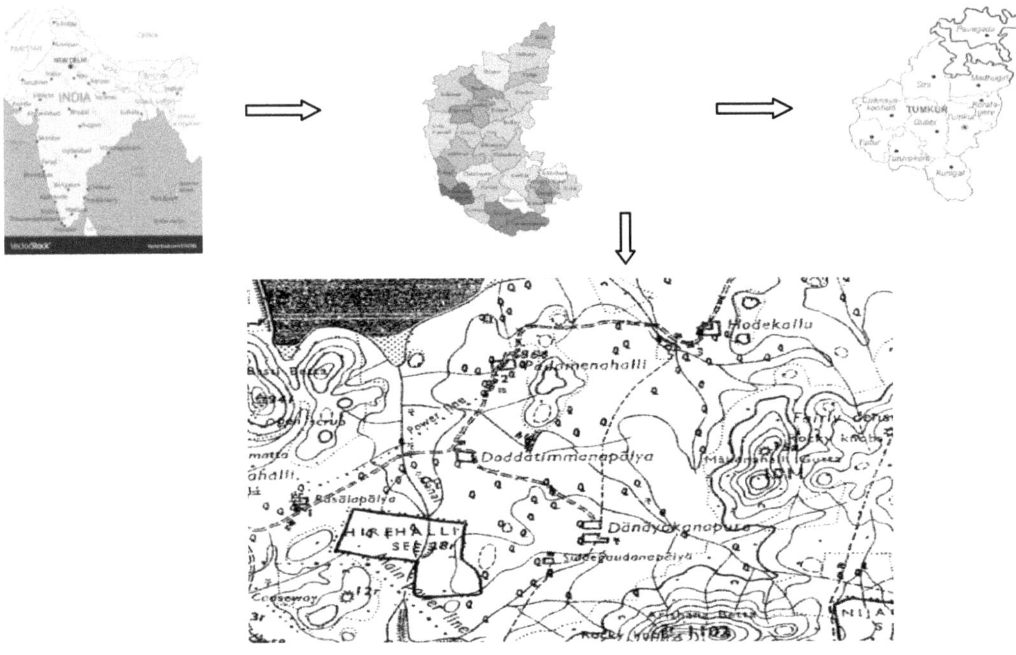

Figure 1 Location Map of Study Area.

Methodology

Samples of groundwater are collected from bore wells & hand pumps ,during the pumping of water is taking place in a 1-liter plastic can and they were brought to the laboratory. Total of 20 groundwater samples were collected from various places of study area, Samples were drawn with a precleaned plastic polyethylene bottle. Prior to sampling, all the sampling containers were washed and rinsed thoroughly with then groundwater. All the parameters were analyzed as per BIS procedure. Soil samples from fifteen locations were collected in polythene bags and analyzed for pH, EC, organic content, Sulphur content, available Phosphorous, Potassium, Calcium, Magnesium, nitrogen and micronutrients like Copper, zinc, Iron and manganese.

Water quality index is one of the most effective tools to monitor the surface as well as groundwater pollution and can be used efficiently in the implementation of water quality upgrading programmes. The objective of an index is to turn multifaceted water quality data into simple information that is comprehensible and useable by the public. Water quality index provide information on a rating scale from zero to hundred. Higher value of WQI indicates better quality of water and lower value shows poor water quality. The concentration of various ions as obtained from chemical analysis of groundwater samples were converted to milli equivalent/litre (meq/l) and used to derive certain parameters and graphical representations. They are Sodium Adsorption Ratio (SAR), percentage sodium (%Na) with the help of these parameters U.S.Salinity diagram, Wilcox diagrams is plotted to determine the suitability of groundwater for agricultural purposes.

Results and Discussion

The understanding of groundwater quality is important because it is the main factor which decides its suitability for domestic, agricultural and industrial purposes. The results of the analysis are compared with Indian standards (BIS) prescribed for drinking water in Tables 1.

Table 1 Summary statistics of different water quality parameters

Parameter	Min (mg/l)	Max (mg/l)	Mean (mg/l)	SD (mg/l)
Calcium	32.4	109.01	67.62	23.43
Chloride	27.99	129.95	73.02	30.53
Magnesium	23.22	77.14	41.34	13.89
Sodium	1.15	59.39	23.84	19.47
Potassium	1.87	5.68	3.25	1.02
Iron	0.01	0.12	0.0445	0.0339
Nitrate	0	4.3	0.45	0.93
Fluoride	0.02	0.35	0.179	0.081
Sulphate	10	71	37.5	12.29
TDS	143	361	226.3	62.78
Total hardness	204	388	282.6	56.36
pH	6.93	8.26	7.72	0.427
EC (µs/cm)	222	472	369.55	69.24

The values of pH in the groundwater samples collected from the study area varied from 6.93 to 8.26, indicating a slightly acidic to slightly basic nature. The Electrical Conductivity (EC) of groundwater in the study area varies widely and ranges between 222 and 472 µS/cm and all the samples are within the permissible limit of 2000 µS/cm. The TDS values varied between 143 and 361 mg/l and all the samples are within the permissible limit of 2000 mg/l. The total hardness (as CaCO3) values range between 204 and 388 mg/l and all the samples are within the permissible limit of 600 mg/l. The concentration of Sodium and Potassium ranged from 1.15to 59.39 and 1.87 to 5.68 mg/l, respectively. The concentration of Calcium and Magnesium was in the range of 32.4 to 109.01 and 23.22 to 77.14 mg/l, respectively. Out of 20 samples 10 samples and 15 samples was having higher Calcium and Magnesium content in comparison to their BIS permissible limit of 200 and 100 mg/l, respectively. In the area of investigation, the Chlorides are in the range of 27.99 to 129.95mg/l and it was found that all the samples was having Chloride values are within the permissible limit of 1000 mg/L. The Sulfate content value varies from 10 to 71 mg/l, all the samples are well within the permissible limit of 400 mg/l. The Nitrate concentration in the region ranges from 0 to 4.3 mg/l all the samples are below the permissible limit of 45 mg/l. Fluoride content values ranges from 0.02 to 0.35mg/l and all the samples are within the permissible limit of 1.5mg/l. Iron value rages from 0.01 to 0.12mg/l and 10samples are above permissible limit of 0.3mg/l.

Table 2 Results of soil analysis

Soil parameters	Min.	Max.	Avg.	SD
pH	7.23	8.61	7.78	0.43
CALCIUM	107.4	172.6	135.5	18.36
SULPHUR	1.98	21.45	7.813	4.34
POTASSIUM	51.46	115.31	84.64	15.85
OC (%)	0.12	1.77	0.79	0.39
IRON	1	5.025	2.0288	1.1358
COPPER	0.2	0.7	0.3536	0.1087
ZINC	1.25	7.925	2.97	1.82
MANGANESE	2.58	13.06	5.07	2.2
PHOSPHORUS	0.98	9.182	4.8	2.3
NITROGEN	19.44	286.7	128.29	64.4
EC	0.01	0.24	0.0547	0.0591

Analysis of soil is carried out to assess the role of soil contamination and its effect on altering the quality of groundwater. The pH values of the soil samples are in the range of 7.23 to 8.61 with an average value of 7.78.The electrical conductivity of water is a measure of the concentration of ionized substance in it. It may also be related to the problems like excessive hardness or mineral contamination. The values of EC varied from 0.01 to 0.24 µS/cm. The values of calcium in the samples varied from 107.4 to 172.6 mg/kg with an average value of 135.5 mg/kg. Sulphur value ranges from 1.98 to 21.45mg/kg with an average of 7.813mg/kg. Organic content in the samples varied from 0.12 to 1.77 mg/kg with an average value of 0.79 mg/kg. The value of potassium varies from 51.46 to 115.31mg/kg. Available iron content in soil sample varies from 1 to 5.025 mg/kg. The copper concentration in soil sample varies from 0.2 to 0.7 mg/kg. Zinc value varies from 1.25 to 7.925 mg/kg with an average value of 2.97 mg/kg. Manganese values vary from 2.58 to 13.06 mg/kg. Phosphorous values vary from 0.98 to 9.182 mg/kg. Nitrogen values varies from 19.44 to 286.7mg/kg.

Water Quality Index Analysis

The WQI values of the study area of various samples are calculated separately. WQI has been calculated based on fourteen selected hydro chemical parameters given in Table below.

Table 3 Water quality classification based on WQI value

Sl. No.	WQI value	Status	% of samples
1	0-25	Excellent	80
2	26-50	Good	20
3	51-75	Fair	0
4	76-100	Poor	0
5	101-150	Very poor	0
6	Above 150	Unfit for drinking	0

The distribution pattern of water types based on WQI indicates that drinking quality of water in the study area is categorized under ''Excellent'' & "Good" category as shown in the above table.

For irrigation purpose, percentage sodium (%Na), Sodium Absorption Ratio (SAR), were determined and rated according to standard. The values of SAR and sodium percentage are calculated for all the samples by the following equation given by.

$$SAR = (Na^{+} / (\sqrt{(Ca^{2+} + Mg^{2+})/2}))$$

$$\% Na = ((Na^{+} + K^{+}) X 100) / (Ca^{2+} + Mg^{2+} + Na^{+} + K^{+})$$

where the concentrations of all the cations, i.e. Na+,K+, Ca2+ and Mg2+, are taken in meq/l.

In the present study the SAR values range from 0.03 to 1.49 with an average value of 0.58.The percentage of Sodium values range from 2.67 to 26.57 with an average value of 13.33.

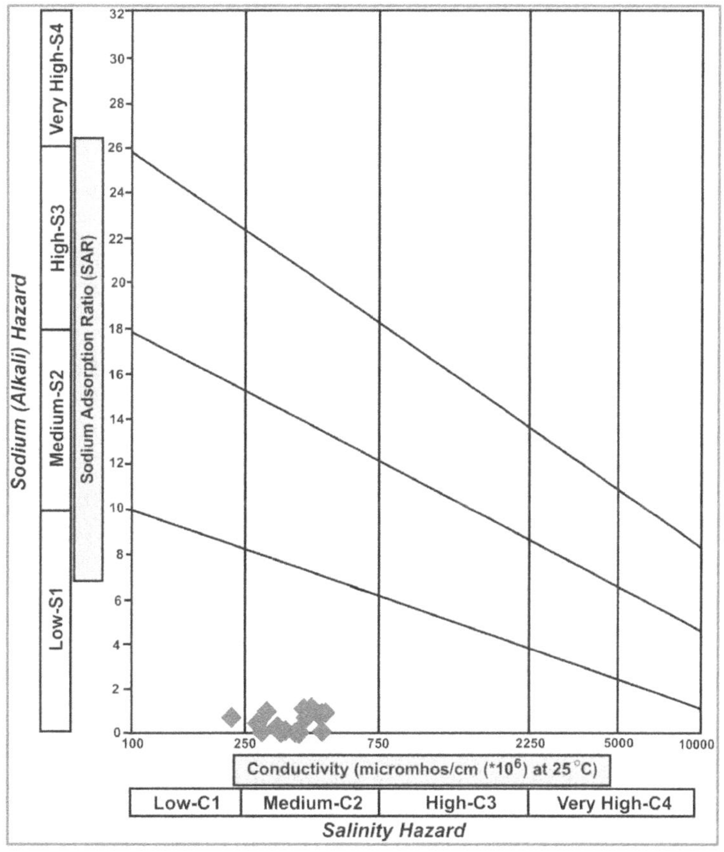

Figure 2 Salinity classification of irrigation water samples using USSL (1954) diagram

This diagram is used in interpreting the analysis for irrigation and the water can be grouped into 16 classes. It uses Sodium Absorption Ratio (SAR) in vertical axis and conductance along horizontal axis (Fig. 5). All concentration values are expressed in equivalents per million. Salinity, sodicity and toxicity generally, need to be considered for evaluation of the suitability of groundwater for irrigation (Todd, 1980, Shainberg and oster, 1976). Sodium absorption ratio is also used to determine the suitability of groundwater for irrigation as it gives a measure of alkali/sodium hazard to crops. If calcium and magnesium are dominant, the hazard is low. As per the U.S. Salinity diagram the water samples fall in the C2-S1 and C1-S1 classes, and hence groundwater in the study area is suitable for irrigation.

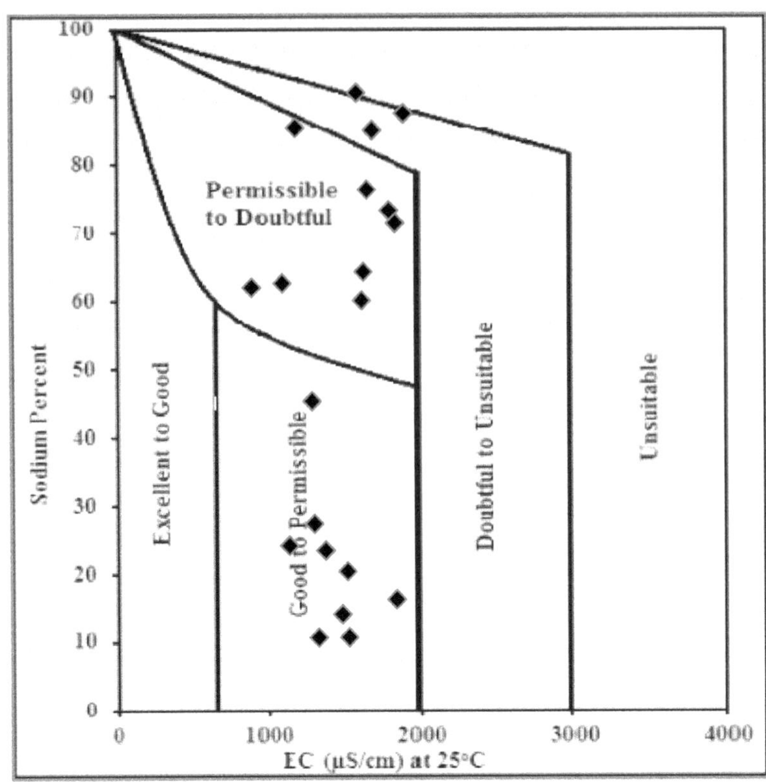

Figure 3 Wilcox's diagram for irrigation water classification

The chemical quality of groundwater samples was studied by plotting analytical data (Wilcox 1948) relating EC and sodium percent shown in above figure is used for classification of irrigation waters. X-axis represented by electrical conductivity and Y-axis represented by soluble sodium percentage. Most of the samples are in "good to permissible" & "permissible to doubtful" category, 3 samples are in "Doubtful to unstable" category".

Conclusion

The study provides significant information on the groundwater quality in Dodda timmayanapalya of Tumkur district. The groundwater in the study area is found to be slightly acidic to basic (pH ranging from 6.93 to 8.26) and hard to very hard (Total Hardness ranging from 204 mg/l and 388 mg/l) in nature.WQI indicates many of the samples are in the range of Excellent to Good category hence collected groundwater samples are suitable for drinking. Based on the classification of irrigation water according to SAR and % Sodium values, all the sample locations are suitable for irrigation purposes. Sodium Adsorption Ratio (SAR) values are categorized as 'Excellent' as per U.S. Salinity diagram collected groundwater samples fall in C1-S1 and C2-S1 classes, and hence can be considered as suitable for irrigation. Based on the Wilcox classification, 15% of the water samples belong to doubtful to unsuitable category for irrigation use. Hence, it is suggested that suitable measures in terms of enhancement of drainage has to be made in areas were high Sodium content is observed for satisfactory crop growth.

References

A. Dhafer, N. Al-Jassim, T. Kenda and P. Hong, "Assessing the Groundwater Quality at a Saudi Arabian Agricultural Site and the Occurrence of Opportunistic Pathogens on Irrigated Food Produce." *Int. J. Environ. Res. Public Health* 2015, *12*, 12391-12411; doi: 10.3390/ijerph121012391.

BIS, Bureau of Indian Standards, Specifications for Drinking Water, IS: 10500, New Delhi, India, 2012.

BIS (1998) Drinking Water Specifications (Revised 2003). IS:10500, Bureau of Indian Standards, New Delhi.

Brown E., Skougstad M.W. and Fishman M.J., Method for collection and analysis of water sample for dissolved minerals and gases, US Department of Interior, 1974, Book No. 5.

G. R. Kalpana, D. P. Nagarajappa, K. M. Sham Sundar, B. Suresh, "Determination of Groundwater Quality Index in Vidyanagar, Davanagere City, Karnataka State, India" *International Journal of Engineering and Innovative Technology (IJEIT)*, Volume 3, Issue 12, June 2014.

J. Delleur, "The Handbook of Groundwater Engineering". CRC Press LLC, USA, 1999.

S. K. Dhameja, "Environmental Studies" S. K. Kataria & Sons, New Delhi. First edition, 2004, 403pp.

Nagaraju, A., Muralidhar, P. and Sreedhar, Y. (2016) Hydrogeochemistry and Ground water Quality Assessment of Rapur Area, Andhra Pradesh, South India. Journal of Geoscience and Environment Protection, 4, 88-99.

O. M. Omorogieva, O. I. Imasuen, M. I. Isikhueme, O. A. Ehinlaye, B. Anegbe 2 and M. O. Ikponmwen Hydrogeology and Water Quality Assessment (WQA) of Ikhueniro and Okhuahe Using Water Quality Index (WQI) *Journal of Geography, Environment and Earth Science International 6(3): 1-10, 2016; Article no. JGEESI. 25615.*

S. Singh, N. J. Raju, Ch. Ramakrishna, "Evaluation of Groundwater Quality and Its Suitability for Domestic and Irrigation Use in Parts of the Chandauli-Varanasi Region", Uttar Pradesh, India". *Journal of Water Resource and Protection,* 2015, 7, 572-587.

Z. T. Zewdu, "Ground Water Quality Determination of former Lake Haramaya, Haramaya District, Eastern Haranghe Zone, Oroma Regional State, Ethiopia." *J. Appl. Sci. Environ. Manage. Sept.,* 2012, Vol. 16 (3) 245–252.

Extreme Rainfall Indices and Climate Change of Vadodara City

Chirayu Pandit[1], Sanskriti Mujumdar[2] and Akash Patel[3]
[1]*Lecturer,* [2]*Associate Professor,* [3]*Research Assistant,*
Department of Civil Engineering, Faculty of Technology & Engineering,
The M.S.University of Baroda, Vadodara
pandit.chirayu-polyced@msubaroda.ac.in, sanskritimujumdar@yahoo.co.in,akash.patel-ced@msubaroda.ac.in

Theme: Effect of Climate change on Water, Environment and its consequences

ABSTRACT

Climate can be defined from average of long term meteorological parameters of the region but to understand climate change extreme values are more helpful. World Meteorological Organisation has identified indices related to extreme rainfall which have impacts on health, agriculture and food security, water resources and hydrology sectors. Vadodara (22.28 N, 73.19E) is the third largest city of Gujarat and eighteenth largest city of India with urban population greater than 1.6 million. Daily rainfall data for Vadodara station from 1961-2016 is collected from State Water Data Centre, Gandhinagar. Annual trends for heavy precipitation days (R10mm, R20mm), extreme precipitation days (R95P), Contribution of extreme precipitation in annual rainfall (R95PT) annual total wet-day precipitation (PRCPTOT), consecutive dry days (CDD), consecutive wet days (CWD), one day maximum rainfall (Rx1), maximum consecutive 5-day precipitation (Rx5) and Standardized Daily Intensity Index (SDII) for the study period were calculated with Mann Kendall test. Generally increasing trend is observed for all indices except CWD, CDD and R95p where no trend is observed. Observed annual magnitudes of one day & five days maximum rainfall (Rx1day and Rx5days) are increasing significantly (p value < 0.05). Daily rainfall intensity (SDII) also shows significant increasing trend. All the indices indicate that high rainfall with shorter duration is increasing for Vadodara city. The observed trends of extreme indices confirm the impact of climate change on rainfall pattern of Vadodara.

Introduction

Cities concentrate wealth, people, and productivity, but they also concentrate vulnerability to natural disasters and to long-term changes in climate. Rising sea levels will affect millions of people living in coastal cities. Similarly, migration, changes in land use, and spatial development are likely to increase the vulnerability of populations to changes in weather and climatic conditions. Adaptation to climate change is therefore an imperative for cities, as it is for the world at large. The urgency of this challenge is also evident when considering the massive investments in buildings and infrastructure that cities in developing countries will undertake in the coming years, which will lock in urban form and structure for many decades thereafter. (Hoornweg, et al. 2011)

Synthesis report of IPCC (Pachauri, et al. 2014) concluded that the changes in many extreme weather and climate events have been observed since 1950. Some of these changes have been linked to human influences, including a decrease in cold temperature extremes, an increase in warm temperature extremes, an increase in extreme high sea levels and an increase in the number of heavy precipitation events in a number of regions. Standardization of extremity of the weather event can help in uniform assessment of climate change. To quantitatively reflect an extreme precipitation event, it should be complemented by characterization with the metrics like magnitude, duration, severity and extent. (L. V. Alexander 2015)

Study of extreme rainfall during south west monsoon all over India (Pattanaik and Rajeevan 2010) during 1951-2005 observed that the occurrence of extreme rainfall events (rainfall >124.4 mm in a day) over India during the southwest monsoon season shows spatial variability with preferred regions of occurrence over the entire west coast of India, parts of central India and parts of northeast India also the frequency of extreme rainfall shows a significant increasing trend over the Indian monsoon region 129 rain gauge stations across India were analysed over the period of 1910 to 2000 by (Sen Roy and Balling 2004). annual time series of seven different indices of extreme precipitation events, including total precipitation, largest 1, 5, and 30 day totals, and the number of daily events above the amount that marks the 90th, 95th, and 97.5th percentiles of all precipitation at each station. The analysis concluded that there was increase in the frequency of extreme precipitation events over the study period.

Study area

Vadodara (22.28 N, 73.19E), the third largest city of Gujarat and eighteenth largest city of India, is situated at Altitude 35.5 m. above mean sea level on the banks of river vishvamitri in Gujarat state. (Figure 1) The topography of the city is relatively flat, with the ground mildly sloping from northeast to southwest. Area of Vadodara city is around 159.95 Sq.Kms having population of 16.6 lakh while Vadodara district is spread over 7794 Sq. Kms Area with 36.5 lakh population. Entire region falls under semi-arid climate. Climate of the Vadodara city is characterized by hot and aside from Southwest Monsoon season Dry throughout the year. Average annual Rainfall is 806 mm. Average annual Rainy days are 37. Mean monthly highest rainfall observed in July is 267 mm. Mean monthly lowest rainfall observed in February is 1 mm. Ever recorded heaviest rainfall in 24 Hrs. is 297 mm on 01 July 2005. Rainfall during the southwest monsoon season is about 94% of Annual rainfall.

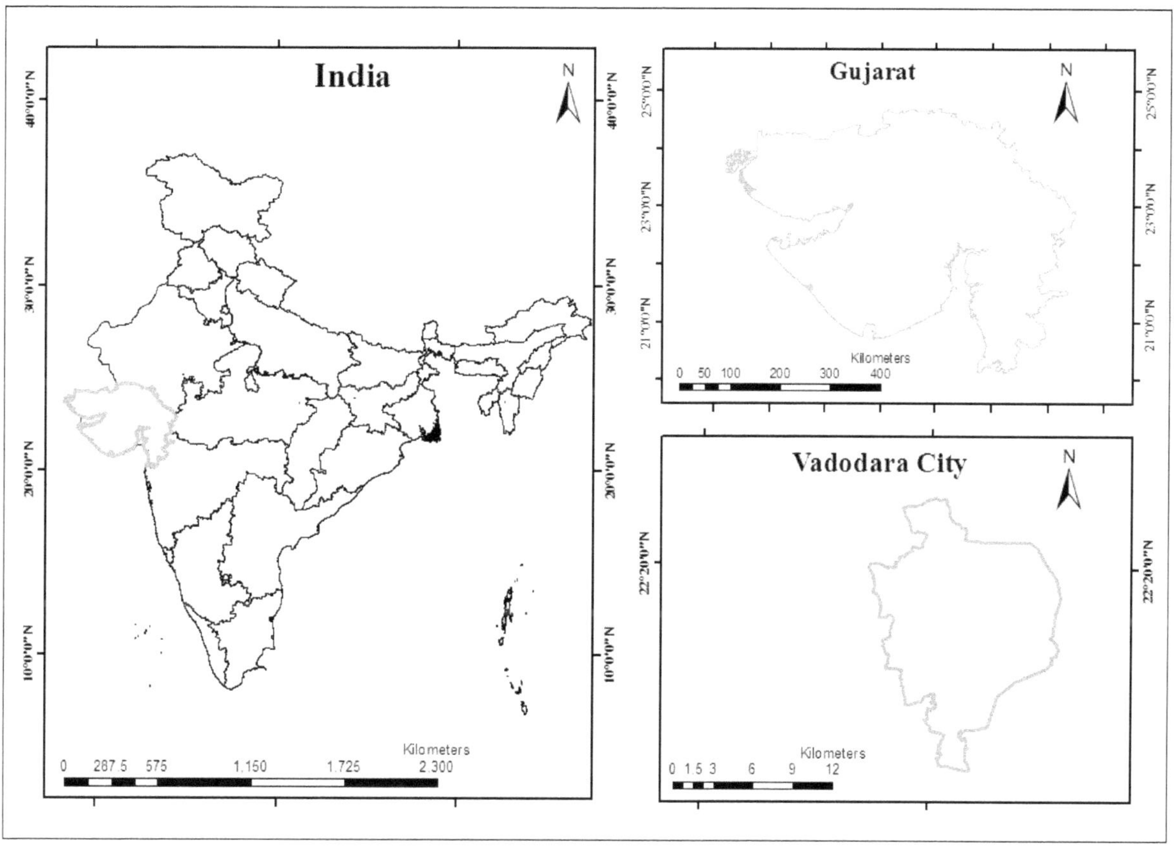

Figure 1 Study Area Vadodara City

Data

It may be noted that for extreme rainfall analysis for studying the behavior of changes in the extreme events using of real or actual station data is more realistic than the gridded data as in the later case extreme events will be missed in most of the occasions due to interpolation or averaging scheme used in gridding. This even mislead regarding the signals of hydrological extremes for better disaster management. (Guhathakurta, Menon, et al. 2010) Data of daily rainfall of Vadodara station for data length 1961-2016 is collected from State Water Data Centre (SWDC) Gandhinagar.

Method

The Expert Team on Sector-specific Climate Indices (ET-SCI) indices related to daily precipitation characteristics are analyzed in the present study. Climate indices provide valuable information contained in daily data, without the need to transmit the daily data itself. For this reason indices derived from daily data are an attempt to objectively extract information from daily weather observations to answer questions concerning aspects of the climate system that affect many human and natural systems with particular emphasis on extremes (Alexander and Herold 2015). Indices related to precipitation and analyzed for Vadodara are described in Table 1.

Table 1 ET-SCI extreme precipitation indices

Code	Indicator name	Description	Unit
R10	Number of heavy precipitation days	Annual count of days when rainfall ≥ 10 mm	Days
R20	Number of very heavy precipitation days	Annual count of days when rainfall ≥ 20 mm	Days
RX1day	Maximum 1-day precipitation amount	Annual maximum 1-day precipitation	mm
Rx5day	Maximum 5-day precipitation amount	Annual maximum consecutive 5-day precipitation	mm
R95p	Very wet days	Annual total PRCP when daily rainfall > 95th percentile	mm
PRCPTOT	Annual total wet-day precipitation	Annual total precipitation in wet days (daily rainfall ≥ 1 mm)	mm
SDII	Simple daily intensity index	Annual total precipitation divided by the number of wet days (defined as daily rainfall ≥ 1.0 mm) in the year	mm day −1
R95PT	Annual contribution from very wet days	(R95p/PRCPTOT) × 100	%
CDD	Consecutive dry days	Maximum number of consecutive days with rainfall < 1 mm	Days
CWD	Consecutive wet days	Maximum number of consecutive days with rainfall ≥ 1 mm	Days

Trend Analysis

Temporal trend determination is an important part of data analysis. Trend analysis will indicate whether sample values are increasing or decreasing over time. In addition, an estimate of the trend's magnitude can help to determine whether a statistically significant trend is of practical concern. Trend over the study period of all indices were calculated by Mann kendall Test.

Mann Kendall Test

Mann (1945) presented nonparametric test for randomness against time which constitutes a particular application of Kendall's test for correlation (1975) known as Mann kendall or Kendall t – test. Non-parametric Mann-Kendall method is widely used to analyze the trends of the environmental time series, which is recommended by World Meteorological Organization (WMO). In the Mann-Kendall test, missing values are allowed and the data need not conform to any particular distribution. In the computation of this statistical test the normal approximation (Z statistics). For time series with 10 or more data points the normal approximation is used. The number of annual values in the studied data series is denoted by n. Missing values are allowed and n can thus be smaller than the number of years in the studied time series.

$$S = \sum_{i=2}^{n} \sum_{j=1}^{i-1} sign(x_i - x_j) \quad .....(1)$$

where sign(xi- xj) is

-1 for xi- xj <0

0 for xi- xj =0

1 for xi- xj >0

First, the variance of S is computed by the following equation, which takes into account that ties may be present:

$$VAR(S) = \frac{1}{18}\left[n(n-1)(2n+5) - \sum_{p-1}^{q} t_p\left(t_p - 1\right)\left(2t_p + 5\right)\right] \qquad .....(2)$$

Here q is the number of tied groups and tp is the number of data values in the pth group. The values of S and Var(S) are used to compute the test statistic Z as follows

$$Z = \begin{Bmatrix} \frac{(S-1)}{\sqrt{Var(S)}} & S > 0 \\ 0 \quad if\ S = 0 & \\ \frac{(S+1)}{\sqrt{Var(S)}} & S < 0 \end{Bmatrix} \qquad .....(3)$$

The presence of a statistically significant trend is evaluated using the Z value. A positive value of Z indicates an upward trend while negative value denote downward trend. The statistic Z has a normal distribution. Sens's slope method is used to derive magnitude of the slope.

Analysis and Results

Absolute indices represented maximum or minimum values of weather parameters within a season or year. The absolute precipitation indices are maximum 1-day precipitation amount (RX1day), maximum 5-day precipitation amount (RX5day) for a year. The result shows significant (0.05 level) positive trend for Rx1day and Rx5days with p value 0.046 and 0.008 respectively. Sen's slope indicates that there is an annual rise of 0.92 mm and 1.74 mm in Rx1day and Rx5days respectively. (Figure 2 c,d)

Threshold indices are defined as the number of days at which temperature or precipitation value falls above or below a fixed threshold number of heavy precipitation days >10 mm (R10), very heavy precipitation days > 20 mm (R20) and extremely wet days when rainfall is above 95 percentile rainfall for Vadodara the threshold value is 74 mm. Positive non significant trend is observed for R10 with annual rise of 0.093 days (Figure 2-a). Similarly very heavy precipitation days (R20) increases non significantly with the rate of 0.077 days per year (Figure 2-b). No trend is observed for extremely wet days. (Figure 2-e)

Table 2 Extreme Rainfall Indices for Vadodara City

Indices Code	Kendall's tau	p-value	Sen's slope
R10 (Days)	0.135	0.152	0.093
R20 (Days)	0.143	0.132	0.077
Rx1day (mm)	0.185	0.046	0.929
Rx5days (mm)	0.246	0.008	1.744
R95p (Days)	0.100	0.312	0.000
R95PT (%)	0.107	0.249	0.167
PRCPTOT (mm)	0.114	0.216	4.021
SDII (mm/day)	0.304	0.001	0.133
CDD (Days)	0.034	0.721	0.000
CWD (Days)	-0.060	0.530	0.000

Duration indices consist of consecutive dry days (CDD) and consecutive wet days (CWD). The CDD index is the length of the longest dry spell in a year while the CWD index is defined as the longest wet spell in a year; No increasing or decreasing trend is observed for CWD and CDD. (Figure 2- i, j)

Other indices include indices of annual precipitation total (PRCPTOT), Contribution of Extreme rainfall in total rainfall (R95PT) and simple daily intensity index (SDII). Though no trend is observed in extremely wet days, the contribution of extreme rainfall is increasing non significantly at the rate of 0.16% annually. (Figure 2-f) Annual total rainfall (PRCTOT) shows non significant increasing trend (Figure 2-g) at the rate of 4.02 mm /year. Simple daily intensity index (SDII) observed significant increasing trend with p value 0.001. SDII increases 0.133 mm/day per year. Calculated Kendall's tau, p value and Sen's slope for all indices can be referred from Table 1Table 2. The total rainfall, contribution of extreme rainfall and daily intensity increasing over the study period of 56 years which point out towards the climate change at local level.

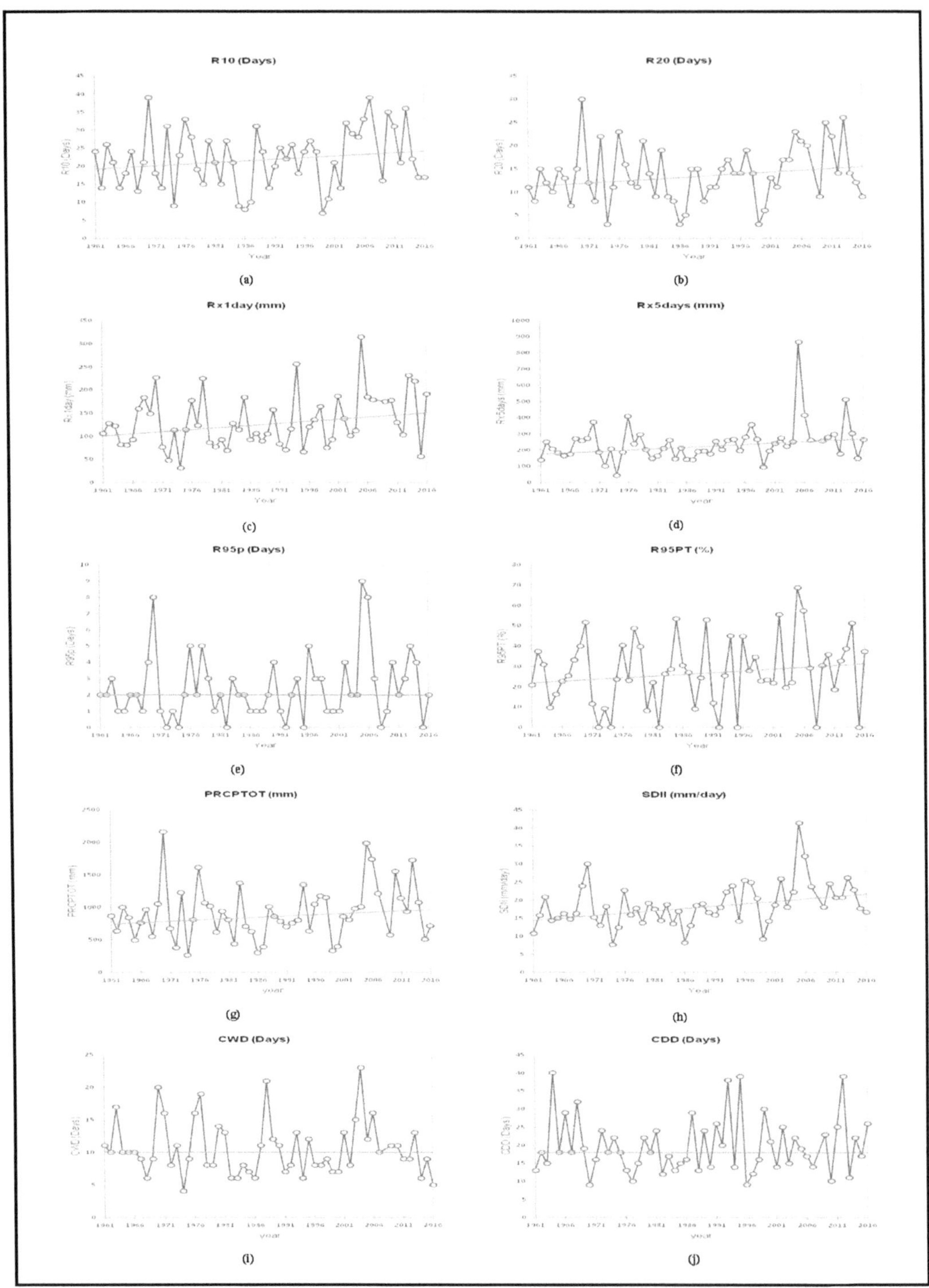

Figure 2 Extreme Rainfall Indice Trends for Vadodara

Conclusion

Extreme daily precipitation characteristics of Vadodara city are explored through the indices suggested by the Expert Team on Sector-specific Climate Indices (ET-SCI). Trend for each index was found with non parametric Mann kendall test and Sen's slope method. Annual rainy days with heavy rainfall (R10 & R20) are increasing non significantly, no trend is observed for rainy days with extreme rainfall (R95). Consecutive dry days (CDD) show decreasing non significant trend while no trend is observed for consecutive wed days (CWD). Total annual rainfall (PRCPTOT) shows non significant increasing trend, similarly contribution of extreme rainfall in total annual rainfall (R95PT) is also showing increasing trend. Observed annual magnitudes of one day & five days maximum rainfall (Rx1day and Rx5days) are increasing significantly. Daily rainfall intensity (SDII) also shows significant increasing trend. All the indices indicate that high rainfall with shorter duration is increasing for Vadodara city. The observed trends of extreme indices confirm the impact of climate change on rainfall pattern of Vadodara. Policy makers and town planners should take this impact of climate change in to consideration while planning for flood mitigation, rain water harvesting and disaster management.

References

Alexander, L., and N. Herold. "ClimPACTv2 Indices and Software. A document prepared on behalf of the Commission for Climatology (CCl) Expert Team on Sector-Specific Climate Indices (ET-SCI)." 2015.

Alexander, Lisa V. "and Climate Extremes, Weather and Climate Extremes." 2015.

Chisanga, Charles Bwalya, Elijah Phiri, and Vernon R. N. Chinene. "Trends of extreme events in precipitation and temperature during the 1963-2012 period at Mt Makulu, Zambia." J Sci Res Rep 15 (2017): 1-19.

Data, Climate. "Guidelines on analysis of extremes in a changing climate in support of informed decisions for adaptation." World Meteorological Organization, 2009.

Guhathakurta, P., and M. Rajeevan. "Trends in the rainfall pattern over India." International Journal of Climatology (Wiley Online Library) 28 (2008): 1453-1469.

Guhathakurta, P., Preetha Menon, A. B. Mazumdar, and O. P. Sreejith. "Changes in extreme rainfall events and flood risk in India during the last century." National Climatic Centre, RR, 2010.

Hoornweg, Daniel, Mila Freire, Marcus J. Lee, Perinaz Bhada-Tata, and Belinda Yuen. Cities and climate change: responding to an urgent agenda. The World Bank, 2011.

Pachauri, Rajendra K., et al. Climate change 2014: synthesis report. Contribution of Working Groups I, II and III to the fifth assessment report of the Intergovernmental Panel on Climate Change. IPCC, 2014.

Pattanaik, D. R., and M. Rajeevan. "Variability of extreme rainfall events over India during southwest monsoon season." Meteorological Applications (Wiley Online Library) 17 (2010): 88-104.

Sen Roy, Shouraseni, and Robert C. Balling. "Trends in extreme daily precipitation indices in India." International Journal of Climatology 24 (2004): 457-466.

Sheikh, Muhammad, et al. "Trends in extreme daily rainfall and temperature indices over South Asia." International Journal of Climatology 35 (6 2014).

Development and Comparison of Rainfall-Runoff Models: Case Study of Mixed Urban Catchment of Western Maharashtra, India

Rushikesh Satpute[1], Kshirsagar M.P.[2], Khare K.C.[3], Dixit P.R.[4] and Janhavi Mahale[5]
[1,5]M.Tech Student, [4]Associate Professor, [2]Assistant Professor, [3]Professor,
[1,4,5]Department of Civil Engineering, VIIT, Pune, [2,3]Department of Civil Engineering, SIT, Pune,

ABSTRACT

For sustainable water resource management, main objective is to estimate the runoff volume from precipitation. In India many of the cities are facing problems because of uncontrolled and unplanned expansion around them, leads to mix urban catchments. Urbanization influences the hydrological cycle due to change in land use pattern, results elevated runoff volume. Therefore, the study aims to find the accurate approach considering this mixed urbanization to calculate runoff volume. This study is an attempt to calculate runoff using different rainfall-runoff models for a mixed urban catchment in western Maharashtra. For ungauged basin, water balance equation was used to calculate observed runoff values. From the comparison of these it was found that the calculated runoff values were underestimated. Geographic condition of the region changes with time resulting in decreased efficacy of the established rainfall runoff models. Hence there is a need to develop a new empirical model for ungauged catchment, which takes into accounts the geographical conditions and its allied changes over the years. Precipitation (rainfall) is taken as primary parameter for calculation of runoff coefficients which takes in to consideration the current scenario of mixed urban catchment. From the results, it is concluded that, the derived equation gives better results for estimation of runoff for the study area when compared to other rainfall-runoff models.

Introduction

Hydrology is the systematic study of the movement, distribution and quality of water on earth including water resources, water cycle and environmental watershed sustainability. Precipitation is the most important part of hydrological cycle. The term precipitation represents all forms of water that reacheson earth from atmosphere. There are different forms of precipitation such as rainfall, snowfall, hail, dew and frost. Rainfall is the major form of precipitation in India due to its tropical metrology. From rainfall to runoff water goes through several losses such as evaporation, initial loses, infiltration and detention storage.

In India, major river basins or catchmentsare gauged for the determination of hydrological variable but small and medium size catchment remains ungauged.Forsuch catchments different methods are developed. Transformation ofRainfall into runoff over a catchment is a complex phenomenon as the process is highly nonlinear, time varying and spatially distributed. If the long range data is available, a model can be established. There are many modeling approaches used for rainfall runoff modeling, such as empirical modeling, numerical modeling and data driven techniques. Many other empirical equation are developed for Indian conditions such as Inglis and Dsouza, Coutagine relationship, Khosla's, Department of irrigation, India and Strange's table method etc.Inglis and Dsouza gives two different formulae for Deccan plateau and for Ghat regions of western India. These empirical equations are applied fordifferent catchments in Maharashtra such as Chaskaman Catchment and Nira Deoghar Catchment, Pune. The results demonstratesthat significant deviations were found between the measured and calculated runoff (Khopade and Oak, 2014). The empirical relationships are generally applicable for particular region, where relationships are developed by taking into account data of that region atthe time. Geographic condition of the region changes with time resulting in decreased efficacy of theseempirical equations. Hence there is a need to develop a new empirical equation for ungauged catchment, which takes into accounts the geographical conditions and allied changes.

New urban cities have been developed by converting small forests, agricultural lands into industries, research centers, big apartments and malls etc. This rapid growth of cities resulted in increased water runoff, reduced groundwater table and reduced width of natural drains (Ministry Of Water Resources Government Of India Maharashtra, 2014). The term 'urbanization' means the increasing share of a nation's population living in urban areas. The growth of urbanization in India is basically due to migration of people from rural area to cities or small cities to big metro cities for employment, improvised education and for better living condition or comfort.Census produced by government of India in 1901 only 10 percent population living in urban area, it increases 17.3 percent in 1951 and reaches 31.2 percent in 2011.(http://www.censusindia)

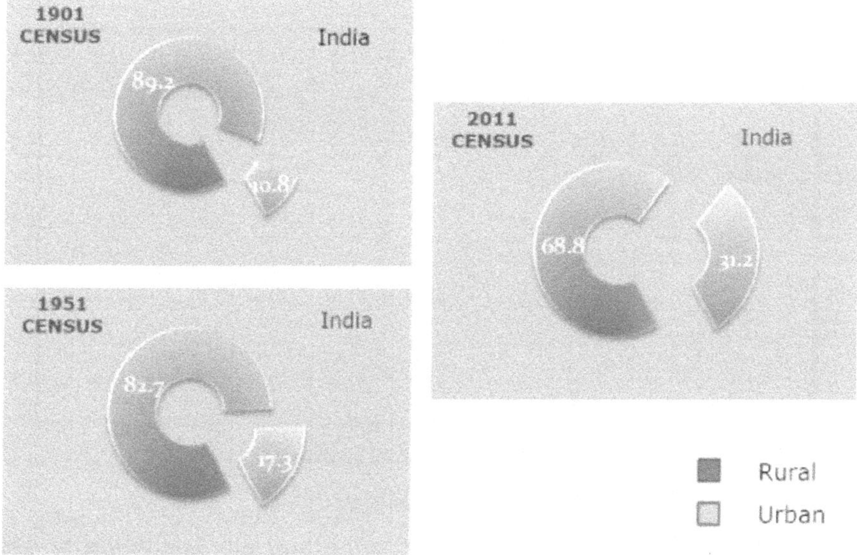

Figure 1 Trends in Rural Urban Distribution of Population- India (in %) (1901, 1951, 2011)(http://www.censusindia)

Many of these cities are facing problems due to uncontrolled and unplanned expansion around them. Urbanization influences the hydrological cycle due to change in land use pattern. It result in elevated runoff volume. Runoff occurs quickly from paved areas, concrete roads, building roofs and other impervious areas, results in less evaporation losses due to less contact time and low infiltration, causing peak runoff with very less lag time causing flooding in many urban areas. Due to change in land use pattern, rainfall-runoff response of catchments changes. To account these changes new method is developed. This study is attempt to find an accurate approach considering urbanization to calculate runoff volume. Observed Runoff values are calculated using water balance equation using rainfall, evaporation and infiltration data. Evaporation and infiltration are heavily dependent on land use pattern that is urbanization.

Material And Methods

Study area and data collection

The Indian city of Pune, witnessed rapid unplanned and uncontrolled radial expansion in the last four decades, leads to mix urban catchment. In mix urban catchments there are no separate zones such as residential, commercial, industrial and agricultural like planed urbanized areas. Rapid urbanization significantly changes the land use pattern of Pune such as agricultural land and open grounds converted into big apartments, malls, industries paved areas that is increase in impervious areas. Table 1 shows the land use/ land cover change of Pune city in four decades (Mundhe et al., 2014).

Table 1 land use/ land cover changes between four decades of Pune city. (Mundhe et al., 2014).

LU/LC Class	1973 Area in(Sq. km)	1992 Area in(Sq. km)	2001Area in(sq. km)	2011Area in(Sq. km)
Built-up	28.50	62.13	130.03	155.99
Agricultural Land	14.42	13.27	20.11	16.82
Vegetation	11.30	11.13	17.98	15.62
Fallow Land	9.97	10.52	17.18	15.89
Scrub Land	67.69	45.60	54.26	35.30
Rivers and Lakes	4.29	1.82	2.72	2.66
Canal	2.59	1.53	1.56	1.56
Total Area	138.76	146.00	243.8	243.8

This accelerated increase in impervious(built-up) area has increased the runoff to a great extent (Gogate& Rawal, 2015).

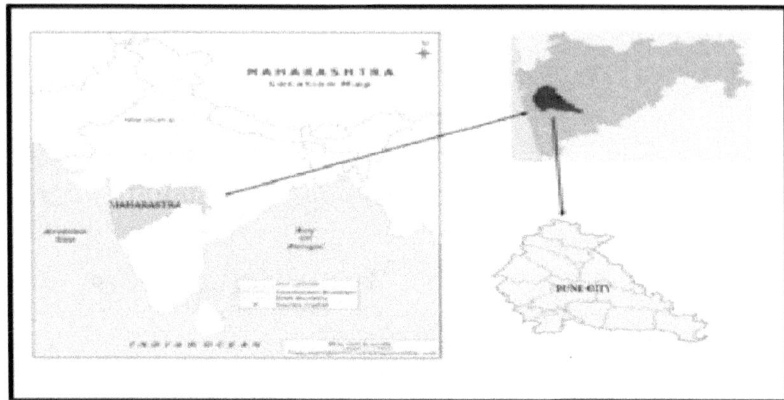

Figure 2 Location Sketch

Pune Municipal Corporation divides Pune district into 23 basins from A to W on basis of topography, out of which G basin is selected which is on the peripheral boundary of Pune city and typically depicts the urbanization trend of Pune city. Figures 4 and 5, presents the rapid changes in built up area of G basin from year 2004 to 2018.

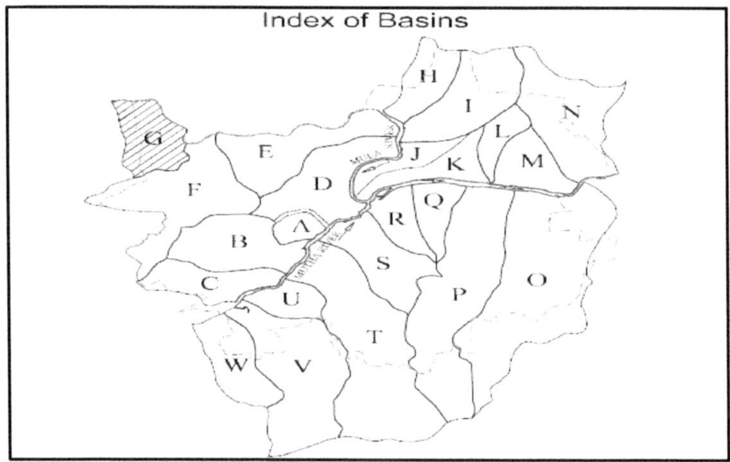

Figure 3 Pune drainage Map-showing study area G basin (PMC).

Figure 4 Satellite view of G basin in 2004.

Figure 5 Satellite view of G basin in 2018.

The rainfall and evaporation data (1982 to 2018) is taken from the India Metrological Department Pune. Hydrological data such as observed runoff data is not available for G basin. Water balance equation is used to calculate observed runoff values.

Estimation of runoff using different empirical methods for Indian catchment

Many of the past engineers have developed empirical equation for runoff estimation. However these equations areapplicable for specific time and specific region, whenever they were developed. These equations arenothing but the rainfall- runoff relationship with addition of third or fourth parameter accounts for climatic or catchment characteristics. Some of the commonly used empirical equation in various part of India are as given below:

Rational Method

The rational method is based on the following empirical relation between the peak runoff rate, Q_p and the rainfall intensity i.

$$Q_p = CAi \quad .....(1)$$

Where,

Q_p is peak runoff.

A is area of catchment.

i is rainfall intensity.

C is runoff coefficient represents surface characteristics and ranges 0 to 1. For the study area 0.75 is taken.

Khosla's Formula (1960)

In this method, the amount of monthly runoff is calculated by following formula (Subramanya, 2008):

$$R_m = P_m - L_m \quad .....(2)$$

$$L_m = 0.48 * T_m \quad .....(3)$$

Where,

Rm is monthly runoff of watershed in cm,

Pm is monthly precipitation in cm and Tm is mean monthly temperature in °C.

Inglis and DeSouza formula (1929)

Inglis and DeSouza (1929) evolved two regional formulae between annual runoff R and annual rainfall P in cm as follows (Subramanya, 2008):

For Deccan plateau

$$R = \frac{P}{254} * (P - 17.8) \qquad .....(4)$$

For Ghat regions of western India

$$R = 0.85 * P - 30.5 \qquad .....(5)$$

Department of irrigation, India

Management of Reihand plan presented the following relationship between the amount of annual precipitation and runoff of Reihand River (Gupta, 1992):

$$R = P - 1.17 * P^{0.86} \qquad .....(6)$$

Where, P is annual precipitation in cm and, R is annual runoff in cm.

Development of Empirical Model

Empirical equations developed like Inglis and DeSouza formula (1929), Khosla's Formula (1960), Department of irrigation, India etc. about 50 to 70 year old. When these equation were developed, the condition at that time was different than present conditions like urbanization.Hence in previous times, perviousarea were more, indirectly infiltration rate was more results low runoff volume.

In present condition due to rapid urbanization the pervious lands are converted into impervious lands such as concrete roads, building roofs, paved areas etc. These results in low infiltration and accelerates the runoff, because of this low evaporation occurs. Low evaporation and infiltration consequences high runoff volume. So to account this changes due to rapid urbanization new equation is developed.

Annual rainfall and runoff data (form water balance) of 1982 to 2001 was taken for trainingthemodel. Precipitation (rainfall) is taken as primary parameter for development of equation. In present many new software tools are available to solve many simultaneous equation in very less time, for present case study MATLAB is used. MATLAB code is written to derive equation taking into consideration the current urbanization scenario, isgiven below,

$$R = 0.0038 * P^2 + 0.4028 * P \qquad .....(7)$$

Where,

P is annual precipitation in cm and,

R is annual runoff in cm.

For validation of empirical model rainfall and runoff data from 2009 to 2018 is used.

Results and Discussion

Various empirical methods and relationships have been developed using past rainfall and runoff data. The first popular method was the Rational Method published by Thomas Mulvaney in 1851 and it is still applied, investigated and interpreted (Grimaldi & Petrosell, 2015).The main limitation in rational formula is calculation of coefficient of runoff and area for different land use.

In this study Rational method, Inglis and DeSouza methodand Department of irrigation, India are used to estimate the runoff of G basin using the rainfall data.

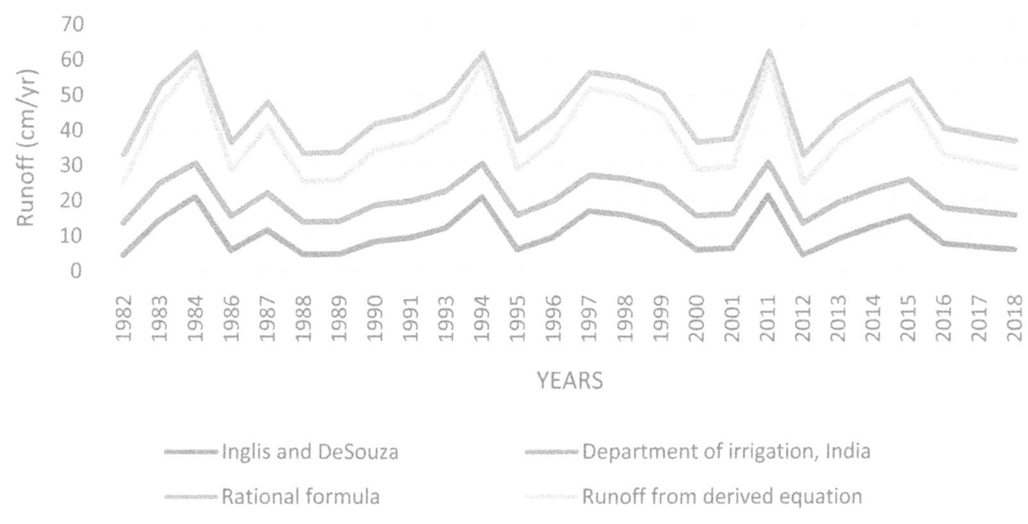

Graph 1 Variation between estimated runoff by various empirical methods and derived equation.

Table 2 Results of statistical methods used for empirical models in G basin.

METHOD	Inglis and DeSouza	Department of irrigation, India	Runoff from derived equation
MAD	34.40	24.46	6.58
MSE	1200.72	613.80	46.13
RMSE	34.65	24.77	6.79
MAPE	77.86	54.91	15.94

When Runoff values estimated were compared withrational method values using two empirical methods, it was found that runoff values are underestimated. The reason behind this is, when this empirical equations was developed the condition at that time was totally different than present condition. Inglis and Desouza develop its formula around 1929. At that time no such development and urbanization take place that is reverse condition at present. Mainly it contain agriculture land, forest, small houses and roads due to this the pervious area is more, results high infiltration. Due to high infiltration runoff volume decreases, results underestimation of estimated runoff values.

Derived equation is easy to use, because it required only one input parameter that is precipitation. From the results and statistical report of the study it was revealed that, derived equation gives better result for estimation of runoff in G Basin in present condition, than other empirical equations.

References

Groundwater Surveys and Development Agency (2014), "Report On The Dynamic Ground Water Resources Of Maharashtra." Ministry for Water Resources Government of India Maharashtra.

Pradip Dalavi, S.R. Bhakar,H.N. Bhangeand B.K. Gavit (2018), "Assessment of Empirical Methods for Runoff Estimation in Chaskaman Catchment of Western Maharashtra, India"International Journal of Current Microbiology and Applied Sciences.7(5):1511-1515.

Khopade, D.K. and Oak, P.R.A. 2014. "Estimation of runoff yield for Nira Deogharcatchment using different empirical equations" The International Journal of Engineering and Science. 3(1): 75-81.

Khosravi, K., Mirzai, H. and Saleh, I. (2013),"Assessment of Empirical Methods of Runoff Estimation by Statistical test (Case study: Banadak Sadat Watershed, Yazd Province)" International Journal of Advanced Biological and Biomedical Research. 1(3): 285-301.

Mundhe et al (2014), "Impact of urbanization on land use/land covers change using Geo-spatial techniques" International journal of geomatics and geosciences Volume 5.

Praveen Kumar, B. J., Pradeep, H., Lokesh, A., Akarshraj, K. H., Surendra, H. J. and Avinash S. D. (2016),"Estimation of Runoff using Empirical Equations and Fuzzy Logic method: A case study" International Journal of Scientific & Engineering Research. 7(5): 28-36.

Salvatore Grimaldiand Andrea Petroselli (2015), "Do we still need the Rational Formula? An alternative empiricalprocedure for peak discharge estimation in small and ungauged basins" Hydrological Sciences Journal, 60:1. Subramanya (2008), "Engineering Hydrology" McGrow hill education.

Dam Break Analysis Using HEC-RAS

Harshwardhan Nikam[1], Shreenivas Londhe[2] and Mrunal Joshi[3]
[1]*Post graduate student, Water Resources and Environmental Engineering,* [2]*Professor in Civil Engineering, Vishwakarma Institute of Information Technology, Pune.*
[3]*Assistant Professor in Civil Engineering NBN Sinhgad College of Engineering, Pune.*
harshwardhan.nikam@gmail.com, shreenivas.londhe@viit.ac.in, mrsmrunaljoshi@yahoo.com

ABSTRACT

Dams offers numerous benefits to the society, however floods causing from dam break leads to huge loss of inhabitants and properties. Dam failures generally occur by overtopping of dam because of insufficient capacity of spillway and freeboard through large inflows from the reservoirs, piping due to internal erosion or seepage. Construction of dams in rivers can provide desired benefits such as supply for drinking water, irrigation, power generation, etc, but there would be catastrophic disasters due to dam failures. Simulation of dam break and causing floods are critical for characterizing and minimizing threats due to possible dam failures. Development of effective emergency action plans needs exact prediction of inundation levels and flood wave arrival time at a specified location. Therefore there is need of dam break analysis before it fails. This study mainly focuses on post effects of dam break phenomenon. In this paper dam break analysis using HEC-RAS 2-D model is being carried out for Khadakwasla dam Pune, which will highlight high flood zones on downstream side of dam and preparation of inundation map. Through which we can assess the type of emergency facility to be provided.

Keywords: Flood, Flood Mapping, Dam Break, HEC-RAS 2-D Model

Introduction

Dams are the important hydraulic structures which are constructed as storage reservoir. It provides a large number of benefits to humanity such as they provides water for domestic, industrial and irrigation purposes but also at the same time its failure results in catastrophic disaster event which is characterized by the sudden uncontrolled release of impounded water (Ekowahyudi, 2004). Dam stores large amount of water which causes significant flooding in downstream region and also affects the economy. Several reasons behind dam failure are high water pressure during floods, inadequate spillwaycapacity, poor maintenance, high pore pressure, embankment slips, defective material, erosion and earthquakes (Basheer, 2017).

But the foremostreason behind the failure of dam is overtopping and piping. Heavy rainfall causes overtopping of dam which firmly discharges the water over the crest which further leads to dam breach. Water also seeps from the bottom of the dam or through walls of the dam which weakens the surrounding material results in breaking of structure. The Great Levee is ruined by ant nest. Basically it notifies people that small cases might result in greater difficulties. However, it always reminds water resource engineers the probability of dam failure. (Xiong2011)

Dam safety is an important issue in today's date. To plan for preventive measures after dambreak is necessary, so that disaster will not lead to loss of lives of population living in downstream areas (Ekowahyudi, 2004). Therefore Dam break analysis is required before it fails, so that large amount of lives or properties can be saved. Hence this study aims the dam break analysis of Khadakwasla Dam by using the HEC-RAS 2D Model.

HEC-RAS is a software developed by Hydrogic Engineering Center of U.S. Army Corps of Engineers River Analysis System which allows to perform one-dimensional, two-dimensional steady and unsteady flow, river hydraulics calculations, sediment transport-mobile bed modeling, and water temperature analysis. The HEC-RAS software supersedes the HEC-2 river hydraulics package, which was a one dimensional, steady flow water surface profiles program (HEC-RAS User's Manual, version 5.0, 2016)

Methodology

Study Area

Many of the cities in developing countries like India are facing the problems of devastating floods which leads to termination of basic Infrastructure. Pune is one of the most sprawling city in the western Indian state of Maharashtra (PMC 2009). Location of Pune City in Maharashtra is as shown in Figure 1. Pune grows into an educational, cultural powerhouse, economical purposes hence the urbanization has expanded up to the khadakwasla village which located near khadakwasla Dam. Therefore Dam Break Analysis for Khadakwasla Dam is required on priority basis, to prevent the city from flooding due to dam failure.

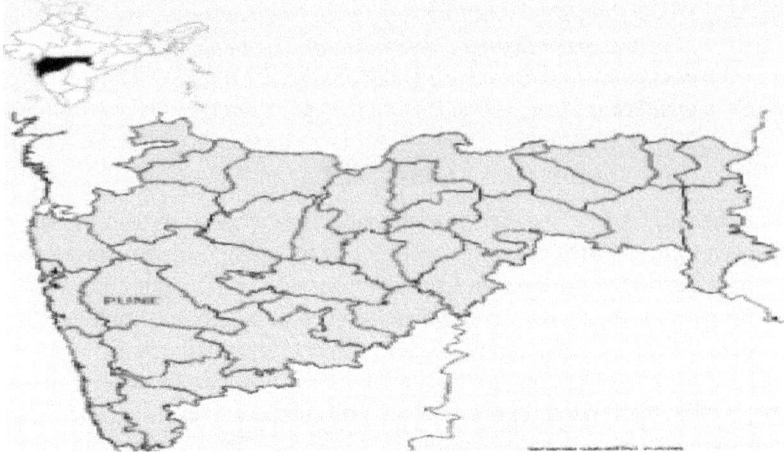

Figure 1 Location of Pune in Maharashtra.

Khadakwasla Dam is a major dam on the MuthaRiver which is located at 21 km from the centre of the city of Pune, Maharashtra. The dam created a reservoir known as Khadakwasla Lake which is the main source of water for Pune and its nearby villages. The length of Khadakwasla Dam is 1539 meter. The dam has constructed on Mutha River, which arises from the convergence of the rivers Mose and Ambi on which the Varasgaon and Panshet Dams are constructed respectively. Mose River has been stated as Mutha River. The length of the Khadakwasla backwaters is nearly 22 km and the width varies from 250 to 1,000 m (820 to 3,280 ft). The dam has 11 radial type sluice gates and six irrigation outlets, flowing into two canals (PMC 2018). Location of Khadakwasla dam in Pune is as shown in figure 2.

Figure 2 Location of Khadakwasla Dam

Data Collection

Khadakwasla dam is constructed across Mutha River in Pune, Maharashtra.

1. Discharge data of khadakwasla dam from SinchanBhavan, Irrigation Department of Pune.
2. Area Capacity/Area Volume Curves Data from SinchanBhavan, Irrigation Department of Pune.
3. Digital Elevation Model (DEM) from Earth Explorer (www.earthexplorer.usgs.gov), Bhuvan (www.nrsc.gov.in) with 30m resolution.

4. Land Use Land Cover classified Map of study area from Maharashtra Remote Sensing Application Centre (MRSAC), Pune.
5. Geometrical Data of Mutha River- Khadakwasla Dam from Divisional Office, Pune and Sub-divisional office, Khadakwasla, Pune.

KhadakwaslaDam Details:

- River- Mutha
- Basin Name- Krishna
- Dam Type- Earthen/ Gravity
- Dam Height - 36 meter
- Type of Spillway- Ogee Crested Spillway
- Length of Spillway – 169 m
- Dam Length - 1539 meter
- Number of Gates – 11
- Gate Type – Radial Gate
- Gate Size – 12.19 x 4.27 meter.

Dam Break Analysis for KhadakwaslaDam

Digital Elevation Model (DEM) from Earth Explorer (www.earthexplorer.usgs.gov) with 30 m resolution and WGS 84 datum and UTM projection (Joshi 2017) is used for creating the Terrain Model in HEC-RAS which is shown in the figure 3.

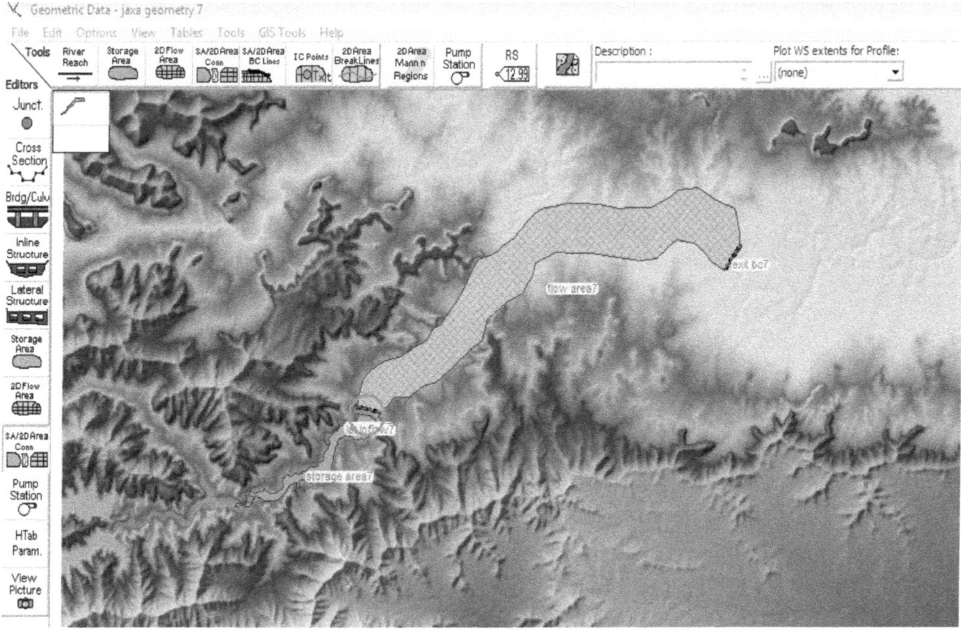

Figure 3 Terrain Model of Khadakwasla Dam

Once terrain model was created; storage area, 2D flow area and connection was marked on terrain mode along with boundary conditions and mesh formation was done for 2D Flow area are shown in figure no. 4

Boundary Conditions

1. Storage Area- Lateral Flow Hydrograph
2. Exit Condition- Normal Depth
3. Connection- TS gate opening.

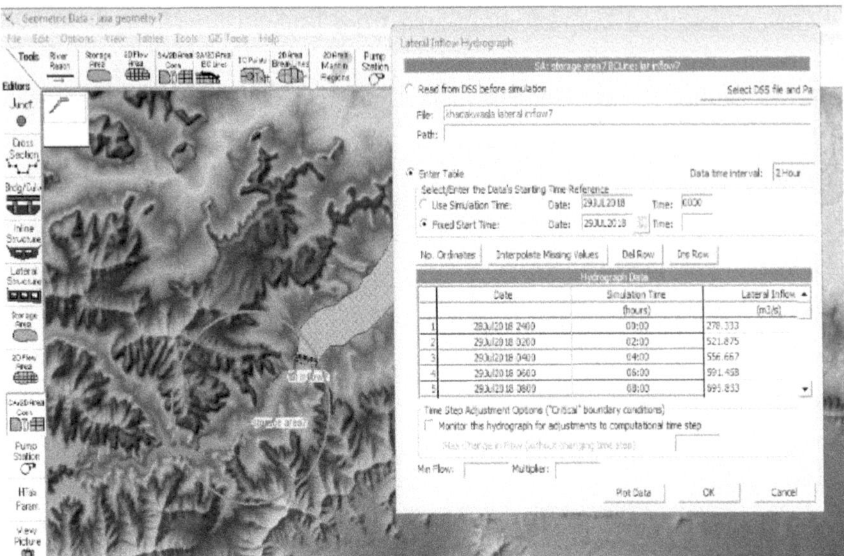

Figure 4 Inflow HydrographKhadakwasla Dam

Breach Parameters

MacDonalds approach was used to find breach parameters that are shown in figure no.5

- Breach Width- 1040 m
- Breach Time- 20 minutes
- Breach Left Side Slope- 0.5
- Breach Right Side Slope- 0.5

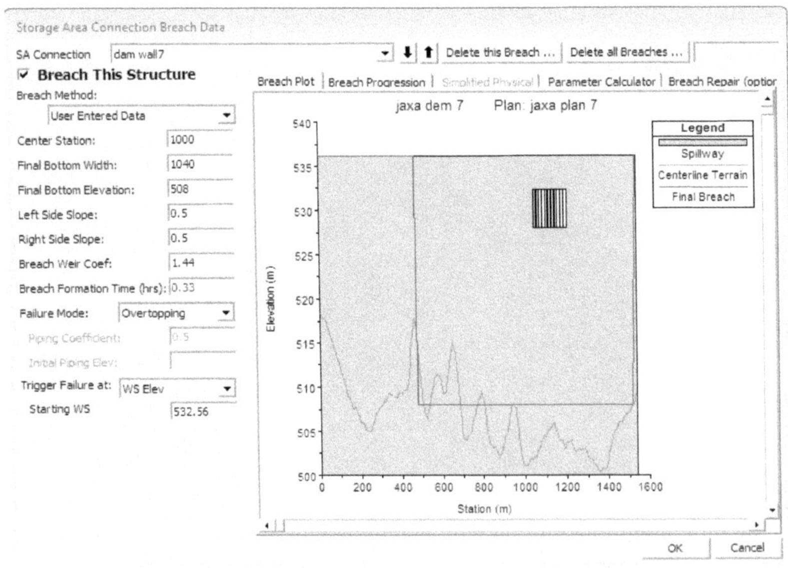

Figure 5 Breach ParametersKhadakwaslaDam

Results

Unsteady flow simulation is worked for dam break study of khadakwasla dam considering computational interval as 10 seconds and Hydrograph output interval as 30 minutes.Khadakwasla dam failed at lateral flow discharge of 3100 m^3/s.

Output maps are generated in RAS Mapper window showing the spread area for the selected boundary of flow area. Inundation map of terrain is shown in figure no. 6 below.

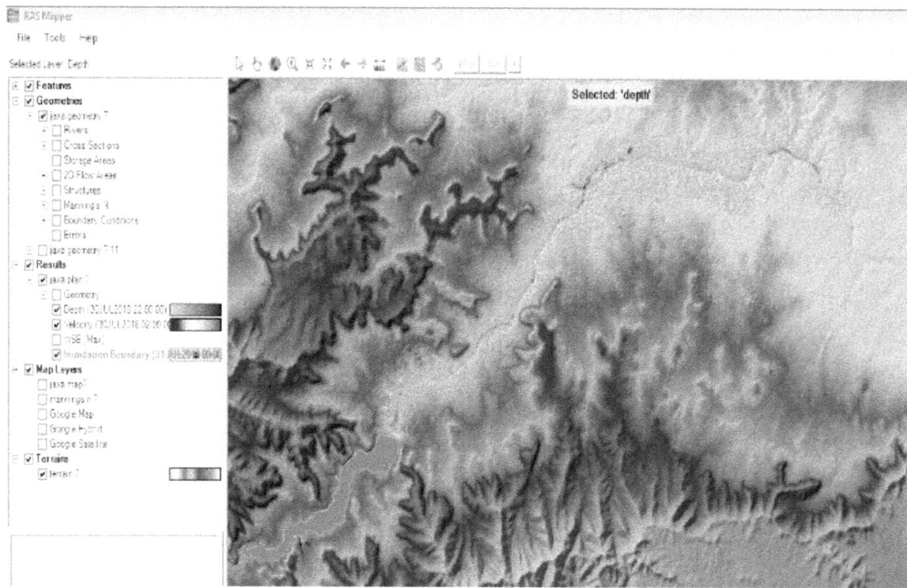

Figure 6 Inundation map (terrain model)

Major part of the city near about the 1 to 1.5 km along the river boundary is under water. Which is shown on inundation map projected on google maps in figure no. 7. All the bridges in the city area over Mula-Mutha River were also under water. Foremost part of the following important places like Vittthalwadi, DeenanathMangeshkar Hospital, PMC STP, GarwareCollege, Poona Hospital, BharatiVidhyapeeth, DeccanGymkhana, Narayan Peth, PMC Bulding, RTO Office and WTPSangamwadi is under water.

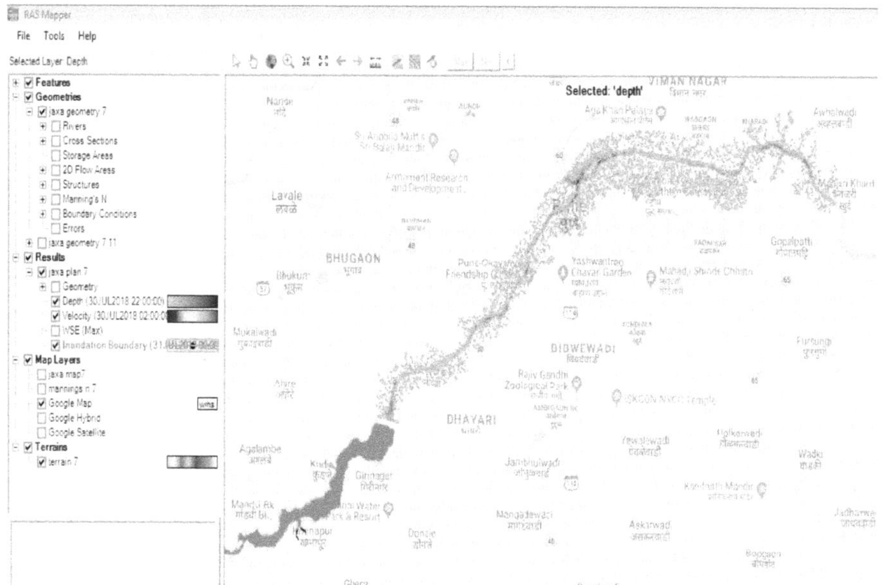

Figure 6 Inundation map (Flow boundary plot over google maps)

Conclusion

Dam break analysis has been carried out for khadakwasla dam by using HEC-RAS 2D modelling software. Majorly affected areas by heavy flood were found out. Depth, velocity, water surface elevation at all the desired cross section along the river reach for finding the affected areas and inundation map was high-lightened.

Structures such as dikes, gabion walls, levees, etc. can be used to minimize the flood impact and protect the important structures. Adverse effects of floods can be minimized and managed by city town planning department by predicting occurrence and magnitude of flood and its travel time well in advance. Thus such flood predicting studies are required to prepare emergency action plans to safeguard the property, life and economy of the city.

References

Talal A. Basheer, (2017); "Dam Breach Parameters and Their Influence on Flood Hydrographs for Mosul Dam

HEC-RAS user manual (2016); "HEC-RAS River Analysis System Hydraulic Reference Manual. United State of America: U.S Army Corps of Engineer.

HEC-RAS user manual (2016); "HEC-RAS River Analysis System Hydraulic User's Manual"

Mrunal Joshi and S. S. Shahapure (2017); "Study of Two Dimensional Dam Break Analysis Using HECRAS for Veer Dam"

S. R. Kulkarni, S. K. Ukarande (2016); "Dam Break Analysis - A Case Study"

PMC Report of 2009 and 2018

Ekowahyudi, (2004); "Dam Break Analysis, A Case Study"

Yi (Frank) Xiong (2011); "A Dam Break Analysis Using HEC-RAS"

Fread, D.L., 1988 (revised 1991); "BREACH: An Erosion Model for Earthen Dam Failures".

Sensitivity Analysis of Wind Factor on Oil Spill Trajectory

Sagar. G. Kharche[1], Prabhat Chandra[2], V. B. Sharma[3] and P. R. Dixit[4]

[1]PG student, [2]Scientist "E", [3]Scientist "B", [4]Associate Prof,

[1,4]Vishwakarma Institute of Information Technology, Kondhwa, Pune

[2,3]Central Water and Power Research Station, Khadakawasla, Pune

sagark9693@gmail.com, pchandra_2003@yahoo.co.in,vipin_bihari@yahoo.com

ABSTRACT

A study has been carried out to determine impact of oil spill in the vicinity ofMormugao. In this regard, different scenarios were considered to determine the impact of wind on spread of oil spill. The currents nearMormugao Port are of the order of 0.3m/s and flow is unidirectional for most of the period. MIKE-21 HD software has been used to simulate the flow conditions. In order to incorporate wave induced currents, radiation stresses were supplied to MIKE-21 HD model and model was calibrated for field observed currents. The output of MIKE-21 HD model was supplied to Oil spill model MIKE – 21/3 OS. The present study is for development of POL jetty in the Vasco bay indicated that within 12hrs of oil spill; about 27% oil gets evaporated. The percentage graphs of evaporation and dispersion was obtained from ADIOS.

Keywords: MIKE-21/3, Oil SpillTrajectory,ADIOS

Introduction

Oil is the fuel that runs the global economy. Half of the world's production of crude oil is transported by sea (Clark, 1992). Ships among the production sites, refineries and point of consumption transport huge quantity of crude oil. Oil production includes crude oil, shale oil and Natural Gas Liquids (NGL). Oil shipments accounts to almost 30% of the global trade. India relies more than 80% on imports to meet its regular oil demands. India has imported 213.93 million tons of crude oil in the year 2016-2017.

In recent years, there is a growing concern over the increased contamination of water bodies and adjacent coastline areas by oil spills. The activities like oil exploration and transport, oil storage facilities, etc. have increased the risk of oil spill accidents. These oil spills directly affect the marine ecosystem, beaches, fish culture, etc. making the life in coastal areas worse. To prepare for such accidents, many government agencies research institutes have prepared contingency and mitigation plans. An important aspect of these plans is the use of numerical method to predict the oil spill and distribution of oil particle concentrations.The oil spills have different sources like ship borne accidents, transfer to downstream, seepage from natural bed, etc. The maximum oil spillage is caused due to accidents i.e. 73%, 13% due to oil transfers, 11% due to miscellaneous reasons and 3% due to ruptures. The numerical modeling approach has emerged to be a useful tool in encountering this oil spill accidents. Spill modeling is important to predict the trajectories and implementing the mitigation plans.

Study Area

Goa (India) has been identified as one of the major tourist destinations of the world because of its beautiful beaches and associated recreational activities. The economy of Goa depends on the tourism largely. The coastal population as well as the State government are very particular about the activities, which are harmful in the coastal zone. Oil spill accidents have taken place in past in vicinity of Goa region. In case spill-remains arrive at the beaches or islands, there will be a huge public disquiet and fall in the ongoing recreational activities (bathing, surfing, boating, diving, etc). Therefore, in this context, it is necessary to have pre-validated trajectories of spill scenarios for effective implementation during the accidents.

Mormugoa Port Trust is developing POL jetty at Vasco Bay. This study is essential to act appropriately in case of any mishaps.

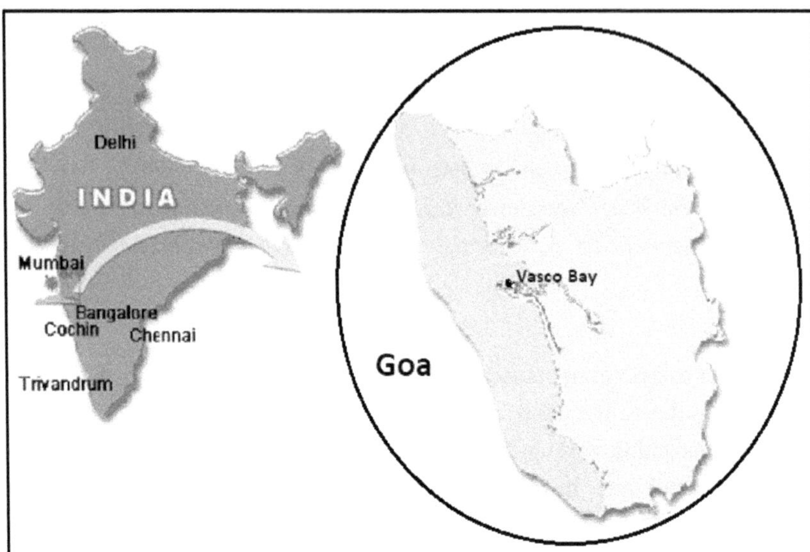

Figure 1 Location map- Goa, India

Environmental Conditions

The mean maximum and minimum temperatures of Goa region over the past 30 years is 35 and 25 degree Celsius. The winds in this region are recorded at Dona Paula.

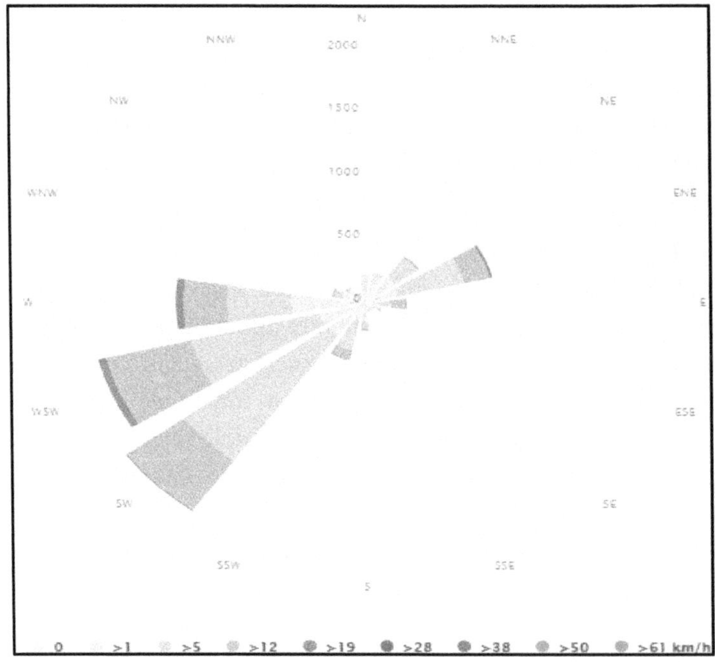

Figure 2 Wind rose diagram for Goa coast

The prevailing winds on the Goa coast are predominantly from the SW direction. The magnitude of these winds vary according to season. The tide data was obtained from C-map.Hence, in the present study we have considered three wind scenarios, 2m/s, 12m/s and 21m/s (extreme wind scenario).

Methodology

Hydrodynamic behavior of costal area in the vicinity of Port area will be simulated using MIKE – 21 HD module which solves the Shallow Water Equations (SWE) using finite volume method. In order to incorporate wave induced currents MIKE – 21 SW module will be used and its output will act as the input for the MIKE – 21 HD module. Further MIKE – 21 OS module will be used to simulate the fate of oil spill.

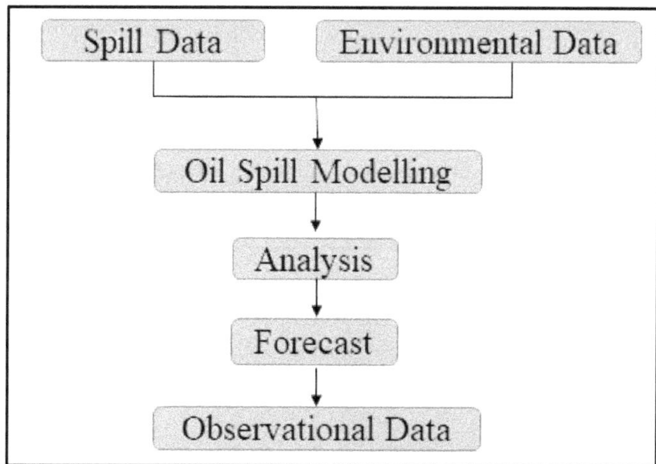

Figure 3 Methodology

Spill Modeling

Bathymetry required for spill modeling, was prepared for the area near Goa coast for a depth of varying upto 100m. Bathymetry was prepared using classical mesh, water data and land data (c-map data) in MIKE ZERO. The tide data was received in time series format from C-map. The wind data was obtained from IMD. MIKE HD module was run and the results obtained were given as input for MIKE 21/3 oil spill module. The results were derived using plot composer, which is a tool in MIKE ZERO. 85 tons of heavy fuel oil was spilled during the oil spill accident, which took place on 23rd March, 2005 due to collision of two vessels. This case was taken as a reference and the hypothetical cases were derived.

Results

SCENARIO 1:

Oil type: Iranian Heavy Oil,API index: 31°, Wind speed: 2 m/s

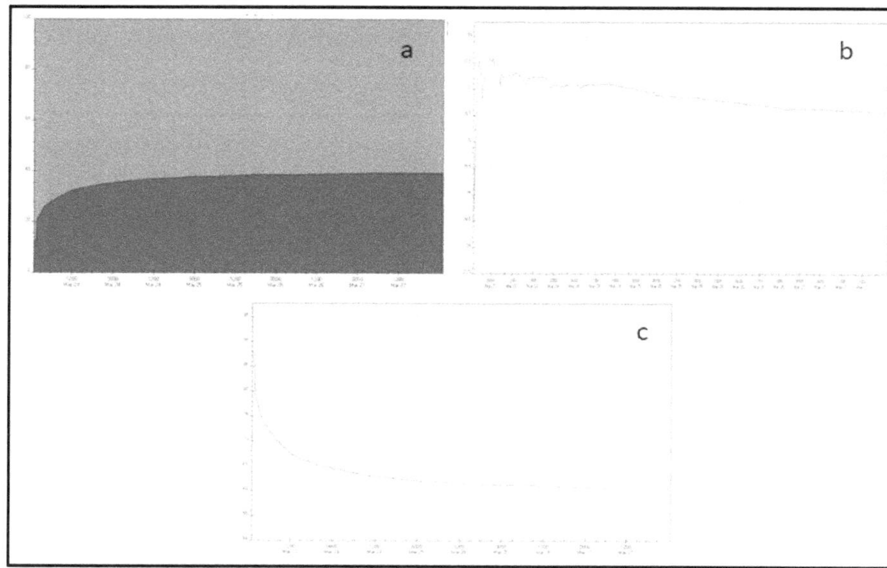

Figure 4 Weathering of oil spill (wind 2m/s)

In the above figure, we can see the effect on oil spill after 5 days. In figure a), the oil spill undergoes the evaporation action. No dispersion of oil spill takes place when the winds are of 2 m/s magnitude. The figure b), indicates that almost 40% of oil gets evaporated and the 60% remains (figure c).

SCENARIO 2:

Oil type: Iranian Heavy Oil ,API index: 31°, Wind speed: 12 m/s

In the figure below, we can see that oil spill pattern changes when wind speed is increased to 12m/s. It is found that, 35% of oil gets evaporated (figure b) and 34% of oil gets dispersed (figure c) and the remainder is 31% (figure d).

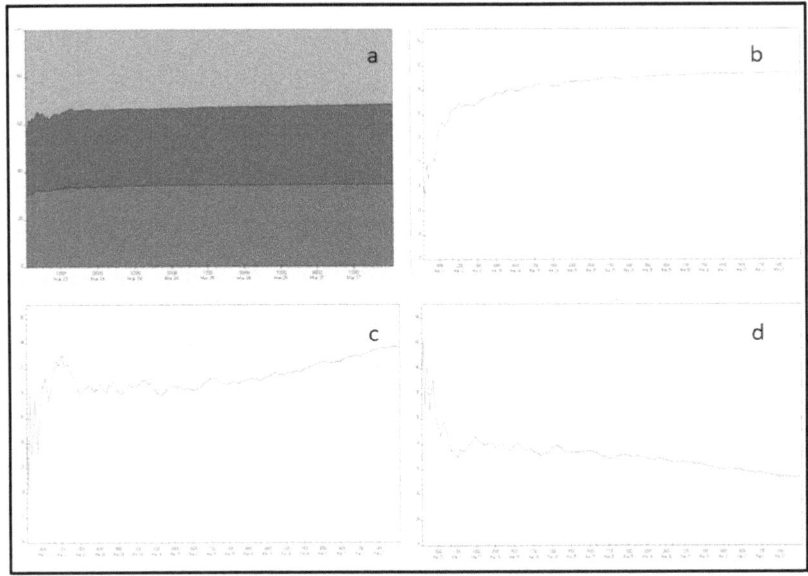

Figure 5 Weathering of oil spill (wind 12m/s)

SCENARIO 3:

Oil type: Iranian Heavy Oil ,API index: 31°, Wind speed: 21 m/s

In the above below, we can see that oil spill pattern changes when wind speed is increased further more to 21 m/s. It is found that, 27% of oil gets evaporated (figure b) and 71% of oil gets dispersed (figure c) and the remainder is 2% (figure d).

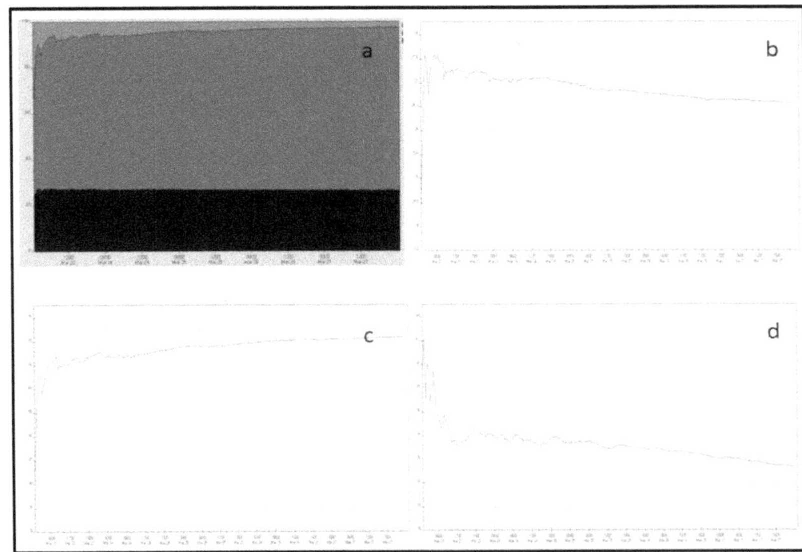

Figure 6 Weathering of oil spill (wind 21 m/s)

Conclusion

Prevailing wind and its magnitude in the region affect the oil spill weathering. Almost 35% of oil is evaporated within span of 24 hours of the spill accident. The average winds (within 5m/s) do not affect cause the dispersion phenomenon while those winds above this cause the evaporation and dispersion in a combination. It was observed that when extreme wind conditions prevail (21m/s), 98% of oil gets evaporated (27%) and dispersed (71%). These condition can be seen through the above graphs which shows the percentage wise evaporation and dispersion value.

References

Xiaobo Chao (2000), "Two- and three-dimensional oil spill model for coastal waters"; Ocean Engineering 28 (2001) ; 1557–1573

Dagmar Schmidt Etkin (2002), "Analysis of Past Marine Oil Spill Rates and Trends For Future Contingency Planning" ; Arctic and Marine Oil spill Program (AMOP) Technical Seminar

M.G.Psaltaki (2005), "Modeling the behavior of an oil spill in the marine environment";International Conference on fluid dynamics & aerodynamics, Corfu, Greece, August 20-22, 2005

R.S. Kankara et.al (2007), "Oil Spill Sensitivity Analysis and Risk Assessment for Gulf of Kachchh, India, using Integrated Modeling"; Journal of Coastal Research; 23/5

P Vethamony et.al (2007), "Trajectory of an oil spill off Goa, eastern Arabian Sea: field observations and simulations"; Research Journal of marine science

William B. Samuels (2013), "Development of a Global Oil Spill Modeling System"; Earth Science Research; Vol. 2, No. 2; 2013 ISSN 1927-0542

R. Mohan et.al (2014), "Oil spill trajectory modelling of Chennai coast, east coast of India"; INCHOE 2014

Gurumoorthi K et.al (2016), "Simulation of hypothetical oil spill trajectory model for southern tip of India"; IJSIT; Volume 5; Issue 5

MIKE 21/3 OS Manual by DHI

Community Problems and Technological Solutions: A Case Study of Harobelavadi-Dharwad - India

Munnoli P M[1], Jadi S[2], Rachana A[3], Kulkarni P[4] and Pramod K[5]

Professor[1], Final Year students [2,3,4,5], Department of Civil Engineering,

S D M College of Engineering and Technology Dharwad,

Corresponding author:munnolipm@gmail.com

ABSTRACT

India is a agrobased country with its 70% population living in rural areas. The schemes like Sansad Adarsh Gram Yojana (SAGY) and Unnat Bharat abhiyan (UBA) are flag ship Programs of Government of India, aiming at " Developed Village Developed India".

Harobelavadi a village situated on Dharwad Savadatti road (KA SH 34) is being covered under Sansad Adarsh Gram Yojana and Unnat Bharat Abhiyan undertaken by S D M College of Engineering and Technology Dharwad. The village survey based on UBA format is carried out with a specific objective of collecting data on rural sanitation/waste disposal and problems faced by the community in general. The data revealed that even though there are arund 90% construction of rural toilets by the Village Panchayat, but the functional utility is yet to be fully realised. The community people are reluctant to use the toilets, or used as a storage place for agricultural wastes.

The paper presents the details of problems faced by the communities and identified through peoples participation in survey along with possible technological solutions to be included in the village development plan prepared using ARC GIS software, leading to developed India.

Keywords: Unnat Bharat Abhiyan, SAGY, rural sanitation, GIS

Introduction

India is a agro based country, with 600000 villages out of which 125000 villages are backward or underdeveloped. Even though there are governmental programs in the five year plans for the developmental activities regularly. The impacts of these programs are far reaching the societal needs. There is need for designing and building the villages in India to grow to the full potential so that they get all the urban facilities without shifting to the urban areas (A P J Abdul Kalam, 1998; 2002). The present Government of India has planned schemes like Sansad Adarsh gram Yojana (SAGY) and Unnat Bharat abhiyan (UBA) for the inclusive development of villages and entrusted these works to the Engineering colleges for identifying problems and provide technological solutions. One of the goals of the SDM College of Engineering and Technology is to develop community interaction and provide solutions to the communities through analytical abilities of students. Therefore the paper presents the work carried out in Harobelavadi Village under UBA in conducting the survey, identifying the problems and providing appropriate technological solutions.

Village Harobelavadi

Harobelavadi situated at 20 km from Dharwad is a peri-urban area facing diversified problems on drinking water, sanitation, waste disposal, drainage.

Table 1 Village Data

Population	**Males(1512), Females(1445)**
Total area of village	Cultivable(2139.02), Non cultivable(31.12)Ha
No. of households	609
Livestock	Cows(778),Buffalo(254),sheep's (255)
Anganwadis	2 No
Primary and Secondary School	1 No
Youth Club	1 No

Survey Work

A field survey of village survey was carried out with a well designed questionnaire as given by Unnat Bharat Abhiyan. The data colllectd had two parts 1) Village survey 2) House hold survey.

The survey was meticulously carried out by the students who were traind by the faculty for the community interaction. The format for survey was adopted from UBA.

Results and Discussion

The survey work revealed that the village is facing the problems as presented in the Table 2. There is also a need for developing village development plan incorporating the above problems.

Table 2 Technological Problems and solutions

Problems of community	Solutions
No safe drinking water	Design of water filter
Disposal of solid waste	Design of Composting and Vermi-composting units individual and community (Munnoli P M,2005; 2010)
Disposal of drainage water	Design of open sewers
Commutation to Dharwad	Direct bus to Dharwad from Harobelavadi for students.
Bus shelter	Design of bus shelter
Cow sheds	Modern cow shed with slope to drain out urine and washing to compost or vermi-compost unit. With stall feeding facility.
Toilets	Construction of two pit toilets
Bio Gas Plant maintenance	Installation of new plants and maintenance (Mohanty M K and Sorayan V P S,2003)
Self Help groups	Participatory approach in developing self help groups.(Chambers 2009)
Road between Kabbenur and Harobelavadi presently takes 9 Km round	Design of flexible pavement between Kabbenur and Harobelavadi straight 2.5KM.

These are the limited to the survey format and our interaction with the VP and Communities if more time is spent in understanding the community and resolve to participatory approach (Mukharjee A, 2004; Desai A R 2002), more number of real life problems can be identified from health, economics, education, child marriage, child labor, problems of widows etc. The identified problems and solutions are utilized for designing the village development plan using ARCH GIS showing existing details is placed in figure 1.0 and developmental activities is placed in figure 1.2

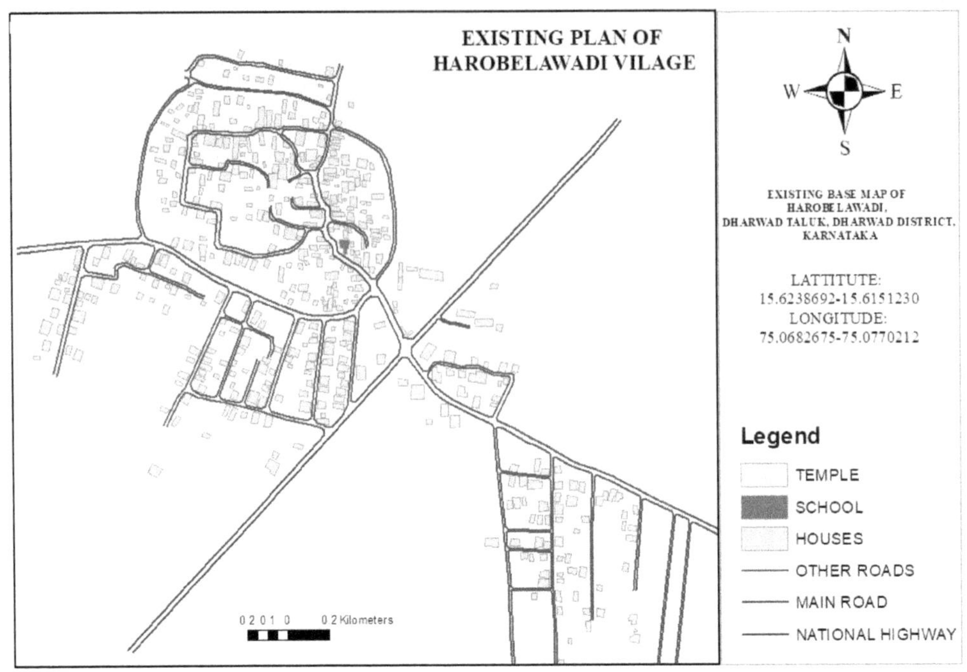

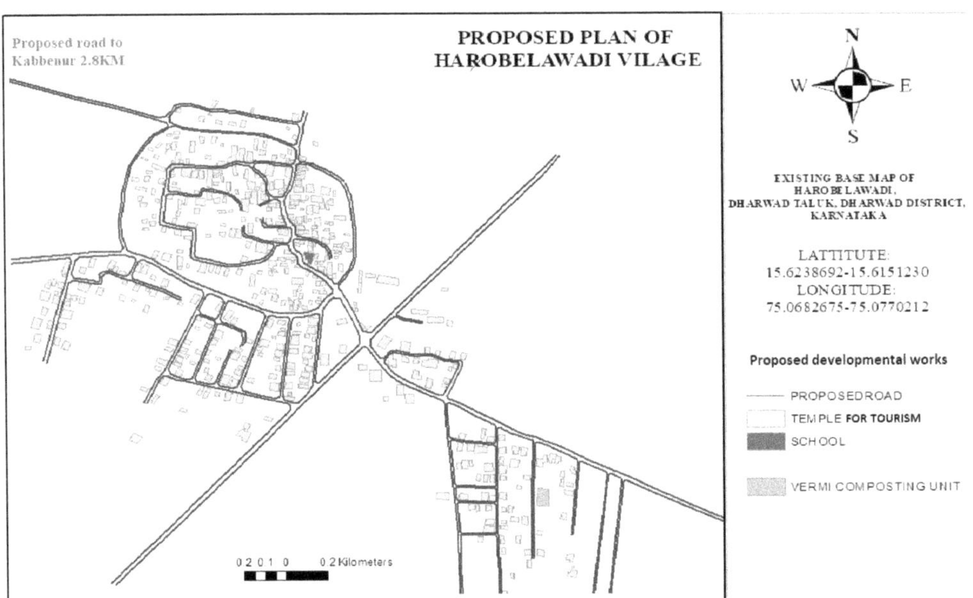

References

Munnoli P M (2010), Dynamics of the Soil-Earthworm-Plant Relationship: A Review, Dynamic Soil, Dynamic Plant, Global 1-21

Mohanty M K and Sorayan V P S, 2003, Bio gas production and utilization in India, In:Proce. Nat. sem.on Downsizing Technology for Rural development, Ed.Khuntia S, Parida A, Misra V N, Allied Pub Pvt. Ltd, pp 331-337

Desai A R (1998) Rural Sociology in India, Popular Prakshan Bombay, pp 167-171

A P J Abdul Kalam, (1998) INDIA 2020: A Vision for the New Millennium, Penguin India, p 1-300.

A P J Abdul Kalam, (2002) Ignited Minds, Penguin India, p 1-200.

Chambers, R, 1994. Participatory rural appraisal (PRA): analysis of experience. *World development, 22*(9), pp.1253-1268.

Munnoli P M, (2005) Rural waste management through training in vermi compost technology: a case study of Malcornem village; Goa" In: national meeting on Vocational Educational and Training for sustainable rural livelihood conducted jointly by Dr. Harisingh gaur university, Sagar Madhya Pradesh and Pandit Sundarlal Sharma Institute of Vocational Education Bhopal a constituent of NCERT New Delhi at Dr. Harisingh Gaur University, Sagar Madhya Pradesh from 23rd to 25th February 2005.

Mukharjee N, (2003) Participatory Rural Appraisal Methodology and applications, Concept pub New elhi, p-1-157.

Rivera F G and Erlich J L (1998).Community organizing in a diverse society, 3rd ed., Allan and Bcon Publishers, London, pp-25-60.

Garg S K (2010)Water supply and Sanitary Engineering Khanna Publisher New Delhi, pp 1-150.

Duggal K N (2012) Environmental Engineering, S Chand Publisher New Delhi, pp-20-45.

Gandhi's Views& Work for Village Development Panchayat Raj, Vithal nagar, Maharashtra , 18-1-2012

Solar power energy solutions for Yemeni rural villages and desert communities http://smartvillage.nic.in

www.gram.org.in

www.unnat.iitd.ac.in

Flood Lines i.e. Red and Blue Line Marking on Downstream of Kurnoor Medium Project on Bori River

Shital Shinde[1], Shreenivas Londhe[2] and Preeti Kulkarni[3]
[1]PG Student, [2]Professor, [3]Associate Professor
Vishwakarma Institute of Information Technology Pune,
[1]shitalkumarshinde21@gmail.com,[2]shreenivas.londhe@viit.ac.in,[3]preeti.kulkarni@viit.ac.in

ABSTRACT

Flood mapping is the process of defining the area covered by water in flood event. Flood information in a graphical format (flood map) enables us to better prepare for potential flood conditions. It is important to have easy to read graphical information so as to mitigate the consequences of flood.

In this report, study area is 25 km downstream of Kurnoor medium project on Bori River in Naldurg Tahsil of Maharashtra State. Flow hydrograph of 25 year and 100 years return flood is prepared based on CWC manual of synthetic unit hydrograph. Using digital elevation model and ArcGIS, river geometry is generated. Manning's 'n' value is extracted from land use and land cover map with help of ArcGIS tool. Unsteady flow modeling is done in HECRAS and based on results flood lines for Restrictive i.e red line and Prohibitive zone i.e blue line are marked. Further, flow spread area due to return period 25 years, 50 years and 100 year is compared.

Keywords: ARCGIS, HECRAS, HEC-geoRAS, floodplain mapping.

Introduction

India receives precipitation in the form of rainfall which causes flood in some areas and drought at other. To know the flood spread in advance to prevent the damages due to overflow of adjoining river banks, the flood line marking plays an important role. Flood is an unusual high stage of a river overflowing its bank and inundating the marginal lands. This is due to severe storm with unusual meteorological combination on the catchment. This may also due to shifting of the course of the river, earth quake causing bank erosion, or blocking of river, or breaching of the river flood banks. Flooding is one of the serious natural hazards in the world. Flood zone marking is very important to ascertain likely inundation area in aspect of cities, town and villages along the river banks. The flood zone marking helps in preventing loss of life, property and environment by regulating land usage in prohibitive and Restrictive zones. The marking of flood zones helps in removal of encroachments to keep required waterway clear for safe passage of floods. The Prohibitive Zone is classified as, the river channel required for passing a 25 years return period flood or a flood equivalent to 1.5 times the flood carrying capacity of established river channel whichever is higher. It is commonly denoted as Blue line.[1] The Restrictive Zone is classified as, area required to pass the spillway design flood of the reservoir or 100 years return flood from free catchment, excluding area of prohibitive zone. It is commonly denoted as Red line.[1]

Necessity of Work

Flood causes loss of lives, property wealth and overall prosperity of nature. In last year many states such as kerala, Assam Bihar have affected with flood. In Maharashtra also flood had caused severe damages. In order to mitigate flood consequences and proper monitoring, the National Green Tribunal has verified many sites in Maharashtra and Suggested to prepare flood maps. Kurnoor medium project is one of these sites.

Objectives

(a) To Estimate peak discharge rate for flood of return period 25, 50 and 100 years for study area.

(b) To calculate and compare flow spread area for different return floods.

(c) To prepare flood map i.e mark blue and red line for 25 km downstream of kurnoor medium project for 25 and 100 years return flood.

Study Area and Data Used

Kurnur Medium Project is located on Bori river in at Motyal in Akkolkot town of Muncipal council in Solapur District with its latitude and longitude as 17'62'' N and 76'21'' E respectively. In the present study 25 km

Downstream of Kurnur Medium project is considered for marking of Flood Zones. The location of study area is shown in Figure 1. [4]

The data required for this study is obtained from different sources. DEM of 32m resolution is downloaded from Bhuvan site using CARTOSAT-1 satellite data. The required study area is clipped from DEM which the help of shape file. The CWC Manual on ''Flood Estimation report for Krishna and Pennar Sub zone 3h'' is used for preparation of flood hydrograph of 25, 50 and 100 years return flood. The land cover and land use data is obtained from website of (ORNL DAAC) is one of the NASA Earth Observing System Data and Information System (EOSDIS) data centers.

Figure 1 Location Map

Methodology

In this study the HEC-RAS model has been used to get the surface water profiles of of area downstream of kurnoor project. Frequency analysis of discharge data was carried out Using synthetic Unit Hydrograph. Secondly, geometry data were prepared from the DEM downloaded from Bahuvan site of 32m resolution. Digitization of selected stretch of study area on DEM is done using Hec-GeoRAS and ARCGIS Tool. The geometrical data of a river is prepared using a tool called HEC-GeoRAS which assists in preparing input file as well as post processing of the HEC-RAS results in GIS environment The important layers that are created are the stream centerline, Flow path centerlines, main channel banks and cross section cut lines as RAS layers. These parameters are used to establish series of cross-sections along the stream. The results of HEC-GeoRAS are exported to HEC-RAS in .sdf format. Unsteady flow modeling is carried out in HECRAS. The results obtained are exported back to ARCGIS and flood delineation maps are prepared for different return floods. The below tree diagram gives general idea about the steps involved in the project.

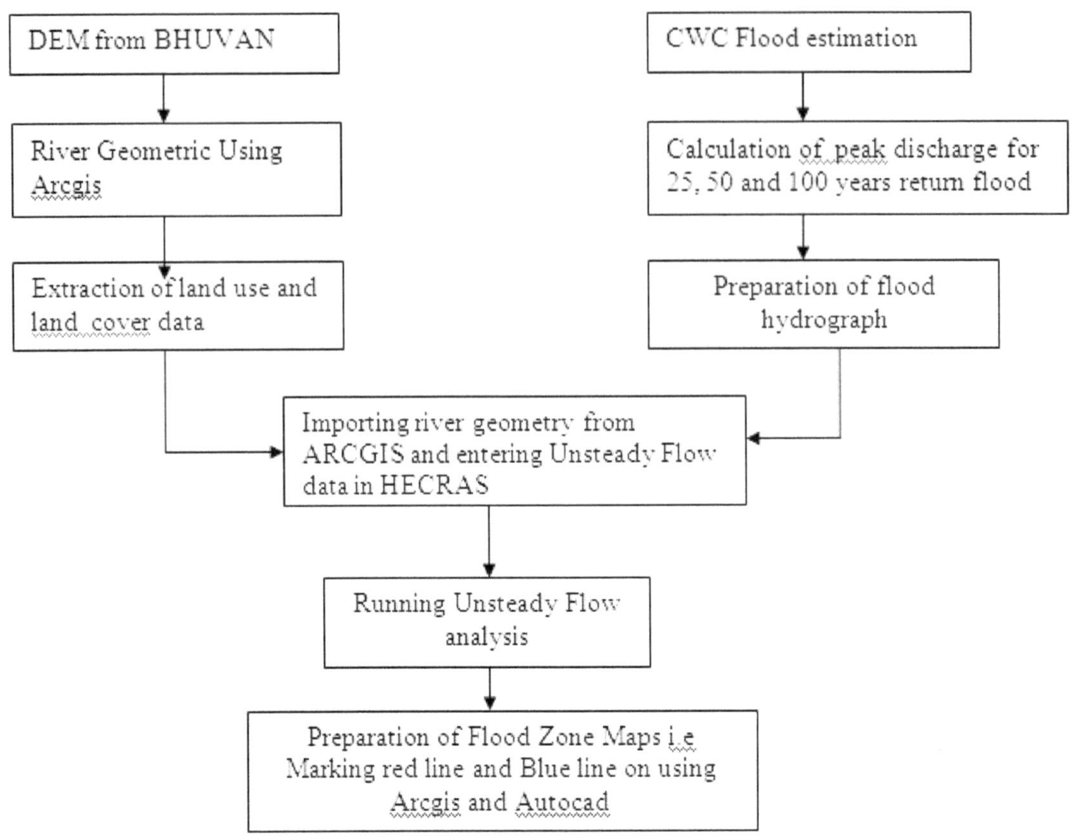

Figure 2 Flow chart of Methodology

In this study, Tools like ARCGIS, HECRAS, HECGEORAS is used for various processing's. HEC-RAS is a computer program that models the hydraulics of water flow through natural rivers and other channels. For unsteady flow, HEC-RAS solves the full, dynamic, 1-D Saint Venant Equation using an implicit, finite difference method. HEC-RAS is a computer program for modeling water flowing through systems of open channels and computing water surface profiles.

ArcGIS is a geographic information system (GIS) for working with maps and geographic information. It is used for creating and using maps, compiling geographic data, analyzing mapped information, sharing and discovering geographic information, using maps and geographic information in a range of applications, and managing geographic information in a database.

Results and Discussion

HEC-RAS model determines the water profiles for various return periods by using flow hydrograph. . The peak discharge for various return periods are tabulated in Table No 1. These flood hydrographs are used in hydraulic model to determine the water surface profiles with respect to various return periods to know flood extent.

Table 1 Peak discharge of different return flood

Sr. No.	Return period	Peak Discharge
1	25 years	1354.49 m3/sec
2	50 years	1508.38 m3/sec
3	100 years	1829.03 m3/sec

The peak discharge is calculated based on synthetic unit hydrograph. 1 hour synthetic unit hydrograph is shown in Figure 3.

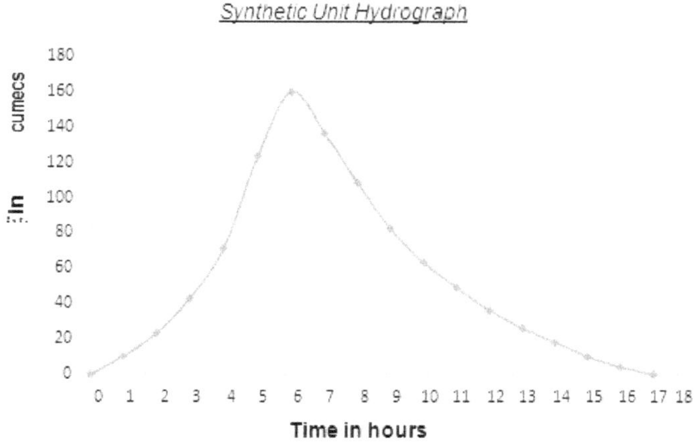

Figure 3 1 hr synthetic unit hydrograph of different return floods

The figure 4 shows flood hydrograph of 25, 50 and 100 years return flood. Figure 5, 6 and 7shows flood delineation maps for return period floods of 25, 50 and 100.

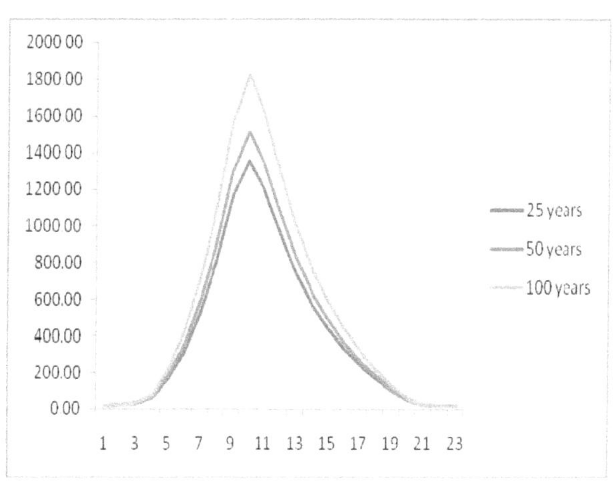

Figure 4 Flood hydrograph of different return floods

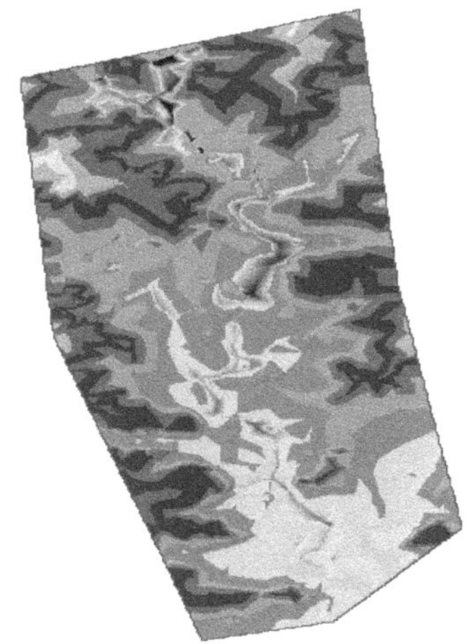

Figure 5 Flood delineation map 25 years

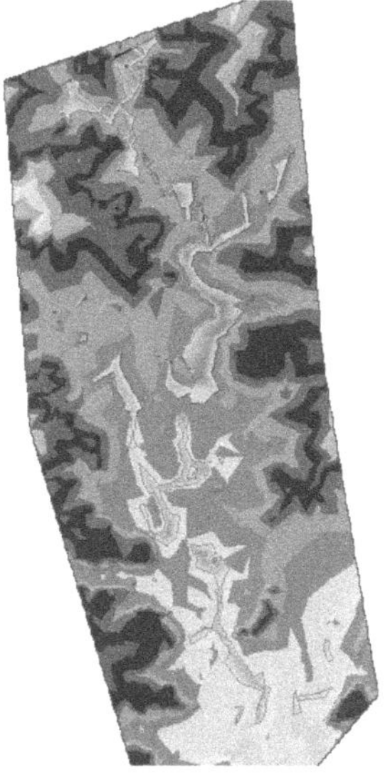

Figure 6 Flood delineation map 50 years

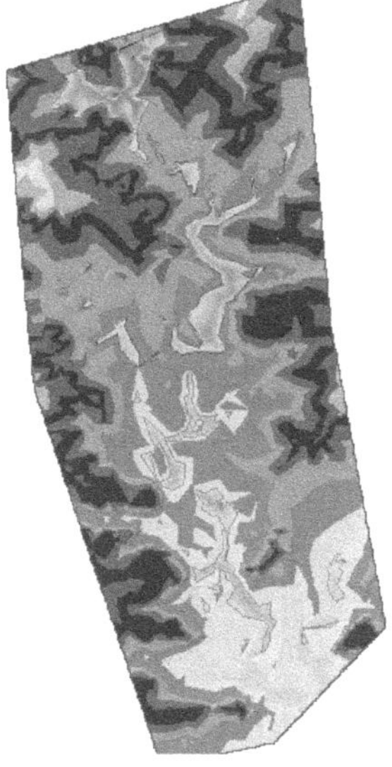

Figure 7 Flood delineation map 100 years

Unsteady flow model is used in HECRAS to calculate flow spread area for different return floods. Using ARCGIS, total 57 cross sections are taken across the river. HECRAS has calculated flow area at all cross-sections. Output of HECRAS is exported to ARCGIS for preparation of flood delineation map. These maps are further exported to AutoCAD to prepare Flood maps showing blue and red lines.

Also, at eleven cross section stations cumulative flow area from one station to another is tabulated as shown below. The water spread area is compared at these cross section stations.

Table 2 Flow spread area due to different return floods.

Sr. No.	River Station in KM	25 years flow area in M2	50 years flow area in M2	100 years flow area in M2
1	From 26721.45 to 25203.04	10550.01	17001.92	46944.33
2	From 24837.8 to 23493.41	4230.4	11242.5	11607.57
3	From 23216.1 to 21555.62	2390.92	3650.66	4476.99
4	From 21308.19 to 18945.91	5317.04	4796.67	5000.51
5	From 18659.33 to 17090.51	6723.64	7175.56	7568.78
6	From 16648.43 to 14922.8	14875.31	15092.67	15680.46
7	From 14472.27 to 12826.03	11254.76	11500.7	12110.14
8	From 12387.3 to 9645.094	5349.59	5420.19	5509.24
9	From 9100.302 to 6617.284	5415.36	5441.18	5532.19
10	From 6328.955 to 4550.959	2300.4	3229	4042.28
11	From 4082.557 to 1435.066	4042.35	3484.09	3763.62

Conclusion

ARCGIS, HEC-RAS and DEM are useful in inundation mapping, watershed and flood plain delineation. The integration of HEC-RAS and ARCGIS helps to save time that required by Physical Survey of river. Synthetic unit hydrograph is very useful for calculation of peak discharges and hydrographs for unguaged river channels. From results it is to state that, up to 7 km downstream of Kurnoor medium project there is much variance in flow area of different return periods. After that the variance is not more for this study area. The flood map will help us to suggest certain measures to mitigate flood during flood season reducing the impact of flood on human and damages.

References

Technical circular on marking of Prohibitive (Blue) and Restrictive (Red) flood zones from MERI, Nashik.

Central Water Commission's ''*Flood Estimation Report for Krishna and Pennar Subzone*'', September 2000.

Manual on Estimation of Design Flood by Hydrology studies organization by Central water Commission, March 2001.

Maharashtra Krishna Valley Development Corporation Pune,'' *Integrated state water Plan for Lower Bhima Sub basin (K-6) of Krishna Basin Report*''.

Sandhyarekha and A. V. Shivapur, ''*Floodplain Mapping Of River Krishnana Using Hec-Ras Model At Two Streaches Namely Kudachi And Ugar Villages Of Belagavi District, Karnataka* '' Irjet, Volume 4,2017 Page 1524-1529.

Sunilkumar P (2017) conducted as study on ''*Flood Modelling Of Mangalam River Using Gis And Hec-Ras*'' IRJET, Volume 7, 2018 Page 1523-27.

Sania Modak and PrashantNagarnaik ''*Flood Control and Prediction of Flood Using HEC-RAS – A Review*'' IJSR, volume, volume 6, issue 3, march 2017.

Morphometric Analysis of Kinnerasani Basin, India

L. Mani Teja[1] and M.V.S.S Giridhar[2]
M.Tech Student, Professor, Centre for Water Rsources, IST, JNTUH

ABSTRACT

The analysis of morphometric parameters with the help of geographic information system (GIS) would prove a viable method of characterizing the hydrological response behaviour of the watershed. It is also well observed that remote sensing satellite data is emerging as the most effective, time saving and accurate technique for morphometric analysis of a basin. Sub-watersheds are also delineated within this basin to calculate the selected morphometric parameters. Morphometric parameters viz; stream order, stream length, bifurcation ratio, drainage density, stream frequency, form factor, circulatory ratio, etc., are calculated. This technique is found relevant for the extraction of river basin and its stream networks through SRTM (DEM). Study area is Kinnerasani watershed, geographically located between 80^0 9'10.84" – 80^0 41'28.82"E longitudes, and 17^0 39'21.3"- 18^0 9' 12.3"N latitudes located in Telangana state in India. The GIS based Morphometric analysis of this drainage basin revealed that the is 4th order drainage basin and drainage pattern mainly in sub-dendritic to dendritic type thereby indicates homogeneity in texture. This study would help the local people to utilize the resources for sustainable development of the basin area.

Keywords: Drainage morphometry. Kinnerasani basin. DEM GIS

Introduction

Geographic information system (GIS) would prove a viable method of characterizing the hydrological response behaviour of the watershed by delineation, and morphometric analysis of drainage basin. Various important hydrologic phenomena can be correlated with the physiographic characteristics of drainage basins such as size, shape, slope of drainage area, drainage density, size and length of the tributaries etc. (Rastogi et al., 1976). Remote sensing data can be used in conjunction with conventional data for delineation of ridgelines, characterization, priority evaluation, problem identification, assessment of potentials and management needs, identification of erosion prone areas, evolving water conservation strategies, selection of sites for check dams and reservoirs etc.,(Dutta et al.,2002). The present paper describes the drainage characteristics of Kinnerasani watershed area in Telangana obtained through RS GIS based morphometric analysis. It is felt that the study will be useful to understand hydrological behavior of basin. It is well established that the influence of drainage morphometry is very significant in understanding the landform processes, soil physical properties and erosional characteristics.

Study Area

The present study area is located in the Telangana. Kinnerasani River basin has natural boundaries of the Godavari river. The Kinnerasani River tributary of Godavari River mainly drain the study area. The climate of the district is tropical. The average rainfall of the area is 970mm and temperature goes up to 46°C in summer and comes down to 16°C in winter season. The Kinnerasani Basin area is geographically located between 80^0 9'10.84" – 80^0 41'28.82" E longitudes, and 17^0 39'21.3"- 18^0 9' 12.3"N latitudes.

Methodology

Automatic extraction techniques have been used for evaluating the morphometric parameters of a basin, i.e., extraction of River basin/watershed boundary and extraction of drainage/stream network from the Kinnerasani River basin using SRTM DEM. The extracted basin and stream networks are projected to the regional projection (WGS-1984, UTM zone 44 N). The different morphometric parameters have been determined with the help of various geo-processing techniques in ArcGIS-10.5.

Morphometric parameters under linear and shape are computed using standard methods and formulae (Horton 1932, 1945; Smith 1954; Strahler, 1964). The values of morphometric parameters namely; stream length, stream length ratio, bifurcation ratio, basin length, basin area, relief ratio, elongation ratio, drainage density, stream frequency, form factor and circulatory ratio, etc., have been analysed using the standard mathematical formulae suggested by Horton (1945), Miller (1953), Schumn (1956), Strahler(1964), Nookaratm (2005) .

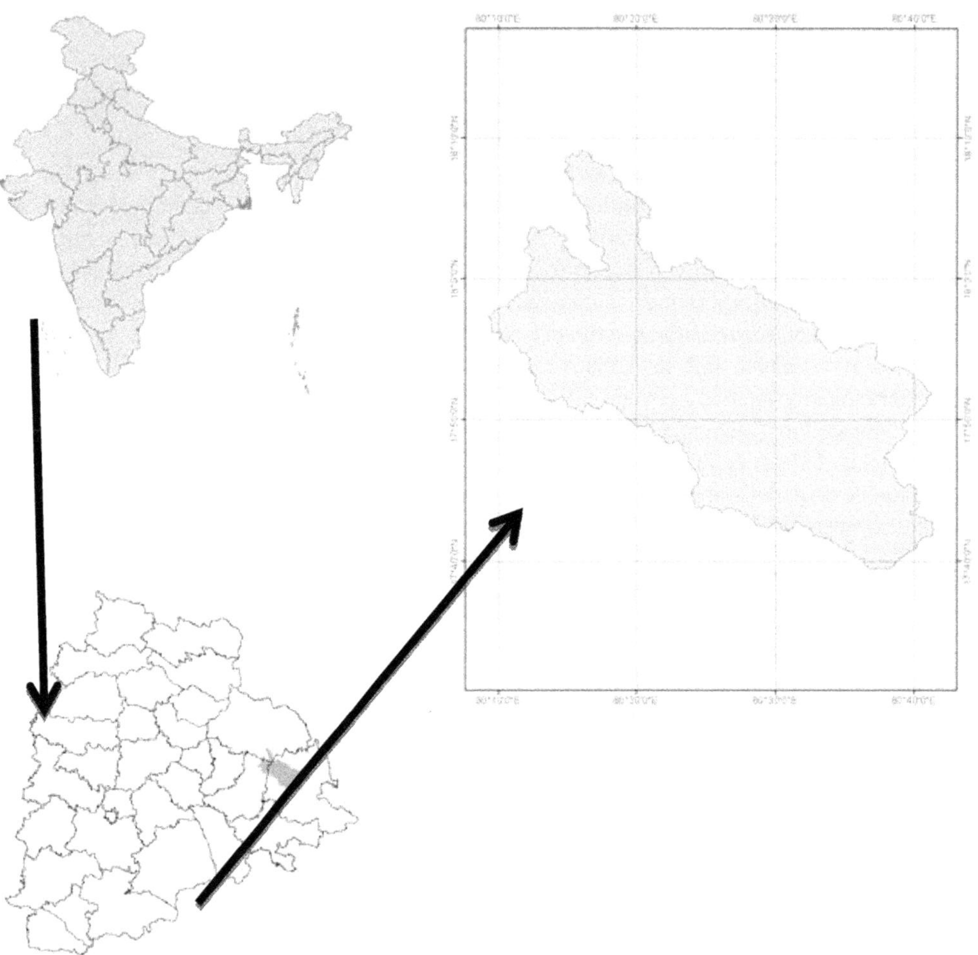

Figure1 Study Area –Kinnerasani Watershed.

Morphometric analysis of basin

The morphometric parameters of Kinnerasani watershed have been calculated and the results are given.

Linear Aspects

The linear aspects of morphometric analysis of basin include stream order, stream length, stream length ratio and bifurcation ratio.

Stream Order (U)

Strahler's (1964) system, which is a slightly modified of Horton system, has been followed for Kinnerasani basin. As 1st order, the confluence of two 1st order channels give a channels segments of 2nd order, two 2nd order streams converge to form a segment of 3rd order and so on. When two channels of different order join then the higher order is attained.

Stream Length (Lu)

The stream length (Lu) has been computed based on the law proposed by Horton. The total length of individual stream segments of each order is the stream length of that order. The stream of relatively smaller length is characteristics of areas with larger slopes and finer textures. The observation of stream order verifies the Horton's law of stream number i.e. the number of stream segment of each order forms an inverse geometric sequence with order number.

Table 1 Formulae for calculation of various watershed parameters

Parameters	Formulae	References
Linear Aspects		
Stream Order (U)	The smallest permanent streams are called "first order". Two first order streams join to form a larger, second order stream; two second order streams join to form a third order, and so on. Smaller streams entering a higher ordered stream do not change its order number.	Strahler (1964)
Stream Length (Lu)	The average length of streams of each of the different orders in a drainage basin tends closely to approximate a direct geometric ratio.	Horton (1945)
Stream Length Ratio (RL)	RL= Lu/(Lu-1)	
Bifurcation Ratio (Rb)	Rb = Nu/(Nu+1)	Horton (1932)
Areal Aspects		
Stream Frequency (Fs)	$Fs = \sum Nu/A$	Horton (1945)
Drainage Density (Dd)	Dd = Lu/A	Horton (1945)
Drainage Texture (T)	T = Dd x Fs	Smith (1950)
Elongation ratio (Re)	$Re = 1.128*\sqrt{A}/L$	Schumm (1956)
Circularity ratio (Rc)	$Rc = 4\prod A/P^2$	Strahler (1964)
Form factor (Ff)	$F_f = A/L^2$	Horton (1945)
Relief Aspects		
Relief (R)	R = H – h	Hadley and Schumm (1961
Relief Ratio (Rr)	Rr = R/L	Schumm (1963)

Stream Length Ratio (RL)

The Stream length ratio is the ratio of mean or average length of segment of order 'u' to the mean or average length of order 'u-1'.The stream length ratio can be defined as the ratio of the mean stream length of a given order to the mean stream length of next lower order and has an important relationship with surface flow and discharge (Horton, 1945). The RL values between streams of different order in the basin reveal that there are variations in slope and topography.

Table 2 Stream Analysis of Kinnerasani Watershed

Sub basins	Area (km2)	No of streams of each order							Stream length of each order(Lu) in km					
		1	2	3	4	5	6	total	1	2	3	4	5	6
1	179.254	404	200	71	59	55	0	789	138.3125	85.13871	20.441	23.30753	20.70231	0
2	128.2634	253	132	39	59	34	0	517	102.6541	54.0908	16.38701	18.14227	12.49636	0
3	58.03264	109	48	32	1	32	2	224	54.05279	21.49541	14.34908	0.264792	11.21269	0.037366
4	92.44267	172	89	61	8	2	35	367	68.93739	39.68766	26.50222	6.049234	0.086122	10.7278
5	46.8846	110	58	48	23	2	0	241	42.35825	16.67335	13.57929	8.470454	0.048748	0
6	137.8645	237	123	82	19	45	3	509	108.4614	57.97545	38.83362	13.47594	17.35344	0.133029
7	61.84192	108	60	30	11	0	30	239	51.04803	26.5221	13.28938	6.050551	0	7.195756
8	69.41407	99	31	47	10	5	1	193	52.86288	14.12331	26.18651	8.393232	2.987032	0.028796

Sub basins	Area (km2)	No of streams of each order							Stream length of each order(Lu) in km					
		1	2	3	4	5	6	total	1	2	3	4	5	6
9	116.3148	229	111	54	43	2	0	439	87.09937	48.38626	26.4322	19.73833	0.41594	0
10	90.339	172	72	41	14	1	43	343	68.24352	39.86938	17.57263	5.200747	0.042567	13.44996
11	103.8155	223	104	34	41	26	0	428	84.22795	44.1524	13.68483	17.97826	7.123145	0
12	9.863232	23	5	0	0	2	24	54	8.87315	2.475141	0	0	0.057042	5.306629
13	32.9277	78	35	16	0	5	34	168	27.16275	10.83047	4.505378	0	0.478461	10.21777
14	76.64242	173	72	60	16	24	2	347	61.51216	27.66486	22.72515	3.660473	5.60358	0.135719
15	130.0753	316	143	82	34	0	61	636	112.082	61.36118	28.49382	12.07475	0	13.92732
Total	1333.975752	2706	1283	697	338	235	235	5494	1067.888	550.4465	282.9821	142.8066	78.60744	61.16013

Bifurcation Ratio (Rb)

Bifurcation ratio is the ratio of a number of the stream segments of specified order and a number of streams in the next higher order. Bifurcation ratio (Rb) may be defined as the ratio of the number of stream segments of given order to the number of segments of the next higher order (Schumn 1956). Horton (1945) considered the bifurcation ratio as an index of relief and dissections

Table 3 Bifurcation ratio of Kinnerasani Watershed

Sub basins	No of streams of each order(Nu)							bifurcation ratio of each order(rb)				
	1	2	3	4	5	6	total	Nu/ Nu+1	Nu/ Nu+1	Nu/ Nu+1	Nu /Nu+1	Nu/ Nu+1
1	404	200	71	59	55	0	789	2.02	2.816901	1.20339	1.072727	0
2	253	132	39	59	34	0	517	1.916667	3.384615	0.661017	1.735294	0
3	109	48	32	1	32	2	224	2.270833	1.5	32	0.03125	16
4	172	89	61	8	2	35	367	1.932584	1.459016	7.625	4	0.057143
5	110	58	48	23	2	0	241	1.896552	1.208333	2.086957	11.5	0
6	237	123	82	19	45	3	509	1.926829	1.5	4.315789	0.422222	15
7	108	60	30	11	0	30	239	1.8	2	2.727273	0	0
8	99	31	47	10	5	1	193	3.193548	0.659574	4.7	2	5
9	229	111	54	43	2	0	439	2.063063	2.055556	1.255814	21.5	0
10	172	72	41	14	1	43	343	2.388889	1.756098	2.928571	14	0.023256
11	223	104	34	41	26	0	428	2.144231	3.058824	0.829268	1.576923	0
12	23	5	0	0	2	24	54	4.6	0	0	0	0.083333
13	78	35	16	0	5	34	168	2.228571	2.1875	0	0	0.147059
14	173	72	60	16	24	2	347	2.402778	1.2	3.75	0.666667	12
15	316	143	82	34	0	61	636	2.20979	1.743902	2.411765	0	0
Total	2706	1283	697	338	235	235	5494	2.109119	1.840746	2.06213	1.438298	1

Aerial Aspects

The aerial aspects of the drainage basin such as Drainage Density (Dd), Stream Frequency (Fs), Drainage Texture (T), Elongation Ratio (Re), Circularity Ratio (Rc) and Form Factor Ratio (Rf) were calculated.

Drainage density (Dd)

Drainage density (Dd) is a measure the total stream length in a given basin to the total area of the basin (Strahler 1964). The drainage density indicates the closeness of spacing of channels, thus providing a quantitative measure of the average length of stream channel for the whole basin. the drainage density (Dd) is an important indicator of the linear scale of land form elements in stream eroded topography. It is the ratio of total channel segment length cumulated for all order within a basin to the basin area, which is expressed in terms of Km/Km^2. High drainage density is the resultant of weak or impermeable subsurface material, sparse vegetation and mountainous relief. Low drainage density leads to coarse drainage texture while high drainage density leads to fine drainage texture (Strahaler, 1964).

Stream Frequency (Fs)

Stream frequency is directly related to the lithological characteristics. The number of stream segments per unit area is termed as Stream Frequency (Fs) or Channel Frequency or Drainage Frequency (Horton 1945).

Drainage Texture (T)

Drainage texture may be defined as the total number of stream segments of all order in a basin per perimeter of the basin (Horton, 1945). It is important to understand geomorphology which indicates the relative spacing of drainage lines.

Elongation Ratio (Re)

Elongation ratio (Re) may be defined as the ratio of diameter of a circle of the same area as the basin to the maximum basin length (Schumm 1956). It is a measure of the shape of the river basin and it depends on the climatic and geologic types. A circular basin is more efficient in runoff discharge than an elongated basin (Singh and Singh 1997).

Circularity Ratio (Rc)

Circularity Ratio is the ratio of the area of a basin to the area of circle having the same circumference as the perimeter of the basin (Miller, 1953). The value of circularity ratio varies from 0 (in line) to 1 (in a circle). Circulatory ratio is influenced by the length and frequency of streams, geological structures, land use/land cover, climate, relief and slope of the basin.

Form Factor (Ff)

Form factor (Ff) is defined as the ratio of the basin area to the square of the basin length. This factor indicates the flow intensity of a basin of a defined area (Horton, 1945).

Table 4 Areal aspects of Kinnerasani Watershed

Sub Basins	Area (km2)	Drainage density of each order(Dd)							Stream frequency	drainage texture
		1	2	3	4	5	6	Total	Fs=∑Nu/A	T=Dd*Fs
1	179.254	0.7716	0.4750	0.1140	0.1300	0.1155	0.0000	1.6061	4.40158	7.06942
2	128.2634	0.8003	0.4217	0.1278	0.1414	0.0974	0.0000	1.5887	4.03077	6.40363
3	58.03264	0.9314	0.3704	0.2473	0.0046	0.1932	0.0006	1.7475	3.85990	6.74518
4	92.44267	0.7457	0.4293	0.2867	0.0654	0.0009	0.1160	1.6442	3.97003	6.52736
5	46.8846	0.9035	0.3556	0.2896	0.1807	0.0010	0.0000	1.7304	5.14028	8.89485
6	137.8645	0.7867	0.4205	0.2817	0.0977	0.1259	0.0010	1.7135	3.69203	6.32635
7	61.84192	0.8255	0.4289	0.2149	0.0978	0.0000	0.1164	1.6834	3.86469	6.50589
8	69.41407	0.7616	0.2035	0.3773	0.1209	0.0430	0.0004	1.5066	2.78042	4.18908
9	116.3148	0.7488	0.4160	0.2272	0.1697	0.0036	0.0000	1.5653	3.77424	5.90797

Sub Basins	Area (km2)	Drainage density of each order(Dd)							Stream frequency	drainage texture
		1	2	3	4	5	6	Total	Fs=∑Nu/A	T=Dd*Fs
10	90.339	0.7554	0.4413	0.1945	0.0576	0.0005	0.1489	1.5982	3.79681	6.06802
11	103.8155	0.8113	0.4253	0.1318	0.1732	0.0686	0.0000	1.6102	4.12270	6.63848
12	9.863232	0.8996	0.2509	0.0000	0.0000	0.0058	0.5380	1.6944	5.47488	9.27647
13	32.9277	0.8249	0.3289	0.1368	0.0000	0.0145	0.3103	1.6155	5.10209	8.24244
14	76.64242	0.8026	0.3610	0.2965	0.0478	0.0731	0.0018	1.5827	4.52752	7.16570
15	130.0753	0.8617	0.4717	0.2191	0.0928	0.0000	0.1071	1.7524	4.88948	8.56813
Total	1333.975752	0.8005	0.4126	0.2121	0.1071	0.0589	0.0458	1.6371	4.11852	6.74254

Table 5 Areal aspects of Kinnerasani Watershed

Sub basins	Area(km^2)	perimeter(km)	circularity ratio	length(km)	Elongation ratio	formfactor
			Rc=4∏A/P2			
1	179.254	89.3096	0.2824	36.91	0.409165947	0.1315772
2	128.2634	62.383	0.4142	35.28	0.362102812	0.1030494
3	58.03264	44.968	0.3606	19.64	0.437525913	0.150449
4	92.44267	46.896	0.5282	18.57	0.584027687	0.2680703
5	46.8846	37.045	0.4293	20.76	0.372046187	0.1087866
6	137.8645	64.54	0.4159	30.06	0.440601883	0.1525719
7	61.84192	39.243	0.5046	16.43	0.539899616	0.2290909
8	69.41407	56.333	0.2749	20.15	0.466399208	0.1709611
9	116.3148	62.215	0.3776	27.58	0.441095195	0.1529137
10	90.339	50.013	0.4539	22.3	0.480774999	0.1816626
11	103.8155	53.864	0.4497	24.2	0.474924784	0.1772685
12	9.863232	14.31	0.6053	7.18	0.493394472	0.1913244
13	32.9277	27.036	0.5661	11.67	0.554649903	0.2417796
14	76.64242	45.026	0.4751	19.96	0.494746998	0.1923748
15	130.0753	64.105	0.3978	24.96	0.515420795	0.2087881
Total	1333.975752	757.2866	0.0292	298.7400	0.137908073	0.0149472

Relief Aspects

Basin Relief

Basin relief is the elevation difference of the highest and lowest point of the valleyfloor. Basin Relief plays a significant role in landform development, drainage development, surface and subsurface water flow, permeability and erosional properties of the terrain.

Relief Ratio

The relief ratio, (Rh) is ratio of maximum relief to horizontal distance along the longest dimension of the basin parallel to the principal drainage line (Schumm, 1956).

Table 6 Relief aspects of Kinnerasani Watershed

Sub Basins	Relief Aspect				
	H	H	Relief	Length(km)	Relief ratio
1	0.504	0.234	0.27	36.91	0.007
2	0.582	0.238	0.344	35.28	0.010
3	0.465	0.203	0.262	19.64	0.013
4	0.574	0.201	0.373	18.57	0.020
5	0.641	0.197	0.444	20.76	0.021
6	0.443	0.181	0.262	30.06	0.009
7	0.417	0.164	0.253	16.43	0.015
8	0.434	0.171	0.263	20.15	0.013
9	0.423	0.123	0.3	27.58	0.011
10	0.418	0.143	0.275	22.3	0.012
11	0.407	0.123	0.284	24.2	0.012
12	0.242	0.12	0.122	7.18	0.017
13	0.562	0.122	0.44	11.67	0.038
14	0.433	0.143	0.29	19.96	0.015
15	0.418	0.088	0.33	24.96	0.013
Total	6.963	2.451	4.512	298.7400	0.015

Conclusion

Morphometric analysis of drainage system is prerequisite to any hydrological study. GIS and Remote sensing techniques have proved to be accurate and efficient tool in drainage delineation. the drainage orders, the Kinnerasani watershed has been classified as sixth order basin. The drainage density (Dd) of study area is 1.45 km/km^2 which indicating impermeable surface materials. The drainage texture of watershed falls under the category of fine drainage texture (6-8) Drainage density, texture ratio, circulatory ratio and elongation ratio shows that texture of basin is fine and shape of basin is elongated, these streams are situated on hilly terrains and volume of water discharge through river tributaries in the basin is more due to presence of large number of streams of lower order present in Kinnerasani watershed. It is observed that the regions with steep slope contributes to high erosion and surface runoff.

References

Martin D. and Saha S.K. (2007) Integrated approach of using Remote Sensing and GIS to study watershed prioritization and productivity. Jour. Indian Soc. Remote Sensing, v.35(1), pp21-30.

Horton, R. E. (1932). Drainage-basin characteristics. Transactions, American Geophysical Union, 13, 350–361.

Horton, R.E. (1945) Erosional development of streams and their drainage density: hydrophysical approach to quantitative geomorphology. Geol. Soc. Amer. Bull., no.56, pp.275-370.

Melton, M.A. (1957)An analysis of the relations among elements of climate, surface properties and geomorphology. Dept. Geol., Columbia University, Technical Report, 11, Proj.NR389-042.off. of Nav. Res., New York.

Strahler, A.N. (1964) Quantitative geomorphology of drainage basins and channel networks. In: V.T. Chow (Ed.), Handbook of Applied Hydrology. McGraw-Hill, New York, pp.4.39-4.76.

Strahelr AN (1952) Hypsometric (area-altitude) analysis of erosional topography. Bull Geol, Soc Am 63

Strahler AN (1957) Quantitative analysis of watershed geomorphology.

Trans Am Geophys Union 38:913–920

Schumm, S.A. (1956) Evolution of drainage systems and slopes in Badlands at Perth Amboy, New Jersey. Bull. Geol. Soc. Amer., v.67, pp.597-646.

M.L.Waikar and Aditya P. Nilawar Morphometric Analysis of a Drainage Basin Using Geographical Information System: A Case study

P. Dinagara Pandi, T. Thena, B. Nirmal, M. R. Aswathy, K. Saravanan & K. Mohan (2017) Morphometric analyses of Neyyar River Basin, southern Kerala, India, Geology, Ecology, and Landscapes.

Praveen Kumar Rai, Kshitij Mohan, Sameer Mishra, Aariz Ahmad & Varun Narayan Mishra A GIS-based approach in drainage morphometric analysis of Kanhar River Basin, India

Assessment of Water Quality Parameters at Kapuluppada Dumpsite in Visakhapatnam, Andhra Pradesh

K. Syamala Devi

Asst. Prof. of ES, G. Narayanamma Institute of Technology of Science, Shaikpet, Hyderabad.

ABSTRACT

Land filling is the preferred method of Municipal Solid Waste (MSW) disposal. However, poorly designed land fill leads to contamination of ground water, soil and air. As water percolates through the landfill, contaminants are leached from the solid waste. Leachate is produced when moisture enters the refuse in a landfill, extracts contaminants into the liquid phase. Leachate is generated in a landfill as a consequence of the contact of water with solid waste. Leachate contains dissolved or suspended material associated with wastes disposed off in the land fill, as well as many by-products of chemical and biological reactions. Leachate tend to migrate in surrounding soil may result in contamination of underlying ground water and soil. The rate at which it percolates depends on the soil, texture and depth of the aquifer. It is established that the ground water was contaminated by the leachate in many cities not only in India and but in many other countries. The peri-urban areas face severe problem of ground water contamination from MSW dumpsites. Owing to the importance of the topic, the present work has been carried near Kapuluppada dumpsite in Visakhapatnam, Andhra Pradesh to study the physico – analysis of water quality parameters at selected stations and results were discussed.

Keywords: MSW, physico- chemical analysis, refuse, leachate

Introduction

The generation of solid waste is not a new phenomenon. It is directly related to human civilization. The first organized dump of solid waste is reported to have been set up outside the Athens in 500 B.C. (Krishnamoorthy, 2001). The problems associated with solid waste management in all major cities of developing countries are more acute than in the major cities of developed countries. The rapid growth in population and urbanization, adds greatly to the volume of waste being generated in the cities which also adds to demand for waste retrieval service in municipal areas. On the hand, an increase in population is not matched with an equal increase in revenue for the municipal authorities for proper waste management.

The pressure on ground water has been increasing as the urbanization is gaining momentums. There is a general notion that ground water is safer than surface water. Several studies have proven this is fabe, as the earlier studies in many places reported the contamination of ground water by various sources. Owing to the importance of the topic, the present work has been carried near Kapuluppada dumpsite in Visakhapatnam, Andhra Pradesh to study the physico – analysis of water quality parameters at selected stations and results were discussed.

Visakhapatnam is the largest city in Andhra Pradesh with a population of 20 lakhs. An area of 100 acres at Kapuluppada village was selected as the dumpsite in 2004 for Landfilling located 15 Km away from the city. The present study has been carried out with the following objectives:

1. To study the physico-chemical characteristics of ground water in and around the MSW dump site at Kapuluppada, Visakhapatnam.
2. To assess the potability of ground water in six selected stations by comparing the results of the present study with standards specified by BIS (2012).
3. To study the seasonal changes in the quality of ground water at the six selected stations.
4. To assess whether there is any contamination of ground water by leachate from the MSW dumpsite.
5. To recommend measures to abate contamination of ground water by the MSW dumpsite.

Results

The present study on the “Impact of Municipal Solid Waste Dumpsite on the ground water quality near Kapuluppada at Visakhapatnam, A.P., India” has been carried out at Visakhapatnam from 2011-12 to 2014-15 (March, 2011 to January 2015). Kapuluppada was the landfill site with an area of 100 acres and in operation since 2004 in Visakhapatnam. Dumping is still actively going on. The rapid urban sprawl has been encircling the dumpsite area. Several residential colonies have come up in the area in which 80% of the housing was in the form of apartments and only 20% of the houses were independent. As result, the exploitation of ground water has increased manifold since 2010. Though there is no scarcity of ground water, the ground water quality has been deteriorating with excess of TDS, Total Hardness, Chlorides etc., affecting the potability of water. No systematic studies were taken up in these areas about the impact of MSW dumpsite on the ground water.

Visakhapatnam city generates 980 tonnes of garbage every day. The landfill site is located at a distance of 15 km to the west of the city, spread across 100 acres of land having an elevation of 45 m. The overall climate of the region was tropical with a maximum temperature of 45°C in summer, the winters are colder with 10°C temperature. Red loamy soils predominate. The soils are poor textured and easily drained. Sandy loam soils come next. In general, the soils are low in organic content and phosphorous. The prominent lineaments are trending in NE-SW, NW-SE and ENE-WSW.

The annual mean rainfall was 1191.5 mm. Major source of rainfall is south west monsoon (48.4%) followed by north east monsoon (18%) and the rest of the rainfall is contributed by occasional rains in winter and summer. The Visakhapatnam district experiences drought conditions too often, as no major irrigation system exists to cushion the vagaries of the monsoon.

Study Area of all six stations

Station V was a bore well located within the dumpsite at Kapuluppada. Dumping was started in 2004 at this site. In order to establish the formation and dispersion of leachate, and its impact on ground water, this station was selected. The bore well water is not in use by public. The quality of ground water is assessed. The pH of ground water ranged between 7.1 and 7.4 with a combined mean of 7.4. The water at this station was slightly alkaline and the pH was within the permissible range of 6.5 – 8.5 (BIS-2012) specified for drinking water.

Total Dissolved Solids (TDS) have exceeded the acceptable limit of 500 mg/l but within the permissible limit of 2000 mg/l. The lowest concentration of TDS were recorded during 2012-13 (1490 mg/l) and highest was during 2013-14 (1550 mg/l) with a combined mean of 1532 mg/l during the 4 year study period. The TDS will affect the palatability beyond 500 mg/l and may cause gastrointestinal irritation. The water at this site at the reported levels of TDS is not fit for drinking.

The concentrations of Chlorides have exceeded the acceptable limit of 250 mg/l. The lowest annual mean recorded in the study was 495 mg/l and highest was 520 mg/l with a combined mean for the study period of 509 mg/l. The chlorides beyond 250 mg/l will interfere with taste, palatability and cause corrosion. However a maximum permissible limit upto 1000 mg/l under no alternate source was given by BIS (2012). Though the ground water at Kapullupada dumpsite has high chloride concentration than the acceptable limit, the values were lesser when compared to the maximum permissible limit of 1000 mg/l. Hence the ground water at this station can be used only in the absence of alternate source.

The total hardness was very high at Station V when compared with the acceptable limit of 200 mg/l. The hardness recorded in this station varied between a minimum annual mean of 785 mg/l and a maximum of 814 mg/l with a combined mean of 810 mg/l. The hardness at this station not only exceeded the acceptable limit but also has exceeded the maximum permissible limit (600 mg/l). When hardness is considered, the water is unfit for drinking and other domestic uses. The nitrate concentration were within the acceptable limit of 45 mg/l and fairly low. The annual means during the study period ranged from a minimum of 10.86 mg/l and a maximum of 12.22 mg/l with a combined mean of 11.65 mg/l. At these levels, the nitrates in ground water at this station does not interfere with potability of water. However, this bore well has not been in use for a long time.

The concentrations of sulphates were within the permissible limit of 200 mg/l (BIS, 2012). The lowest annual mean recorded during the 4 year study period was 135 mg/l (2011-12) and the highest was 152 (2013-14) with a combined mean of 146 mg/l. Sulphates generally cause gastrointestinal irritation when magnesium or sodium was present. Though presently the ground water has lesser sulphates there is a chance of sulphate concentrations exceeding the acceptable limit in near future as the current mean is about 75% of the acceptable limit.

The iron concentrations were far below the acceptable limit of 0.3 mg/l (BIS, 2012). The annual means of the study period varied between 0.058 mg/l and 0.066 mg/l in 2014-15 and 2011-12 respectively, with a combined mean of 0.062 mg/l. The currently there is not chance of contamination of ground water by iron, unless a specific source of iron were added at the dump and leached into ground water.

The concentrations of zinc were fairly low in the bore well at dumpsite. The Zinc concentration was always below 0.05. The lowest annual mean recorded was 0.024 mg/l and 0.032 mg/l in 2013-14 and 2012-13, respectively, with a combined mean of 0.027. This shows that the MSW does not bear any source of Zinc and the zinc at the recorded concentrations does not cause any astringent taste to ground water. The phenols and cyanides were present in below detectable limits while lead, copper, Nickel, Chromium and cadmium were not detected in any of the samples. The Total Dissolved Solids, Chlorides and total hardness were higher than the acceptable limits. The TDS was higher than acceptable limit but lesser than maximum permissible limits, while total hardness was higher than the maximum permissible limits. All other parameters were within the permissible limits in the ground water at Kapuluppada Municipal Dump site.

Table 1 Annual Mean Concentrations of different water quality parameters compared with BIS Standards at Station V (at Kapuluppada MSW dumpsite) during the study period.

Sl. No.	Parameter	2011-12	2012-13	2013-14	2014-15	Mean	BIS STANDARD
1	pH	7.1	7.4	7.3	7.2	7.4	6.5-8.5
2	Total Dissolved Solids	1506	1490	1550	1538	1532	500 mg/l
3	Chlorides	495	520	506	512	509	250 mg/l
4	Total hardness as $CaCO_3$	785	814	802	806	810	200 mg/l
5	Nitrates Nitrogen (as NO_3-N)	10.94	11.62	12.22	10.86	11.65	45 mg/l
6	Sulphates	135	144	152	146	146	200 mg/l
7	Phenols	BDL	BDL	BDL	BDL	BDL	0.001 mg/l
8	Cyanide	BDL	BDL	BDL	BDL	BDL	0.05 mg/l

Sl. No.	Parameter	2011-12	2012-13	2013-14	2014-15	Mean	BIS STANDARD
9	Lead as Pb	ND	ND	ND	ND	ND	0.05 mg/l
10	Iron as Fe	0.066	0.061	0.064	0.058	0.062	0.3 mg/l
11	Zinc	0.028	0.032	0.024	0.030	0.027	5 mg/l
12	Nickel	ND	ND	ND	ND	ND	-
13	Copper	ND	ND	ND	ND	ND	0.05 mg/l
14	Chromium	ND	ND	ND	ND	ND	0.05 mg/l
15	Cadmium as Cd	ND	ND	ND	ND	ND	0.01 mg/l

Discussion

The generation of solid waste is not a new phenomenon. It is directly related to human civilization. The first organized dump of solid waste is reported to have been set up outside the Athens in 500 B.C. (Krishnamoorthy, 2001). The problems associated with solid waste management in all major cities of developing countries are more acute than in the major cities of developed countries. The rapid growth in population and urbanization, adds greatly to the volume of waste being generated in the cities which also adds to demand for waste retrieval service in municipal areas. On the hand, an increase in population is not matched with an equal increase in revenue for the municipal authorities for proper waste management.

The present study has been carried out in Visakhapatnam from March 2011 to January, 2015. Visakhapatnam city is spread in 540 Sq. Km with a solid waste generation of 980 Tonnes Per Day (TDP) municipal waste is sent to Kapuluppada Dump Site located at 15 Km to the west of Visakhapatnam. The landfill site is spread across 100 acres of land having elation of 45 m.

The ground water at Visakhapatnam occurs in all the geological formation. Ground water occurs under unconfined and semi confined conditions in the hard rock formations while it occurs under unconfined to confined conditions in soft formations.

The composition of MSW and leachate change from place to place and a seasonal change of composition was also reported. The dumped solid wastes gradually release its initial interstitial water and some of its decomposition by products gets into water moving through the waste deposit. This complex liquid is known as leachate. This leachate accumulates at the bottom of the landfill and percolates through the Soil. Areas near to landfill have a greater possibility of ground water contamination. Such contaminations of ground water contamination pose a substantial risk to local resource user to the environment (Zhu *et al.,* 2008).

The ground water quality was tested for 4 years in and the parameters analyzed were: pH, TDS, Chlorides, Total Hardness, Nitrate Nitrogen, Sulphates, Phenols, Cyanide, Lead, Iron, Zinc, Nickel, Copper, Chromium and Cadmium. The analysis results were compared with the standards prescribed by Bureau of Indian standards (BIS, 2012).

Conclusion

The results of the present study revealed that the total dissolved solids, total hardness and chlorides were in very concentrations. These parameters, though not directly affect the human health, indicate the increase of ions in the ground water through leachate contamination. The Kapuluppada dumpsite is a landfill and not scientifically designed. Hence, to prevent future contamination of heavy metals and organic and inorganic materials, the following recommendations are made:

A sanitary landfill site shall be planned, designed and developed with proper documentation of construction plan as well as closure plan in a phased manner. Waste processing units should be permitted in the vicinity.

A state of the art incineration plant shall be established to minimize solid waste problem. A buffer zone of no development shall be maintained around solid waste dumping and disposal facility. Special attention must be paid not to allow the biomedical waste to mix with MSW. Steps must be taken to segregate e-waste from MSW.

A proper leachate collection system should be installed with immediate effect. Approach roads must be either concretized or black topped to minimize the fugitive dust. The Landfilling must be compacted using heavy compactors A vermicomposting unit must be developed as an integral part at the edge of the dump site. Steps must be taken to prevent entry of stray cattle which feed on the vegetable waste and paper/cellulose material packed in polythene bags. Proper drainage system shall be provided to divert run-off away from the active cell of the landfill. A monitoring system should be established at the dump for analysis of landfill gases and leachate quality. Preferably a land fill gas recovery system be established. Care must be taken not to permit open dug wells in the surrounding villages, as the shallow waters are prone early for contamination. A Permeable Reactive Barriers (PRB) Technology for in situ ground water remediation shall also be considered, that passively capture a plume of contaminants and removes or breakdowns the contaminants, finally releasing uncontaminated water.

References

Abbasi, S. A. and Vinithan, S. 1999. Water Quality in and around an industrialized suburb of Pondicherry. The Indian Journal of Environmental Sciences and Health. 41(4): 253-263.

Abdullah, S., Iqbal, M. A., Ilyas, M., Shejule K.B. and Rehana, A. 2012. "Ground water Pollution at Beed, Maharastra as an Effect of MSW Dumping". Bull. Env. Pharmacol. Life Science. 1 (12): 43-46.

Abedi, J. and Bagheri. M. 2001. The effects of infiltrated sewage on ground water. Congress of the Impacts of Sewage in Ground Water and Surface Water, Tehran, Iran.

Adams, J. O. and Adetoro, E. O. 2014. Quality Assessments of Water in Selected Waste –Dumping Site Areas. International Journal Of Scientific Research And Education. 2(11): 2399-2404.

Adekunle, I. M., Adetunji M. T., Gbadebo A. M. and Banjoko O. B. 2007. Assessment of groundwater quality in a typical rural settlement in southwest Nigeria. International journal environmental research and public health. 1661-7827.

Adeolu O.A., Ada V.O., Gbenga A. A. and Adebayo A. O. 2011.Assessment of groundwater contamination by leachate near a municipal solid waste landfill. African Journal of Environmental Science and Technology. 5(11): 933-940.

Aderemi, A.O., Oriaku, A.V., Adewumi, G.A. and Otitoloju, A.A., 2011. Assessment of ground water contamination by leachate near a municipal solid waste landfill. Afr J Environ Sci Technol. 5(11): 933-940.

Adeyemo, O.K., Ayodeji, I.O., and Raji and Aiki CO. 2002. The Water Quality and Sanitary Conditions in a Major Abbatoir (Bodija) in Ibadan, Nigeria. African Journal of Biomedical Research. Ibadan Biomedical Communications Group. 1(2): 51 – 55.

Agrawal, A., Pandey, R. and Agrawal, M.L. 2011. Impact of Solid Waste Leachate on Ground Water Sources-A Case Study. International Journal of Chemical and Environmental Engineering. 2(2): 113-118.

Ahmed, K. and Ali, W. 2000. Evaluation of Ravi River Water Quality. Pakistan J. Drainage Water Manage. 4: 10-15.

Alslaibi, T.M., Mogheir, Y.K. and Afifi, Samir., 2011. Assessment of Ground water Quality due to Municipal Solid Waste Landfills Leachate. Science Alert. 4: 419-436.

Amar, M.D., Pawar, C.B., Pardeshi, P.B. and Patil, D.A. 2008. Municipal Solid Waste disposal in Pune city – An analysis of air and ground water pollution. Current Science. 95(6): 773-777.

Anilkumar, A., Dipu Sukumaran., Salom Gnana. and Thanga Vincent. 2015. Effect of Municipal Solid Waste Leachate on Ground Water Quality of Thiruvananthapuram District, Kerala, India. Applied Ecology and Environmental Sciences. 3(5): 151-157.

Annapoorani, A., Murugesan, A., Ramu, A. and Renganathan, N.G. 2013. Hydrochemical Characteristics of Groundwater Quality in Chennai City Using GIS – A Case Study. International Journal of Research in Chemistry and Environment. 3 (4): 39-47.

Anwar, K.M. andAggarwal, V. 2014. Analysis of Groundwater Quality using Statistical Techniques: A Case Study of Aligarh city (India). International Journal of Technical Research and Applications. 2 (5): 100-106.

Arafath, Y., Praveen Kumar., Vigneshwaran and Banupriya, 2014. Analysis of Pollutants present in the ground water due to Leachates at Thuraipakkam Dumpyard, Chennai. Civil and Environmental Research. 6 (6): 55-60.

Aremu, M. O., Gav, B. L., Opaluwa, O. D., Atolaiye, B. O., Madu, P. C. and Sangari, D. U. 2011: Assessment of physico-chemical contaminants in waters and fishes from selected rivers in Nasarawa state, Nigeria. Research J. of Chemical Sciences. 1(4): 6-17.

Bahar, M. and Reza, M.S. 2010. Hydro chemical characteristics and quality assessment of shallow groundwater in a coastal area of Southwest Bangladesh. Environ. Earth Sci. 61: 1065-1073.

Bandara, N.J.G.J and Hettiachchi, P.J. 2003. Environmental impacts associated with current disposal practices in a municipality, Srilanka- A case study. Proc. of the work shop on sustainable landfill management, Chennai, India. 19-26.

Behera, B., Mira Das and Rana, G. S. 2012. Studies on ground water pollution due to iron content and water quality in and around, Jagdalpur, Bastar district, Chattisgarh, India. Journal of Chemical and Pharmaceutical Research. 4(8): 3803-3807.

Bhalla, G., Swamee, P.K., Arvind Kumar and Ajay Bansa. 2012. Assessment of groundwater quality near municipal solid waste landfill by an Aggregate Index Method. International Journal of Environmental Sciences. 2(2): 1492-1503.

Blight, G. 1996. Standards for landfills in developing countries. Waste Manage. Res. 14: 399-408.

Bohlke, J.K. and Denver, J.M. 1995. Combined use of groundwater dating, chemical, and isotopic analyses to resolve the history and fate of nitrate contamination in two agricultural watersheds, Atlantic coastal plain, Maryland. Water Resources Research Journal. 31: 2319–2339.

Borah, K.K., Bhuyan, B. and Sarma, H.P., 2009. Heavy Metal Contamination of Groundwater in the Tea Garden Belt of Darrang District, Assam, India. E-Journal of Chemistry. 6(S1):S501-S507.

Bundela, P.S., Anjana Sharma, Akhilesh Kumar Pandey, Priyanka Pandey and Abhishek Kumar Awasthi. 2012. Physicochemical Analysis of Ground water Near Municipal Solid Waste Dumping Sites Jabalpur. International Journal of Plant, Animal and Environmental Sciences. 2 (1): 217-222.

Butt, Ibtisam ., Ghaffar and Abdul. 2012. "Ground water quality assessment near Mehmood Boti Landfill, Lahore, Pakistan", Asian Journal of Social Sciences & Humanities. 1(2): 13-24.

Chain, E. S. K and Walle, F. B.D. 1976. "Sanitary Landfill Leachates and their Treatment", Journal of Environmental Engineering Division. 10(2): 411-431.

Chatterjee, R. 2010. Municipal Solid Waste Management in Kohima City- India, Iran Journal of Environmental Health Sciences and Engineering, 7 (2): 178 – 180.

Chavan, B. L. and Zambare, N. S. 2013. "Assessment of Ground Water Quality from Wells Located Near Municipal Solid Waste Dumping Sites of Solapur City, Maharashtra". International Journal of Research in Sciences. 2 (1): 01-07.

Chavan, B. L. and Zambare, N. S. 2013. "A case study on Municipal Solid Waste management in Solapur city, Maharashtra, India" International Journal of Research in Civil Engineering, Architecture & Design. 1(2): 46-53.

Chavan, B.L. and Zambare, N.S., 2014. Ground Water Quality Assessment near Municipal Solid Waste Dumping Site, Solapur, Maharashtra, India. International Journal of Research in Applied, Cheboterev. II., 1955. Metamorphism of natural waters in the crust of weathering-I. Geochimica et Cosmochima Acta.8(1-2): 22-48.

Christensen, T.H., Kjeldsen, P., Bjerg, P.L., Jensen, D.L., Christensen, J.B., Baun, A., Albrechtsen, H.J. and Heron, G., 2001. "Biogeochemistry of landfill leachate plumes", Applied

Clarke, R, Lawrence, A.R. and Foster, S. S. D. 1995. Groundwater a threatened resource. UNEP Environ. Library, 15.

Das, K., Das, S.N., and Dhundasi, S.A., 2008. Nickel, its adverse health effects & oxidative stress. Indian J Med Res. 128: 412-425.

Dhale, D.A. and Pachkore, G.L., 2012. Assessment of Physico-Chemical Status of Ground Water Samples of Parbhani District (M.S., INDIA). International Journal Pharmaceutical Sciences and Research. 3(5):1367-1370.

Dharmarathne, N. and Gunatilake, J., 2013. Leachate Characterization and Surface Groundwater Pollution at Municipal Solid Waste Landfill of Gohagoda, Sri Lanka. International Journal of Scientific and Research Publications. 3(11): 1-7.

Dhok, R. P., 2016. Studies on Groundwater Quality and Its Suitability for Drinking Purpose in Baramati City, India. Journal of Geography, Environment and Earth Science International 5(1): 1-11.

Dhok, R.P., Patil, A.S., and Ghole, V.S., 2011. Quality of Ground water and its suitability for drinking water purpose, Water Research and Development, 1(1): 27-35.

Dhok, R.P., Patil, A.S., and Ghole, V.S., 2013. Hardness of Groundwater Resources and its Suitability for Drinking Purpose. International Journal of Pharmaceutical and Chemical Sciences, 2(1): 169-172.

Fadel, M.E., Findikalis, A.N. and Leckie, J.O., 1997. "Environmental impacts of solid waste landfilling," Journal of Environmental Management, 50 (1): 1-25.

Goyer, R., 1991. Toxic effects of metals. In: Amdur MO, Doull JD, Klaassen CD, editors. Casarett and Doull's toxicology, 4thed. New York: Pergamon Press. 623-80.

Hamazah, A, Addullah, M.P., Sarmani, S. and Johari, M.A. 1997. Chemical and Bacteriological monitoring of drinking water from an urbanized water catchment drainage basin. Environmental Monitoring and Assessment. 44: 327-328.

Harmsen, J. 1983. Identification of organic compounds in leachate from a waste dump. Journal of Water Research. 17(6): 699-705.

Hauser, B. A. 2001. "Drinking Water Chemistry: A Laboratory Manual" Lewis publishers, A CRC Press Company: Boca Raton, FL. 71.

Hem, J. D. 1959. Study and interpretation of the chemical characteristic of natural water. Water Supply Paper, 2254, USGS, USA.

Howarth, R.W. and Marino, R. 2006. Nitrogen as the limiting nutrient for eutrophication in coastal marine ecosystems: Evolving views over three decades. Journal of Limnology and Oceanography. 51: 364–376.

Hughes, J. L. 2004. Evaluation of groundwater Degradation Result from Waste Disposal to Alluvium Near Barstow, California. U.S. Geological survey professional Paper 878.

Jadhav, S.D., Sawant, R.S.,.Godhate, A.G. Patil, S.R. and Patil, R.S. 2012. Assessment of Ground Water Quality of Ajara Tahsil from Maharashtra. Rasayan J. Chem. 5(2): 246-249.

Jain, C. K. 2004. Ground Water Quality of District Dehradun, Uttaranchal. Ind, J. Env. & Ecoplan. 8 (2): 475-484.

Jayanthi, M., Duraisamy, P., Sharma, K.K. and Paramasivam, K, 2012. Potential impacts of leachate generation from urban dumps on the water quality of Pallikaranai Marsh – the only surviving fresh water Wetland of Chennai City in India. Indian Journal Innovations Dev. 1(3): 186-191.

Jhamnani, B. and Singh, S. K., 2009. Groundwater Contamination due to Bhalaswa Landfill Site in New Delhi. International Journal of Civil and Environmental Engineering. 1(3): 121-125.

Jibrin,N.A., Mairami, F.M. and Husseini, H.U. 2015. Assessment of Groundwater Quality in Some Rural Areas of the Federal Capital Territory, Abuja, Nigeria. International Journal of Scientific Research and Education 3 (5): 3363-3376.

Johnson, C. J., Bonrud, P. A., Dosch, T. L., Kilness, A. W., Senger, K. A., Busch, D. C., and Meyer, M. R., 1987, Fatal Outcome of Methemoglobinemia in an Infant, Journal of the American Medical Association. 257: 2796-2797.

Johnson, C.A, Richner, G.A., Vitvar, T., Schittli, N. and Eberhard M. 1998. Hydrological and geochemical factors affecting leachate composition in municipal solid waste incinerator bottom ash part-I, The hydrology of landfill Lostorf, Switzerland. J. Contam. Hydrol. 33: 361-376.

Joseph.O., Adams Ezekiel and Adetoro. A. 2014. Quality Assesments of Water in Selected Waste – Dumping Site Areas. International Journal of Scientific Research and Education. 2 (11): 2399-2404.

Kale, S.S., Kadam, A.K., Kumar, S. and Pawar, N.J. 2010. Evaluating pollution potential of leachate from landfill site, from the Pune metropolitan city and its impact on shallow basaltic aquifers. Environ. Monit. Assess. 162: 327-346.

Kamboj, N. and Choudhary, M. 2013. Impact of Solid waste disposal on ground water quality near Gazipur dumping site, Delhi, India. Journal of Applied and Natural Science 5(2): 306 – 312.

Karanth, K. R. 1987. Groundwater Assessment Development and Management. New Delhi: Tata McGraw Hill Publishing Company Ltd. 725.

Kassenga, G.R. and Mbuligwe, S.E. 2009. Impacts of a Solid Waste Disposal Site on Soil, Surface Water and Ground water Quality in Dares Salaam City, Tanzania. Journal of Sustainable Development in Africa. 10(4): 73-94.

Katsari, A. K., Bodogianni, A. and Chritoulas, D. 1999. Prediction of leachate quality from sanitary landfills. Journal of Environmental Division. 125: 950-957.

Kelly, W. E. 1976. "Groundwater pollution near a landfill". Journal of Environmental Engineering Division. 102 (6): 1189-1199.

Khahlown, M.A., Tahir, M.A. and Ashraf. M. 2002. Water Quality Issues and Status in Pakistan. In Proceedings: Seminar on Strategies to Address the Present and Future Water Quality Issues. Pakistan Council of Res. in Water Resources. Islamabad.

Krushnamoorthy, B. 2001. Environmental Management. Abhinav kumar Publications, Mumbai. 80-81.

Kugali, N.M., Ankalagi, R. F. and Yadawe, M. S. 2013. Estimation of nitrate, nitrite, arsenic and other physico- chemical properties of water. International J. of Plant, animal and Environmental Sciences. 3(14): 132 – 136.

Kulikowska, D. and Klimiuk, E. 2008. "The effect of landfill age on municipal leachate composition". Bioresour. Technol. 99(13): 5981-5985.

Kumar, A., .Bhawsar, N.G., Khandelwal, S., Sakir, S.S., Ahuja, S., Pathak, M., Hanote, R.K., Gatakwad, S.R. and Khasdeo. K. 2013. Physico-Chemical Parameters Apply to Analysis of Drinking Water From Some Selected Area of Betul District, Madhya Pradesh. International Journal of Pharmaceutical, Chemical and Biological Sciences. 3(4): 1109-1114.

Kumar, D. and Alappat, B. J., 2003. "Analysis of Leachates contamination Potential of a municipal landfill using leachate pollution index", Workshop on sustainable Landfill Management, India. 147-153.

Kumar, D., Khare, M. and Alappat, B.J. 2002. "Threat to groundwater from the municipal landfills in Delhi, India", Proceedings of the 28th WEDC Conference on Sustainable Environment Sanitation and Water Services, Kolkata, India. 377-380.

Kumar, N. and Sinha, D.K., 2010. Drinking water quality management through correlation studies among various physic-chemical parameters. International Journal of Environmental Sciences. 1(2): 253-259.

Kumar, P. 2013. Environmental Effect/ Impact Assessment of Industrial Effluent on Ground Water. Oriental Journal of Chemistry. 29 (3): 1243-1249.

Kumar, S. and Gaikwad, S. A. 2004. Municipal solid waste management in Indian urban centres: An approach for betterment in urban development debates in new millennium. Atlantic publishers and distributors, New Delhi: 101-111.

Kumaraswamy N., Subramanyam B. and Kotaiah B. 2002. "Movement of Ground water Pollution in and around a solid waste disposal site – A Case Study", Indian Journal of Environment Protection, 20: 206-208.

Kumaravel, B., Gopalaswamy, V. and Kanakasabai, V. 2003. Groundwater contamination due to municipal garbage dumping –A comparative study. Poll. Res. 22(1): 67-71.

Lee G.F. 2002. "Solid Waste Management: USA Lined Landfill Reliability, Natural Resources Forum", A United Nations Journal, New York.

Longe, E.O. and Balogun, M.R. 2010, Groundwater quality assessment near a municipal landfill, Lagos, Nigeria, Department of Civil and Environmental Engineering, University of Lagos, Nigeria. Research Journal of Applied Sciences, Engineering and Technology. 2(1): 39-44.

Longe, E.O. and Enekwechi, L.O. 2007. Investigation on potential groundwater impacts and influence of local hydrogeology on natural attenuation of leachate at a municipal landfill. Int. J. Environ. Sci. Technol. 4(1): 133-140.

Lu., J.C.S., Eichenberger, B. and Stearn, R.J. 1985. Leachate from municipal landfills – Production and management. Noyes, Park Ridge, NJ. 453.

Meenakshi and Maheswari, R. C. 2006. Fluoride in drinking water and its removal. J Haz Matter. 137: 456 –463.

Meenakshi Khajuria and Datta. S. P. S. 2010. Hydro-chemistry of underground water drinking water adjacent to river Tawi, Jammu (J&K). Aqua.Biol. 25 (2): 62-69.

Mishra, U.K., Tripathi, A.K., Tiwari, S. and Ajay Mishra, A. 2012. Assesment of Quality and Pollution Potential of Groundwater around Dabhaura Area, Rewa District, Madhya Pradesh, India. Earth Science Research. 1 (2): 249-261.

Mohiuddin, M. 2015. Assessment of Ground Water Quality in Gokunda Taluka Kinwat of Nanded District, Maharashtra (INDIA), International Recognized Double – Blind Peer Reviewed Multidisciplinary research Journal, research Directions. 2(12): 1-9.

Mor, S., Ravindra K., Dahiya R.P. and Chandra A. 2006. "Leachate Characterization and Assessment of Groundwater Pollution near Municipal Solid Waste Landfill Site". Environmental Monitoring and Assessment. 118 (1-3): 435-456.

Nagarajan, R., Subramani, T., Elango, L. 2010. Ground Water Contamination due to Municipal Solid Waste Disposal – A GIS Based study in Erode City. International Journal of Environmental Sciences. 1(1): 39-55.

Nagarajan, R., Thirumalaisamy, S. and Lakshumanan, E. 2012. Impact of leachate on groundwater pollution due to non-engineered municipal solid waste landfill sites of erode city, Tamil Nadu, India. Iranian Journal of Environmental Health Sciences & Engineering. 9: 35-42.

Nandwana R. and Chhipa, R C. 2014. "Impact of Solid Waste Disposal on Ground Water Quality in Different Disposal Site at Jaipur, India", International Journal of Engineering Sciences and Research Technology. 3 (8): 93-101.

Navneet Kumar and Sinha D.K. 2010. Drinking water quality management through correlation studies among various physicochemical parameters: A case study", International Journal of Environmental Quality. 22: 392–402.

Niloufer, S., Dr. Swamy, A.V.V.S., and Devi, K.S. 2014. Gaseous Emissions from MSW dumpsites in Vijayawada. American International Journal of Research in Science, Technology, Engineering and Mathematics. 6(1): 67-73.

Niloufer, S., Dr. Swamy, A.V.V.S., and Devi, K.S., 2014. Seasonal Variations in the Leachate Characteristics in MSW sites. Journal of International Academic Research for Multidisciplinary. 2(8): 479-487.

Nubi, O.A. and Ajuonu, N., 2011. Impacts of industrial effluent and dumpsite leachate discharges on the quality of groundwater in Oyo state, Nigeria. Journal of Biodiversity and Environmental Sciences. 1(3): 13-18.

Palmer, C.D. and Wittbrodt, P.R., 1991. Processes Affecting the Remediation of Chromium-Contaminated Sites. Environmental Health Perspectives. 92: 25-40.

Pande, G., Sinha, A. and Agarwal, S. 2015. Impacts of Leachate Percolation on Ground Water Quality: A Case study of Djamnad City. Global NEST Journal. 17(1): 162-174.

Parveen R. S., Arjun B.B. and Vyankatesh, Y. 2012. The impact of landfill on soil and ground water quality of the Nanded city, Maharashtra. Researcher. 4(7): 56-63.

Patil, S. R., Sawant, R. S., Patil, S. S., Sathe, T. V. and Patil, R. S. (2013). Avian fauna and Physico-chemical Parameters of Gajargaon Pond of Ajara Tahsil, Kolhapur (M. S.). Rasayan J. Chem. 6(1):76-79.

Patil. A, Krishnan. S. 2012. "Groundwater situation in urban India: Overview, Opportunities and Challenges," IWMI – Tata Water Policy Program.

Paul, M.K. and Sujata, S. 2012. The Occurrence of TDS and Conductivity of Domestic Water in Lumding Town of Nowgong District of Assam, N.E. India. Current World Environment. 7(2): 251-258.

Perwak, J., Bysshe, S., Goyer, M., Nelken, L. and Scow, K., 1980. Exposure and Risk Assessment for Copper. Washington, DC, USA. Office Water Regul. Stand., Environmental Protection Agency. ICA: 477.

Pramod Kumar. 2013. Environmental effect /impact assessment of industrial effluent on ground water. OJCHEG. 29(3): 1243-1249.

Prasad, G.B. 2003. "Evaluation of water quality in Tadepalli mandal Guntur Dist. A.P." Natl. Environ. Poll. Technol. 2(3): 273-276.

Pushpendra, S. B., Anjana, S., Pandey, A. K., Pandey, P. and Kumar, A. A. 2012. "Physicochemical Analysis of Ground Water near Municipal Solid Waste Dumping Sites in Jabalpur". International Journal of Plant, Animal and Environmental Sciences. 2 (1): 217.

Rajagopal, R. 1978. Impact of land use on groundwater quality in Grande Tranverse Bay, Michigan, Journal of Environmental Quality. 7: 93-98.

Rajan, A.P., Raja, A.S., Prathipa, V. and Dheenadayalan, M.S. 2015. A Study of Ground Water Quality on the Bank of Santhanavarthini River, Dindigul Town, Tamil Nadu, India. Journal of Environmental Science and Pollution Research. 1(1): 15-16.

Raman, N. and Sathiyanarayana, D. 2011 Quality Assessment of Ground Water in Pallavaram Municipal Solid Waste Dumpsite Area Nearer to Pallavaram in Chennai, Tamilnadu. Rasayan J. Chem. 4(2): 481-487.

Ray, K. L. 1988. "Hydrology of water quality" McGraw Hill Book Co. Inc. Hill Book Co. Inc. New York.

Reddy, V.H., Prasad, P.M.N., Reddy A.V.R. and Reddy, Y.V.R., 2012. Determination of heavy metals in surface and ground water in and around Tirupati, Chittoor (Di), Andhra Pradesh, India. Der Pharma. Chemica. 4: 2442 – 2448.

Robertson, W.D., Russell, B.M. and Cherry, J.A. 1996. Attenuation of nitrate in acquitter sediments of southern Ontario. Journal of Hydrology. 180: 267–281.

Rowe, R.K., 1995. Leachate characteristics for MSW landfill. Proc. Sardinia. 95, 5th International Symposium on Sanitary landfills, CISA, Italy. 2: 327-344.

Saarela, J. 2003. "Pilot Investigations of surface parts of three closed landfills and factors affecting them", Environment Monitoring Assessment. 84: 183-192.

Saidu, M. 2011. Effect of refuse dumps on ground water quality. Advances in Applied Science Research. 2 (6): 595-599.

Sairy, A., Mohammed, A.I. and Mohammed, I.F. 2010. "Physicochemical analysis of freshwaterat Kundlika Dam Upli, Dist. Beed (MS) India". Global Jr. Envnmtl. Res. 4(1):01-05.

Salami, L. and Susu, A.A. 2012. Leachate Characterization and Assessment of Ground water Quality: A case of Soluos Dumpsite in Lagos State, Nigeria. Greener Journal of Science, Engineering and Technological Research. 3(2): 042-061.

Sarala, C. and Ravi Babu, P. 2012. Assessment of Groundwater Quality Parameters in and around Jawaharnagar, Hyderabad. International Journal of Scientific and Research Publications. 2(10), 1-6.

Saravanan, R., Sundaram, V.L.K., Gopalakrishna A.N. and Ravikumar, G. 2010. Modelling of Ground water Pollution due to Solid Waste Dumping – A Case Study. International Journal of Earth Sciences and Engineering. 3(2): 176-187.

Shahidullah, S.M., Hakim, M.A., Alam, M.S. and Shamsuddoha, A.T.M. 2000. Assessment of groundwater quality in a selected area of Bangladesh. Pak. J. Biol. Sci. 3: 246-249.

Sharma, A. 2005. Physico – Chemical analysis of tap water of Chandigarh (UT), India. INT J Curr Sci, 14: 51-54.

Sharma, P., Dubey, A. and Chatterjee, S.K., 2014. Pre-monsoon physic – chemical analysis of surface and ground water of Abhanpur block in district Raipur, (C.G). Recent Research in Science and Technology. 6(1):157-161.

Sharma. A. 2015. Physico chemical analysis of ground water of Chandigarh (UT), India. Int J Curr Sci. 14: E51-54.

Singh, G., Dapinder Deep Singh, Prof. S.K. Sharma. 2013. "Effect of Polluted Surface Water on Groundwater: A Case Study of Budha Nullah". IOSR Journal of Mechanical and Civil Engineering. 5 (5): 1-8.

Singh, R.K., Datta, M. and Nema, A.K. 2009. A new system for groundwater contamination hazard rating of landfills. Journal of Environmental Management. 91(2): 344-357.

Singh, V. and Dev, P. 2009. Chemical Quality Evaluation of Groundwater for Human Domestic Consumption in Saharanpur and Adjoining Area, Uttar Pradesh, India. Int. J. Econ. Env. Geol. 2(1): 1-10.

Subba Rao, N. 2003. Ground water quality: Focus on flouride concentration in rural parts of Guntur District, Andhra Pradesh, India. Hydro. Sci. Jour, -des sciences Hydrologiques. 48 (5): 835-847.

Tabatabaei, S.H., Lalehzari, R., Nourrmahnad, N. and Khazaei, M. 2010. Groundwater Quality and Land Use Change (A Case Study: Shahrekord Aquifer, Iran). Journal of Research in Agricultural Science. 6: 39-44.

Taiwo, A.M. Adeogn, A.O. Olatunde, K.A. and Adegbite, K.I. 2011. Analysis of groundwater quality of hand-dug wells in Peri-Urban area of Obantoko, Abeokuta for selected physico-chemical parameters. Pecific Journal of Science and Technology. 12(1): 527-534.

Tengrui, L., AlHarbawi, A.F., Lin, M.B., Jun, Z. and Long, X.U. 2007. Characteristics of nitrogen removal from old landfill leachate by sequencing batch biofilm reactor. American Journal of Applied Sciences. 4(4): 211-214.

Tesoriero, A.J., Liebscher, H. and Cox S.E. 2000. Mechanism and rate of denitrification in an agricultural watershed: Electron and mass balance along groundwater flow paths. Water Resources Research Journal. 36: 1545–1559.

Tiwari, R.N. 2011. Assessment of ground water quality and pollution potential of Jawa Block Rewa District, Madhya Pradesh, India. Proceedings of the International Academy of Ecology and Environmental Sciences. 1(3-4): 202-212.

Toxic link. 2003. Scrapping the Hitech myth: computer waste in India. http://www.toxicslink.org/ovrvw.

Ullah, R., Malik, R.N. and Qadir, A. 2009. Assessment of groundwater contamination in an industrial city, Sialkot, Pakistan. African Journal of Environmental Science and Technology. 3 (12): 429-446.

Umar, M., Aziz, H. A. and Yusoff, M. S. 2010. Variability of Parameters Involved in Leachate Pollution Index and Determination of LPI from Four Landfills in Malaysia. International Journal of Chemical Engineering. Hindawi Publishing Corporation. 10(6): 1-6.

Ustad, I. 2015. Evaluation of Groundwater Quality and its Suitability for drinking use in parts of Khultabad, District Aurangabad, MS, India. Journal of Environmental Science, Computer Science and Engineering & Technology. 4 (2): 452-457.

Vasanthi, P., Kaliappan, S. and Srinivasaraghavan, R. 2008. Impact of Poor Solid waste Management on ground Water. Journal of Environment Monitoring and Assessment, Springer, Netherlands. 143(1-3): 227-238.

Walton, W. C. 1970. Groundwater Resources Evaluation. New York: Mc-Graw Hill Book Company, New York. 664.

WHO, 2010. Guidelines for Drinking Water Quality. 3rd Edn., World Health Organization, Geneva, Switzerland.

World Health Organization (WHO), 1997. Guidelines for Drinking Water Quality, 2nd edition Volume 2, Health criteria and other supporting information, World Health Organization, Geneva. 940-949.

Yilmaz, E. and Koc, C. 2014. Physically and chemically evaluation for the water quality criteria in a farm on Akcay. J. Water Resour. Prot. 6: 63-67.

Zerboc, O. 2003. “Urban Solid Waste Management: Waste Reduction in Developing Nations”. Michigan Technological University Accessed Online at website 15th 2006. http://www.cee.mtu.edu/peacecorps/documents, july03/Waste_ reduction.

Zhu, D., Asnani, P.U., Zurbrugg , C., Anapolsky , S., Mani, S. 2008. Improving municipal solid waste management in India. A source book for policymakers and Practitioners. The World Bank.

Impact of PMKSY-Watersheds on Dairying in Anantapur District of Andhra Pradesh - An Ex-Post Analysis

S. Ramana Reddy[1], N. Polappa[2], B. V. Ramana Kumar[3], K.Chaithanya Chakravarthy[4] and R.Jhansi Rani[5]

[1]*I.R.P.S, Director watersheds,* [2]*Joint Commissioner (Watersheds), State Level Nodal Agency, Amaravati, Andhra Pradesh.*
[3]*Chief Executive Officer,* [4]*Project Manager,* [5] *Project Manager - Remote Sensing Instruments, Hyderabad.*

ABSTRACT

An ex-post analysis was conducted to find out the Impact of Pradhan Mantri Krishi Sinchayee Yojana-Watersheds on dairying in Anantapur district of Andhra Pradesh by randomly selecting 3 out of 14 Batch I projects. Primary data was collected from randomly selected 60 households belonging to 6 micro watersheds for both pre and post periods and analysed the costs of maintenance of milch animals, incomes from it duly considering with and without imputed family labour cost. The analysis revealed that the cost of maintenance of milch animal/day was more economical without family labour compared to the with-family labour cost. However, in the project area majority of the operations in maintenance of the milch animals were attended by the family labour only and rarely by the hired labour. Similarly the incremental gross returns, net incomes and benefit cost ratios from pre to post periods were higher when we do not account for family labour cost. The increase in incremental average milk yield and gross incomes were to the extent of 28% and 77% respectively in the project area. Further due to the promotion of livestock interventions in the study area respondents started purchasing the milch animals so as to increase the milk production though the productivity was not significant but the incomes are at desired levels because of the availability of green fodder, conduct of animal health camps, artificial insemination and travices construction for cattle drinking purpose. The same was expressed by the majority of the households during the opinion survey.

Introduction

India ranks first in the world milk production with a record production of 165.4 million tons according to 2016-17 census. In India Andhra Pradesh state occupies fifth place with a milk production of 9.65 million tons. Moreover, 80% of the milk comes from landless, marginal and small farmers, the weaker sections of the society in rural areas, so any improvement in milk production and livestock would have a tremendous social impact. The milk production constitutes around two percent of Gross Domestic Product of Indian economy. The dairy sector has also been identified as one of the "Engines of Growth" in the state. The milk yield of dairy animals is very low in India as compared to other countries. Consequently, the country has to maintain a much larger stock to produce the required quantity of milk. However, the country has progressed from being deficient in milk production at 20 million tons in 1970 to becoming the world's largest producer at 164.5 million during 2016-17 tons accounting for around 19% of global milk production. In a developing country like India, it is important to weigh the strategies for productivity enhancement against the alternatives before investing resources on technological development which requires a substantial capital investment (Karjana et al., 2012). The projected milk production by 2012-22 is 254.5 million tons as per the vision 2022 document. In order to address this the Govt. of India launched special programme called PMKSY-Watersheds erstwhile Integrated Watershed Management Programme (IWMP) as one of the developmental programmes especially in drought prone areas as majority of the area is still under rained conditions. One of the major components of their programme is the promotion of livestock activities in the selected areas during the implementation period of the project by providing milch animals and water saving infrastructure facilities in the rural villages to vulnerable sections of the society thus increasing the milk production and productivity there by reaping the profits of the farming community. Of late the PMKSY Watersheds of Batch-I (2009-10) projects have completed their period of operation and now it is the time for ex-post evaluation of the projects duly comparing with baseline period data pertaining to milch animals. In view of the above, the present study has been conducted with the following objectives.

(i) To analyse the incremental costs and incremental returns of maintenance of milch animals in the study region.
(ii) To document the benefits derived in milk production from the project.

Data and Analytical Tools

Data used: The study was based on the primary and secondary data. Primary data was collected through personal interview using pre-tested schedule and secondary data were obtained from published documents and records of the concerned project officers and staff responsible for project implementation.

Sampling frame: A multistage random sampling procedure was followed for assessment of dairy farmers belonging to Batch-I projects of PMKSY-Watersheds. Anantapur, a backward district of Andhra Pradesh was selected where the projects have completed their evaluation phase and the Govt. support was withdrawn.

Again from Batch-I PMKSY-Watersheds three projects namely Jakkalacheruvu, Kammavaripalli and Lakshmampalli were selected randomly out of 14 for the present analysis. From each selected project two micro-watersheds randomly were chosen for primary data collection. Further from each selected micro-watershed project 10 dairy household families mostly landless, marginal and small were selected randomly thus 20 from each watershed and total sample beneficiary households were 60.

Table 1 Sample frame for the analysis

S.No.	Name of the district	Name of PMKSY-Watershed	Name of Micro-Watershed	Number of sample respondents
1	Anantapur	Jakkalacheruvu	Jakkalacheruvu	10
2			Kothapeta	10
3		Kommavaripalli	Kasireddipalli	10
4			Ragamekalapalli	10
5		Lakshmampalli	Cherlopalli	10
6			Kairevu	10
	Total	**3**	**6**	**60**

Method of collection: Survey method was followed in addition to focussed group discussions for collection of primary data.

Data period: The reference period i.e., base year was 2009-10 and post-project period was 2016-17. One of the major components of the PMKSY-Watersheds programme is Entry Point Activities (EPA). These were organized to have rapport with the community and to create awareness about implementing procedures of various activities under the project during the initial stages in the form of provision of water trough for cattle drinking purpose, travices, artificial insemination and animal health camps that would enhance livestock productivity and enrich soil fertility for milk production as well as nutritious consciousness.

Another component of the programme is Production System Improvements (PSI) where they provide milch animals to landless group, conversion of forest land to cultivable land, creation of water harvesting structures under the income generation activities.

Analytical tools: The simple statistical tools like mean, percentages were applied for better presentation of the results. Further, the per animal cost of maintenance for milch animals was estimated by using following method:

Δ cm=ØTP-ØTC

Where Δ cm = Incremental cost of maintenance of milch animal

ØTP = Cost of maintenance of milch animal of project beneficiaries during ex-post period

ØTC = Cost of maintenance of milch animal of beneficiaries during ex-ante period.

Similarly incremental benefit-cost ratio (ΔBCR) was calculated by:

$\Delta BCR = \Delta B/\Delta cm$ Where ΔB = Incremental benefits

Δcm = Incremental costs

Further to compare the difference between two means of key variables i.e., cost of maintenance of milch animal, productivity and production of respondents during ex-ante and ex-post periods, 't' test was used as follows:

$$t = \frac{\overline{x_1} - \overline{x_2}}{\sqrt{\frac{s_1^2}{n_1} + \frac{s_2^2}{n_2}}}$$

Where $\overline{x_1}$=Mean of key variables of ex-post period.

$\overline{x_2}$=Mean of key variables of ex-ante period.

S_1=Standard deviation of key variables for ex-post group.

S_2=Standard deviation of key variables for ex-ante group.

n_1=Sample size of ex-post group

n_2=Sample size of ex-ante group

Results and Discussion

The results are discussed under the following headings:

- Socio-Economic profile of the sample respondents
- Cost of maintenance of milch animal
- Yield of milch animals
- Income/returns from milch animals
- Opinion on usefulness of benefits

Socio-Economic profile of the sample households

The average age of the respondents selected in the watershed projects were more than 50 and highest being 53.12 in Jakkalacheruvu project. Most of the beneficiaries had primary school education and around 18% of the sample studied up to secondary level education. The average family size ranged from 5.47 in Lakshmampalli Watershed to 5.83 in Kammavaripalli watershed. Almost all the selected respondents had more than 20 years of experience in maintenance of milch animals. The selected dairy farmers possessed more than 2 animals/household on an average and in some cases up to 5 animals including cows and buffaloes.

Tables 2 Socio-Economic profile of the respondents

S.No.	Particulars	PMKSY-Watersheds			Average
		Jakkalacheruvu	Kommavaripalli	Lakshmampalli	
1	Age of the respondent (Year)	53.12	51.26	48.17	50.85
2	Years of schooling	4.78	4.31	4.66	4.58
3	Family size (Number)	5.69	5.83	5.47	5.66
4	Experience (Years)	26.78	24.04	21.31	24.04
5	Size of animal (Number)	2.16	2.71	2.48	2.45

Cost of maintenance of milch animal: The details of costs of maintenance of milch animals/day are furnished in Table-3 for both pre and post project periods. The costs included fodder, concentrates, labour and miscellaneous costs. In the study area indigenous/non-cross breed cows and buffaloes are predominant. The cross breed animals are rarely available in the drought prone study area. The analysis was carried at two intervals one with imputed family labour cost and another without cost of family labour.

The costs of maintenance of milch animal/day was more when we consider imputed value of family labour both in pre and post periods compared to without family labour. The average cost of maintenance was in the order of Rs. 22.48, Rs. 21.8 and Rs. 21.72 for Lakshmanpalli, Jakkalacheruvu and Kammavaripalli watersheds respectively. However average total costs without imputed family labour was highest in Kammavaripali and the lowest being in Jakkalacheruvu watershed with Rs. 9.16/day. The amount spent on green fodder and concentrates were higher during post-period compared to before. This could be due to the infrastructure created in the micro-watersheds resulting to availability of water in water storage structures even in rabi season and some cases during summer season also. Thus the availability of green fodder and concentrate in the form of rice bran increased the cost of maintenance during post-project. Generally, in the study region dairy farmers don't incur money purchasing these items, but utilise their form produced products for maintenance of animals.

Table 3 Cost of maintenance of milch animals among the respondents between pre and post periods (Rs/animal/day)

S. No.	Parti-culars	Watershed Projects							
		Pre-project				Post project			
		Jakkalacheruvu	Kommavaripalli	Lakshmampalli	Average	Jakkalacheruvu	Kommavaripalli	Lakshmampalli	Average
1	Green fodder	2.93	3.38	3.05	3.12	4.28	5.03	4.67	4.66
2	Dry Fodder	4.59	4.31	4.84	4.58	6.13	6.21	6.33	6.22
3	Concent-rates	0.68	0.72	0.83	0.74	1.24	1.39	1.51	1.38
4	Total feed cost	8.2	8.41	8.72	8.44	11.65	12.63	12.51	12.26
5	Family labour	12.64	11.36	12.29	12.10	21.77	22.08	20.16	21.34
6	Hired labour		0.71	0.38	0.36	1.65	3.49	3.68	2.94
7	Total labour cost	12.64	12.07	12.67	12.46	23.42	25.57	23.84	24.28
8	Misc. expense	0.96	1.24	1.09	1.10	1.29	1.95	2.07	1.77
9	Total costs with family labour	21.8	21.72	22.48	22	36.36	40.15	38.42	38.31
10	Total costs without family labour	9.16	10.36	10.19	9.9	14.59	18.07	18.26	16.97

Yield and incomes of milch animals: In all the selected watersheds the average milk yield increased from pre to post period i.e., 1.36 l/day to 1.74 l/day due to the availability of green fodder as well as cultivation of crops during rabi season (Table-4). Similarly the cost of production/litre has been enhanced from before to after the project period. Some of the uncultivable lands and barren lands were also brought under cultivation duly utilising the infrastructure generated by the project officials. The increase in the yield resulted in realising additional incomes i.e., gross and net incomes from pre to post periods. The increase in incomes of the beneficiaries was more when we do not consider the imputed value of family labour. In the study area, only family members generally attend to maintenance of milch animals. The average net income/day/milch animal without family labour cost increased from Rs. 15.83 to Rs. 28.53 a better rewards for family labour. However, other variables like yield and gross income with and without imputed value of family labour costs were not shown considerable impact in the study area.

Incremental cost of maintenance and yield of milch animals: The incremental change in cost of maintenance from pre to post periods in the selected watersheds of drought prone district of Anantapur with and without family labour was Rs. 16.31 and Rs. 7.07 per day/milch animal respectively as indicated in Table-5. Similarly the incremental change in average milk yield was 0.37 l/day/milch animal which was low and also not significant. It implies that project is yet to derive its benefits on milk yield.

Incremental change in incomes from milch animals: Per day/animal net and gross incremental incomes derived from milch animals in the selected watersheds from ex-ante and ex-post are presented in Table-6. It was observed in Lakshmampet watershed the incremental net income/day/animal was more (Rs. 4.93), when we include the family labour cost while the same without the cost of family labour was highest in Kommavaripalli watershed with Rs. 14.15/day/animal. This might be due to the availability of green fodder in the project area and creation of a sort of employment potential. The net incremental incomes without imputed value of family labour for kommavaripalli and Lakshmampalli watersheds were significant at 5 percent and one percent level of probability respectively. The overall additional net income in the study area was also significant.

Percentage change in incremental costs and incomes: Further the percent increase in incremental costs and incomes were calculated by including and excluding family labour cost and furnished in Table-7. A perusal of this revealed that there was an increase in incremental average milk yield and gross income to the tune of around 28% and 77% respectively. This increase was considered to be the impact of the project in the selected areas of the district, due to the implementation of various interventions by the project staff.

Incremental Benefit-cost Ratio of milch animals (ΔBCR): The incremental BCR was calculated for the selected watershed projects (Table-8) and found that the ΔBCR was more beneficial when the imputed value of family labour was not included in maintenance cost of milch animal in all the projects. It means for every rupee investment on maintenance would benefit the household to an extent of Rs. 2.80 on an average in the project area; whereas the same with family labour was Rs. 1.21/day/animal.

Table 8 Incremental benefit-cost ratios of milch animals (Rs/day/animal)

S.No.	Name of the Watershed	Incremental maintenance cost (Rs)		Incremental Gross income (Rs)	Benefit cost ratio	
		With family labour	Without family labour		With family labour	Without family labour
1	Jakkalacheruvu	14.56	5.43	16.20	1.11	2.98
2	Kommavaripalli	18.43	7.71	21.86	1.19	2.84
3	Lakshmampalli	15.94	8.07	20.87	1.31	2.59
	Average	16.31	7.07	19.77	1.21	2.80

Table 4 Yield of milch animals per day

S.No.	Particulars	Watershed Projects							
		Pre-project				Post project			
		Jakkalacheruvu	Kommavaripalli	Lakshmampalli	Average	Jakkalacheruvu	Kommavaripalli	Lakshmampalli	Average
1	Yield (Liter/day)	1.26	1.64	1.19	1.36	1.52	2.03	1.66	1.74
2	Dung value (Rs.)	0.95	0.88	0.92	0.92	1.33	1.42	1.47	1.41
3	Cost of production/litre								
(a)	With family labour	17.3	13.24	18.89	16.48	23.92	19.78	23.14	22.28
(b)	Without family labour	7.27	6.32	8.56	7.38	9.6	8.9	11	9.83
4	Sale price/milk (Rs/Lit)	17.81	18.22	18.68	18.24	25.17	25.22	25.63	25.34

S.No.	Particulars	Watershed Projects							
		Pre-project				Post project			
		Jakkalacheruvu	Kommavaripalli	Lakshmampalli	Average	Jakkalacheruvu	Kommavaripalli	Lakshmampalli	Average
5	Milk income (Rs/animal)								
(a)	Gross income (Milk yield X Sale price + Dung value)	23.39	30.76	23.15	25.73	39.59	52.62	44.02	45.50
(b)	Net income with family/labour	1.59	9.04	0.67	3.73	3.23	12.47	5.6	7.19
(c)	Net income without family/labour	14.23	20.40	12.96	15.83	25.00	34.55	25.76	28.53

Table 5 Incremental cost of maintenance and yield of milch animals under different watershed projects (Rs/animal/day)

S.No.	Name of the Watershed	Cost of maintenance						Average milk yield		Difference
		With family labour			Without family labour					
		Pre	Post	Δ	Pre	Post	Δ	Pre	Post	Δ
1	Jakkalacheruvu	21.80	36.36	14.56	9.16	14.59	5.43	1.26	1.52	0.26
2	Kommavaripalli	21.72	40.15	18.43	10.36	18.07	7.71	1.64	2.03	0.39
3	Lakshmampalli	22.48	38.42	15.94	10.19	18.26	8.07	1.19	1.66	0.47
	Average	22.00	38.31	16.31	9.9	16.97	7.07	1.36	1.74	0.37

Table 6 Incremental gross and net income from milch animals under different watershed projects (Rs/animal/day)

S.No.	Name of the Watershed	Gross income		Difference (Δ)	Net income					
					With family labour			Without family labour		
		Pre-project	Post-project		Pre-project	Post-project	Δ	Pre-project	Post-project	Δ
1	Jakkalacheruvu	23.39	39.59	16.2	1.59	3.23	1.64	14.23	25	10.77
2	Kommavaripalli	30.76	52.62	21.86	9.04	12.47	3.43	20.4	34.55	14.15**
3	Lakshmampalli	23.15	44.02	20.87	0.67	5.60	4.93	12.96	25.76	12.80*
	Average	25.73	45.50	19.77	3.73	7.19	3.46	15.83	28.53	12.70*

* Significant (P<0.05) ** Significant (P<0.10)

Table 7 Percentage change in incremental cost of maintenance, milk yield, gross income and net income under different watershed projects (%)

S.No.	Name of the watershed	Incremental cost of maintenance		Incremental milk yield	Incremental gross income	Incremental net income	
		With family	Without family			With family labour	Without family labour
1	Jakkalacheruvu	66.79	59.28	20.63	69.26	103.15	75.69
2	Kommavaripalli	84.85	74.42	23.78	71.07	37.94	69.36
3	Lakshmampalli	70.91	79.20	39.50	90.15	735.80	98.77
	Average	74.14	71.41	27.94	76.84	92.76	80.23

Average number of milch animals and production: Due to the promotion of livestock interventions like artificial insemination to induct composition of improved breed, animal health camps, disease management, green fodder production, water troughs for cattle drinking in the selected areas respondents were tempted to purchase more of livestock in order to enhance their incomes. As a result the average possession of milch animals which was 1.28 at the base period has been increased to 2.45 animals/household by the end of the project period thus an incremental increase was 1.17/household, in addition to an average increase in milk production/animal/day was 2.66 litres though there was no substantial increase in productivity, but the production of milk contributed significantly, hence the respondents preferred to have more animals thus increase in total production leading to increase in incomes (Table-9). This is a good sign. In years to come there will be change in stock of the milch animals as well as productivity in the district as the project has started yielding the benefits, thus contributing their might to the Gross Value Added (GVA) of the state.

Table 9 Average number of milch animals and milk productivity between two periods for the respondents

S.No.	Name of the project	No. of milch animals			Milk production		
		Ex-ante	Ex-post	Δ	Ex-ante	Ex-post	Δ
1	Jakkalacheruvu	1.07	2.16	1.09	1.35	3.28	1.93
2	Kommavaripalli	1.32	2.71	1.39	2.16	5.50	3.34
3	Lakshmampalli	1.45	2.48	1.03	1.73	4.12	2.39
	Average	1.28	2.45	1.17	1.75	4.30	2.55

Opinion on the benefits of the project interventions: The opinions of the respondents on the usefulness of the interventions implemented during the project period was obtained by conducting focus group discussions and the same were arranged on 3 point scale highly satisfied, satisfied and not satisfied. An examination of Table-10 showed that among the interventions, 30% of the respondents perceived the usefulness of artificial insemination as highly satisfied followed by animal health camps (23.33%) and disease management (20%). Majority of the households (75%) satisfied with the green fodder availability and travices (71.67%) in their regions as these would solve the problem of fodder and drinking water. Yet around 43% of the respondents were not satisfied with the incremental milch yield in the study region inspite of these interventions. However, this group would be benefitted shortly like their counterparts in realizing sufficient incremental income. More than 93% of the total households expressed that availability of green fodder availability has benefitted in obtaining the incomes in the project area.

Table 10 Opinion on usefulness of activities implemented/derived from the project (%)

S.No.	Benefits	Highly satisfied	Satisfied	Net satisfied
1	Animal health camps	14 (23.33)	28 (46.67)	18 (30.00)
2	Artificial insemination	18 (30.00)	33 (55.00)	9 (15.00)
3	Disease management	12 (20.00)	26 (43.33)	22 (36.67)
4	Green fodder availability	11 (18.33)	45 (75.00)	4 (6.67)
5	Trevices	8 (13.33)	43 (71.67)	9 (15.00)
6	Incremental yield	6 (10.00)	28 (46.67)	26 (43.33)

Conclusion

The foregoing analysis revealed that the cost of maintenance of milch animal/day was more economical without family labour compared to the with-family labour cost. However, in the project area majority of the operations in maintenance of the milch animals were attended by the family labour only and rarely by the hired labour. Similarly the incremental gross returns, net incomes and benefit cost ratios from pre to post periods were higher when we do not account for family labour cost. The increase in incremental average milk yield and gross incomes were to the extent of 28% and 77% respectively in the project area. Further due to the promotion of livestock interventions in the study area respondents started purchasing the milch animals so as to increase the milk production though the productivity was not significant but the incomes are at desired levels because of the availability of green fodder, conduct of animal health camps, artificial insemination and travices construction for cattle drinking purpose. The same was expressed by the majority of the households during the survey.

References

Census of India, 2014-District Census Hand Book of Anantapur: Village and Town wise Primary Census Abstract.

Hand Book of Statistics-2014: Anantapur: 2010-2011, Chief Planning Officer, Anantapur.

Government of India-2011, Common Guidelines for Watershed Development Projects, National Rainfed Area Authority (NRAA), India.

Nagaraja B and Ekambaram G-2015, A Critical Appraisal of Integrated Watershed Management Program in India, IOSR, Journal of Humanities and Social Sciences; 20(6); 17-23.

Priyanka Lal and B.S.Chandel (2017) Total Factor Productivity in Milk Production, Agricultural Economics Research Review, Vol 30 (No.2) PP 279-284.

Livelihood Security through Resource Conservation and Management in the Semi arid Region

S. Ramana Reddy[1], N. Polappa[2], B.V. Ramana Kumar[3], K. Chaithanya Chakravarthy[4] and R. Jhansi Rani[5]

[1] I.R.P.S, Director watersheds, [2]Joint Commissioner (Watersheds), State Level Nodal Agency, Amaravati, Andhra Pradesh.
[3]Chief Executive Officer, [4] Project Manager, [5] Project Manager - Remote Sensing Instruments, Hyderabad.

ABSTRACT

An integrated watershed development programme was taken up in the rainfed district of Ananthapur, Andhra Pradesh during 2009-10 to 2016-17. The programme focussed on natural resource conservation and optimum utilization through a variety of measures. We conducted an impact assessment study on the impact of the programme on the changes in the availability of surface and groundwater resources in the project area. The study revealed that consequent to crop demonstrations, rain water conservation, and various livelihood opportunities created through the watershed projects, groundwater was recharged, which prompted the farmers to go in for new bore wells, expand areas under irrigation, and adopt diversified cropping systems including horticultural area expansion. This resulted in increased crop productivity and production, and increased household income in case of all farmer categories (65% in landless, 58% in marginal and large farmers, and 57% in small farmers) across the watershed projects.

Keywords: Watershed, conservation, rain water conservation.

Introduction

Natural Resources Management (NRM) refers to the sustainable utilization of these natural resources for providing vital ecosystem services, by focussing on the interactions between people and natural landscapes (CGIAR, 2003). In addition to the dynamic and inter-linked ecological and hydrological cycles that occur in nature, NRM also deals with various stakeholders and their needs, policies and economic implications. Therefore, 'people' is central to the concept of NRM, and any management strategy that limits itself to the conservation of natural resources without considering their use by people, and how it influences livelihoods, is not likely to be sustainable. This has resulted in the evolution of various participatory approaches in NRM (Catacutan *et al.*, 2001) and decentralisation of natural resource management based on community participation replacing top-down natural resource management (Akbulut and Soylu, 2012). Further, evaluation and funding agencies emphasises not only participatory development and implementation of natural resource conservation plans, but wants participatory impact assessment wherein local stakeholders play an integral part in the assessment process, assisted by experts (Vaidya and Mayer, 2014).Participatory watershed approach has resulted in new opportunities by supporting agricultural intensification processes, and the subsequent increases in crop intensity and productivity has led to the creation of labour opportunities.

In order to address the constraints of rainfed farming, GOI launched a special programme to develop rainfed areas through integrated approach by converging all the developmental programmes under different departments into a comprehensive action plan for conserving natural resources and to promote their productive use considering resources on one hand and the demands of the communities on the other using watershed as a unit for development, so as to stabilize rural dynamic economy. This programme, known as Integrated Watershed Management Programme (IWMP) was implemented from April 2009. The major activities undertaken as part of project implementation were: soil and water conservation measures in cultivated and degraded lands; afforestation including block and avenue plantation; horticulture development to provide resilience to agricultural income; drainage line treatment with engineering and bio-engineering structures; crop demonstration with improved cultivation practices; repairs, renovation and up-gradation of existing common property assets ; and promotion and propagation of energy conservation devices. With the prime objectives of improving livelihood security of farmers through increased crop productivity, conservation of natural resources and their optimum utilization, reduced migration and increased income-generating activities, 14 watershed projects in Ananthapur district of Andhra Pradesh were implemented from 2009-10 to 2016-17. Bio-physical and biological interventions were made to conserve resources and increase bio-diversity. Alternative livelihood action plans were prepared and implemented to provide additional income to landless and vulnerable groups. Final impact evaluation of the watershed management programmes on the water resources developed in the district was conducted as per standard methodology.

Methodology

The study was conducted in Ananthapur district of Andhra Pradesh, which lies in the southern part of the state of AP between latitudes of 13° 41' to 15°14'N, longitudes 76° 47'to 78° 26' E, and at an altitude of 670 m above mean sea level (MSL) with a gradual decline towards North and North-East. Its elevation towards south is about 660 m with gradual decline to about 300m at Gooty in the north and 270 at Tadipatri in the northeast. The total area of the district is 19,13,586 ha with about 1200 villages grouped into 63 *mandals*. The district receives the lowest quantum of rainfall (550mm) among the districts of Andhra Pradesh, and second lowest among the Indian districts. The maximum temperature varies through the year between 25 to 42° C.

Agricultural lands occupy 64.3% of the total area while waste land and forests occupy about 21.4 and 10 per cent, respectively of the total area. Area under water bodies such as rivers, tanks etc. is 4.1% and residential area is about 1.4%. The soils can be classified as sandy loam (31%), clayey (24%), loamy sands (14%), sandy clay loams (13%) and rocky land (12%). Area is undulating with ridges and valleys. Black cotton soils occur sporadically in certain areas. Groundnut, paddy, sorghum, pigeon pea, chick pea and sunflower are the major crops grown in the district. Out of the total rainfall received, only 10 to 15% is utilizable for agriculture, the remaining is lost as runoff through streams. Soil degradation is rampant across the district due to excessive biotic pressure coupled with absence of resource conservation measures. Groundwater levels in the district are alarmingly low. Further, degeneration of existing marginal and degraded forests during the last four decades has resulted in the acute scarcity of drinking water, fodder and fuel in every alternate year which is a serious concern.

Fourteen watershed projects in the district covered under the IWMP (Batch-I) were covered under the study (Table 1, Figure 1). The 14 projects are spread across 14 *mandals* and 75micro-watersheds, comprising of 170 villages in the district and have 28,724 households, ranging from 1,012 in Muttala project to 4,537 in Vajrakarur project. The major activities undertaken as part of IWMP implementation are: soil and water conservation measures in cultivated and degraded lands; afforestation including block and avenue plantation; horticulture development to provide resilience to agricultural income; drainage line treatment with engineering and bio-engineering structures; crop demonstration with improved cultivation practices; repairs, renovation and up-gradation of existing common property assets to obtain sustained benefits from previous public investments; and promotion and propagation of energy conservation devices. The impact of these activities on the change in household income in different watersheds of the district has been presented in this paper.

Table 1 Details of Projects Covered under IWMP Batch-I in Anantapur District

S. No.	Watershed Project	Name of *mandal*	Area	Number of		
			in ha	Micro-watersheds	Villages	Households
1	Bandlapalli	Narpala	2557	4	8	2141
2	JakkalaCheruvu	Gooty	3449	5	7	2866
3	Kammavaripalli	Gorantla	4723	6	22	1500
4	Lakshmampalli	Settur	4460	6	6	1758
5	Muddinayanapalli	Kalyandurg	4153	7	8	1517
6	Muradi	D. Hirehal	2244	2	4	1103
7	Muttala	Atmakur	2535	4	4	1012
8	Nallamada	Nallamada	4030	7	19	2539
9	Puttaparthi	Puttaparthi	3670	5	13	2200
10	Tavalam	Tanakal	4027	9	53	2517
11	Thagarakunta	Kanaganapalle	4318	3	3	1015
12	Thumbiganoor	Kanekal	4301	4	5	1639
13	Vajrakarur	Vajrakarur	4974	6	6	4537
14	Varadayapalli	PeddaPappuru	5562	7	12	2380
		Total	55003	75	170	28724

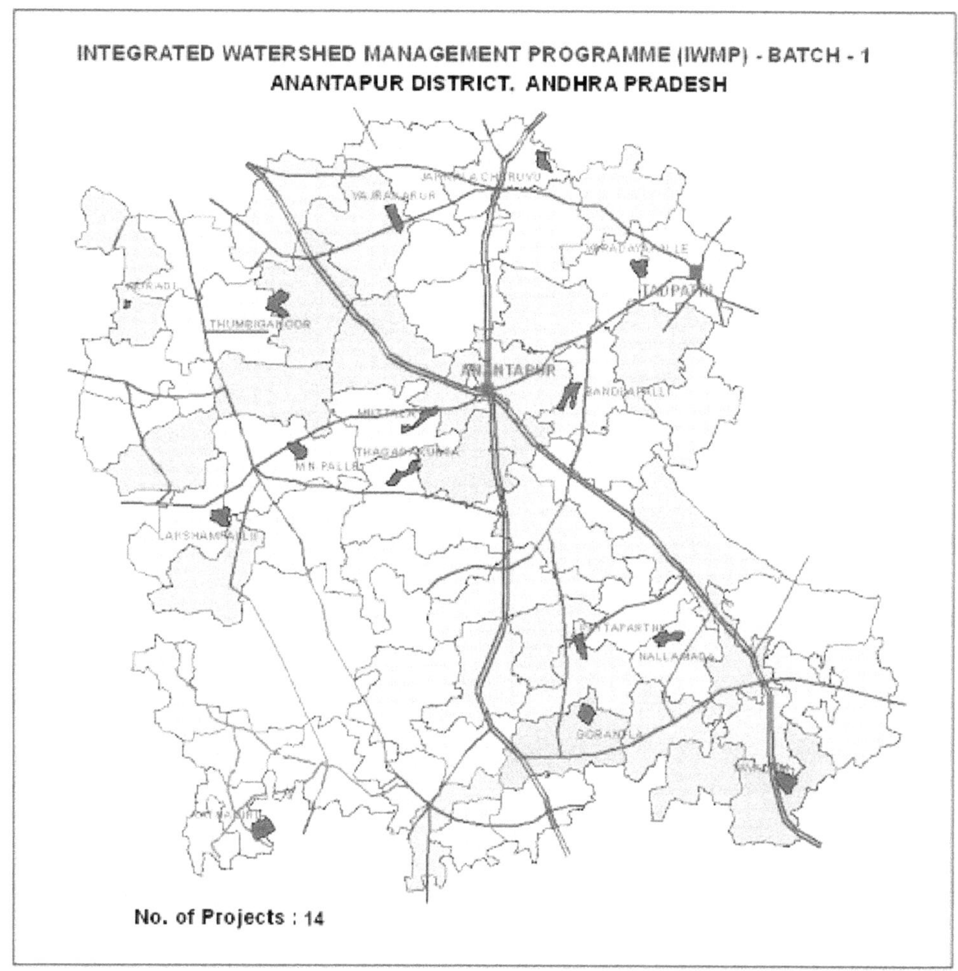

Figure 1 Location of Projects Selected under IWMP Batch-I in Ananthapur District

Results and Discussion

Impact of watershed programme on household income

Consequent to rain water conservation in the watersheds, the groundwater recharge increased and the water availability period increased varying from 3 to 9 months in a year. This prompted the farmers to go in for new bore wells as well as expansion of area under irrigation by the existing bore wells. As a result, the irrigated areas increased during post-project period by 33% across 14 projects. Area under horticulture, for ensuring economic stability to the farming community, has increased by 88%. In view of the increased availability of soil moisture more remunerative crops including vegetables are being cultivated as compared to those cultivated during pre–project period. Consequent to such developments in the agriculture sector, the total production of crops registered growth. Due to promotion of livestock activities, the total milk production increased by 73%. All the above indicate that the watershed development increased the average household income leading to economic stability at the household level. Average household income which varied from Rs. 17,476/- to Rs. 30,677/- before the project shot up varying from Rs 29,277/- to Rs. 49,198/-.

Implementation of various rain water harvesting techniques across the fourteen watershed projects resulted in an overall increase in household income of all categories of farmers (Tables 2 and 3). Among the different farmer categories, the household income in landless improved from Rs 17,736/- to Rs 29,335/-. In case of marginal farmers, it increased from Rs 22,198/- to Rs 35,003/- and for small farmers the increase was from Rs 25,589/- to Rs 40,112/-. The increase in household income ranged from 13.1 to 92.3%across all categories and watershed projects. The highest overall growth in household income was observed in case of Muradi watershed project (80%), while the lowest growth occurred in case of Muttala project (25%).

Table 2 Change in household income (in INR) of landless and marginal farmers across the 14 watershed projects.

Change in household income in Rs./Annum							
		Landless farmers			Marginal farmers		
S. No.	Watershed Project	Pre-project	Post-project	Change (%)	Pre-project	Post-project	Change (%)
1	Bandlapalli	17050	28224	65.5	21112	35700	69.1
2	JakkalaCheruvu	17659	33317	88.7	23425	40280	72.0
3	Kammavaripalli	18966	33980	79.2	23926	38842	62.4
4	Lakshmampalli	18508	32547	75.9	22437	39928	78.0
5	Muddinayanapalli	16799	26614	58.4	20077	30423	51.5
6	Muradi	17593	33825	92.3	23083	41120	78.1
7	Muttala	16812	21270	26.5	21317	25846	21.2
8	Nallamada	19053	32865	72.5	23853	36441	52.8
9	Puttaparthi	19685	32320	64.2	26145	35107	34.3
10	Tavalam	19220	33275	73.1	23980	35462	47.9
11	Thagarakunta	14600	22500	54.1	18483	26524	43.5
12	Thumbiganoor	16562	28410	71.5	23713	40822	72.2
13	Vajrakarur	20626	27254	32.1	19184	33772	76.1
14	Varadayapalli	15168	24289	60.1	20042	29781	48.6
Total		17736	29335	65.4	22198	35003	57.7

Table 3 Change in household income (in INR) of landless, marginal and small farmers across the 14 watershed projects.

Change in household income in Rs./Annum							
		Small farmers			Large farmers		
S. No.	Watershed Project	Pre-project	Post-project	Change (%)	Pre-project	Post-project	Change (%)
1	Bandlapalli	24301	41207	69.6	28493	49545	73.9
2	JakkalaCheruvu	27147	45181	66.4	31536	54300	72.2
3	Kammavaripalli	28363	45767	61.4	45376	51299	13.1
4	Lakshmampalli	25376	44995	77.3	29307	50800	73.3
5	Muddinayanapalli	23231	35070	51.0	27705	47487	71.4
6	Muradi	26768	43093	61.0	29127	54450	86.9
7	Muttala	21132	28116	33.1	29018	34500	18.9
8	Nallamada	26915	40956	52.2	31161	48726	56.4
9	Puttaparthi	28655	40865	42.6	33262	46986	41.3
10	Tavalam	27251	40927	50.2	32399	49571	53.0
11	Thagarakunta	24008	32545	35.6	29480	48000	62.8
12	Thumbiganoor	27762	47290	70.3	32120	48000	49.4
13	Vajrakarur	22354	39943	78.7	29153	55962	92.0
14	Varadayapalli	24983	35610	42.5	28348	49250	73.7
Total		25589	40112	56.8	31178	49205	57.8

Conclusion

Fourteen watershed development programmes were undertaken in Ananthapur district of Andhra Pradesh under IWMP with the objective of improving agricultural production through conservation of natural resources and their optimum utilization. Adequate attention has been given for inclusiveness and establishing economic equity by ensuring that the benefits of investments reach all sections of the society. Consequent to rain water conservation, the groundwater recharge increased and the water availability period increased varying from 3 to 9 months in a year. This prompted the farmers to go in for new bore wells as well as expansion of area under irrigation by the existing bore wells. This, in turn resulted in increased area under irrigation to both field crops and horticulture, thereby increasing crop productivity and production. The positive impact of all these indicators resulted in the increased income among the farmer groups thereby ensuring their livelihood security on a sustainable basis.

References

Akbulut, B. and Soylu, C. (2012) An inquiry into power and participatory natural resource management. Cambridge Journal of Economics, 36: 1143-1162.

Catacutan, D.C., Duque, C. and Garrity, D.P. (2001) Replicating models of institutional innovations for devolved, participatory watershed management. Paper presented to the 1st Annual In-House Review for Lantapan-Based R&D Projects, MKAVI, Alanib, Lantapan, Bukidnon, Philippines, 25 March 2001. 9 p.

CGIAR (2003) Research towards Integrated Natural Resources Management-Part 2: Towards integrated natural resource management research (Harwood, R.R. and Kassam, A.H. eds.) CGIAR Interim Science Council, CGIAR, Rome, 2003.

Vaidya, A. and Mayer, A.L., 2014. Use of the participatory approach to develop sustainability assessments for natural resource management. International Journal of Sustainable Development & World Ecology, 21: 369-379.

Application of Remote Sensing and GIS Application for Monitoring & Evaluation of Amidala Watershed Project in Anantapur District, Andhra Pradesh, India

S.Ramana Reddy, N. Polappa, B.V. Ramana Kumar and K. Chaithanya Chakravarthy

[1] *I.R.P.S, Director watersheds,* [2]*Joint Commissioner (Watersheds), State Level Nodal Agency, Amaravati, Andhra Pradesh.*
[3]*Chief Executive Officer,* [4] *Project Manager,* [5] *Project Manager - Remote Sensing Instruments, Hyderabad.*

ABSTRACT

Monitoring and evaluation of the resultant progress due to the treatment of the watershed projects are important both for tracking the implementation and for assessing the effectiveness of the programme implemented. Satellite remote sensing data is utilized for impact assessment to establish the net impact of the project on productivity, environment and related indicators. In this regard remote sensing satellite images captured during pre and post treatment periods enable an assessment of the changes in the watershed landscape over time. Essentially analysis of such remote sensing temporal images provide changes like reclamation of wasteland and increased cultivation, change in forest and horticulture assets, cropping pattern, construction of moisture and water harvesting structures, post environmental/ ecological conditions, etc., The satellite images have the advantages of providing synoptic view and large area coverage which helps in obtaining the proverbial "bird's eye view" of the features on the earth. Thus satellites orbiting around the earth, act as a vantage point to observe, measure, map and monitor the earth's natural resources. They depict accurate picture of earth surface and there is no need to work in the field as in the conventional method.The research paper explains about the Thus, satellite remote sensing data plays a vital role in this connection by depicting the status of watershed for the monitoring and evaluating the watershed projects.

Introduction

The satellite images have the advantages of providing synoptic view and large area coverage which helps in obtaining the proverbial "bird's eye view" of the features on the earth. Thus satellites orbiting around the earth, act as a vantage point to observe, measure, map and monitor the earth's natural resources. They depict accurate picture of earth surface and there is no need to work in the field as in the conventional method. Measuring and evaluating changes in a landscape over time is an important application of remote sensing. So, Impact evaluation of watershed projects has been carried by satellite remote sensing data supplied by the National Remote Sensing Centre (NRSC) of ISRO at Hyderabad. Multispectral Resourcesat-1/2, LISS-IV images of 5.8m resolution for both kharif and rabi seasons have been procured from NRSC for the pre treatment period (2010-11) and post treatment period (2017-18). These imageries are interpreted depicting the area under various types of land use/ land cover units for these two time periods (pre-treatment and post treatment) using Remote Sensing and GIS techniques. Image interpretation is the act of examining images for identifying objects and judging their significance. It will be carried out by either visual interpretation techniques or digital interpretation techniques. In this project the mapping has been carried out by visual interpretation. In the visual interpretation, remotely sensed data has been studied by the elements of Image Interpretation and attempted through logical processes in detecting, recognising and identifying, analysing, classifying, measuring and evaluating the significances of physical and cultural objects, their patterns and special relationship. The extent of area under each unit is derived from the generated map by GIS Techniques and impact evaluation has been carried by comparing the areas of the same unit for two time periods (pre treatment and post treatment). The impact evaluation has been carried by comparing the areas of the same unit between two time periods (pre treatment and post treatment) for the indicators like vegetation cover, water spread area, additional area brought under cropped area, shift from annual crops to perennial crops, reclamation of wastelands, cropping intensity, changes in different land use/ land cover categories like double crop (crop in both kharif and rabi season), single crop, fallow lands etc.

Study area

Amidala Watershed Project is located in Uravakonda Mandal, Anntapur district of Andhra Pradesh as shown in the Figure 1 & Figure 2. It is covering an area of 10,346.42ha in Amidala, Lathavaram, Chinnamusturu, Peddamusturu, Mopodi, Indravathi and Shaikshanipalli revenue villages. The treatment area of this watershed project is 4,126 ha.for the five years.

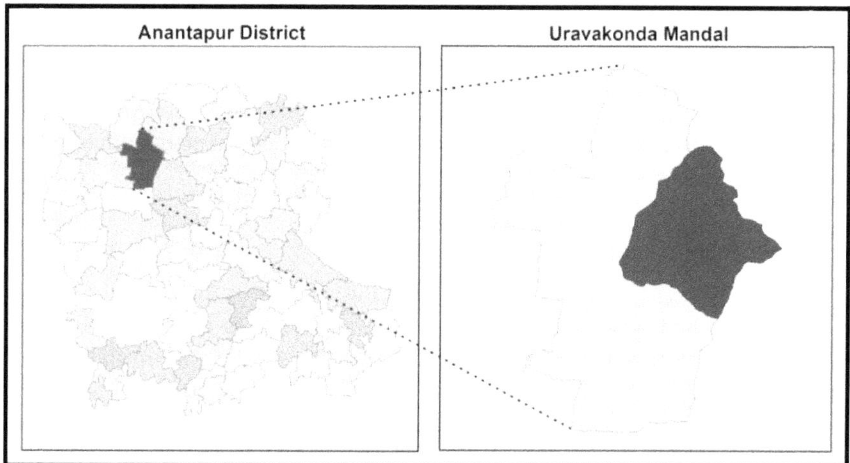

Figure 1 Location Map

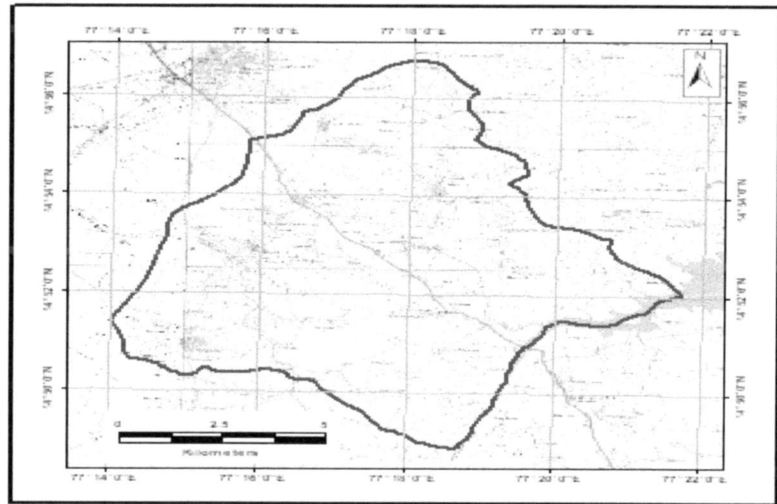

Figure 2 Topographical Map

Data used

Below given aspects have been used for the monitoring and evaluation of Amidala watershed projects through the remote sensing and GIS application.

Toposheets

Toposheets on 1:50,000scale, pertaining to Amidala Watershed Project, have been procured from the Survey of India. These procured toposheets have been scanned and converted to Geo Tiff format.

Satellite Data

The IRS data used for a diverse range of applications such as crop acreage and production estimation of major crops, drought monitoring and assessment based on vegetation condition, flood risk zone mapping and flood damage assessment, hydro-geo-morphological maps for locating underground water resources, irrigation command area status monitoring, snowmelt run-off estimation, land use and land Path, Rows and date of pass of the RESOURCESAT-1 (IRS-P6) scenes used.

Methodology

Below given methodologies have been used for the analaysis of Amidala Watershed project monitoring and evaluation.

Image Processing of Satellite Data

Image processing has been carried out by using Erdas Imagine 9.1 software. In this process, correction for Distortions, Degradations and Noise, removing of radiometric errors, contrast enhancement etc., has been carried out.

Geo-Referencing of Satellite Data

In this Amidala watershed project, geo-referencing has been carried out with the help of GCP's (Ground Control Points). For this purpose we have been selected four to five GCP's covering each CARTOSAT scene. These GCP's are selected on the basis of, that they must be identifiable both on the image and on the ground and covering all the corners of the scene. The co-ordinates of the each GCP have been collected by the Leica 1200, DGPS instrument. With the help of the co-ordinates of the GCP's collected, all the CARTOSAT scenes have been geo-referenced by using Erdas Imagine 9.1 software. Then RESOURCESAT (LISS IV MAX) scenes have been rectified, by using the geo-referenced CARTOSAT scenes,

Mosaicking

The joining of the adjacent images is called mosaicking. As the Amidala Watershed Project covers in more than one scene, all the RESOURESAT scenes are mosaicked. While mosaicking the scenes, care has been taken such that the features should be continuous in edge matching of side-by-side scenes, as they come from different paths and time.

Image Interpretation

Image interpretation is defined as the act of examining images for identifying objects and judging their significance. In this Amidala project the mapping has been carried out by visual interpretation.

Visual Interpretation

In the visual interpretation, interpreters study remotely sensed data by the elements of Image Interpretation and attempt through logical processes in detecting, recognizing and identifying, analyzing, classifying, measuring and evaluating the significances of physical and cultural objects, their patterns and special relationship.

Ground Truth / Ground Survey

The ground truth has been carried out by the subject specific thematic specialists to collect information in the field regarding all doubtful areas noted down in the pre-field interpretation and also to conform the pre-field interpretation. Necessary corrections and additions have been noted in the field dairies. Representative photographs for each class in different themes were taken and their precise location recorded through handheld GPS receivers. In the case of forest details like forest type, vegetation type, forest condition, major species, tree age class, regeneration status, soil type and condition etc., ground information has been noted in the prescribed formats. Then all the thematic maps have been finalized by incorporating modifications noted in the field for each specific theme. The mapping of each theme has been described below.

(A) Base Mapping

The base details of each JFM site mapped by onscreen visual interpretation of RESOURCESAT data and ground survey is follows.

- Transport network with roads (metalled / unmetalled / cart tracks / footpaths etc.) and railway line.

- All the drainage system containing rivers perennial streams, dry streams and canal.
- All the water bodies like ponds, reservoirs and tanks.
- Villages, hutments and tribal settlements with names.
- District and forest boundaries

(B) Land Use/Land Cover Mapping:

Land use indicates the usage of land, i.e., manmade usage of the earth's surface like built-up lands, agriculture, etc., and land cover is the natural coverage like forests, water bodies, snow etc. Spatial information on land use/land cover is a necessary prerequisite in planning, utilizing and management of natural resources. The land use /land cover units, mapped for each JFM site, by onscreen visual interpretation of RESOURCESAT data are described below.

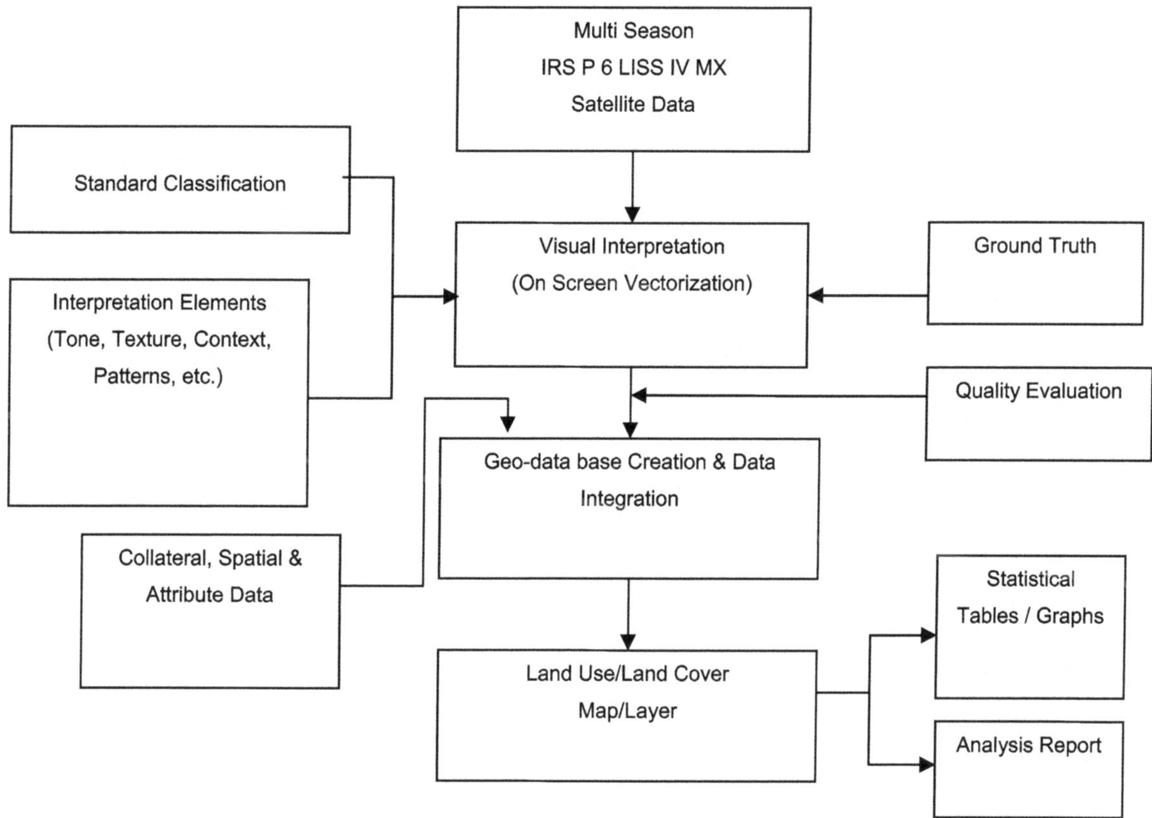

Flow Chart 1 The land use /land cover units, mapped for each JFM site, by onscreen visual interpretation of RESOURCESAT data

Results

Distribution and its Changes in Vegetation Cover

Normalised Difference Vegetation Index (NDVI) is required for analysis of changes in vegetation cover. So, it has been generated for kharif season of pre and post treatment periods by using NIR and Red bands of the Resourcesat LISS-IV MX data. These NDVIs has been classified into different vegetation vigor classes like dense, moderate, open and no vegetation. These vegetation cover maps prepared for kharif season are shown in the Figure 3.

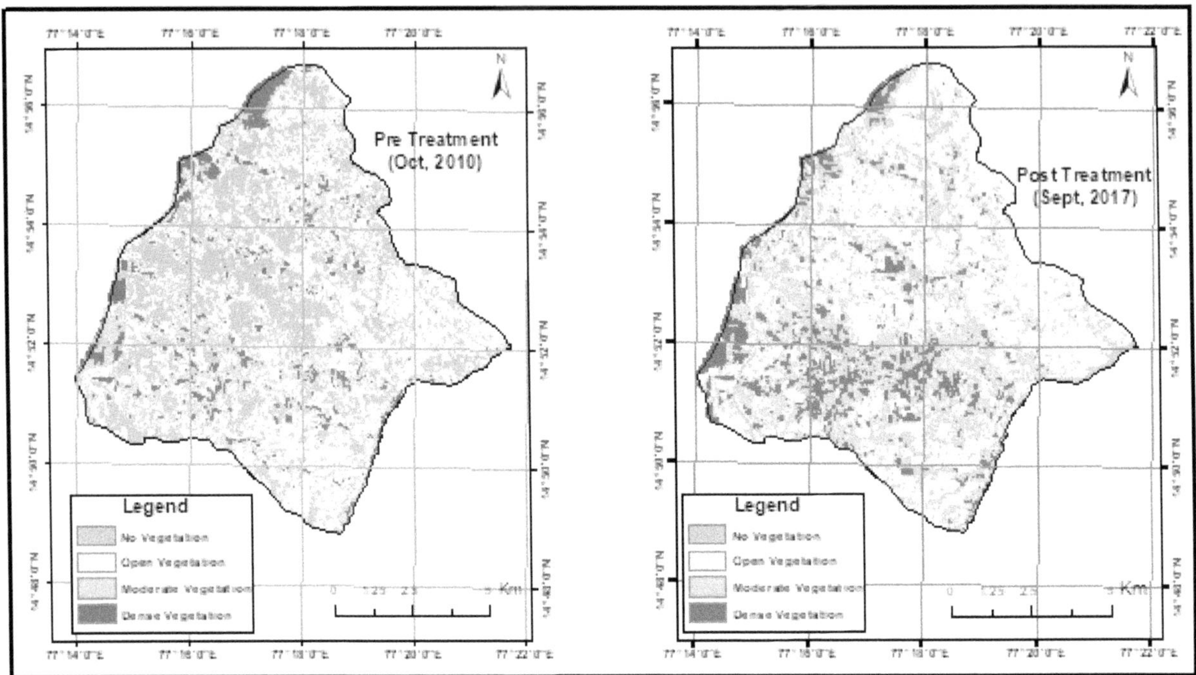

Figure 3 Vegetation Cover Maps generated from NDVI

Table 1 Details of Vegetation Cover

Sl. No.	Category	Pre Treatment (Oct-2010)		Post Treatment (Sept-2017)		Change	
		Area (in Ha.)	**%**	**Area (in Ha.)**	**%**	**Area (in Ha.)**	**%**
1	No Vegetation	3,013.35	29.12	2,632.21	25.44	- 381.14	- 12.65
2	Open Vegetation	3,543.33	34.25	3,623.02	35.02	79.68	2.25
3	Moderate Vegetation	2,343.45	22.65	2,398.79	23.18	55.35	2.36
4	Dense Vegetation	1,446.30	13.98	1,692.40	16.36	246.10	17.02
	Total	10,346.42	100.00	10,346.42	100.00		

From the above Figure 1 and Table 1 it has been observed that dense vegetation has been increased by 246.10ha (17.02%) i.e., 1,446.30ha in pre treatment period to 1,692.40ha in post treatment period.

Distribution and its Changes in Land use/Land Cover

The acquired satellite data for kharif and rabi seasons of the years 2010-11 and 2017-18 have been interpreted and different land use/land cover categories have been delineated. The spatial distribution of land use land cover units have been shown in the Figure 4 and their statistics are given in the Table 2. Distribution of land use for pre and post treatment has been shown in the Figure 2.

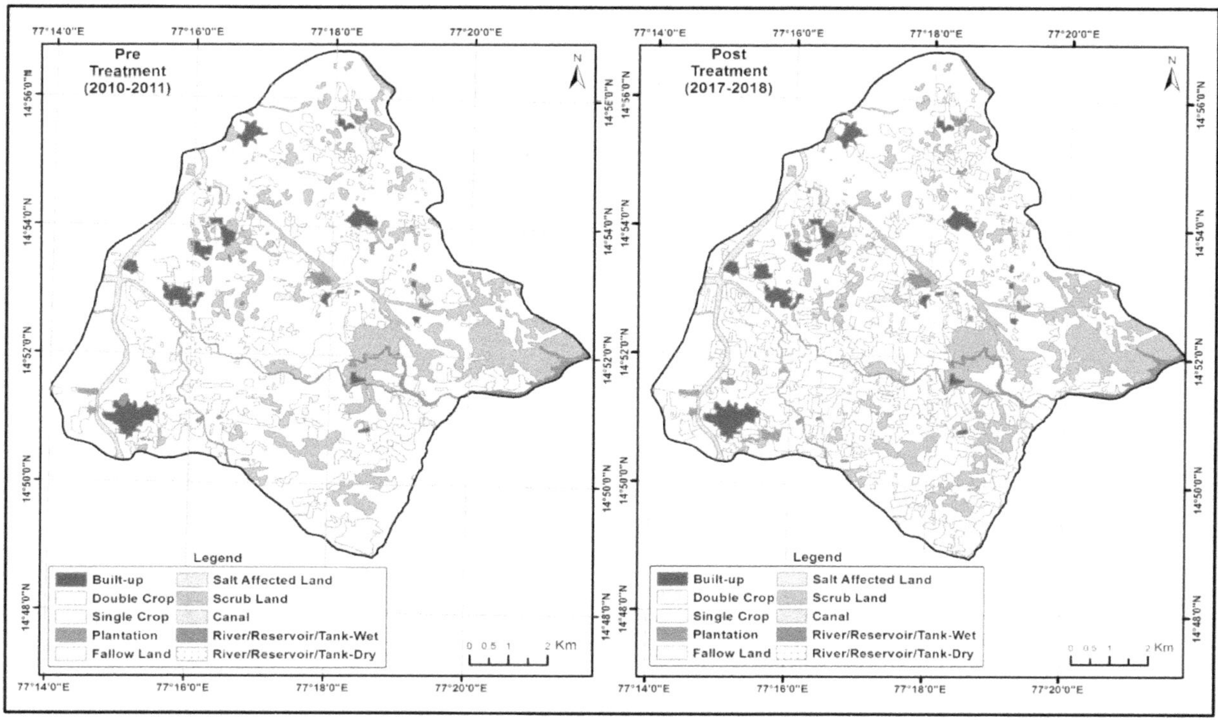

Figure 4 Land Use/Land Cover Maps

Table 2 Details of Land Use/Land Cover

Sl. No.	Category	Sub-Category	Pre Treatment (2010-11)		Post Treatment (2017-18)		Change	
			Area (in Ha.)	%	Area (in Ha.)	%	Area (in Ha.)	%
1	Built-up	Built-up	282.05	2.73	298.59	2.89	16.54	5.86
2	Agricultural Land	Double Crop	564.12	5.45	2,353.19	22.74	1,789.07	317.14
		Single Crop	6,481.45	62.64	5,218.43	50.44	-1,263.02	-19.49
		Plantation	26.86	0.26	42.44	0.41	15.58	57.99
		Fallow Land	1,068.36	10.33	526.69	5.09	-541.67	-50.70
3	Waste	Salt Affected Land	62.02	0.60	53.40	0.52	-8.62	-13.89
	Lands	Scrub Land	1,480.65	14.31	1,466.65	14.18	-14.00	-0.95
4	Water Bodies	Canal	173.96	1.68	173.96	1.68	0.00	0.00
		River/Reservoir/ Tank-Wet	160.78	1.55	164.47	1.59	3.69	2.30
		River/Reservoir/ Tank-Dry	46.17	0.45	48.60	0.47	2.43	5.27
Total			10,346.42	100.00	10,346.42	100.00		

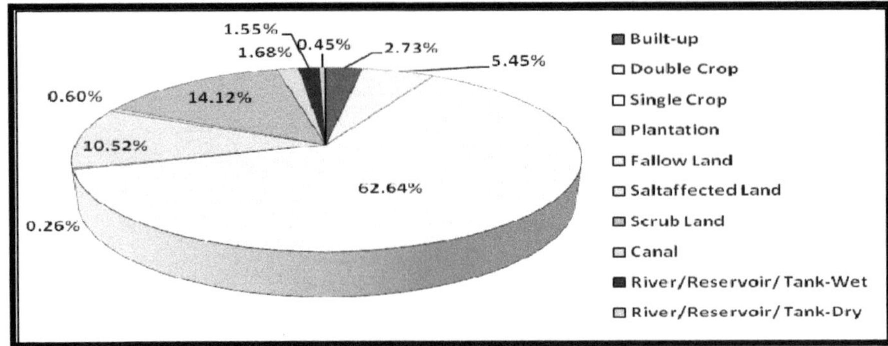

Figure 5 Distribution of Land Use/Land Cover (Pre Treatment)

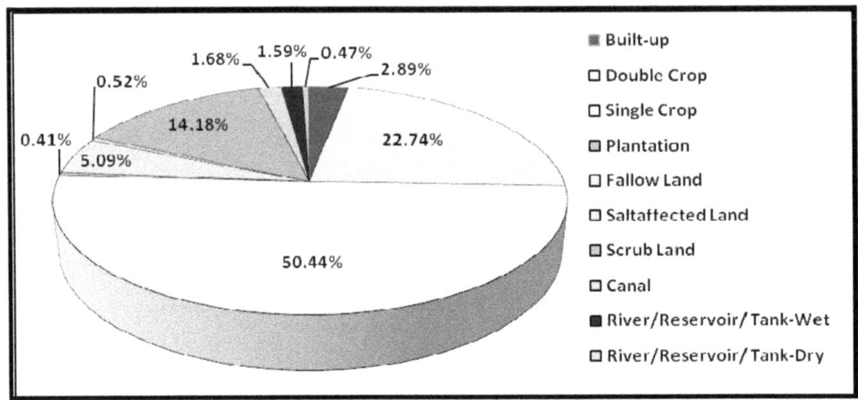

Figure 6 Distribution of Land Use/Land Cover (Post Treatment)

From the above Table 2 and Figure 6 it has been observed that double crop has been increased by 1,789.07ha (317.14%) i.e., 564.12ha in pre treatment period to 2,353.19ha in post treatment period.

Changes in Water Spread Area

An increase in water spread area has been observed in this project for an area of 3.69ha (2.30%) i.e., from 160.78ha to 164.47ha. These changes are shown in the Figure 7. This increase is due to the number of water conservation activities carried during treatment period.

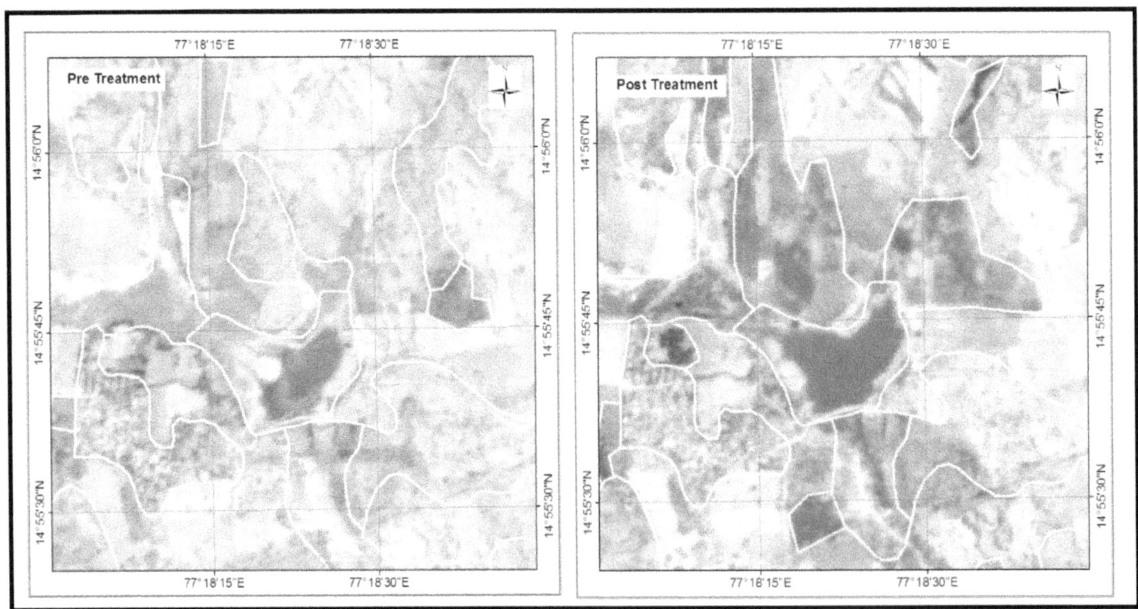

Figure 7 Comparison of Water Spread area between Pre and Post Treatment

Additional Area Brought Under Cultivation

Due to soil and water conservation activities additional area of 541.62ha (7.66%) has been brought under cultivation. This additional cropped area is mainly came from waste lands and fallow lands by adoption of soil and water conservation measures during the treatment of this watershed project.

Shift from Annual Crops to Perennial Crops

Plantations have been increased for an area of 15.58ha (57.99%) i.e., 26.86ha in pre treatment period to 42.44ha in post treatment period.

Reclamation of Waste Lands

Various wasteland reclamation activities have been implemented during the treatment period in this watershed project. Due to these activities waste lands have been decreased by 22.61ha (1.47%), i.e., 1,542.66ha to 1,520.05ha from pre to post treatment period.

Cropping Intensity

Cropping intensity is the ratio of the area under crops to the cultivable area (agricultural land) operated by the farmer in each season. In this watershed project, the cropping intensity has been increased by 6.7% (86.9% to 93.5%) in kharif season and 22.2% (7.3% to 29.4%) in rabi season from pre to post treatment period.

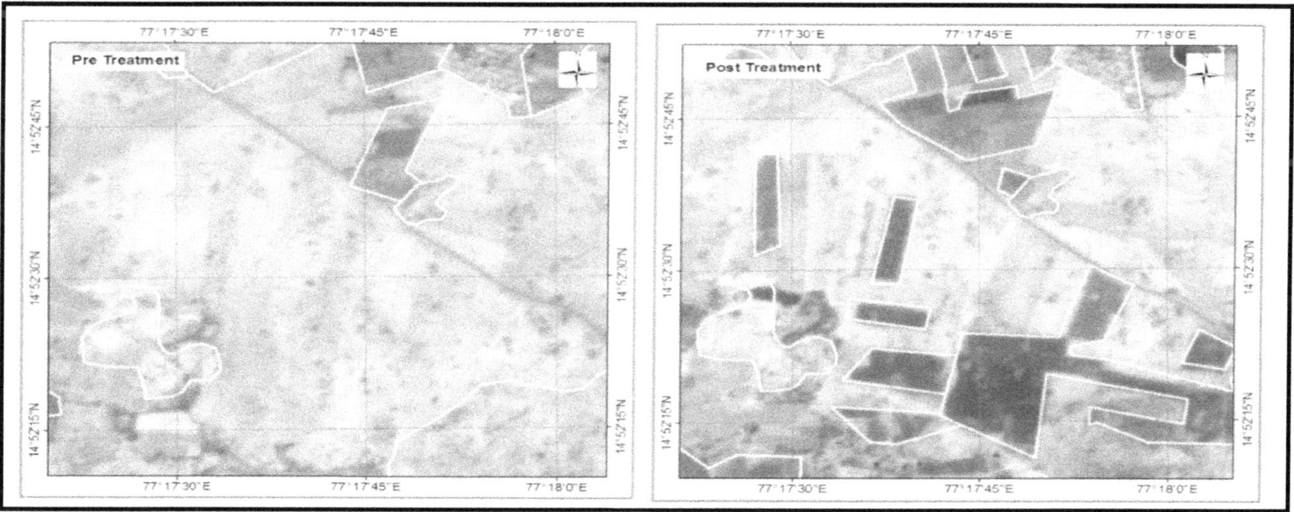

Figure 8 Changes in Cropping Pattern

Greening of Hillocks

Greening of hillocks has been carried in this watershed project during the treatment period. As it is not clearly visible in the LISS-IV images, Google image of the same is shown in the following Figure 9.

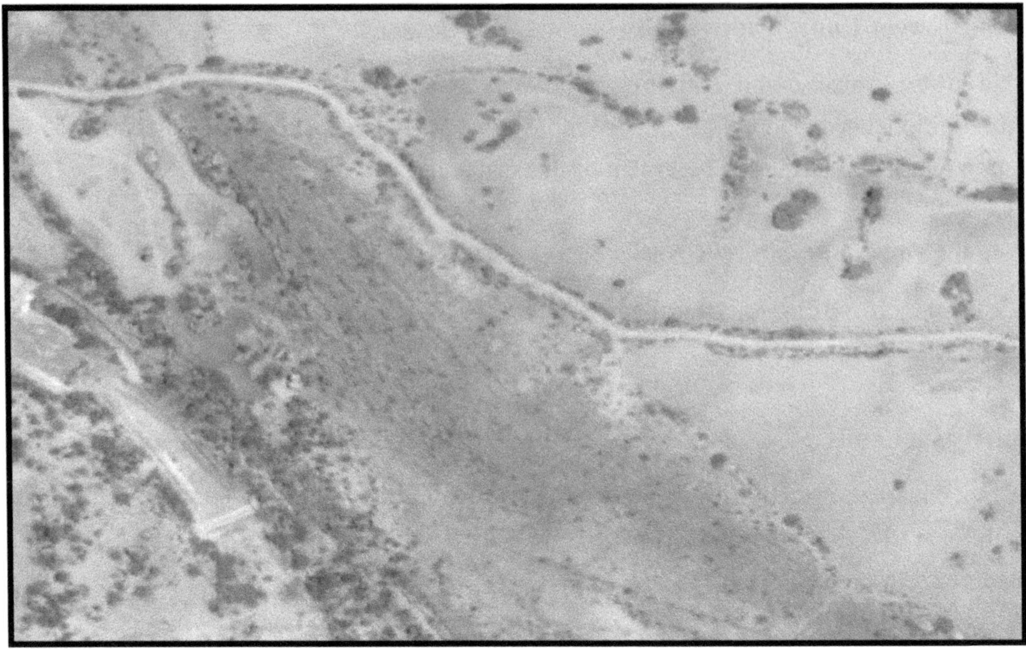

Figure 9 Google Image showing Greening of Hillocks

Conclusions

After the analysis of the LISS-IV MX satellite data of kharif and rabi seasons of the pre and post treatment period, following changes have been observed. Area under double crop has been increased by 1,789.07ha (317.14%) i.e., from 564.12ha to 2353.19ha. Plantations have been increased for an area of 15.58ha (57.99%) i.e., from 26.86ha to 42.44ha. An area of 541.62ha under fallow lands and wastelands has been brought into cultivation. Fallow lands have been decreased by 541.67ha (50.70%) i.e., from 1,068.36ha to 526.69ha. Waste lands have been decreased by 22.61ha (1.47%) i.e., from 1,542.66ha to 1,520.05ha. Water spread area has been increased for an area of 3.69ha (2.30%) i.e., from 160.78ha to 164.47ha.

References

Census of India, 2011-District Census Hand Book of Ananthapur District: Village and Town wise Primary Census Abstract, 2014.

Hand Book of Statistics: Ananthapur: 2010-2011, Chief Planning Officer, Ananthapur, 2014.

GoI, Common Guidelines for Watershed Development Projects, National Rainfed Area Authority (NRAA), India, 2011.

Detailed Project Report of Amidala Project, Ananthapur District, DWMA.

Environmental Impact Assessment Study of a Pharma Industry

P. Pallavi[1] and V. Hanumantha Rao[2]
[1]Research Scholar, JNTUK and [2]Research Scholar, Andhra University
pallavichemico@gmail.com, hrvelivala@gmail.com

ABSTRACT

Now-a-days the importance is focusing on burning issues like environmental sector because public health is in great threat for environmental decaying due to improper procedures of industries. Industries mainly pharmaceutical industry is one of those which signify not only for its revenue but also for the safety and proper medication depends on it as long as the health issue is concerned. This paper was aimed to assess the present conditions of pharmaceutical industries and tried to evaluate the present scenario and figure out some of the mitigation measures of the existing problems through the concept of EIA. The bulk drug manufacturing process utilizes various process equipment and chemical methods. The scientific research and development in API manufacturing are focused on increasing the yields and reducing the toxicity of wastes and consumption of solvents, using alternative manufacturing methods etc. Accordingly, Ministry of Environment, Forests & Climate Change, Government of India mandatory require environmental clearance for synthetic organic chemicals manufacturing units. The Environmental Impact Assessment studies is conducted to identify the negative and positive impacts and to delineate effective measures to control the pollution and to mitigate the environmental pollution. Primary and secondary data are collected initially. Subsequently, monitoring has commenced for collection of data on meteorology, ambient air quality, surface and ground water quality, soil characteristics, noise levels, flora and fauna at the specified location. The assessment concluded with some rational mitigation measures which would be firmly related to the public and manufacturer to protect the environment from pollution.

Keywords: EIA, sampling locations, API, mitigate, pharmaceutical industry.

Introduction

The pharmaceutical industry is one of the fastest growing segments of the Indian economy and has experienced rapid and sustained expansion since the second half of the 20th Century. In recent decades India's pharmaceutical industry has scaled new heights in step with a steady rise in population and thanks to its reputation as a low-cost manufacturing destination for multinational drug companies [1]. The pharmaceutical industry is based primarily upon the scientific research and development (R&D) of medicines that prevent or treat diseases and disorders. Combined with the growth in contract manufacturing and outsourcing by multinational companies (MNCs) to low-cost Indian suppliers led to the rapid development of India's generic drugs sector, often described as the "backbone" of the country's pharmaceutical industry. Today, India is one of the world's leading suppliers of generic drugs, which account for approximately 75 per cent of its market by volume[2]. Liquid effluents, air emissions and solid wastes generated are the major pollutants from the process operations of bulk drug manufacturing. Sources of emissions include dryers, reactors, distillation units, storage and transfer of materials, filtration, extraction, centrifugation, and crystallization. Waste streams generated are numerous and complex due to the raw materials used and the varied nature of operations. Unconverted reactants, byproducts, and residual product in a solvent or aqueous base, as well as acids, bases, metals, etc. So, in order to assess the feasibility of the project EIA studies should be conducted. Environmental Impact Assessment (EIA) is a process to assess the environmental consequences of any project and design proper mitigation plans to minimize the possible adverse impacts. EIA certainly has a vital role to play in addressing environmental issues surrounding project development [3,4]. EIA studies involve three basic components i.e., identification, prediction and evaluation of impacts.

EIA is considered the starting point in the process of implementing sustainable development agendas. In terms of benefits, it has identified EIA as the most effective tool for integrating environmental concerns in development planning and implementation. EIA also provide a good example on how a combination of 'top-down' and 'bottom-up' approaches could improve democracy and service delivery [5]. EIA is used to identify, predict, evaluate and mitigate the environmental, social and other potential impacts and consequences of projects prior to major decisions being taken and commitments made to recommend suitable mitigation measures and to decrease possible adverse impacts. It is a good management tool to predict the type, magnitude and probability of environmental and social changes likely to occur as direct or indirect result of a plan or policy and to decrease adverse impacts [6]. The need for EIAs has become increasingly important and is now a statutory requirement in many developing countries.

Objectives of the Study

The objectives of the present environmental impact assessment study briefly described by sustaining goals of environmental protection and development. For Pharma Projects EIA has been used as an effective tool in decision making process for production that affect\environmental factors. To predict environmental impacts of Pharma projects due to land use change or modifications. Decision makers considering the costs and benefits before projects are started. Reduce adverse impacts during the production phase. To avoid, minimize or offset the adverse significant biophysical, social and other relevant effects on environment. To protect the productivity and capacity of natural systems. Optimize resource use as well as management opportunities in future.

Study of the Project Location

EIA report has been structured covering various aspects like project description, baseline conditions, environmental impacts, mitigation measures, environmental management plan. For the establishment of any project "Environmental Clearance" is compulsory. The location of the project area is identified along with survey numbers, latitudes and longitudes. As soon the location is identified, the collection of primary and secondary data will be implemented. Monitoring for collection of data on meteorology, ambient air quality, surface and ground water quality, soil characteristics, noise levels, flora and fauna, land use and land cover and socio-economic status of community are studied. Ambient air quality data are obtained from India Meteorological Department (IMD). The geology, hydrogeology nature of the soil was analysed. The base line data for a circular radius of 10 km radius around the industrial site was collected. The nearby means of transportation, waterbodies, national parks and wildlife sanctuary along with ecological sensitive areas are identified within the 10 km radius. The location of plant area was represented on a topo map.

Process Description

In this segment the pharmaceutical products that are manufactured are represented by calculating the total amount of water required and the types of emissions that are released during the process.

Water pollution

Based on the plant production capacity the water requirements and effluent stream released are noted. The characteristics of the streams are identified and effluent management system is developed to ensure Zero liquid discharge by establishing a Multiple effect evaporator (MEE), ETP plant. The organic bottom is sent to cement plants as material of co-incineration.

Air Pollution

Different sources of air pollutants like compounds of Sulphur, nitrogen and particulate matter are notified which can be controlled using bag filter. The different types are emissions that are released are treated by using various sources and released into the atmosphere so as to maintain pollution free. The industry should maintain ambient air quality standards of PM10 (size less than 10μm)-100 μg/ m^3, PM 2.5 (size less than 2.5μm)- 60 μg/ m^3, SO_2-80 μg/ m^3, NO_x-80 μg/ m^3outside the factory premises.

Solid waste

The solid waste that is generated from various segments of the manufacturing process is recovered as the incineration material for the cement industry and the waste that cannot be recovered are sent to TSDF facility for secured landfill.

Noise pollution

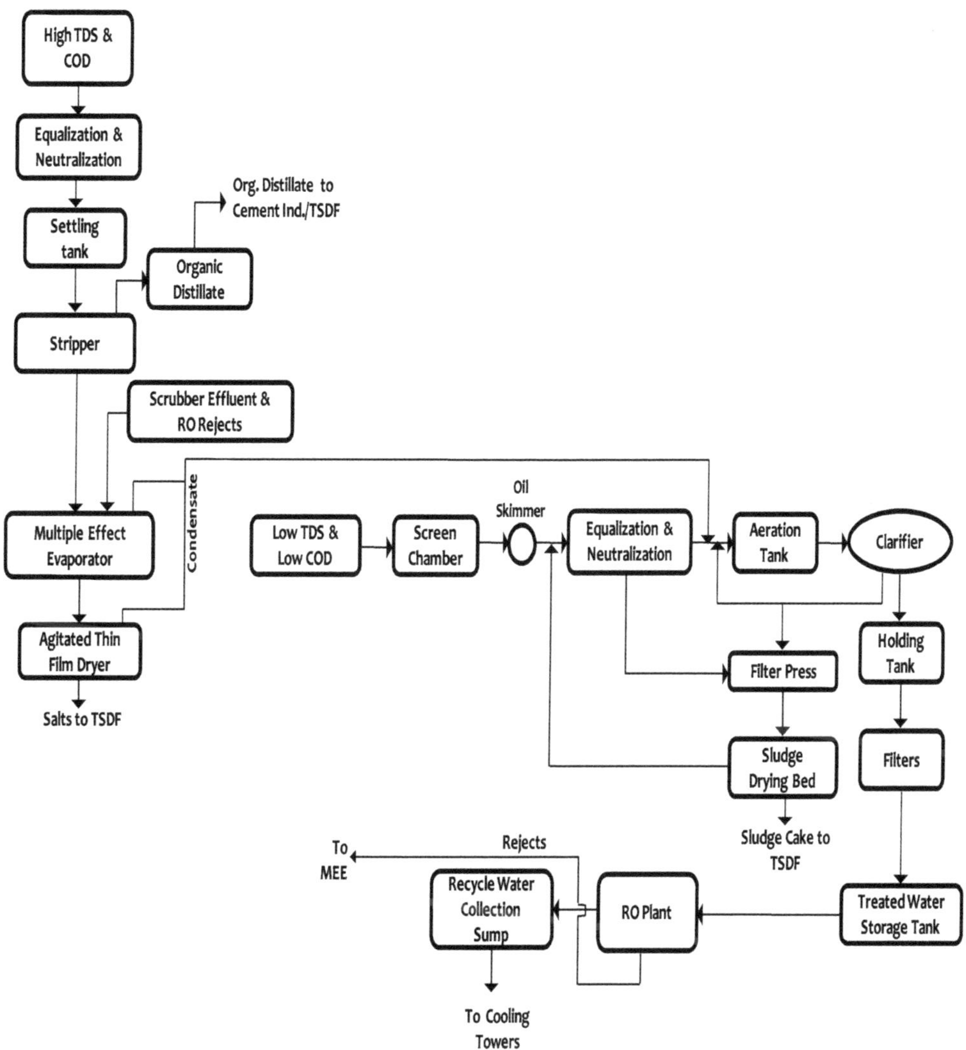

Figure 1

The equipment that generate noise pollution should be guarded and shall be mounted to ensure reduction of noise and vibration. According to ambient air quality standards during day time (6 AM to 10 PM)-75 dB and night time (10 PM to 6 AM)- 70 dB of sound levels should be maintained. Identification of impacts is one of the basic analytical steps in EIA for subsequent prediction and evaluation of impacts. The impacts are signified as either negative or positive, second as direct, indirect or cumulative and thirdly a scale is use to determine the severity of effect like low, medium and high. The ware house should be provided with adequate fire fighting facilities as per the norms. The raw materials for the process are transported by different forms like solid, liquid and crystalline. Depending on the nature of materials, they are handled in drums, bags, storage tanks, cylinders etc., and transported through roadways. Rain water harvesting shall also be adopted by providing rain water harvesting structures along the drains, storm water storage sump and tanks [7]. Cold water and chilled brine were identified as the best medium to mitigate VOC concentration in ambient air. The usage of high-pressure water jet in place of fill and vacate method of washings is chosen as the best alternative of equipment washing. Hazard analysis involves the identification and quantification of the various hazards that exist in the plant. Risk analysis deals with the identification and quantification of risks. Hazard and risk analysis involve very extensive studies, and require a very detailed design and engineering information. Hazard identification and risk assessment involve a critical sequence of information gathering and the application of a decision-making process [8].

Health monitoring of the employees is monitored regularly. The employees working in the organization should be provide with gumshoe, helmet, masks and googles. Social activities are planned for the betterment of neighborhood through awareness and welfare programs. Green belt development around the plant area is proposed.

The ETP plants are used widely in pharmaceutical industry to remove the effluents from the bulk drugs. The treatment of effluents in pharmaceutical industry is essential to prevent pollution of the receiving water. The effluent treatment plants are used in the removal of high amount of organics, debris, dirt, grit, pollution, toxic, nontoxic materials, polymers etc. from drugs and other medicated stuff.

Effluent Treatment Plant

Conclusion

Environmental Impact Assessment (EIA) can broadly be defined as a Process, providing an anticipatory and preventive mechanism for environmental management and protection to achieve sustainable development. It is a study of the effects of a proposed project, plan or program on the environment. In other words. EIA is an administrative process that identifies the potential environmental effects of any proposal along with its advantages and disadvantages on environment. Positive effects are maximized whereas; adverse effects are minimized to greatest possible extent. To identify and assess environmental, social, economic, and cultural consequences of a proposed activity and to delineate plans for the mitigation of adverse impacts resulting from the proposed activity.EIA certainly plays a vital role in assessing the environmental impacts of surrounding developmental project To provide for the involvement of the public, Government department and Government agencies in the review of the proposed activities. Thus, by using EIA tool both environmental and economic benefits can be achieved, such as reduced cost & time of project design & implementation, avoided treatment/clean-up costs, impacts of laws and regulations, etc.

References

The Hague. Ministry of Housing, Spatial Planning and the Environment of the Netherlands, Strategic Environmental Assessment - status, challenges and future directions.by B. R. Sadler, and Verheem, (1996).

The Environmental Impact of Buildings. Building Research Establishment (BRE) Information Paper, BRE, UK, 18-19 by P.B. Barlett, and J.J. Prior, (1991),

An integrated fuzzy AHP–ELECTRE methodology for environmental impact assessment. Expert Systems with Applications by T. Kaya, and C. Kahraman, 38(7), (2011), 8553–8562.

Environmental and social impact assessment. New York: Wiley, by F. Vanclay, and D. A. Bronstein, (1995).

Environmental Impact of projects, planning and policy issues, New Delhi, India. Institude of Economic Growth by Chopra, K., and *et al.* (1993).

Environmental Impact Assessment (EIA) and Construction. Int. Res. J. Environmental Sci., by Dutta, A.B. and Sengupta, I. 3 (1), 58–61 (2014)

A Review of Low-Cost Housing Technologies in India. In Advances in Structural Engineering, by V. Puri and *et al.* 1943-1955, (2015),

An Environmental impact assessment based on the numbers. Expert Systems with Applications by X. Deng and *et al.,* 635–643 (2014).

Annual Water Balance Assessment of a Micro-Catchment in Dryland Agriculture

R. S. Patode, M. B. Nagdeve, K. Ramamohan Reddy[1], Ranee Wankhade, V. V. Gabhane and M. M. Ganvir

All India Coordinated Research Project for Dryland Agriculture

Dr. Panjabrao Deshmukh Krishi Vidyapeeth, Akola (M.S.)

[1]Centre for Water Resources, IST, Jawaharlal Nehru Technological University, Hyderabad, (Telangana)

ABSTRACT

Rainfall is the single input in the Dryland agriculture for fulfilling the crop water requirement. Based on the water balance approach it is possible to make a quantitative evaluation of water resources and their change under the influence of man's activities. Knowledge of the water balance assists the prediction of the consequences of artificial changes in the regime of streams, lakes, and ground-water basins. An understanding of the water balance is also extremely important for studies of the hydrological cycle. With water balance data it is possible to compare individual sources of water in a system, over different periods of time, and to establish the degree of their effect on variations in the water regime. Due to its multiple benefits and the problems created by its excesses, shortages and quality deterioration, water as a resource requires special attention. Hydrological models are useful tools for quantification of water balance components. MIKE SHE model set up was developed and used for simulation of hydrological components in treated as well as untreated micro-catchment at the experimental field of AICRP for Dryland Agriculture, Dr. Panjabrao Deshmukh Krishi Vidyapeeth, Akola in Vidarbha region of Maharashtra state. The results of the hydrological water balance of treated and untreated catchment are presented here. During the year 2018, the annual rainfall was 830mm out of which 62mm surface runoff + losses were occurred in untreated micro-catchment. The evapo-transpiration losses of 462mm and 446mm were occurred in treated and untreated catchments respectively. Around 234mm and 199mm recharge was observed in treated and untreated micro-catchments respectively.

Keywords: micro-catchment, model, water balance.

Introduction

In India main stay of people is on agriculture that's why it is necessary to conserve rainwater in the soil to increase the groundwater table and provide sufficient amount of water to crops in rainfed condition. Efficient conservation of rain water is the central issue in successful dryland farming. Extensive trials conducted by the soil conservation and dryland research centres have led to the identification of number of inter-terrace land treatments besides contour and graded bunds. Farmers have not widely adopted mechanical measures like contour bunds, graded bunds, grassing of waterways and construction of farm ponds without the government support due to financial constraints. However, studies at Hyderabad, Banglore and Anantpur revealed that more than 80 per cent farmers follow simple conservation measures like sowing across the slope, opening of dead furrow and key line conservation.

In hydrology, a water balance equation can be used to describe the flow of water in and out of a system. A water balance can be used to help manage water supply and predict where there may be water shortage. It is used in irrigation, runoff assessment, flood control and pollution control. Several diagnostic measures in hydrology can be used to select and evaluate the performance of water balance models. The groundwater recharge has been routinely estimated as a residual of various components of soil moisture budget viz. surface runoff, evaporation from soil, and transpiration from plants, interception losses, soil moisture store and precipitation. Water balance uses the principles of conservation of mass in a closed system, whereby any water entering a system (via precipitation), must be transferred into either evaporation, surface runoff (eventually reaching the channel and leaving in the form of river discharge), or stored in the ground.

Today, MIKE SHE is an advanced, flexible framework for hydrologic modeling. It covers the major processes in the hydrologic cycle and includes process models for evapotranspiration, overland flow, unsaturated flow, groundwater flow, and channel flow and their interactions. Each of these processes can be represented at different levels of spatial distribution and complexity, according to the goals of the modeling study, the availability of field data and the modeler's choices. The MIKE SHE user interface allows the user to intuitively build the model description based on the user's conceptual model of the watershed. The model data is specified in a variety of formats independent of the model domain and grid, including native GIS formats. At run time, the spatial data is mapped onto the numerical grid, which makes it easy to change the spatial discretisation. (Graham and Butts, 2005).

Study Area

The study was conducted at the experimental field of All India Coordinated Research Project for Dryland Agriculture, Dr. Panjabrao Deshmukh Krishi Vidyapeeth, Akola (M.S.), India. The site is situated between latitude of 20° 43' 05.8" to 20° 43' 09.3" North and Longitude of 77° 02' 43.1" to 77° 02' 46.0" East with the altitude of 307m above MSL. It is located in the Western Vidarbha Zone, a part of the Central Maharashtra Plateau Agro-climatic zone. In the present study three types of soils were identified in the micro-catchment viz. Inceptisol, Entisol and Vertisol. The infiltration rate was determined by double ring infiltrometer. The infiltration rates for the soils under study were found to be 3.77, 4.4 and 1.2 cm hr^{-1} for Inceptisol, Entisol and Vertisol, respectively (Sarda Tariku, 2012).

Methodology

The experimental area of 1.0 ha was divided into two micro-catchments. The micro-catchment-1 (MC-1) is treated with Continuous Contour Trenches (CCTs) and has horticultural plantations of Atemoya (*Anona atemoya*) and Custard apple (*Anona squamosa*). The intercrop has been practiced along the continuous contour trenches in between plantation rows. The adjacent micro-catchment-2 (MC-2) has been kept without continuous contour trenches. This untreated micro-catchment also has horticultural plantations of Atemoya (*Anona atemoya*) and Custard apple (*Anona squamosa*). The intercrop of green gram has been sown in between the plantation rows of Custard apple and Atemoya in MC-1 and MC-2 during kharif season in every year. Model setup of MIKE SHE for treated as well as control (untreated) micro-catchment was prepared. For water movement model set up, different components were considered. Performance of the CCTs existing in the treated micro-catchment was evaluated by comparing the relevant components of the hydrological cycle. The results were used for the impact assessment purpose (Shinde, 2006 and Pendke, 2009). Other than hydrological monitoring, the physical observations and analysis of fruit production was also used for assessing CCT performance. Model setup of MIKE SHE for treated as well as control micro-catchment was prepared. For water movement model set up, saturated zone, unsaturated zone, evapotranspiration and overland flow were included.

Results and Discussions

The treated and untreated micro-catchments were monitored and impact assessment was done based on the results obtained as below.

Soil Moisture

The soil moisture at the depths 0-15, 15-30 and 30-45cm is given in Table 1 and presented in Figure 1. The soil moisture status in CCT treated catchment (T_2) was observed to be better as compared to the untreated catchment (T_1) at 0-15, 15-30 and 30-45cm depth in every recorded month. The prolonged moisture in the CCT treated catchment has enhanced the growth of perennial plantation of Custard Apple (*Annona squamosa*) and Hanumanphal (*Annona atemoya*).

Table 1 Soil moisture content (%) recorded at 0-15, 15-30 and 30-45cm depths in different months.

Treatments	Depth (cm)	Soil moisture content (%)			
		30/07/2018	29/08/2018	29/09/2018	15/10/2018
Untreated catchment, T_1	0-15	29.86	29.76	23.89	19.25
	15-30	30.21	30.08	24.80	20.58
	30-45	31.54	31.34	25.70	21.69
CCT treated catchment, T_2	0-15	31.12	30.80	26.20	22.05
	15-30	31.40	31.59	27.30	22.85
	30-45	31.66	31.78	27.58	23.01

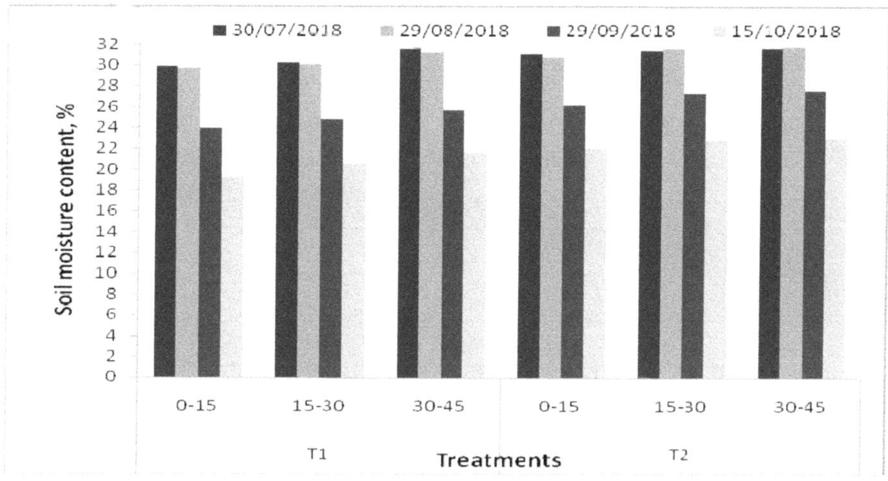

Figure 1 The soil moisture content at different depths in untreated and CCT treated micro-catchment recorded in different months.

Monitoring of ground water levels

The experimental area (1.0ha) was divided into two small catchments. One catchment was treated by constructing continuous contour trenches (CCTs) and other was without continuous contour trenches. The small catchments were again divided into two parts, thus in entire area there are four parts. In each part the observation wells were prepared for monitoring the ground water levels. The observation wells in all the four parts were monitored and the readings during the season are presented in the Table 2. From the table, it is observed that the percentage fluctuations in groundwater levels of CCT treated catchment over non treated catchment was more in the month of July (25.16%) followed by December (22.58%) and November (20.77%). It was also observed that the groundwater levels were more in CCT treated catchment compared to non treated catchment in all the months and this effect was depicted in figure 2 and 3. On an average during the twelve months the observed ground water recharge in the CCT treated catchment was more by 17.15% compared to the non treated catchment and this will clearly indicate the benefits of continuous contour trenches for groundwater recharge in small catchment.

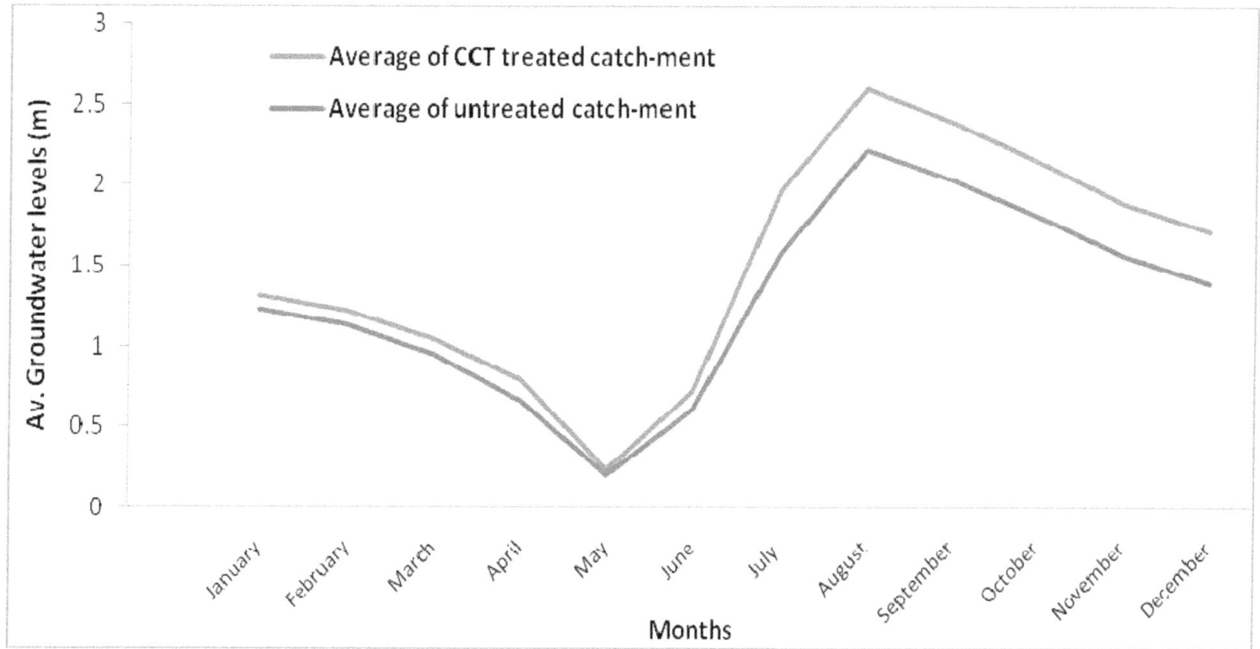

Figure 2 Ground water levels in different months in the micro-catchment

Table 2 Average monthly ground water levels (m) in the observation wells at AICRPDA, Dr. PDKV, Akola during 2018

S. N.	Month	Average ground water levels, m						Increase in GW levels of CCT treated over untreated catchment (%)
		CCT treated catchment (T_2)		Untreated catchment (T_1)		Average of CCT treated catchment	Average of untreated catchment	
		OW-1	OW-3	OW-2	OW-4			
1	January	1.29	1.35	1.20	1.26	1.32	1.23	7.32
2	February	1.19	1.26	1.10	1.16	1.23	1.13	8.41
3	March	1.01	1.08	0.91	0.99	1.05	0.95	10.00
4	April	0.76	0.83	0.62	0.70	0.80	0.66	20.45
5	May	0.21	0.28	0.18	0.23	0.25	0.21	19.51
6	June	0.69	0.76	0.58	0.65	0.73	0.62	17.89
7	July	1.92	2.01	1.52	1.62	1.97	1.57	25.16
8	August	2.55	2.64	2.18	2.25	2.60	2.22	17.16
9	September	2.34	2.43	2.00	2.06	2.39	2.03	17.49
10	October	2.10	2.21	1.75	1.86	2.15	1.81	19.11
11	November	1.81	1.92	1.51	1.62	1.89	1.57	20.77
12	December	1.67	1.78	1.35	1.44	1.71	1.40	22.58

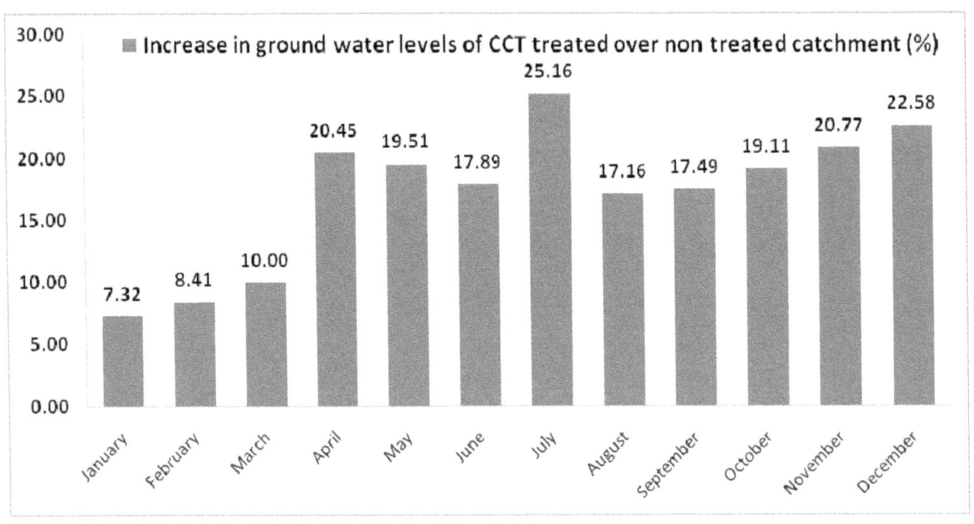

Figure 3 Increase in groundwater levels in CCT treated micro-catchment

Water Balance estimation through CCT treated and untreated micro-catchments

For knowing the CCT performance, the rainfall was main hydrological process thus the rainfall data during year 2018 were collected. Other data in respect of runoff, evaporation, soil moisture, LAI and ground water levels were monitored throughout the year. This year wise data was simulated with calibrated and validated MIKE SHE model. Performance of the CCTs existing in the treated micro-catchment was evaluated by comparing the appropriate components of the hydrological processes of the CCT treated and untreated micro-catchments. The results of water balance obtained with MIKE SHE model for the two micro-catchments, viz., treated and untreated for the year 2018 are given in the Table 3 and depicted in Figure 4. The hydrological water balance for the year 2018 was observed (Figure 4). The annual rainfall was 830mm out of which 62mm surface runoff + losses were occurred in untreated micro-catchment. The evapo-transpiration losses of 462mm and 446mm were occurred in treated and untreated catchments respectively. Around 234mm and 199mm recharge was observed in treated and untreated micro-catchments respectively.

Table 3 Water balance of CCT treated and untreated micro-catchments during the year 2018

Hydrological components (2018)	CCT Treated	Untreated	Increase/decrease over control (%)
Rainfall (mm)	830	830	-
UZ Storage Change (mm)	130	123	5.69
Runoff + Surface Losses(mm)	4	62	-93.55
Evapotranspiration (mm)	462	446	3.59
Recharge (mm)	234	199	17.58

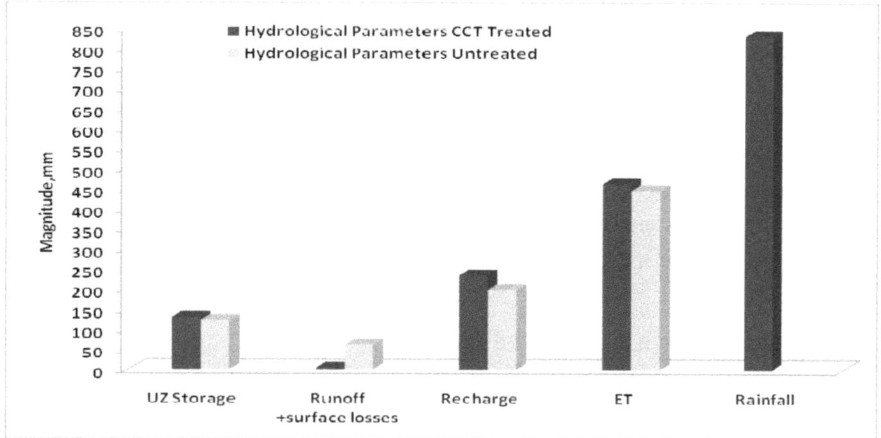

Figure 4 Hydrological water balance components for CCT treated and untreated micro-catchment during 2018

Conclusion

The soil moisture status in CCT treated catchment (T_2) was observed to be better as compared to the untreated catchment (T_1) at 0-15, 15-30 and 30-45cm depth in every recorded month. On an average during the twelve months the observed ground water recharge in the CCT treated catchment was more by 17.15% compared to the non treated catchment.

References

Graham, D.N. and M. B. Butts 2005. Flexible, integrated watershed modelling with MIKE SHE. In Watershed Models, Eds. V.P. Singh & D.K. Frevert Pages 245-272,

Pendke, M. S. 2009. "Qualitative evaluation of soil and water conservation structures in Daregaon watershed." Journal of Soil and Water Conservation, 8(1), 9-13.

Pramana, K.E.R. 2007. "The Design of contour trenches in Vietnam." Master Thesis, Technische Universiteit, Delft, The Netherlands, 1-87.

Sadgir, P. A., Patil, G. K. and Takalkar, V. G. 2006. "Sustainable watershed development by refilled continuous contour trenching technology." National seminar on rainwater harvesting and water management, Nagpur (M.S.), 331-338.

Sarda Tariku 2012. "Modelling one dimensional water flow in Vadose zone using HYDRUS." M.Tech Thesis (Unpublished) submitted to Deptt. of Soil and Water Conservation Engg., Dr. P.D.K.V., Akola.

Shinde, M., Smout, I. and Gorantiwar, S. 2006. "Assessment of water harvesting and groundwater recharge through continuous contour trenches." Recharge systems for protecting and enhancing groundwater resources, UNESCO, IHP-VI, series on Groundwater, 13, 229-235.

Sussman, D. 2007. "Design manual- Contour trenches." Bren school of environmental science and management, University of California Santa Barbara, U.S.A, 1-7.

Thomas, C. George 2010. "Land husbandry and watershed management." 1st Eds. Kalyani publishers, New-Delhi, India, 324-352.

Soil Loss under changing Climatic Scenarios at Mudhole Watershed in Adilabad

R. Sudha Rani[1], R. Rejani[2], G. Manoj Kumar[1], Y. Siva Laxmi[1] and D. Kalyana Srinivas[2]

[1]*College of Agricultural Engineering (PJTSAU), Kandi, Sangareddy, Telangana.*

[2]*ICAR-Central Research Institute for Dryland Agriculture, Hyderabad, India.*

ratnala504@gmail.com

ABSTRACT

The Mudhole watershed in Adilabad district of Telangana is prone to severe erosion and water scarcity. The information on soil loss data is essential and necessary for a watershed to support natural resource management and for planning soil and water conservation interventions etc. In this study, an attempt was made to estimate the soil loss spatially, temporally and under climate change for Mudhole watershed using RUSLE coupled with ARCGIS and its validation was found satisfactorily (R^2=0.802) since the observed and recorded data matches. The spatially estimation of soil loss during the period from 2004 to 2016, that most of the area has soil loss < 5 t ha^{-1} year^{-1} and the soil loss > 20 t ha^{-1} year^{-1} in very less areas. Under the climate change scenarios, soil loss under low emission scenario in 2020's (RCP 2.6) is predicted to increase by 11 t ha^{-1} year^{-1} and 17 t ha^{-1} year^{-1} by 2050's, medium emission scenario in 2020's (RCP 4.5) is predicted to increase by 8.5 t ha^{-1} year^{-1}and 21 t ha^{-1} year^{-1} by 2050's, high emission scenario in 2020's (RCP 6.0) is predicted to increase by 8.5 t ha^{-1} year^{-1} and 19 t ha^{-1} year^{-1} by 2050's and very high emission scenario in 2020's (RCP 8.5) is predicted to increase by 9.6 t ha^{-1} year^{-1} and 29 t ha^{-1} year^{-1} by 2050's. Similar trend was observed during 2080's.

Keywords: Adilabad, climate change, RUSLE, soil loss.

Introduction

The land and water are very essential natural resources, under the tremendous stress due to ever increasing human pressure on the earth. Soil degradation has already rendered 0.3-0.8% of the world's arable land unsuitable for agricultural production every year. In India, out of the total geographical area of 329 million ha, approximately 145 million ha of the total land resource is subjected to various degrees of wind and water erosion, which cause a loss of 5.3 million Mg of soil every year. In eastern Himalayan region of India, soil erosion by water is a major factor causing land degradation. About one third area of the region suffers from various forms of land degradation problems (Sehgal and Abrol, 1994). The soil erosion removes the fertile top soil and organic matter from the soil surface which in turn affect the soil fertility and the reduction of crop yields (Ismail and Ravichandran, 2008). Soil erosion mainly depends on the intensity and duration of rainfall, the slope of the land, soil type, land use land cover, and the land management practices. For estimate soil loss are different models are there like, Universal Soil Loss Equation (USLE), Revised Universal Soil Loss Equation (RUSLE), Water Erosion Prediction Project (WEPP), Soil Erosion Model for Mediterranean Regions (SEMMED), Soil and Water Assessment tool (SWAT), European Soil Erosion Model (EUROSEM), Agricultural Non- Point Source Pollution Model (AGNPS) etc. Among these models, USLE and RUSLE are widely used to predict long term average annual soil loss using rainfall, soil type, topography, crop systems and management practices. These factors key role is better understanding of the dynamics of erosion phenomena and they are crucial for planning soil and water conservation practices.

The impacts of climate change are experienced worldwide, but countries like India are more vulnerable in view of the high population depending on agriculture. The changes in the intensity of rainfall and prolonged dry spells attributes to the climate change effects in Indian agriculture. In India, the surface temperature is predicted to increase by 2 to 4°C (Ranuzzi and Srivastava, 2012). Also, changes in the distribution and frequency of rainfall, decrease in the number of rainy days, increase in rainfall intensities and intensity of cyclonic storms are also projected by 2030. Hence, the present study was taken to estimate the soil loss spatially and under climate change scenarios from a watershed.

Methodology

Study Area

The Mudhole watershed lies in Godavari river basin. This watershed is located in Adilabad district (Figure 1), is having a drainage area of 300 km^2 and extended from 18°56'34" to 18°56'49" N latitudes and 77°52'39" to 77°56'45" E longitudes. The average annual rainfall in the Mudhole area is around 997.5 mm. The South West monsoon sets by the middle of June and withdraws by the middle of October. About 90% of annual rainfall is received during the monsoon months, of which more than 70% occurs during July, August and September.

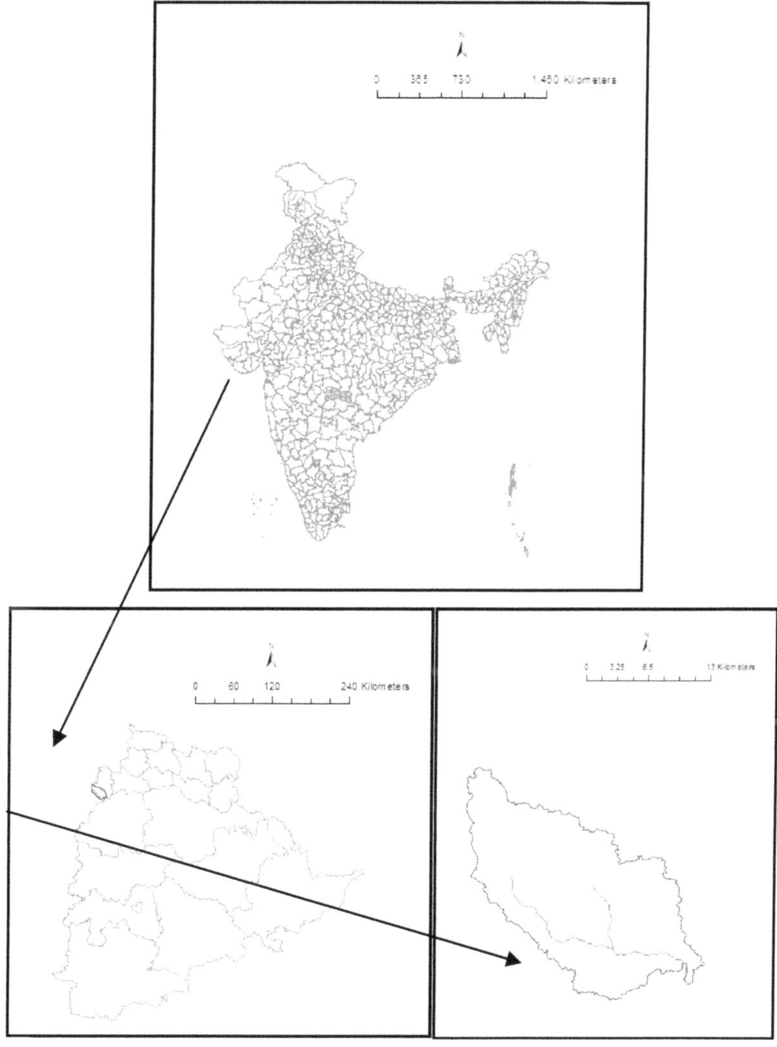

Figure 1 Location Map of the Study Area

The RUSLE equation is a multiplicative function of five factors controlling the rill and inter-rill erosion and can be expressed as:

$$A = R*K*LS*C*P \quad .....(1)$$

Where, A = Annual soil loss (t ha^{-1} year^{-1}); R = Rainfall erosivity factor (MJ mm ha^{-1} h^{-1} year^{-1}); K = Soil erodibility factor (t ha h ha^{-1} MJ^{-1} mm^{-1}); LS= slope length factor (dimensionless); C = Crop cover management factor (dimensionless) and P = Conservation practices factor (dimensionless).

Rainfall erosivity factor was estimated from daily rainfall using equations (Rejani *et al.*, 2016). Soil samples were collected from the field and analyzed for organic matter content and soil texture. Soil erodibility factor was determined from soil data by using equation (2) (Wischmeier et al., 1971). The major portion of the study area was characterized as clay.

$$K = 2.77 \times 10^{-6} x M^{1.14} (12\text{-}a) + 0.043 (b\text{-}2) + 0.033 (c\text{-}3) \quad .....(2)$$

Where, K= the soil erodibility factor (t ha h ha^{-1} MJ^{-1} mm^{-1}), M=particle size parameter (%silt+%very fine sand)*(100-clay), a = organic matter content(%), b = soil structure code (1-very fine granular, 2-fine granular, 3-medium or coarse granular, 4- blocky, platy or massive), c=soil permeability class (1-rapid, 2-moderate to rapid, 3-moderate, 4-slow to moderate, 5-slow, 6-very slow.

Topographic factor using the following equation (3) (Tirkey *et al.*, 2013):

$$LS = \left(\frac{\text{flow accumulation} * \text{cell value}}{22.1}\right)^m (0.065 + 0.045s + 0.0065s^2) \quad .....(3)$$

Where, s is the slope of DEM in degrees, cell value is the resolution of the DEM and m is dimensionless constant which depends upon slope.

Crop cover management factor was estimated from NDVI using equation (4) (Van der knijff *et al.*, 2000)

$$C = exp\left(-\alpha \frac{NDVI}{\beta - NDVI}\right) \quad .....(4)$$

Where α *and* β are the parameters that determine the shape of the NDVI-C curve, $\alpha = 2$ and $\beta = 1$.

The Conservation practices factor (P) lies between 0.03 to 1.00 and it depend upon the soil conservation methods practiced in the area (Reddy *et al.*, 2005).

For climate change estimations, different factors such as erosivity, constant K-factor, LS-factor, LULC (2015) and P-factor was used. In this study, erosion rates were computed for historic, current, and future conditions using RUSLE.

Results and Discussion

Spatial soil loss estimation model using GIS coupled with RUSLE and its validation

The observed soil loss data obtained from the Project Directorate, Vikarabad, Government of Telangana was used for validation of the developed model. It was found that both the observed and recorded data matches satisfactorily (R^2=0.802) (Figure 2).

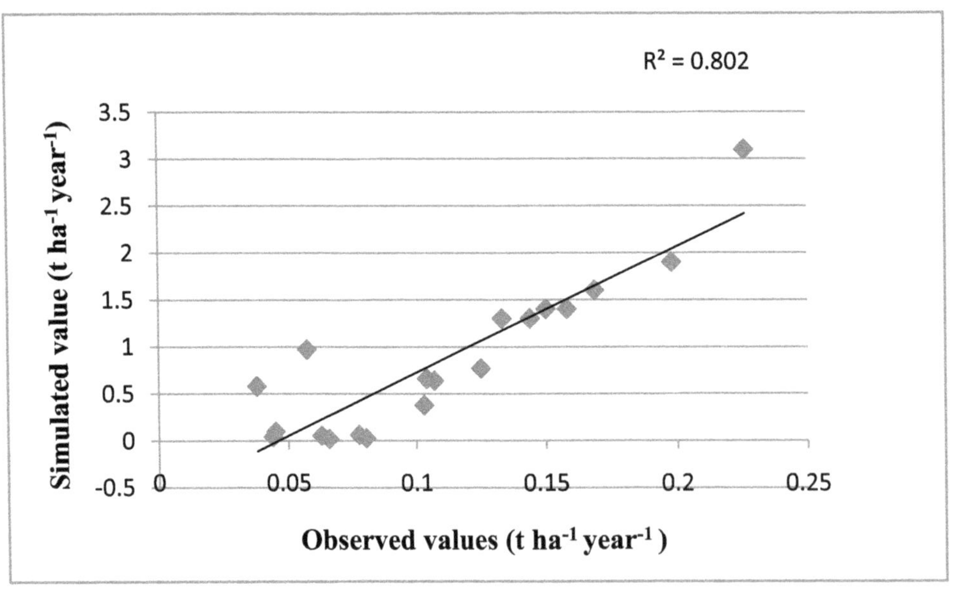

Figure 2 Validation of the Soil Loss Estimation Model (2004 To 2012)

Spatially Estimation of Soil Loss

The annual soil loss from the Mudhole watershed varied considerably (Figure 3). It was noticed that most of the area has soil loss ranged from 0 to 5 t ha^{-1} $year^{-1}$ and in very less area, the soil loss more than 20 t ha^{-1} $year^{-1}$.

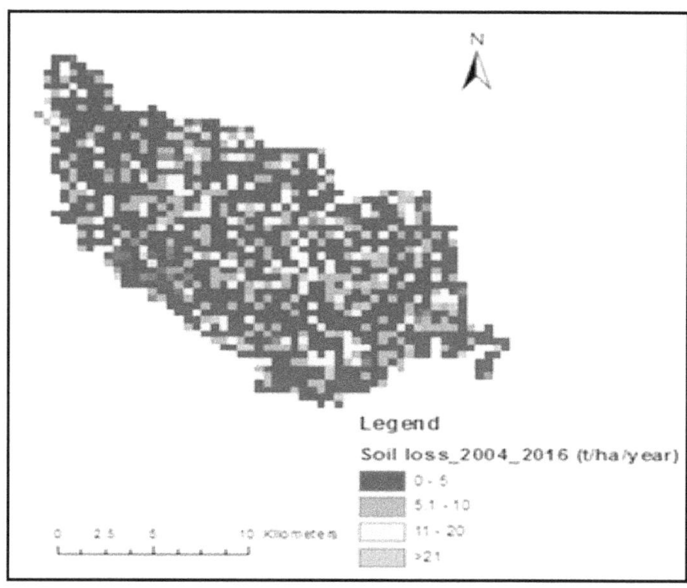

Figure 3 Spatial Variation of Average Annual Soil Loss within the Watershed

Climate Change on Soil Loss

In the base line period (Figure 4), the soil loss from most of the area, varied from 0 to 5 t ha^{-1} $year^{-1}$.

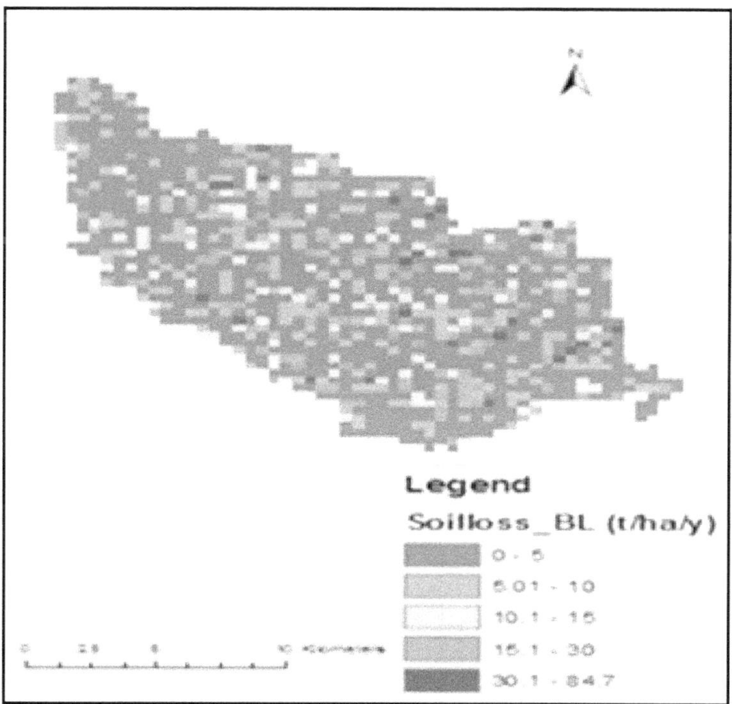

Figure 4 Variation of Soil Loss in Base Line Scenario

During the period of 2020's, the soil loss varied in different area. Soil loss under low emission scenario in 2020's (RCP 2.6) is predicted to increase by 11 t ha^{-1} $year^{-1}$ (Figure 5) and 17 t ha^{-1} $year^{-1}$ by 2050's (Figure 7). Soil loss under medium emission scenario in 2020's (RCP 4.5) is predicted to increase by 8.5 t ha^{-1} $year^{-1}$ (Figure 6) and 21 t ha^{-1} $year^{-1}$ by 2050's (Figure 8). Soil loss under high emission scenario in 2020's (RCP 6.0) is predicted to increase by 8.5 t ha^{-1} $year^{-1}$ and 19 t ha^{-1} $year^{-1}$ by 2050's. Soil loss under very high emission scenario in 2020's (RCP 8.5) is predicted to increase by 9.6 t ha^{-1} $year^{-1}$ and 29 t ha^{-1} $year^{-1}$ by 2050's. Similar trend was observed during 2080's.

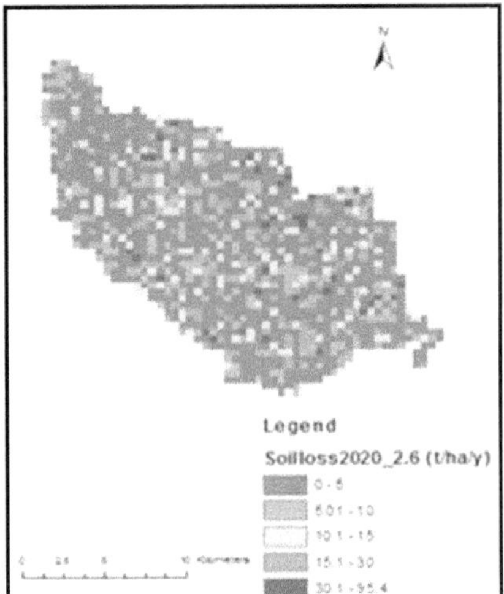

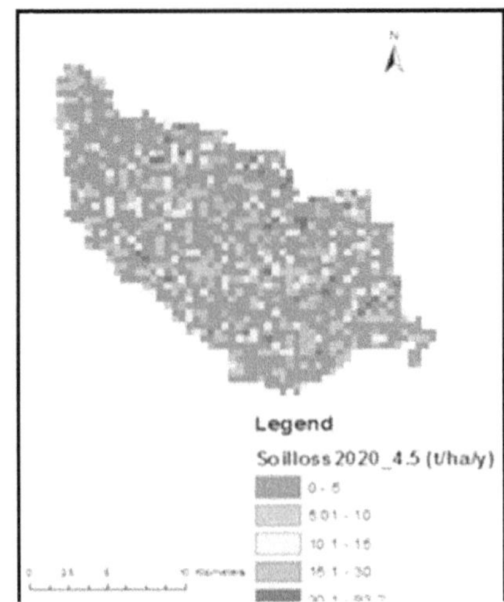

Figure 5 & 6 Variation of Soil Loss under Low Emission RCP 2.6 and Medium Emission RCP 4.5 during 2020's

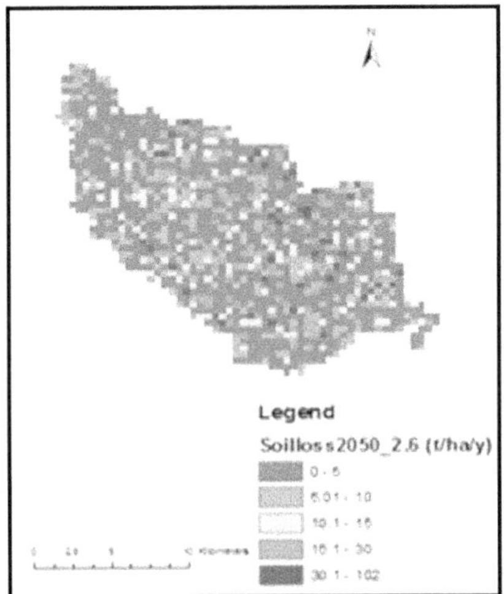

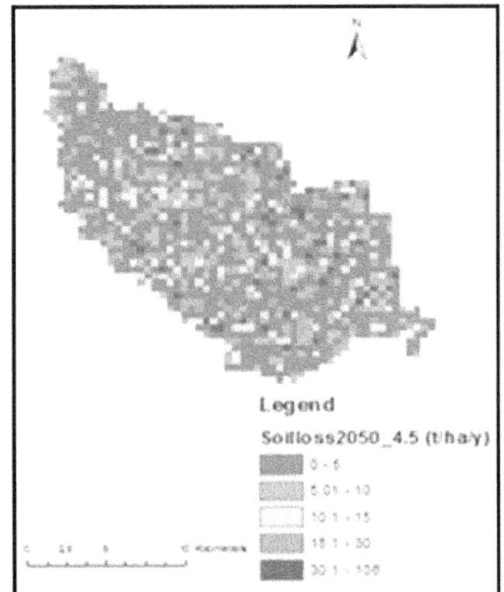

Figure 7 & 8 Variation of Soil Loss under Low Emission RCP 2.6 and Medium Emission RCP 4.5 during 2050's

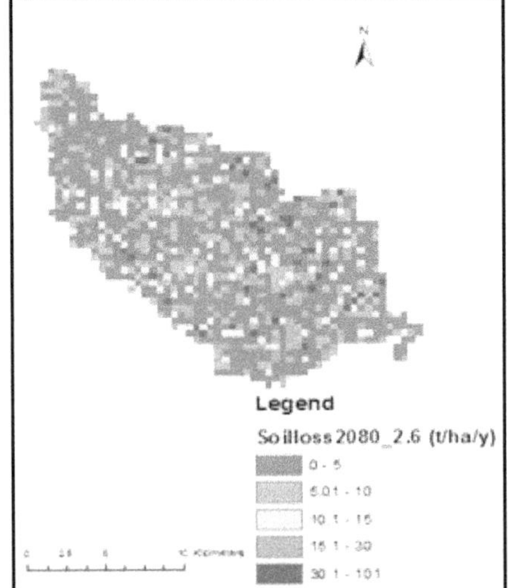

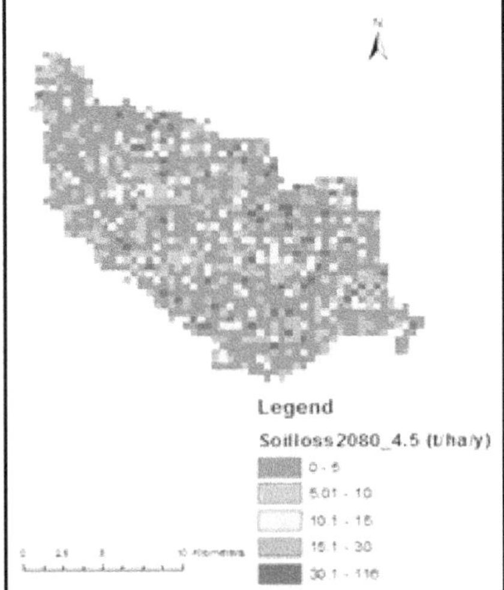

Figure 9 & 10 Variation of Soil Loss under Low Emission RCP 2.6 and Medium Emission RCP 4.5 during 2080's

Conclusion

The spatial estimation of soil loss is most important for planning interventions in Mudhole watershed. During 2004 to 2016, that most of the area has soil loss < 5 t ha^{-1} $year^{-1}$ and in very less areas have soil loss > 20 t ha^{-1} $year^{-1}$. Spatial variation of soil loss and under changing climatic scenarios gives an idea to plan for future. Under climate change scenarios, the soil loss at Mudhole watershed predicted to increase considerably.

References

Ismail, J and Ravichandran, S. 2008. RUSLE2 model application for soil erosion assessment using remote sensing and GIS. *Water Resources Management,* 22(1): 83-102.

Ranuzzi, A and Srivastava, R. 2012. Impact of climate change on agriculture and food security. *ICRIER Policy series*. 16: 33.

Reddy, R.S., Nalatwadmath, S.K and Krishnan, P. 2005. Soil Erosion Andhra Pradesh. NBSS Publication No. 114, NBSS&LUP, Nagpur, India, 76 pp.

Rejani, R., Rao, K.V., Osman, M., Srinivasa Rao, Ch., Reddy K.S., Chary ,G.R., Pushpanjali and Samuel, J. 2016. Spatial and temporal estimation of soil loss for the sustainable management of a wet semi-arid watershed cluster. *Environmental Monitoring and Assessment.* 188(143):1-16.

Sehgal, J.L and Abrol, I.P. 1994. Soil degradation in India: Status and impact. Oxford and IBH Pub co Pvt Ltd, New Delhi, India. 80 pp.

Tirkey, A.S., Pandey, A.C and Nathawat, M.S. 2013. Use of satellite data, GIS and RUSLE for estimation of average annual soil loss in Daltonganj watershed of Jharkhand (India). *Journal of Remote Sensing Technology*, 1(1): 20-30.

Van der Knijff, J.M., Jones, R.J.A and Montanarella, L. 2000. Soil erosion risk assessment in Europe. http://eusoils.jrc.ec.europa.eu/ESDB_Archive/pesera/pesera_cd/pdf/ereurnew2.pdf

Wischmeier, W.H., Johnson, C.B and cross, B.V. 1971. Soil erodibility nomograph for farmland and construction sites. *Journal Soil and Water Conservation*, 26: 189-193.

Annual Rainfall Pattern and Analysis for Eastern and Western Parts of Tamil Nadu

M.Thirumarran[1], Tirupathi Rao[2] and Sakkeel P.T[2]
[1]*Department of Physics, Sathyabama Institute of Science and Technology, Jeppiaarnagar, Chennai, TamilNadu, India*
[2]*Department of Statistics, Ramanujan School of Mathematical Sciences, Pondicherry University, Puducherry, India*
drthirumarranmeganathan@gmail.com, drtrpadi@gmail.com and sakeelptnkt@gmail.com

ABSTRACT

The rainfall and characteristics of annual rainfall in the coastal areas of Tamil Nadu for the districts Chennai, Cuddalore, Thiruvarur, Nagapattinam, Sivaganga, Tanjore and the Western districts Coimbatore, Erode, Dindigul, Namakkal, Nilgiris and Tirunelveli districts are considered for the study. The monthly rainfall data for the above stations are collected for eleven years from 2007 to 2017, to find out the annual rainfall pattern and spatial distribution of rainfall in these eleven districts. In every district five to fifteen rain gauge stations data are taken for large sampling. The spatial distributions of rainfall for all the studying districts were explored. Data analysis is carried out with the Descriptive statistics and analysis of variance has also been made to understand the difference between the mean rainfall and between the districts. The trends of annual rainfall anomalies over these areas are observed and it is seen that annual monsoon rainfall trends in these twelve districts are increasing trend in nature. Also itis observed that the trend of rainfall is increasing over Tamil Nadu and Pondicherry region for the Northeast monsoon rainfall for the period from 1950 to 2014.

Key words: Annual rainfall, Eastern and Western part of Tamil Nadu, Rainfall.

Introduction

Tamil Nadu is the only subdivision of the Indian Union which receives more rainfall in the Northeast monsoon (October to December) season than in the South west monsoon season (June to September). In a previous study (Dhar*etal* 1981) it was found that the Southwest and Northeast monsoon seasons contribute about 34 % and 47 % respectively, to the mean annual rainfall of this region. Rao (1953) observed that in the same year, there is an indication that a good Northeast monsoon rainfall over Tamil Nadu is generally associated with deficient or normal Southwest monsoon rainfall and *vice versa*. Thus, rainfall received in Tamil Nadu during Northeast monsoon is highly variable from Southwest monsoon rainfall and great economic value. Major agricultural activities are normally undertaken during that season. Therefore if the rainfall behavior and distribution in all the districts of Tamil Nadu during annual period could be predicted in advance, then it would go long way towards planning and helping the agricultural and industrial activities in all the districts of Tamil Nadu. The state of Tamil Nadu occupies the southeastern side of Indian peninsula. It was bounded on the East by the Bay of Bengal, in the South by the Indian ocean, in the West by the states of Kerala and Karnataka and in the North by Karnataka and Andhra Pradesh. The latitudinal extent of Tamil Nadu is between 8°4' to 13°25' N. The longitudinal extent between 76°15' E to 80°21' E. The state of TamilNadu covers 31 districts, with area of 1,30,058 sq. km. In the present study an attempt has been made to investigate the Rainfall characteristics over Chennai, Cuddalore, ThiruvarurNagapattinam,Namakkal,Tanjore,Dindigul,Sivaganga,Tirunelveli,Coimbatore, Erode and Nilgiris districts of Tamil Nadu which are situated in the Coromandel coastal region, called East coastal region of Bay of Bengal and western parts of TamilNadu.

There are also several studies relating to the prediction of NE monsoon rainfall using predictors such as El Nino, Southern Oscillation, Indian Ocean Dipole, etc; (Ropelewski&Halpert 1987; Sridharan&Muthuswamy 1990; Singh Chattopadhyay 1998; JayanthiGovindachari 1999; De Mukhopadhyay 1999; MedhaKhole De 2003; Bhanu Kumar *et al* 2003 b). El Nino generally appears in the months of December, January and sometimes February and it persists even in later months and influences the Indian summer as well as winter monsoon activity. Secondly, there is an inverse relationship between NE monsoon rainfall over Tamil Nadu and the preceding March April-May SOl (Singh Chattopadhyay 1998). A significant positive relationship with SSTs of Nino-4 region during June, July and August is also detected. Thus, an El- Nino ENSO episode generally enhances NE monsoon rainfall.

Dhar*et al* (1982) studied the fluctuations in Northeast monsoon rainfall of Tamil Nadu. The average rainfall series of Tamil Nadu for the Northeast monsoon months of October to December and the season as a whole were analyzed for trends, periodicities and variability using standard statistical methods. The trend analysis showed that there are no long-term trends of increasing or decreasing rainfall in the individual months or the season as a whole. Dhar and Rakhecha (1982) studied the association between the southwest and northeast monsoon (October to December) rainfall over Tamil Nadu. They examined the correlation and showed that there is a negative correlation between Southwest monsoon rainfall and Northeast monsoon rainfall for the period of 100 years from 1877 to 1976.Raj (1989) screened several upper-air and surface parameters of representative stations in India and derived linear regression equations to predict winter monsoon rainfall in four sub divisions of South India. Raj *et al* (1996) reported that rainfall from Northeast Monsoon in Tamil Nadu is highly variable with a Coefficient of Variation (CV) of 27% for a normal rainfall of 48 cm. As such the seasonal rainfall manifested year to year fluctuations with as much as 60% deficiency in 1974 and 66% excess in 1946 as the range during the period between 1901-1993.

Ganesan*et al* (2000) studied the nature of distribution of rainfall in various classes for the post monsoon season in respect of one of the most cyclone prone areas of India namely Tamil Nadu coast. The stations chosen in Tamil Nadu coast are Chennai, Cuddalore, Nagapattinam, Adiramapatinam and Pamban. They classified the rainfall (mm) as Light (0.1-7.4), Moderate (7.5-34.4), Rather heavy (34.5-64.5), Heavy (64.5-124.4), Very heavy (124.5) and above. The percentage frequencies of various classes of rainfall in respect of various stations are calculated. Finally they concluded that seasonal rainfalls over coastal stations are above normal rainfall. Manorama et al.in 2007 studied the probability of occurrence of rainfall over Nilgiris in Tamil Nadu by using the thirty years of time series data on rainfall. They analyzed the coefficient of variation and the probability of occurrence of rainfall over Nilgiris in Tamil Nadu. This would help in planning the crop cultivation twice in a year. Thirumarran 2013 studied the rainfall pattern and characteristics of rainfall over the coastal areas of Cuddalore, Thiruvarur and Nagai districts of Tamilnadu and it is observed that the rainfall distribution and pattern over Cuddalore,Thiruvarur and Nagai districts are not spatially homogeneous. Thirumarran 2013 studied the behavior of rainfall and characteristics of rainfall over the land and hilly terrain areas of western Tamil Nadu (Coimbatore, Erode &Nilgiris districts). The trends of rainfall anomalies over these areas are observed. It is observed that the rainfall trends in these three districts are not varying in nature during Northeast monsoon season. It is also observed that the trend of rainfall is partially maximum at Coimbatore district compared to Erode and Nilgiri district and minimum of trend of rainfall is observed over Nilgiri district.

Hence a detailed statistical study is made in this paper to understand the annual rainfall pattern for the period of eleven years from 2007 to 2017 over Eastern parts of Tamilnadu for the selected districts Chennai, Cuddalore, Tiruvarur, Nagapattinam, Namakkal, Tanjore which are located in the Coromandel coastal region and Western parts Dindigul, Sivaganga, Tirunelveli, Coimbatore, Erode and Nilgiris which are located in thewestern region of Tamil Nadu.

Materials and Methods

An attempt has been made to investigate the Rainfall Characteristics over Coastal area(Chennai,Cuddalore, Thiruvarur and Nagapattinam), Agricultural area (Namakkal,Tanjore,Dindigul,Sivaganga&Tirunelveli) and Terrain area(Coimbatore, Erode and Nilgiris) districts of TamilNadu. The monthly rainfall data for the above mentioned districtsare supplied by India Meteorological Department, Surface Ground and surface water Resources Data centre (SGSWRDC), Taramani, Chennai. The rainfall data for the period of eleven years from 2007 to 2017 are taken up for the present study. The monthly data set in each district includes five to sixteen rain gauge stations are taken for large sampling. The cumulative rainfall data over coastal area, agricultural area and terrain area of eleven districts of Tamil Nadu have been collected and for each district, the data is collected on monthly basis and for twelve months in a year for a period of eleven years from 2007 to 2017.

Results and Discussion

The spatial distribution of cumulative annual rainfall over coastal area, agricultural area and terrain area for the period 2007 to 2017 are shown in Figure 1,2 and 3. In the coastal areait is observed that maximum rainfall (3429.94 cm) is observed over Cuddalore district and minimum rainfall (860.74cm) is observed over Chennai district during the year 2015,maximum rainfall (1138.41cm) is observed over Cuddalore district and minimum rainfall (408.63 cm)is observed over Chennai district during the year 2016,maximum rainfall (144.06cm) is observed over Nagai district and minimum rainfall (103.87cm) is observed over Tiruvarur district during the year

2017. In the agricultural area it is clear that maximum rainfall (3780.46cm) is observed over Tirunelveli district and minimum rainfall(319.6cm) is observed over Namakkal district during the year 2015, maximum rainfall(1095.08cm) is observed over Thanjavur district and minimum rainfall (319.6cm) is observed over Namakkal district during the year 2016, It is observed that maximum rainfall(112.7cm)was occurred over Tirunelveli district and minimum rainfall (72.05cm)was observed over Namakkal district during the year 2017. In the terrain area it is clear that maximum rainfall (5378.22cm) is observed over Coimbatore district and minimum rainfall (2430.93cm) is observed over Erode district during the year 2015. Also it was noted that maximum rainfall (3409.92cm)was observed over Coimbatore district and minimum rainfall(897.86cm)was observed over Erode district during the year 2016. It is observed that maximum rainfall(146.9cm) occurred over Nilgiri district and minimum rainfall(71.63cm) is observed over Erode district during the year 2017.

The statistical parameters over Eastern and Western districts of Tamilnadu are given in Table 1. The twelve districts are divided into two regions as Eastern and Western region based on their geographical location. The districts such as Chennai, Cuddalore, Thiruvarur, Nagapattinam, Sivaganga and Tanjore arelocated in the in eastern region and the districts such as Coimbatore, Erode, Nilgiris, Namakkal, Dindugul and Tirunelvelilocated in the western region of Tamilnadu. The measures like mean, median, maximum, Standard Deviation, Coefficient of Variation (CV) and Skewness of rainfall occurred during the period from 2007 to 2017 over twelve districts of Tamilnadu are given in Table 1. The highest annual rainfall 13100.5 mm occurred over Tirunelveli district during the year 2015, the maximum Mean rainfall (730.47mm) is recorded over Nilgiris district and minimum mean rainfall (128.97mm) was recorded over Namakkal district during the time period 2007 to 2017. The maximum Standard Deviation was occurred over Coimbatore district with a value 1929.58mm which indicates that the rainfall was highly dispersed or there was inconsistency in the rainfall pattern over Coimbatore district. The rainfall greater than 10000 mm was occurred four times during the period from 2007 to 2017. It means that rainfall occurred over Tirunelveli district was 13100.5 mm during November 2015, rainfall occurred over Cuddalore district was 12870.8mm during Nov 2015, rainfall occurred over Nilgiris district was 10802.5mm during June 2015 and occurred rainfall value over Coimbatore district was 10441.9mm. The annual rainfall data over the twelve districts are positively skewed and it is significant in each district.

The summary statistics of annual rainfall during 2007 and 2017 aregiven in Table 2& Table 3 and visualized in Figure 4 and Figure 5. The highest median rainfall was reported at Nilgiris district, which is greater than the third quartile of rainfall at Cuddalore, Tanjore, Sivaganga, Dindugul, Namakkal, Tirunelveli, Coimbatore and Erode districts. The highest maximum rainfall (483mm) occurred at Nagapattinam district and the lowest maximum rainfall (161.2mm) occurred at Tirunelveli district. The rainfall at the districts such as Chennai, Cuddalore, Thiruvarur, Nagapattinam, Tanjore, Dindugul,Namakkal, Tirunelveli, Coimbatore, Erode and Nilgiris are positively skewed, but the rainfall over Sivaganga district is negatively skewed (Table 2). During the year 2017 the highest median rainfall was reported at Nilgiris district and lowest median rainfall was reported at Chennai district. The highest maximum rainfall (704.9mm) was occurred at Nagai district and lowest maximum rainfall (184mm) was occurred at Sivaganga district. All the districts are positively skewed except Nilgiris district (Table 3).

The box plots of rainfall over twelve districts without outliers are shown in Figure 4. The rainfall occurred over eastern and western districts for the year 2007 and 2017 are shown in Figure 5 and Figure 6. It is evident that maximum outlier was occurred over cuddalore district and minimum outlier was occurred over Coimbatore district during 2007. It is also seen that maximum outlier was observed over Thiruvarur district and minimum outlier was observed over Tirunelveli district during the year 2017.

In order to verify whether there is any significant difference between the average rainfall of various study area, Analysis of Variance (ANOVA) test has been carried out with the hypothesis (H_0) testing. The results reveal that that there is no significant difference between the average rainfall of various study districts. The results are shown in Table4, Table 5and Table 6for coastal area, agricultural area and terrain area respectively. By observing the results from the Table4 the test critical value $F_{critical}$ (2.838) is significantly greater than the test statistical value ($F_{statistics} = 0.47$) at 95% confidence level. Therefore test hypothesis (H_0) is accepted. Hence it is concluded that difference between the average rainfall is not significant over the coastal districts. In other words, coastal districts are not significantly differed with their average rainfall over coastal area. By observing the results from the Table 5the test critical value $F_{critical}$ (2.64) is significantly greater than the test statistical value ($F_{statistics} = 0.4293$) at 95% confidence level. Therefore test hypothesis (H_0) is accepted. Hence it is concluded that difference between the average rainfall is not significant over the agricultural area. In other words, agricultural districts are not significantly differed with their average rainfall over agricultural area. Also it is observed from the results and from the Table6 the test critical value $F_{critical}$ (3.315) is significantly greater than the test statistical value ($F_{statistics}$ = 0.4640) at 95% confidence level. Therefore test hypothesis (H_0) is accepted. Hence it is concluded that difference between the average rainfall is not significant over the terrain districts. In other words, terrain districts are not significantly differed with their average rainfall over terrain area.

In order to verify whether there is any significant difference between the average rainfall among the Eastern and Western regions of Tamilnadu for the period 2007 to 2017 , Analysis of Variance (ANOVA) test has been carried out with the hypothesis (H_0) testing and the results are shown in Table 7. The results reveal that the probability value p=0.004 is less than 0.05. Therefore test hypothesis (H_0) is rejected which indicates that the mean annual rainfall of the western region and eastern region of Tamilnaduare significantly different and it is evident that the mean annual rainfall of western region is greater than the mean annual rainfall of Eastern region of Tamilnadu.

To verify whether there is any significant difference between the average rainfall over the Eastern and Western regions of Tamilnadu for the year 2007 and for the year 2017, Analysis of Variance (ANOVA) test has been carried out with the hypothesis (H_0) testing and the results are shown in Table8 and Table 9. The results reveal that the probability value p=0.029 is less than 0.05and therefore test hypothesis (H_0) is rejected which indicates that the mean annual rainfall of the western region of Tamilnadu are significantly different during the year 2007. Moreover the probability value p=0.046 for Coimbatore and Nilgiris district and p=0.044 for Erode and Nilgiris districts which are less than 0.05 which also indicates that test hypothesis (H_0) is rejected and can be observed from the Table 9, which indicates that the mean annual rainfall between Coimbatore, Erode and Nilgiris districts of the western region of Tamilnadu during the year 2007 are significantly different. Also it is observed that there is no significant difference in the mean rainfall over Eastern and western regions during the year 2017.

In order to test the data quality two variable t - test is performed for the annual cumulative rainfall and monsoon cumulative rainfall for the twelve districts. The results are given in Table.10 along with standard deviation (SD) values. It can be seen from the Table 10 that the t-values are highly significant at 5 % level for the eleven districts of study area and hence the data quality is established beyond any doubt.

In order to understand the relation between annual rainfall with respect to time over these area regression lines are fitted with respect to time by the method of least squares to study the nature of trend of rainfall behavior during the total annual period. Table 11 shows the district wise regression equations, correlation coefficients, direction of trend and slope value of the trend line over the study area. It is observed that there exists a positive trend of annual rainfall with time for the period from 2007 to 2017 for the eleven districts. It is also observed that there exists aincreasing trend of Northeast monsoon rainfall with time (m=0.0885) for the period of sixty four years from 1951 to 2014 with correlation coefficient(r=0.0461) over Tamilnadu and Pondicherry region.

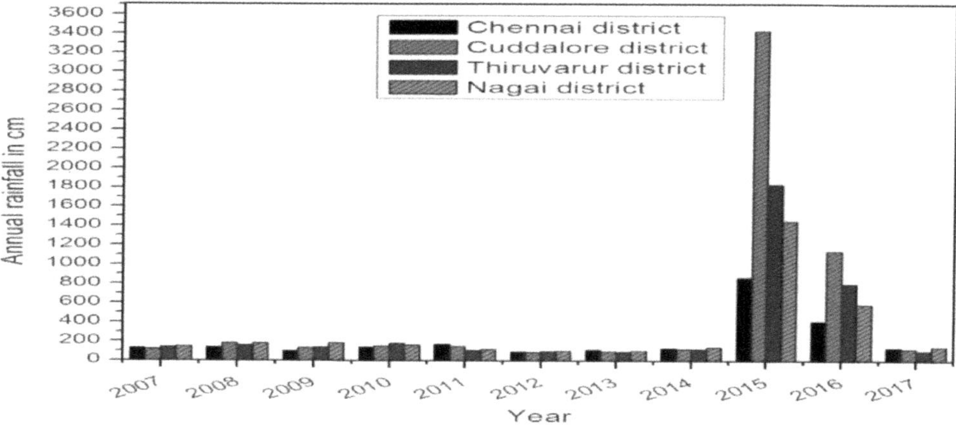

Figure 1 Spatial distribution of annual rainfall over Coastal area

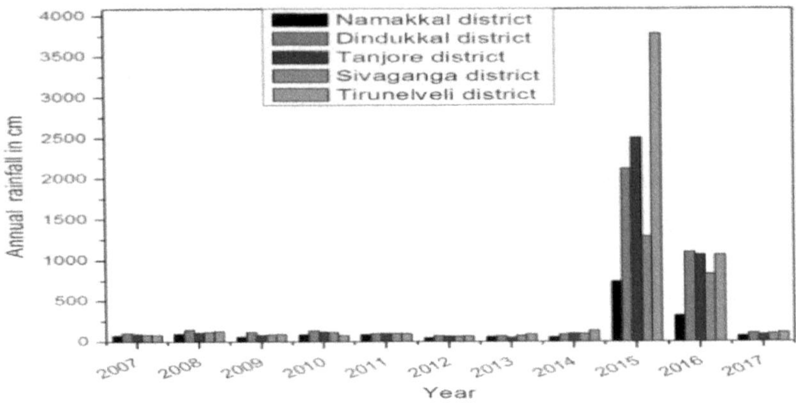

Figure 2 Spatial distribution of annual rainfall over Agricultural area

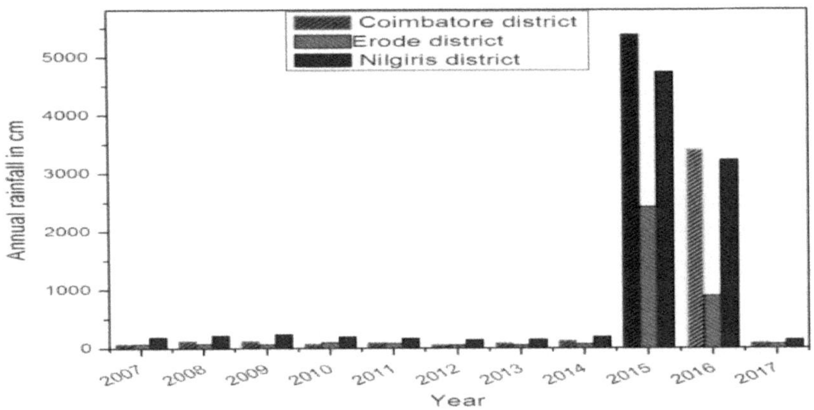

Figure 3 Spatial distribution of annual rainfall over Terrain area

Table1 Analysis of Statistical Parameters over Eastern and Western parts of Tamilnadu

Region	Station	Mean	Median	Maximum	Standard deviation	CV	Skewness
Eastern Tamilnadu	Chennai	183.12	79.6	4157.7	427.92	233.68	6.91
	Cuddalore	435.05	72.45	12870.8	1393.76	320.37	6.93
	Thiruvarur	286.22	66.45	5908.1	724.87	253.26	5.56
	Nagapattinam	251.63	57.65	5527.8	635.86	252.7	6.35
	Sivaganga	224.32	75.6	3457.1	462.9	206.36	3.94
	Tanjore	315.41	57.45	6484.2	802.72	254.5	4.94
Western Tamilnadu	Dindugul	331.23	57.7	6341.3	852.79	257.46	4.37
	Namakkal	128.97	55.3	1574	247.54	191.93	3.84
	Tirunelveli	433.86	59.4	13100.5	1459.09	336.3	6.3
	Coimbatore	729.26	64.95	10441.9	1929.58	264.6	3.26
	Erode	302.76	59.75	7325.6	873.48	288.51	5.38
	Nilgiris	730.47	163.3	10802.5	1788.92	244.9	3.72

Table 2 Summary statistics for the year 2007

Station	Minimum	1st Quartile	Median	Mean	3rd Quartile	Maximum	Skewness
Chennai	0	4.425	99.8	107.79	201.59	253.7	0.022
Cuddalore	0	16.23	45.1	101.87	113.9	422	1.678
Thiruvarur	0	19.25	47.65	117.92	165.53	416.9	1.13
Nagapattinam	0	12.5	45.4	123.7	169.2	483	1.411
Tanjore	0	24.07	51.85	89.37	89.67	326.3	1.604
Sivaganga	1.3	15.8	86.6	73.88	104.55	214.4	-0.213
Dindugul	0	25.88	52.65	76.72	67.88	325.9	2.096
Namakkal	0	20.7	53.75	62.87	83.15	172.8	0.745
Tirunelveli	0.1	17.12	66.25	70.03	114.67	161.2	0.14
Coimbatore	0	29.07	51.15	60.19	72.92	198	1.461
Erode	0.3	12.85	37.1	59.52	81.55	221.4	1.337
Nilgiris	6.4	58.33	142.4	157.88	278.38	335.7	0.029

Table 3 Summary statistics for the year 2017

Station	Minimum	1st Quartile	Median	Mean	3rd Quartile	Maximum	Skewness
Chennai	0	0	31.5	109.1	123.3	583.8	1.829
Cuddalore	0	8.33	67.15	99.94	142.15	391.4	1.279
Thiruvarur	0	8.9	60.5	86.56	124.47	363.9	1.613
Nagapattinam	0	5.85	44.05	120.05	132.43	704.9	2.371
Tanjore	0.2	30.3	85.1	89.76	115.28	237.4	0.722
Sivaganga	1.1	53.4	78.2	80.63	120.1	184	0.26
Dindugul	0	22.55	40.85	71.92	102.33	216	0.862
Namakkal	0	14.18	35.25	60.04	102.3	194.8	0.938
Tirunelveli	5	38.02	64.2	93.92	93.65	344	1.576
Coimbatore	0	20.3	60.05	70.11	90.97	262.7	1.591
Erode	0	11.93	39.15	59.69	92.75	187	0.885
Nilgiris	0	92.3	123.2	122.4	147.3	294.8	-0.882

Table 4 ANOVA over Coastal Area (Annual rainfall in cm)

ANOVA: Single factor						
Summary						
Groups	*Count*	*Sum*	*Average*	*Variance*		
Chennai	11	2417.177	219.7434	52710.91		
Cuddalore	11	5742.63	522.0573	1022263		
Thiruvarur	11	3778.08	343.4618	285653.5		
Nagai	11	3321.5	301.9545	163658.3		
ANOVA						
Source of Variation	*SS*	*df*	*MS*	*F*	*P-value*	*F crit*
Between Groups	537688.3	3	179229.4	0.47033	0.704649	2.838745
Within Groups	15242855	40	381071.4			
Total	15780544	43				

Table 5 ANOVA over Agricultural Area (Annual rainfall in cm)

Anova: Single factor						
Summary						
Groups	*Count*	*Sum*	*Average*	*Variance*		
Namakkal	11	1702.45	154.7681	42750.227		
Tanjore	11	4163.46	378.4963	423850.23		
Dindugul	11	4372.2	397.4727	573411.35		
Sivaganga	11	2961.04	269.1854	163194.87		
Tirunelveli	11	5726.97	520.6336	1252225.7		
ANOVA						
Source of Variation	*SS*	*df*	*MS*	*F*	*P-value*	*F crit*
Between Groups	843200.7	4	210800.173	0.429252635	0.786804	2.557179
Within Groups	24554325	50	491086.498			
Total	25397526	54				

Table.6 ANOVA over Terrain Area (Annual rainfall in cm)

Anova: Single Factor						
Summary						
Groups	*Count*	*Sum*	*Average*	*Variance*		
Coimbatore	11	9626.17	875.1064	3221188		
Erode	11	3996.4	363.3091	531465.8		
Nilgiris	11	9642.16	876.56	2473609		
ANOVA						
Source of Variation	*SS*	*df*	*MS*	*F*	*P-value*	*F crit*
Between Groups	1926339	2	963169.3	0.464084	0.633149	3.31583
Within Groups	62262630	30	2075421			
Total	64188969	32				

Table 7 ANOVA over eastern and western districts of Tamilnadu

Variable	Regions	Mean rainfall in mm	Standard. Deviation	t-value	df	P- Value
Districts	Eastern	282.625	809.11	-2.8732	1298	0.004
	Western	449.757	1343.67			

Table 8 Analysis of variance ANOVA over western districts and eastern districts during the year 2007 &2017

ANOVA (Rainfall in mm)						
		Sum of Squares (SS)	df	Mean Square (MS)	F	P value at 5% significance
Rainfall at western stations (2007)	Between Groups	87275.312	5	17455.062	2.683	0.029
	Within Groups	429419.968	66	6506.363		
	Total	516695.280	71			
Rainfall at eastern stations (2007)	Between Groups	20489.557	5	4097.911	0.259	0.934
	Within Groups	1045316.453	66	15838.128		
	Total	1065806.010	71			
Rainfall at western stations (2017)	Between Groups	35594.269	5	7118.854	1.279	0.284
	Within Groups	367480.678	66	5567.889		
	Total	403074.947	71			
Rainfall at eastern stations (2017)	Between Groups	13364.928	5	2672.986	0.154	0.978
	Within Groups	1145411.291	66	17354.717		
	Total	1158776.219	71			

Table 9 Post Hoc Analysis of rainfall occurred over western districts during the year 2007

Multiple comparisons (2007, Western Tamil Nadu)					
Dependent Variable	(I) station	(J) station	Mean Difference (I-J)	Std. Error	P- Value at 5% Significance
	Coimbatore	Nilgiris	-97.69167*	32.93014	0.046
	Erode	Nilgiris	-98.36667*	32.93014	0.044

Table 10 Test of Significance between Monsoon and Annual Cumulative Rainfall over Study Area (Cumulative rainfall in cm)

District/Region	Rfm	Rfa	SDm	SDa	t level at 5%
Chennai	204.04	219.74	239.94	229.59	1.73
Cuddalore	457.73	522.06	941.52	1011.07	1.73
Thiruvarur	279.88	343.46	449.56	534.47	1.74
Nagapattinam	257.60	301.95	361.92	404.55	1.73
Namakkal	118.76	154.77	161.27	206.76	1.74
Thanjavur	311.60	378.50	549.63	651.04	1.74
Dindugul	284.96	397.47	531.24	757.24	1.75
Sivagangai	208.14	269.19	317.67	403.97	1.74
Tirunelveli	401.44	520.63	917.55	1119.03	1.74
Coimbatore	709.82	875.11	1501.48	1794.77	1.74
Erode	278.31	363.39	570.94	729.02	1.74
Nilgiris	730.52	876.56	1349.43	1572.77	1.74
Tamilnadu & Pondicherry	78.88	94.57	12.7790	14.4338	1.66

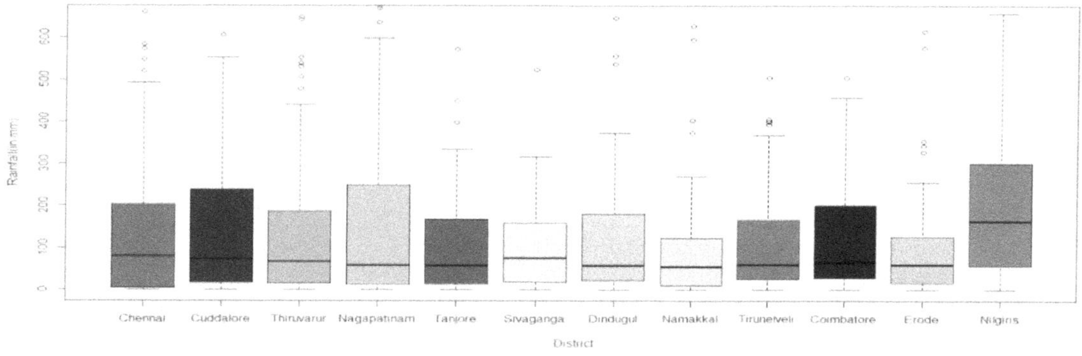

Figure 4 Box plot of rainfall at different districts without outliers

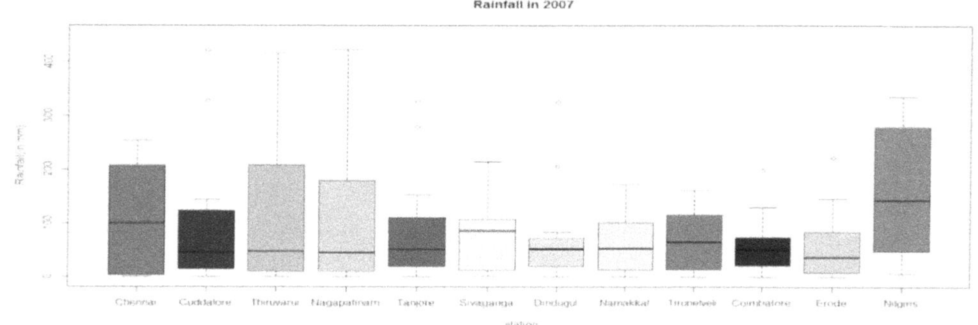

Figure 5 Rainfall at Eastern and Western Districts duringthe year 2007

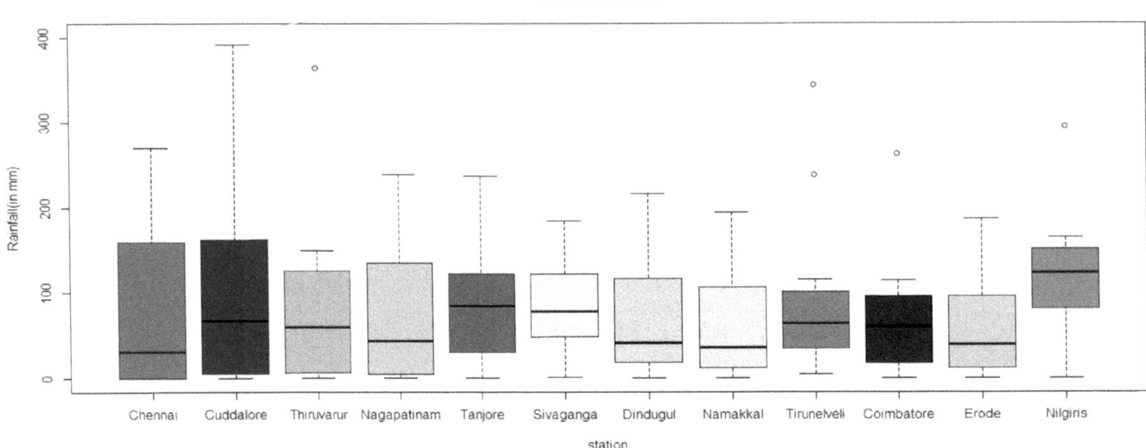

Figure 6 Rainfall occurrence at eastern and western districts in 2017

Table 11 Trend analysis over study districts and study area.(Annual cumulative rainfall in cm)

Region/District	Monsoon	Regression Equation	Slope(m)	Correlation Coefficient(r)	Direction of Trend
Coastal Area					
Chennai	Annual	y=-59579.2659+29.7212x	29.72	0.4294	Increasing
Cuddalore	Annual	y=-2.476E5+123.3204x	123.32	0.4045	Increasing
Tiruvarur	Annual	y=-1.3332E5+66.4313x	66.43	0.4122	Increasing
Nagapattinam	Annual	y=-96995.8047+48.3587x	48.35	0.3965	Increasing
Agricultural Area					
Namakkal	Annual	y = -51429.9853+25.6385x	25.63	0.4113	Increasing
Tanjore	Annual	y=-1.7701E5+88.1634x	88.16	0.4491	Increasing
Dindugul	Annual	y = -2.0021E5 + 99.7064*x;	99.70	0.4367	Increasing
Sivagangai	Annual	y=-1.1712E5+58.3455x	58.34	0.479	Increasing
Tirunelveli	Annual	y=-2.7462E5+136.7497	136.74	0.4053	Increasing
Terrain Area					
Coimbatore	Annual	y=-5.3075E5+264.2261x	264.22	0.4883	Increasing
Erode	Annual	y=-1.8753E5+93.3887x	93.38	0.4249	Increasing
Nilgiris	Annual	y = -4.6096E5+229.5419x	229.5	0.4841	Increasing
Tamilnadu and Pondicherry	Winter	Y=24.4052-0.0108x	-0.0108	-0.0638	Decreasing
	Summer	y=31.7426-0.0093x	-0.0093	-0.0345	Decreasing
	Southwest	y=120.9473-0.0439x	-0.0439	-0.1325	Decreasing
	Northeast	y=-130.4816+0.0885x	0.0885	0.1461	Increasing
	Annual	Y=124.4736-0.0151X	-0.0151	-0.0198	Decreasing

Conclusion

The rainfall pattern and characteristics of Eastern and western parts ofTamilnadu are studied in detail and it is clear that Northeast monsoon is the major contribution of rainfall in these areas. The statistical parameters over Eastern and Western region are studied and it is observed that the statistical parameters are highly significant over the study area and during the year 2007 and 2017. The spatial distribution of rainfall over Tirunelveli district is highly significant over agricultural area and itis also observed that the rainfall occurrence during annual season is maximum over Coimbatore and Nilgiri district. The reason being Coimbatore and Nilgirisdistrict receives more rainfall both in Northeast and Southwest monsoon. Moreover the orography rainfall is playing a major role in the observed rainfall over Nilgiri district.The test of significance values are highly significant for all the districts over the study area. From the ANOVA test analysis it is evident that there is significant difference in the mean rainfall between Eastern and western districts and the mean annual rainfall of western region is greater than mean annual rainfall of Eastern regions of Tamilnadu. It is noted that the annual rainfall trend over the study area for the annual period of rainfall from 2007 to 2017 are increasing trend in nature. Moreover it is also noted that the rainfall trend over Tamilnadu and Pondicherry region are increasing trend in nature during Northeast monsoon rainfall for the period from 1950 to 2014.

Acknowledgement

One of the authors M.Thirumarran want to thank our Honorable Founder chairman Colonel Dr.Jeppiaar, Ph.D., Honourable Chairman and Secretary, Vice chairman and Dean, Department of Science and Humanities, Head of the Physics Department, for their valuable support and encouragement. I owe my special gratitude to faculty members of Physics, Department, Sathyabama Institute of Science and Technology, Chennai-600119, for their support and encouragement. I would like to thank my guide Dr.P.TirupathiRao, Professor, Ramanujan school of Mathematical Sciences, Kalapet, Puducherry for his constant support and encouragement.

References

Bhanu Kumar O S R U, Naidu CV, Rao SRL and Naga Ratna K (2003).An empirical relation for prediction of winter monsoon rainfall over Tamil Nadu.Proc. AP Akademi Sci. 7(2): 151-154.

De U S and Mukhopadhyay R K (1999)The effect of ENSO / Anti ENSO on Northeast monsoon Rainfall. Mausam. 50(4): 343-354.

Dhar O N, Rakhecha R and Kulkarni A K (1984) Fluctuations in Northeast monsoon rainfall of Tamil Nadu.Inter. J. Climatol.2: 339-345.

Dhar O N, Rakhecha P R and Kulkarni A K (1982b) Trends and fluctuations of seasonal and annual rainfall of Tamil Nadu. Proc. Indian Acad. Sci (Earth Planetary Sci.) 91(2): 97-104.

Dhar O N andRakhecha P R (1983) Foreshadowing Northeast monsoon rainfall over Tamil Nadu, India.Monthly weather review, American.Mete. Soci. 111: 109-112.

Ganesan G S, Muthusami A and Ponnusamy A (2000) Heavy rain at Chennai and its relation to cyclonic disturbances. Mausam. 51.1: 17-24.

Ganesan G S, Muthusami A and Ponnusamy A (2001)Various classes of rainfall in the Coastal stations of Tamil Nadu. Mausam. 52.2: 433-436.

Jayanthi N and GovindachariS (1999) EL-Nino and NE monsoon rainfall over Tamil Nadu.Mausam.50 : 2: 217-219.

MedhaKhole and De U S (2003) A study on Northeast monsoon rainfall over India. Mausam. 54:2 : 419-426.

Raj Y E A(1989)Statistical relation between winter monsoon rainfall and the preceding summer Monsoon. Mausam. 41(1): 51-56.

Rao P R K (1953) Rainfall of Madras state with special reference to Tamil Nadu and Rayalaseema.Indian meteorological Department: 1.

Ropelwski C F and Halpert M S (1987) Global and regional scale precipitation patterns associated with the El Nino/ Southern oscillation. Mon.Wea.Rev. 115: 1606-1626.

Raj Y E A (1996) Inter and intra-seasonal variation of thermodynamic parameters of the atmosphere over coastal Tamil Nadu during northeast monsoon.Mausam. 47(3): 259-268.

Singh G P and Chattopadhyay J (1998) Relationship of tropospheric temperature anomaly with Indian southwest monsoon rainfall.International J.Climatol.18 : 759-763.

Sridharan S and Muthuchami A (1990) Northeast monsoon rainfall in relation to El Nino: QBO and Atlantic hurricane frequency. Vayumandal 20(3-4): 1.08-1.11.

Suresh, R. and A.K. Bhatnagar(2005), "Seasonal variability of raindrop size distribution over Cuddalore, a coastal station in Tamil Nadu of southern peninsular India", Mausam, 56(2), 337-342.

Thirumarran. M (2013), "Northeast monsoon rainfall pattern over WesternTamilnadu ", J.Ecotoxicol.Environment. monit.,23 (1),59-67.

Thirumarran. M(2013),"Northeast monsoon rainfall pattern overCuddalore,Thiruvarur and Nagai districts of Tamilnadu", J.Ecotoxicol.Environment. monit.,23(2)107-116.

Tripathy, K.P. and A.K. Sikka(2004), "Rainfall depth class distribution in Nilgiris", Ind.J. Soi.Conser., 32, 239-241.

Effect of Vermicompost on Shear Strength and Erosion Control Parameter of Mining Waste Soil

Munnoli P M[1], Udagatti N[2], Chavan A[3], Harshitha K[4], Joshi I[5]

Professor,Students Deparment of Civil Engineering S D M Engineering College Dharwad

ABSTRACT

Investigations were made on mining soil taken from a vermi-compost set up of to 1: 3 (Mining waste soil: Cow dung) proportion using *Eudrilus Eugenia* at laboratory scale. The vermicompost obtained was studied in a Petri plate system for aggregation and shear strength of mining soil. it was found that found 42.00% of mining soil is retained when incorporated 1g of Vermi-compost on 90 μ sieve and the shear strength experiment showed 118% increase in shear strength (Cohesion value) of mining vermi-compost soil as compared to mining soil alone. Both soil binding and increase in cohesion parameters have played a significant role in erosion control. Suggesting treating in large scale mining soil with *E. Eugenia* and utilizing the treated Vermi-compost in large scale mining waste land development and soil erosion control.

Keywords: Mining waste Soil, Vermicompost, Shear strength

Introduction

The mining industry whose products are metals and minerals on which national economy of any country depends and there has been a progressive increase in the production of metals and minerals over the past 50-60 years has made the progress on one side, ecological imbalance on plants, animals, humans, water and atmosphere on the other side are inevitable (Cooke and Johnson 2002; Anderson 1992; Gowdy and Mc Daniel 1999; Sengupta 1993). The process of being eroded or eroding by wind, water, or other natural agents is known as erosion. As the top soil get depleted due to erosion it becomes necessary to control the erosion. Erosion control is the practice of preventing or controlling the wind or water erosion in agriculture, land development, coastal areas river banks and construction. Erosion can be controlled by using compost, using compost in highly erosive areas can decrease erosion and allow quicker establishment of vegetation. Compost controls erosion by increasing water infiltration in to the soil surface, reducing runoff and soil particle transport in runoff, increasing plant growth and soil cover, reducing soil particle dislodging, increasing water holding capacity, New vegetation can be established directly into compost. (Risse .L.M, and Faucette. L.B 2015). Mining is an economically important industry for many countries but it comes at a cost. Mining can result in the pollution of natural resources, harm to the biodiversity of habitats and erosion of the land. Mining causes erosion when it strips an area of vegetation such as trees and plants. The roots of trees and plants are essential to keeping soil in its place. Without trees and plants, there is nothing to keep the soil from being dispersed and eroded by the wind or from being subject to mud, land, and rockslides that severe weather, like storm events, can cause. The erosion does not just effect the immediate area but can have a negative impact on nearby communities, which can be damaged or destroyed by these large movements of land caused by the denuding of the landscape. Also, the displacement of rich topsoil's by erosion from wind can affect the agricultural aspects of a region. Vermicompost a bio soil has the properties of water holding and erosion control along with enhanced micro and macro nutrients (Munnoli 2007). In the present paper an attempt is made to investigate the shear strength parameters and conduct a separate Petri plate experiment using VC to measure the erosion control of mining soil.

Material and Methods

Procurement of research materials: Earthworms *Eudrilus Eugenia* is procured from UAS Dharwad and maintained in concrete lab. Cow dung: 10 days old procured from fresh local diary. Wheat straw: wheat straw was procured from UAS Dharwad. Mining source soil, straws, cow dung were placed in 3 layers in a plastic tub for study. Earthworms were put into this set up to produce the VC. The whole setup was kept under observation for 90 days so that the compost was ready. The cow dung and straws were removed. The soil was collected after 90 days. The geotechnical properties of the composted soil and source soil were determined. Specific gravity of soil samples were determined according to IS2720 part III-1980.Particle size distribution of soil samples was determined according to IS2720 part IV-1985. Liquid limit of the soil samples were determined by cone penetration method according to IS 2720 part V-1985.Heavy compaction test was carried out for the determination of OMC and MDD of the soil samples according to IS 2720 part VII-1980/87 . CBR test was conducted as per IS 2720 Part-1,2,4,5,8,16,40.

The pH of the soil samples were measured in distilled water after shaking the solution for 30 minutes by means of a pH meter fitted with a glass electrode.

The EC was measured in a sutured solution extract of soil samples using a EC meter.

Bacterial count: Was determined by as per the standard procedure by dilution pour plate method (Khannan 1996).

Petri plate experiment set up-

1g of this VC of MWS having bacterial count 10 x 10^7 which is a combined micro flora of bacteria, fungi, yeasts, and actinmycetes is placed in the Petri dish and by taking a sterilized straight wire the VC was spread around the plate as shown in photograph.

10 g of MWS passing through 90 micron was spread above the VC. The plates observed at every 2 hr interval. The plates 1 and 2 taken after 10 hrs, in vertical and inverted positions presented in Plate 3 and 4.

Particle size distribution of the soil was determined in accordance with IS: 2720(part IV) - 1985 Using standard set of IS Sieves, Samples were machine sieved and percentage assign in various sieves was obtained. Based on the test results, a graph representing particle size distribution prepared on a semi log graph paper.

1g of VC of MWS having bacterial count 10 x 10^7 which is a combined micro flora of bacteria, fungi, yeasts, and actinmycetes is placed and evenly spread. 10 g of MWS passing through90μ is spread on this plate. The observations were made after 30 minutes. The results are placed in Table and figure

Results and Discussion

Properties of source soil

Table 1 Properties of source soil

Property	Value
liquid limit	23
plastic limit	-
IS classification	SM
OMC(%)	8
MDD(g/cc)	2.574
CBR	55.94
Cohesion(n/mm2)	0.11
Angle of internal friction(□)	57

The GT Properties LL 23 indicates that the MWS waste has water holding capacity, however it is non Plastic. The dry density itself indicate that the value of 2.574 is very high and a good for stabilization indicator. The value of CBR 55.95 suggests that the MWS can be directly used as a resource material for filling in engineering structures and sub base for pavements as IRC suggests the range of CBR from to . The soil has a very low cohesion and high angle of internal friction indicating the classification of soil as closer to city sand and can be a very good in taking compressive loads in confined state.

Properties of vermicompost

Table 2 Properties of soil treated with vermicompost

Property	Value
Liquid limit	22
Plastic limit	-
IS classification	SM
OMC(%)	11%
MDD(g/cc)	2.32
Cohesion(N/mm2)	0.2
Angle of internal friction(□)	41

The angle of internal friction as compared with the cohesion of MWS indicated the addition of VC has increased the cohesion to 81.8%. A significant property to controls the binding of soil particles which in turn controls the erosion of this soil.

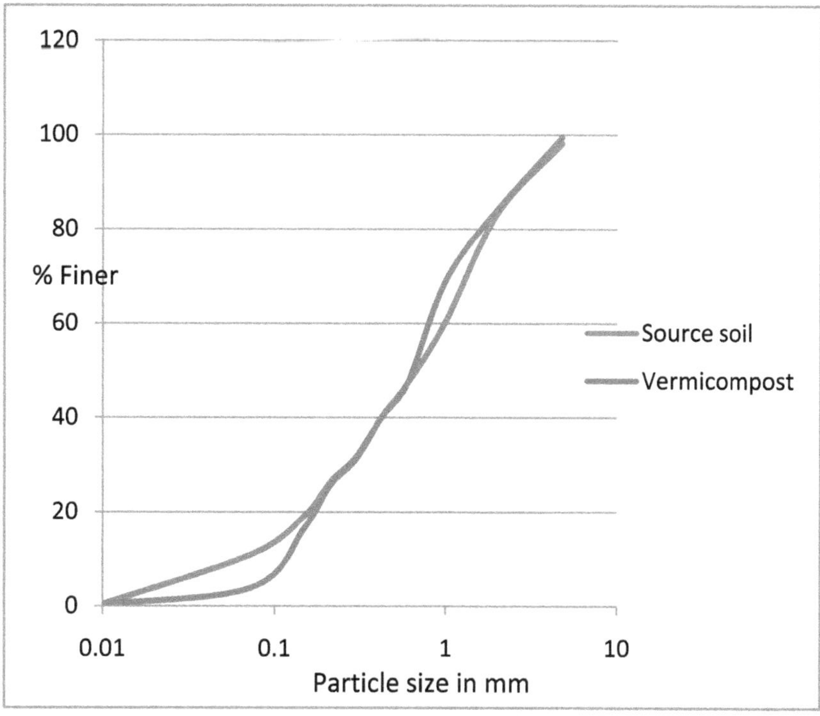

Figure 1 Particle size distribution curve of MWS and VC

The particle size distribution graph shows the MWS and VC treated MWS are showing the similar characteristics and both can be classified as SM.

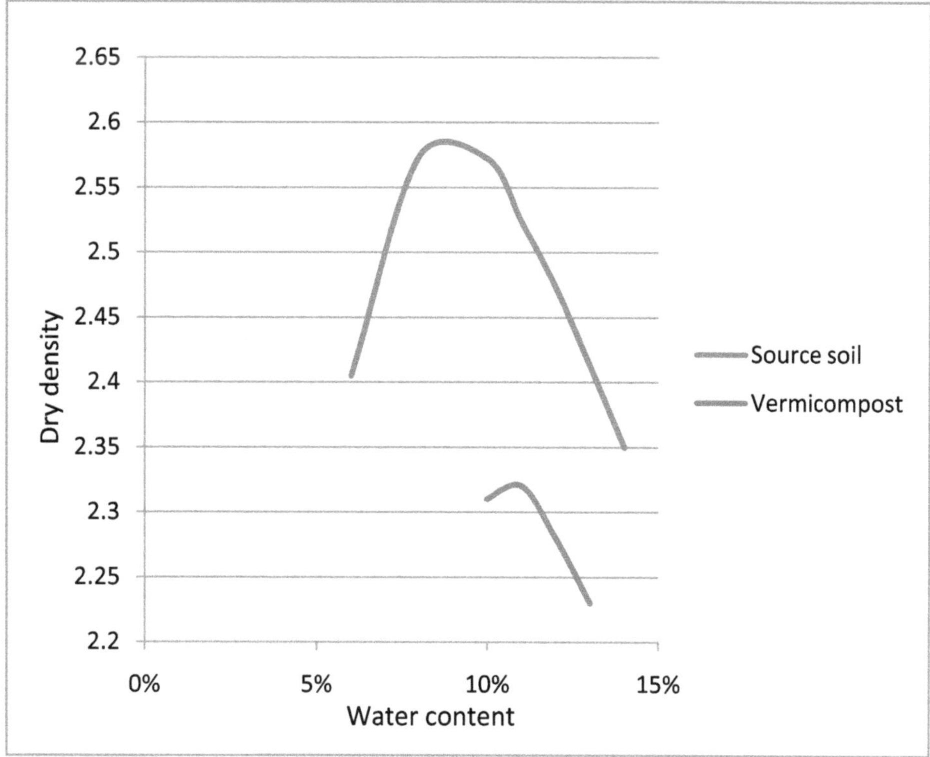

Figure 2 OMC and MDD graph

Bio conversion

Physical observations w.r.t bioconversion

Days	0	10	20	30
Observation for bio conversion	NIL	XX	XXX	XXXX
No of earthworms in 10 g of sample	NIL	4	8	20

Petri plate experiment results:

Plate 1 10 g MSW evenly spread

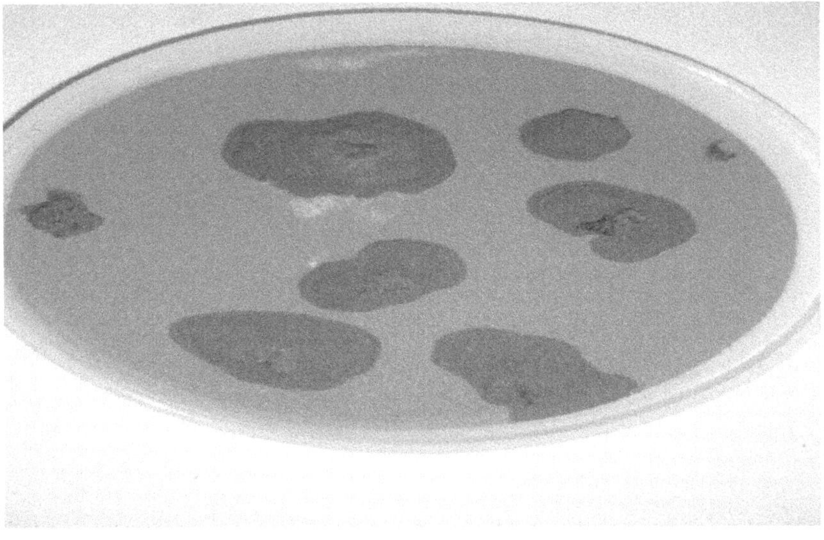

Plate 2 1g of VC spread evenly

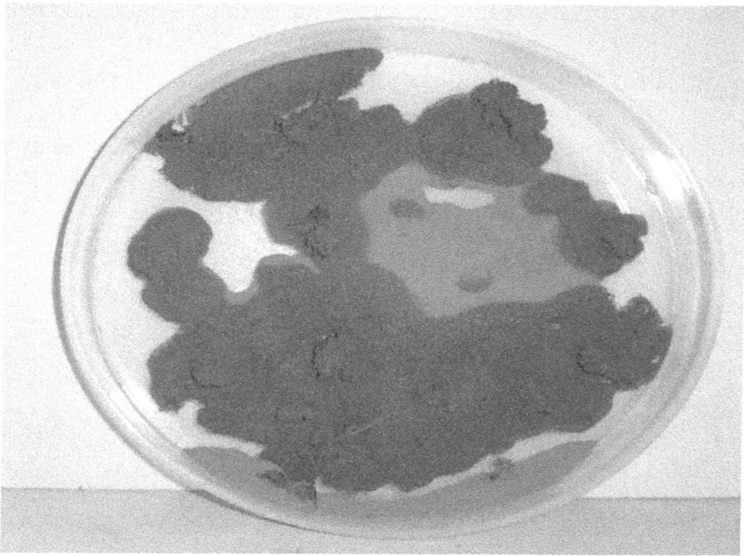

Plate 3 PetriPlate vertically held position

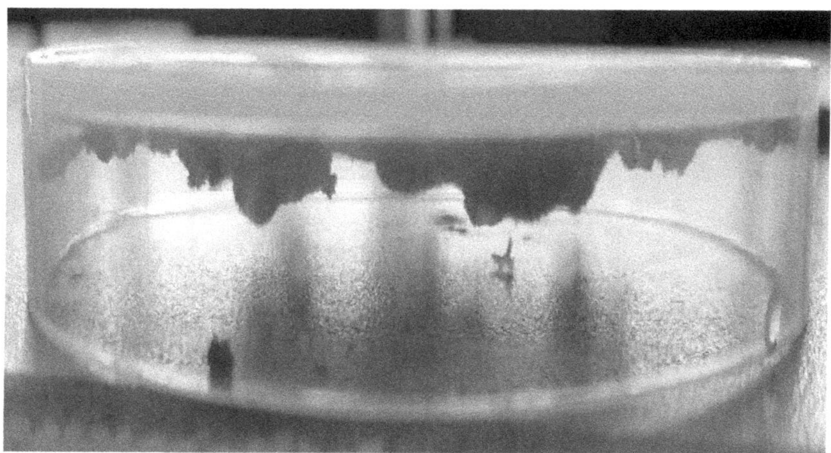

Plate 4 Petri Plate inverted

Conclusion

The particle distribution places the mining waste soil in SM category.

The shear strength of MWS with 1: 3 proportion VC showed a increase of 118%in Cohesion

The MWS passing through 90μ sieve poured on 1 g of VC was held in position in vertical and inverted position suggesting the VC has micro flora responsible for holding the soil together.

The soil properties of cohesion and aggregation are responsible for controlling the erosion in soil.

References

Cooke, J.A., and Johnson, M.S. (2002) "Ecological Restoration of Land with Particular Reference to the Mining of Metals and Industrial Minerals: A review Of Theory and Practice Environmental Revolution, Vol.10, 41-71.

Kannan N (1996) Laboratory Manuel in General Microbiology, Palani Paramount Pub Palani India 1996: 30 – 90.

Munnoli P M (1998)"A study on management of organic solid waste of agro based industries through vermiculture biotechnology", ME Thesis, TIET Patiala, India, 11-30.

Munnoli P M (2011) Water holding capacity of earthworms vermicompost made of sugar industry waste (press mud) in a mono-and polyculture vermi-reactors, The Environmentalist. vol.31 (4), 394-400.

Munnoli, P.M. and Bhosale, S (2008) "Soil aggregation by vermicompost of press mud", Current Science, vol.95 (11), 1533-1535.

Risse .L.M, and Faucette.L.B (2015) "Compost utilization of erosion control", University of Georgia extension, 1-8.

Singh, N.B. (1997)"Dissertation on Development of process package for organic solid waste management through vermiculture Biotechnology, in organic waste generating industries in Punjab" 91-93.

https://study.com/academy/answer/how-does-mining-cause-erosion.html

Vermicompost Acts as Storage Reservoir of Urine Energy

***Munnoli P M, Udagatti N, Chavan A, Harshitha K, Joshi I**
Professor, Deparment of Civil Engineering S D M Engineering College, Dharwad.
**Corrosponding Author: munnolipm@gmail.com*

ABSTRACT

Investigations were carried out on VC of cow dung subjected to urine sample revealed 7.2%: 2.4%: 2% increase as compared to zero day Samples 0.806:0.3:0:43 in N, P K values along with complete removal of odor in 5 days. The bioconversion studies of two pit latrine sludge using *E. fetida* revealed that 29.5%: 23.6%: 16.5% increase as compared to NPK values 1.53%:1.288%:0.32% of VC of cow-dung respectively, suggests the use of urine separated toilets and to harvest energy.

Keywords: Vermicompost, Human Urine, Nitogen, Phosphorus, Potassium

Introduction

India is a developing country with its 70% of population living in rural areas without adequate facility of minimum standards of food, cloth and shelter. Adding to this lack of sanitation facilities like no proper drains, cow sheds, toilets and adopted practices of open defecation has led to the environmental degradation and affecting health of humans and live stocks in general leading to Communicable diseases, Creation of obnoxious and unhygienic conditions, Damage to aesthetics of land, water ,Ground water pollution (Garg S K 2014).,Growth of rodents, rats, cockroaches etc. acting as disease transmitters. Addressing rural sanitation has now been considered as priority by the Ministry of drinking water and sanitation GOI and Karnataka. The present scenario is 33% of villages in Karnataka are without toilets having the entire problem listed above. Even though there is a continuous effort toward rural sanitation either it is viewed secondary or a neglected aspect of life without really knowing the effect of such a sanitation system. Therefore, there is a need that alternative and appropriate methods of designing the rural sanitation systems, like sustainable methods construction of rural toilets, use of alternative materials, economy etc. Further, as a value added product VC of cow dung are subjected to urine treatment and sludge of two pit latrine. With the increase of growing rural population in the villages in concern, the un systematic way of disposal of liquid and solid wastes generated has led to the poor sanitation conditions prevailing, leading to serious environmental degradation and affect on the health of people, especially the health of women is being a serious threat even though she is the one bringing the growth of the whole family. Therefore the Swachh Bharath Mission of Government of India addresses timely both rural and urban sanitations. The Mission aims at achieve a Swachh Bharat by 2019, as a fitting tribute to Mahatma Gandhi on his 150th birth anniversary.

Further with the increase in the cost of chemical fertilizers and there non availability of sufficient quantity and there toxic effects on the land ecosystem the production of alternative organic manure and harvesting urine energy is being considered as simple and economically viable solution. The paper is an attempt to analyze the manure values of urine and two pit latrine sludge treated with VC as a control and with urine as an experiment. Quality of life and human development index is indicated by the sanitation facilities of the country. (WHO 1992). The sanitation practices in general involve the water carriage system started. The overall purposes of sanitation are to provide a healthy living environment for everyone, to protect the natural resources, and to provide safety, security and dignity for people when they defecate or urinate (Duggal K N 2010). Sanitation infrastructure has to be adapted to several specific contexts including consumers' expectations and local resources available (Thomas C et al. 2014). The global sanitation sector is in a severe situation. Over 2.6 billion people are without access to improved sanitation facilities in the world (Seetharam K 2015). (Peal A et al. 2010; Thomas C et al.2014). An estimated 2.5 billion people lack access to improved facilities for the disposal of human excreta, such as a basic pit latrine (Patil S.R et al. 2014). (Srinath P 2013). Note that, by removing the urine at source the volume of solid matter produced is very small and, as moisture is extracted, becomes even smaller, normally leaving only 10% by volume of what comes out of us. (Bremner J M 1982; Jaana BB 2018)

Materials and Methods

VC and Earthworms: Were procured from UAS Dharwad and the same were maintained in the civil engineering laboratory. Sludge: Sample was collected from a two pit latrine situated at Yarikoppa village.

Urine sample: Fresh urine sample at the time of experimentation was collected minutes before inoculating in the experiment from the project team mates.

Experimental setup for removal of odor

Four beakers of 200ml capacity were taken and washed and wiped with clean cloth. VC made out of cow dung collected from University of Agricultural Science, Dharwad was filled in all the four beakers weighing 100gm each.

Fresh urine of 50ml was added in three beakers which act as experimental samples and remaining one beaker containing only VC act as experimental control sample.NPK test was carried out for VC on zero day. Samples were kept under observation till the smell vanishes.

NPK test for experimental samples was carried out after the smell vanished after 6 days, as per the procedure outlined in Jackson M L 1977.

Table 1.0 Nutrient composition of urine

Nutrients	Range
Nitrogen (N)	15 - 19 %
Phosphorous(P)	2.5 - 5%
Potassium(K)	3 - 4.5%

Setting up of Vermicomposting Experiment in a culture box

Two Rectangular wooden boxes of size 30 X 30 X 30 cm were used for maintaining cultures of surface feeders. One box is for experimental sample and another one for experimental control. Sludge sample was collected from two pit toilet from Yarikoppa village. Earthworm surface feeder species (*E.fetida* and *E.eugeniae*) were procured from University of Agricultural Science, Dharwad. The culture box is filled with 5cm thick layers of each of dry leaves followed by black cotton soil and then cow dung was spread evenly. Collected sludge sample was spread into experimental sample box in the form of 5cm thick layer. Experimental setup was covered with gunny bags; this system was kept at moisture level of 70% by sprinkling water daily.

NPK test was carried out for sludge sample on zero days; the same test was carried for samples collected from experimental sample box on 10th and 20th day as per Jackson M L 1958.

Results and Discussion

Table 2.0 Observation for odour

0 day	1 day	2 day	3day	4day	5day
Control	-	-	-	-	-
XXXXX	XXX	XX	X	-	-
XXXXX	XXX	XX	X	-	-
XXXXX	XX	X	X	-	-

Where X - Indicates the odor intensity

The Table 2.0 reveals that there is no odor after 4th day suggesting use of vermicomposting and passage of urine will not cause any problems of odor.

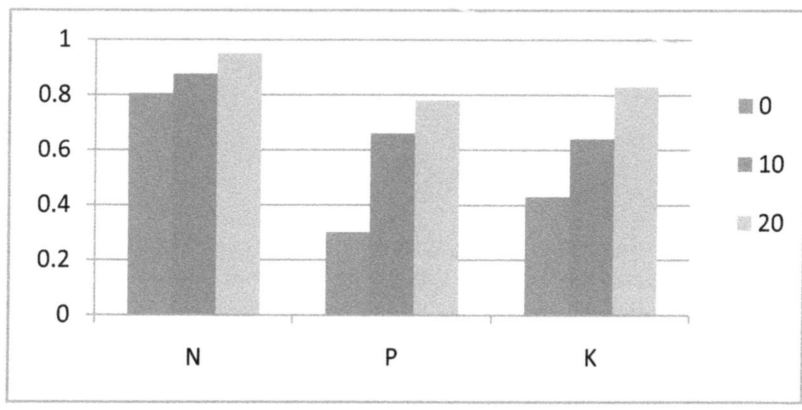

Figure 1.0 Increase in NPK test results of vermicompost with urine

The trends of experiments reveal (Figure1.0) that all the three nutrients have shown increased values. The % increase in Nitrogen is 7.2%; in Phosphorous is 2.4%, in Potassium is 2%. It was observed that urine samples had odor which was also bio absorbed completely and continuous decrease in odor noticed for 5 days, this suggest that if urine is separated and passed through Vermicompost unit leads to increased storage of nutrients in the form NPK and also removes the odor. This can be achieved by simple modification in the design of water closet by providing separate outlet for urine. The urine pipe can be connected to the VC inlet and allowed to stay for 10 to 20 day. The energy rich vermicompost can be harvested and stored also application as manure in farmlands or horticulture. The increasing trend of NPK when urine is passed has also been studied.

NPK test details for Sludge

Table 3.0 Energy values in terms of NPK

Parameter	0 day	10 day	20 day
N%	1.532	1.80	2.12
P%	1.288	1.463	1.76
K%	0.32	0.49	0.65

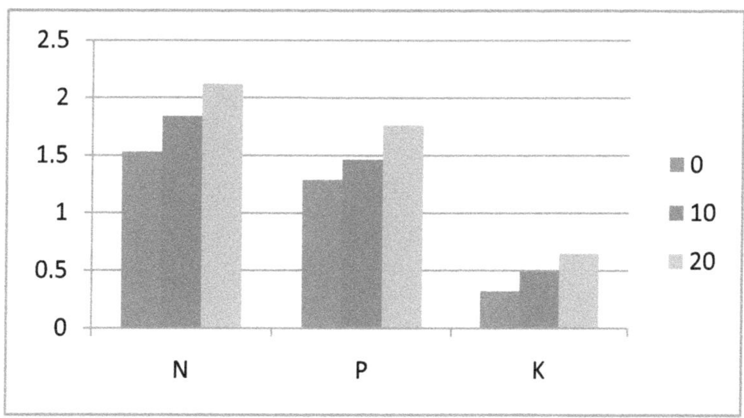

Figure 2.0 Increase in **NPK** test results of vermicompost with Sludge

The suggested method to treat the sludge through VC revealed that there is increase of 29.5 % in nitrogen, 23.6% in phosphorous, 16.5% in potassium (Figure 2.0). The sludge sample drawn from the VC unit of our experiment reveals that the Nitrogen values are increasing continuously from 0 to 20 days; this can be attributed to continuous earthworm activity which releases aerobic bacteria which bio converts our substrate sludge of two pit latrine. The sludge with increased NPK values can directly be utilized as manure.

Conclusion

The odor of urine sample was completely removed in 5 days. The % increase in N: P: K is 7.2%: 2.4%: 2% as compared to the N: P: K 0.806:0.3:0:43 values of VC of cow-dung respectively. The suggested method to treat the sludge through VC revealed that there is increase in N: P: K, 29.5 %: 23.6%: 16.5% as compared to NPK values 1.53%:1.288%:0.32% of VC of cow-dung respectively

References

Bremner JM (1982) Nitrogen-urea. In: Page AL et al. (eds) Methods of Soil Analysis, Part 2, 2nd edition, pp 699-709. Agronomy Series No. 9. Madison, Wisconsin. cited on 23/4/2018 on 3:30 pm.

Duggal, K. N. 2010, Elements of Environmental engineering by S. Chand & Company Ltd., New Delhi, 2010 p :1-50

Garg, S.K. (2009) Environmental Engineering Vol. -Water supply engineering by, Khanna Publishers, Delhi, 2009. p:1-70

Jackson, M. L.: Soil chemical analysis. Verlag: Prentice Hall,

Jana.B.B (2018), Human Urine as an Organic Alternative to Chemical Fertilizers, DREAM 2047 vol 20 no 6 pg no. 29-32.

Patil SR, Arnold BF, Salvatore AL, Briceno B, Ganguly S, et al. (2014) The Effect of India's Total Sanitation Campaign on Defecation Behaviors and Child Health in Rural Madhya Pradesh: A Cluster Randomized Controlled Trial. PLoS Med 11(8): e1001709. doi:10.1371/journal.pmed.1001709

Peal A, Evans B, van der Voorden C. 2010. Hygiene and sanitation software. An overview of approaches. Geneva (Switzerland): Water Supply & Sanitation Collaborative Council.

Seetharam K (2015). Challenges and opportunities for sanitation in developing countries, journal of science policy and governance, ISPG volume 7 issue 1.

Srinath P (2013) Sanitation in Rural India and Karnataka – How has the needle moved?, © 2018 The Transition State,http://catalyst.nationalinterest.in/2013/03/14/sanitation-rural-india- karnataka-2001-2011- improvement/Cited on 22/4/2018 at 10:10 pm.

Thomas C, (2014) Effectiveness of a rural sanitation programme on diarrhea, soil-transmitted helminth infection, and child malnutrition in Odisha, India- a cluster- randomized trial : Lancet Glob Health,

http://dx.doi.org/10.1016/S2214-109X(14)70307-9

WHO, 1992 A Guide to the Development of on-Site Sanitation, World Bank. 1995. Guidelines on community toilets (English). Water and Sanitation Program. Washington, DC : World Bank.

Effect of Geotextile on CBR Strength of Unpaved Road with Soft Subgrade

A. Vittalaiah[1] Akula Prakash[2] and Rathod Ravinder[3]

[1,2,3]*Assistant Professor, Department of Civil Engineering, Gokaraju Rangaraju Institute of Engineering and Technology (Autonomous), Hyderabad*

ABSTRACT

Utilization of Geosynthetics is guaranteed in a given geotechnical application, as it isn't just good yet powerful in improving the dirt properties when suitably set. In this investigation the exhibition of nonwoven geotextile interfaced between delicate subgrade as one and two layers in an unpaved adaptable asphalt framework is done tentatively, using the California Bearing Ratio (CBR) testing course of action. So as to assess the presentation, the fortification proportion is gotten dependent on the CBR load – entrance connection of both one and two layers of nonwoven geotextile for unpaved adaptable asphalt. Correlation of support proportion decided utilizing the CBR quality test demonstrates that the presentation is improved with the consideration of nonwoven geotextile.

Keywords: Geosynthetics, Nonwoven geotextile, Unconfined Compressive Strength, California Bearing Ratio

Introduction

The monetary advancement of a nation is firmly identified with its street transport framework offices accessible. Particularly in an under creating nation, the rustic streets interfacing horticultural towns is imperative in improving the country economy. It is realized that the alternative of unpaved streets are prudent for low traffic volume in such regions, notwithstanding, when unpaved streets laid on delicate subgrade experiences huge disfigurements, where the periodical upkeep of the provincial street is constrained because of cost contemplations, which may upset the administration and influence the capacity of the street. In such circumstances, looking at different techniques, geotextiles can be used to improve not just the presentation of the unpaved street by expanding the existence time, yet in addition, limiting the upkeep cost just as diminishing the thickness of the street.

Need of restoration

The primary utilization of unpaved streets are in low volume traffic and getting to streets. Fundamentally, in farming nations, low volume streets assume a significant job in the provincial economy. At the point when these kinds of streets with delicate establishment soils are built, there is the likelihood for enormous disfigurements to happen, which expands support cost and lead to interference of traffic administration. This contracting and swelling development is the outcome of sporadic street surface and street disintegration, bringing about a requirement for untimely restoration of the asphalt street.

Utilization of Geotextiles

The utilization of geotextile for fortification to improve feeble soil is as of now a mainstream strategy. The elasticity of geotextile and the dirt geotextile communication are the main considerations that impact the improvement of soil. Change in fine substance inside the sand can change the interface conduct among soil and geotextile. Geotextiles have swarmed practically every one of the parts of geotechnical designing with just about a boundless number of uses. With respect to its use as a partition in the unpaved streets, the reason for its application is to give the division or support to the dirt which is debased by time passes and settle the dirt.

Objectives of the Study

1. To assess the adequacy of the geotextile in street development and support.
2. To fuse the geotextile in gathered soil material and evaluate the presentation.
3. To break down the outcomes and make fitting suggestions for ideal use.
4. To assess the presentation, the fortification proportion is acquired dependent on CBR load-infiltration connection.

5. To test the dirt example for soil properties like fluid farthest point, plastic cut off, ideal dampness content, dry thickness and so forth., .and to touch base at CBR load-entrance an incentive for geotextile place in various thickness of soil.
6. And furthermore to learn about the properties of geotextile like elasticity and cut quality and so forth.

Literature Review

Deb, K., Konai, S.,.Bearing Capacity of Geotextile-Reinforced Sand with Varying Fine Fraction (2014). Geomechanics and Engineering, a series of model tests have been carried out and the load settlement curves are obtained. The ultimate load carrying capacity of unreinforced and reinforced sand with different percentages of fines is compared. The interface behaviour of sand and geotextile with various percentages of fines is also studied. It is observed that sand having around 5% of fine is suitable or permissible for bearing capacity improvement due to the application of geosynthetic reinforcement. The effectiveness of the reinforcement in load carrying capacity improvement decreases due to the addition of excessive amount of fines.

P. B. Ullagaddi, T.K.Nagaraj (2013) presented an "Investigation on geosynthetic reinforced two layered soil system", the road laid on bases formed of weaker soil leads to large deformations, causing increases in maintenance cost and interruption of traffic service. And if the subgrade layer of pavement consists of expansive soil (black cotton soil), due to its susceptibility to moisture change and results in subsequent high swelling and shrinkage characteristics. These soils possess less strength and bearing capacity and thereby results in increasing the thickness of pavement. There are many stabilization methods available to improve engineering properties of these types of soft subgrade soil. Use of geosynthetics over a soft subgrade (expansive soil) found to be one of the feasible and economic solutions to strengthen road pavement and thereby increasing service life. This study presents the effect of various geosynthetic products on soft subgrade (black cotton soil) in improving strength parameters.

A.K.Choudhary, K.S.Gill and J.N.Jha (2011) presented on "Improvement in CBR values of expansive soil sub-grades using geo-synthetics" Street asphalts are powerless (feeble) to soil execution in light of the fact that the establishment of the asphalt is a streets most critical component. What's more, if the sub level layer of asphalt comprises of sweeping soil (dark cotton soil), because of changes in dampness content and ensuing shrinkage and swelling, it experiences disappointment. In this way, for the development on such kind of soil it is required to enhance the designing properties of soil or to supplant the dirt itself. Supplanting the current soil won't not be a viable and achievable choice, along these lines it is required to balance out the dirt with suitable stabilizer. However the choice of stabilizer relies on the sort of sub-evaluation soil, kind of soil change sought, accessibility of stabilizer, the required quality and toughness of balanced out layer, different settling methods, natural conditions and the most vital expense component. This paper surveys the work of different scientists on adjustment of soil and utilization of geosynthetic materials in enhancing its quality.

Dhule, S.B., S.S. Valunjkar, S.D. Sarkate and S.S. Korrane (2011), studied thatlocally available sub-grade soil of the road is modified by addition of geotextile material in different percentage i.e.1%, 2% and 2.5% and 3% respectively. Similarly geogrid is been used for improvement in properties of soft murum has also been studied. Also geogrid is used in mix of soil and 2% cement in different proportion to study its effects. With these attempts it will be possible to find out optimum mixes which are to be used for further construction to achieve desired stability and economy in construction. For this purpose different test were performed i.e sieve analysis, liquid limit, Plastic limit, Standard proctor test to find its maximum water content and maximum dry density, specific gravity, Laboratory Unsoaked CBR and Laboratory soaked CBR test to find it resistance to penetration.

Babu, K.K., K.S. Beena and A.K. Raji "Estimation of CBR of Coir Geotextile Reinforced Subgrade, (2011)" Highway Research Journal, Considerable length of roads planned to be constructed in India under various programmes require construction over poor subgrade soils. The performance of a road largely depends on properties of the subgrade soil. One such subgrade soil often encountered is the black cotton (BC) soil. It is inorganic clay of medium to high compressibility, high shrinkage and swelling property, very hard when dry, but lose its strength completely when in wet condition. As a result of wetting and drying process, vertical movement takes place in the soil mass leading to failure of pavement, in the form of settlement, heavy depression, cracking and unevenness. In order to improve the performance of roads on such soils Coir textile has scope as reinforcement. It is expected that with the inclusion of coir geotextile layer below Granular Subbase (GSB) layer would be helpful in restricting the movement of upper pavement layers due to seasonal moisture variation in subgrade expansive, shrinkable soil. A pilot project plan for implementing the idea has been proposed in the paper.

Materials Used

Table 1 Soil Properties from the tests conducted

Test	Results
Liquid limit (%)	38
Plastic limit (%)	21.1
Plasticity Index (%)	16.9
Maximum dry density (gm/cc)	1.648
Optimum moisture content (%)	18.86
CBR (%)	3.04
Unconfined compression strength (KN/m^2)	95
Shear strength (KN/m^2)	47.5

Table 2 Properties of Geotextile

Particulars	Nonwoven geotextile
Mass per unit area (g/m2)	203
Grab Tensile Strength (N)	710
Puncture Resistance (N)	1820

Atterberg Limits and Particle size distribution

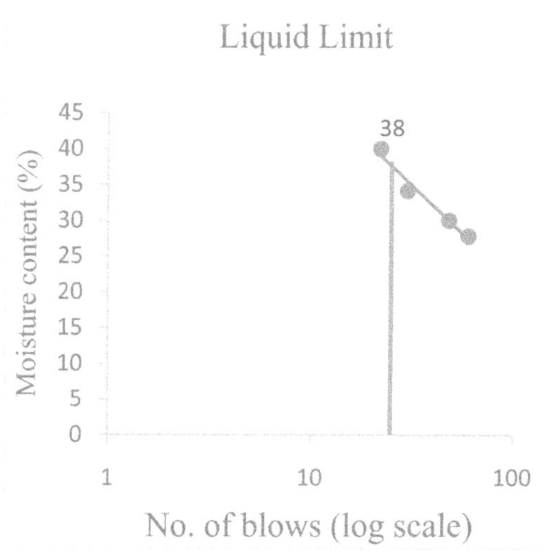

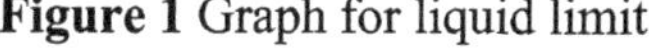

Figure 1 Graph for liquid limit

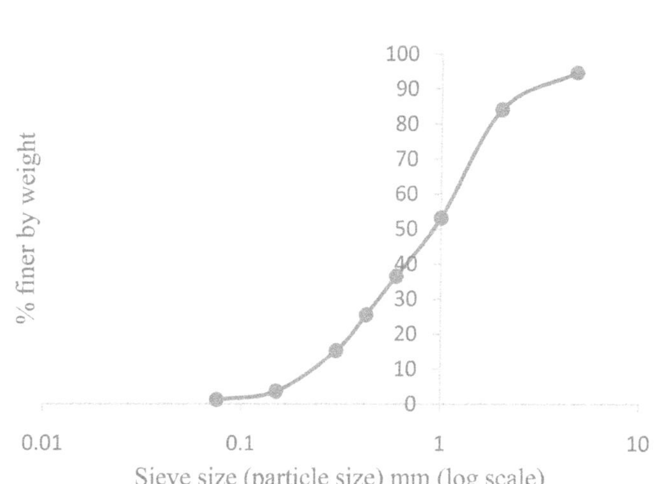

Figure 2 Graph for particle size distribution

- From the above Liquid Limit graph, the obtained Liquid Limit value for the soil sample is 38%.
- From the particle size distribution graph,

  D_{10}= 0.24, D_{30}= 0.54, D_{60}= 1.18

  i.e.., C_C=0.954, C_U=4.91

Therefore, the soil sample we used for our investigation is a well graded soil.

Standard Compaction Test

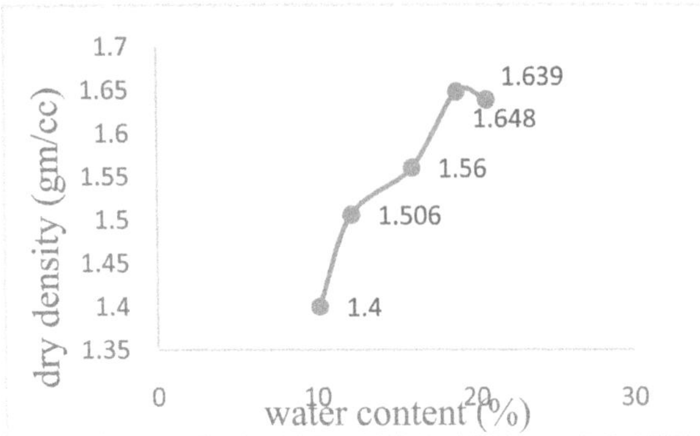

From the graph the Maximum Dry Density (MDD) and Optimum Moisture Content (OMC) of soil are 1.648 gm/cc and 18.86% respectively.

Figure 3 Graph for OMC and MDD

Unconfined Compressive Strength

Table 3. Experimental values for Unconfined Compressive strength

Stress (N/mm^2)	0.0	28.5	33.0	39.8	46.6	57.7	64.2	70.6	74.6	80.7	84.6	88.4	92.1	93.6	95.0	90.0	87.2	80.2
Axial Strain	0.0	0.7	1.4	2.1	2.8	3.5	4.2	4.9	5.6	6.3	6.9	7.6	8.3	9.0	9.7	10.4	11.1	11.8

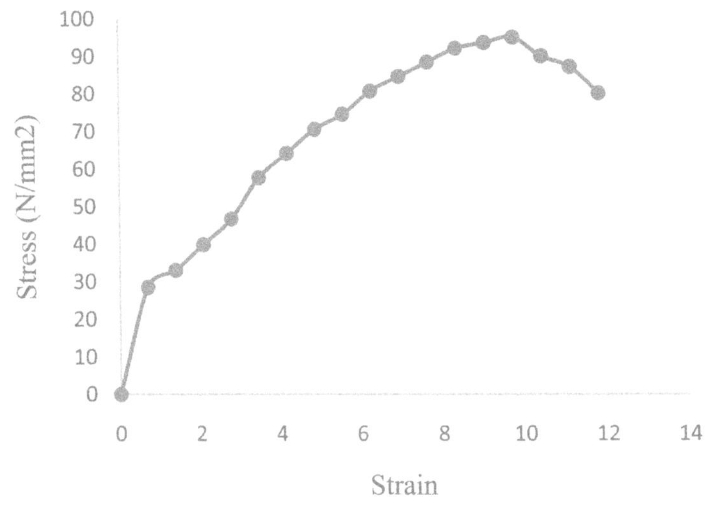

Figure 4 Graph for Unconfined Compressive Strength

From the graph, Unconfined Compression Strength of the Black cotton soil,

$$q_u = 95 \text{ KN/m}^2$$

Shear strength of the Black cotton soil = $q_u/2 = 47.5 \text{ KN/m}^2$

CBR without Geotextile

Table 4 Experimental values of CBR without geotextile

Penetration (mm)	0.5	1.0	1.5	2.0	2.5	3.0	3.5	4.0	4.5	5.0	5.5	6.0	6.5
Load (Kg)	13.9	20.8	24.3	34.8	41.7	41.7	45.2	45.2	48.1	55.6	55.6	55.6	55.6

7.0	7.5	8.0	8.5	9.0	9.5	10.0	10.5	11.0	11.5	12.0	12.5
59.1	59.1	62.6	62.6	66.1	66.1	69.5	73.0	73.0	76.5	80.0	83.4

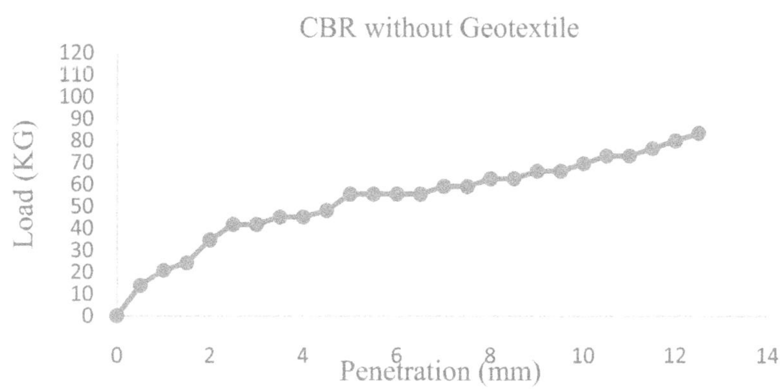

Figure 5 Graph for CBR test for Soil without geotextile

CBR = (Pt/Ps)*100, Where Pt = corrected load, Ps = standard load

- At 2.5mm penetration : CBR = (41.7/1370)*100 = 3.04 %
- At 5mm penetration : CBR = (55.6/2055)*100 = 2.70 %

CBR at 2.5mm penetration is greater than CBR at 5 mm penetration. Therefore CBR of Black cotton soil is 3.04 %.

CBR with Geotextile in one layer

Table 5 Experimental values of CBR with geotextile in one layer

Penetration (mm)	0	0.5	1	1.5	2	2.5	3	3.5	4	4.5	5	5.5	6
Load (Kg)	0.00	17.34	24.27	27.74	38.14	45.08	46.81	52.00	52.01	55.48	65.88	65.88	72.82

6.5	7	7.5	8	8.5	9	9.5	10	10.5	11	11.5	12	12.5
72.82	74.55	74.55	78.02	78.02	81.49	81.49	84.96	86.09	91.89	95.36	100.56	102.29

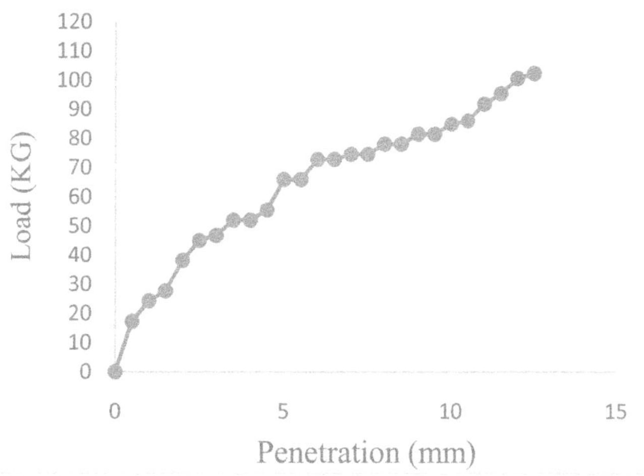

Figure 6 Graph for CBR Test for soil with one layered Geotextile

CBR = (Pt/Ps)*100, Where Pt = corrected load, Ps = standard load

- At 2.5mm penetration : CBR = (45.0788/1370)*100 = 3.29 %
- At 5mm penetration : CBR = (65.8844/2055)*100= 3.18%

CBR at 2.5mm penetration is greater than CBR at 5 mm penetration. Therefore CBR of Black cotton soil is 3.29 %.

CBR with Geotextile in two layers

Table 6 Experimental values of CBR with geotextile in two layers

Penetrartion (mm)	0	0.5	1	1.5	2	2.5	3	3.5	4	4.5	5	5.5	6
Load (Kg)	0.00	19.07	26.01	29.47	39.88	46.81	50.28	55.48	57.22	64.15	69.35	69.35	72.82

6.5	7	7.5	8	8.5	9	9.5	10	10.5	11	11.5	12	12.5
74.55	79.75	79.75	84.96	86.69	91.89	97.09	100.56	100.56	102.29	104.03	109.23	110.96

CBR = (Pt/Ps)*100 : Where Pt = corrected load , Ps = standard load

- At 2.5mm penetration : CBR = (46.8126/1370)*100 = 3.48 %
- At 5mm penetration : CBR = (69.352/2055)*100 = 3.29%

CBR at 2.5mm penetration is greater than CBR at 5 mm penetration. Therefore CBR of Black cotton soil is 3.48 %.

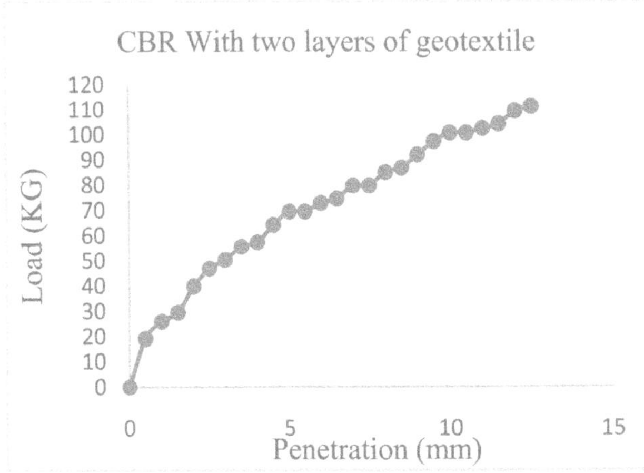

Figure 7 Graph for CBR test for soil with two layered Geotextile

Test Results

By analysing the test results found above, we compared the three different soil samples with and without placing geotextiles, in different thicknesses and observed the improvement in CBR strength in the soil samples. The following are the test results compared based on the CBR test value among three soil samples with geotextile and geotextiles placed at different layers.

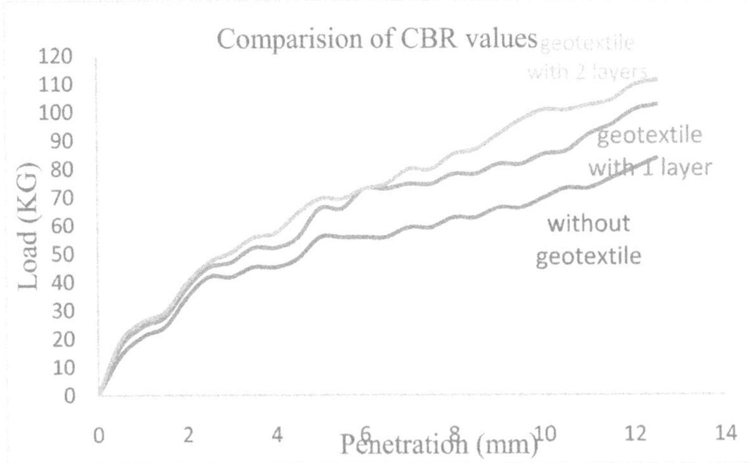

Figure 8 Graph for comparison of CBR values for soil with geotextile at different thicknesses

From the graph, we can clearly observe the variation of the CBR strength with respect to the interfacing of geotextile at different thicknesses of the soil sample in the standard CBR mould.

Conclusions

All test results generally showed that with the utilization of geotextile in the states of one and two layers, the resistance of samples against loading increased appreciably. Interfacing of nonwoven geotextile in an unpaved road especially with soft sub grade, increases the penetration resistance and hence the CBR strength. Therefore, the performance of the unpaved road is better with the inclusion of geotextiles. It is clear that the utilization of geotextiles improves further at larger depth of penetration when used with more number of layers of geotextile

References

Deb, K., Konai, S., 2014.Bearing Capacity of Geotextile-Reinforced Sand with Varying Fine Fraction. Geomechanics and Engineering.Vol. 6, No. 1.33-45.

P.B.Ullagaddi,T.K.Nagara," Investigation on geosynthetic reinforced two layered soil system ", Missouri University of Science and Technology,2013.

A.K.Choudhary, K.S.Gill and J.N.Jha (2011) "Improvement in CBR values of expansive soil subgrades using geo-synthetics" IGC J-233.

Dhule, S.B., S.S. Valunjkar, S.D. Sarkate and S.S. Korrane (2011) "Improvement of Flexible Pavement with Use of Geogrid," Electronic Journal of Geotechnical Engineering, Vol. 16, Bundle C, 269-279

Babu, K.K., K.S. Beena and A.K. Raji (2011) "Estimation of CBR of Coir Geotextile Reinforced Subgrade," Highway Research Journal, Vol. 4, No. 2, 41-47.

R. Ziaie Moayed and M. Nazari,"Effect of Utilization of Geosynthetic of Reducing the Required Thickness of Subbase Layer of a Two Layered Soil",World Academy of Science, Engineering and Technology, International Journal of Environmental, Chemical, Ecological, Geological and Geophysical Engineering Vol:5, No:1, 2011.

Vaishali S. Gor, L. S. Thakur, Dr. K.R. Biyani,"study the typical characteristic of expansive soil with geotextile and cushion materials",Missouri University of Science and Technology,2013.

Madhavi Latha Gali, Asha M Nair (2010), Performance of geosynthetics in unpaved roads, International Journal of Geotechnical Engineering (2010) 4: (337-349).

Lekha, K.R., Kavitha, V., 2006. Coir Geotextile Reinforced Clay Dykes for Drainage of low-lying areas. Geotextiles and Geomembranes 24 (1), 38–51.

Burd, H.J (1995) "Analysis of Membrane Action in Reinforced Unpaved Raods," Canadian Geotechnical Journal, Vol. 32, 946-956.

A Study on Use of Wasteplastic in Bituminous Paving Mixes

Kandlagunta Mounika[1] Akula Prakash[2] and A Vittalaiah[3]

[1]Assistant Professor, [2,3]Assistant Professor, [1]Department of Civil Engineering, CMR Institute of Technology (Autonomous), Kandlakoya, Medchal, Hyderabad [2,3]Department of Civil Engineering, GokarajuRangaraju Institute of Engineering and Technology (Autonomous), Hyderabad

ABSTRACT

Bituminous blends are most generally utilized everywhere throughout the world in adaptable asphalt development. It comprises of black-top or bitumen (utilized as a fastener) and mineral total which are combined, set down in layers and after that compacted. Under typical conditions, ordinary bituminous asphalts whenever structured and executed appropriately perform palatably yet the presentation of bituminous blends is poor under different circumstances. The present asphaltic solid asphalts are relied upon to perform better as they are encountering expanded volume of traffic, expanded loads and expanded varieties in every day or regular temperature over what has been knowledgeable about the past. What's more, the exhibition of bituminous asphalts is observed to be exceptionally poor in dampness prompted circumstances. Thinking about this as a great deal of work has been done on utilization of added substances in bituminous blends and just as on adjustment of bitumen. Research has demonstrated that the expansion of polymers to black-top folios expands the interfacial cohesiveness of the security between the total and the fastener which can improve numerous properties of the black-top asphalts to help fulfil these expanded needs. Notwithstanding, the added substance that will be utilized for adjustment of blend or cover ought to fulfil both the quality prerequisites just as prudent aspects.Plastics are wherever in the present way of life and are developing quickly all through especially in a creating nation like India. As these are non-biodegradable there is a noteworthy issue presented to the general public as to the administration of these strong squanders. Low thickness polyethylene (LDPE) has been observed to be a decent modifier of bitumen. Indeed, the recovered polyethylene initially made of LDPE has been seen to alter bitumen. In the present examination, an endeavor has been made to utilize recovered polyethylene which has been acquired from plastic containers utilized in bundling of a mainstream cooldrinks and water, in dry structure with the totals like a fibre in a bituminous blend. Definite examination on the impacts of these locally squander polyethylene on designing properties of Bituminous cement (BC)

Keywords: Bituminous concrete (BC), Stone mix asphalt (SMA), Dense bound macadam (DBM), plastic, Marshall Properties.

Introduction

Bituminous fasteners are generally utilized by paving industry. By and large asphalts are sorted into 2 gatherings, for example flexible and rigid pavement..

Flexible Pavement: Flexible asphalts are those, which all in all have low flexural quality and are fairly adaptable in their auxiliary activity under burdens. These sorts of asphalt layers mirror the misshaping of lower layers on-to the outside of the layer.

Rigid Pavement: If the surface course of an asphalt is of Plain Cement Concrete then it is called as unbending asphalt since the complete asphalt structure can't twist or redirect because of traffic loads.Pavement plan and the blend configuration are two noteworthy contemplations if there should arise an occurrence of asphalt designing. The present investigation is just identified with the blend plan of adaptable asphalt contemplations. The plan of black-top clearing blends is a multi-step procedure of choosing bitumen and total materials and proportioning them to give a proper trade off among a few factors that influence blend conduct , considering outer factors, for example, traffic stacking and atmosphere conditions.

Principle destinations of bituminous blend configuration are to discover

1. Ideal bitumen substance to guarantee a tough asphalt,
2. Adequate solidarity to oppose shear disfigurement under traffic at higher temperature,
3. Legitimate measure of air voids in the compacted bitumen to consider extra compaction done by traffic,
4. Adequate usefulness, and
5. Adequate adaptability to abstain from splitting because of rehashed traffic load.

Necessities of bituminous blends

Bituminous blend utilized in development of adaptable asphalt ought to have following properties;

1. Stability
2. Durability
3. Flexibility
4. Skid resistance
5. Workability

Destinations of present examination A near report has been made in this examination between SMA, BC, and DBM blends with shifting folio substance (5.5% - 7%) and plastic (bottles) content (0.5% - 1.5%). The targets of this examination are to watch the followings; Study of Marshal Properties of blends utilizing both 1. Stone residue as filler and, 2. Slag as fine total.

Objective of study

1. The impact of plastic as admixture on the quality of bituminous blend with various filler and supplanting some level of fine total by slag
2. The execution of bituminous blend submerged with and without plastic admixture with various filler and supplanting some level of fine total by slag.
3. To examination protection from perpetual misshaping of blends with and without plastic.
4. Evaluation of SMA, BC, and DBM blends utilizing diverse test like Drain down test.

Literature Review

Bindu and Beena (2010) examined how Waste plastic goes about as a settling added substance in Stone Mastic Asphalt when the blends were exposed to execution tests including Marshall Stability, elasticity, compressive quality tests and Tri-pivotal tests. There results showed that adaptable asphalt with superior and sturdiness can be acquired with 10% destroyed plastic.

Fernandes et al. (2008) contemplated Rheological assessment of polymer altered black-top covers by utilizing thermoplastic elastomer styrene butadiene styrene (SBS) and they analyzed the properties of Modified fastener by expansion of both oil shale and sweet-smelling oil to improve their perfectly. The rheological attributes of the SBS PMBs were broke down in a dynamic shear rheometer (DSR) and the morphology gotten to by fluorescence optical microscopy. The outcomes demonstrated that the fragrant and shale oils effectsly affect the microstructure, stockpiling steadiness and viscoelastic conduct of the PMBs. Along these lines, shale oil could be effectively utilized as a compatibilizer specialist without loss of properties or could even supplant the fragrant oil.

Awwad and Shbeeb (2007) showed that the altered blend has a higher security and VMA rate contrasted with the non-changed blends and accordingly emphatically impact the rutting obstruction of these blends. As per them altering black-top blend with HDPE polyethylene upgrades its properties undeniably more than the enhancements acknowledged by using LDPE polyethylene.

Gawande et al. (2012) gave a review on waste plastic use in asphalting street by utilizing both wet and dry strategy. They said that utilization of changed bitumen with the expansion of handled waste plastic of around 5-10% by weight of bitumen helps in improving the life span and asphalt execution with negligible sparing in bitumen use and as per them utilization of waste plastics in the assembling of streets and overlaid material likewise help to expend extensive amount of waste plastics. Along these lines, these procedures are socially exceedingly pertinent, giving better framework.

Khan and Gundaliya (2012) expressed that the procedure of alteration of bitumen with waste polythene upgrades protection from splitting, pothole arrangement and rutting by expanding mellowing point, hardness and lessening stripping because of water, consequently improving the general execution of streets over a significant lot of time. As indicated by them the waste polythene used in the blend shapes covering over totals of the blend which lessens porosity, retention of dampness and improves restricting property.

Materials used in Present Study

For preparation of Bituminous mixes (DDT,BC) aggregates as per MORTH grading as given in Table 1, Table 2 respectively, a particular type of binder and polyethylene in required quantities were mixes as per Marshall procedure. The specific gravity and physical properties of aggregate are given in Table-3, Table - 4 and Table 5.

Table 1 Gradation of aggregates for Drain down test

Sieve sizes(mm)	Percentage passing
19	50
13.2	375
9.5	150
4.75	185
2.36	40
1.18	30
0.6	20
0.075	50

Table 2 Gradation of aggregate for BC

Sieve sizes(mm)	Percentage passing
19	50
13.2	375
9.5	150
4.75	185
2.36	40
1.18	30
0.6	20
0.3	0
0.75	50
PAN(<0.075)	100

Table 3 Specific gravity of aggregates

Types of aggregates	Specific gravity
Coarse	2.75
Fine(stone)	2.6
Fine(slag)	2.45
Filler(stone dust)	2.7
Lime	2-2.3

Table 4 Physical properties of coarse aggregates

Property	Test Result
Aggregate Impact value (%)	20
Aggregate crushing value (%)	22.6
Los Angles abrasion value (%)	21.8
Water absorption (%)	0.1

Table 5 Physical properties of bitumen

Property	Test method	Value
Penetration at 25 c (mm)	IS:1203-1978	38
Softening point	IS:1203-1978	81
Specific gravity	IS:1203-1978	1.03
Ductility	IS:1203-1978	42

Table 6 Physical properties of bitumen with plastic

Property	Test method	Value
Penetration at 25 c (mm)	IS:1203-1978	32
Softening point	IS:1203-1978	64
Ductility	IS:1203-1978	56

Drain down test

The loose asphalt mixtures are prepared for the drain down test. Either of the gradation Indian or Chinese is chosen from the pocket of aggregates, a 1000 g sample is prepared in each case. The analysis was made between no fibre and trail contents of 0.5%, 1% and 1.5% plastic fibre. The feasibility in its application is checked based on specifications of ASTM D6390 and Chinese for Indian and Chinese pavements respectively. The sample performance is tested at 160°C and 170°C for the maximum binder content of 7% and later checked for Optimum binder content (OBC). The draindown test for loose SMA mixes was performed using the basket drainage test as per ASTM D 6390 (2005). The results of the draindown are presented for both the maximum binder content of 7%, and also at the optimum binder contents obtained for the two gradations are tabulated below in Table and compared in the Figure.

Table 7 Draindown values of SMA mix (PLASTIC fibre) (Figure 1)

PLASTIC Content %	Draindown %	
	Draindown at 160°C	MoRT&H Specification
0.5	0.9	0.6 Maximum
1	0.6	
1.5	0.7	

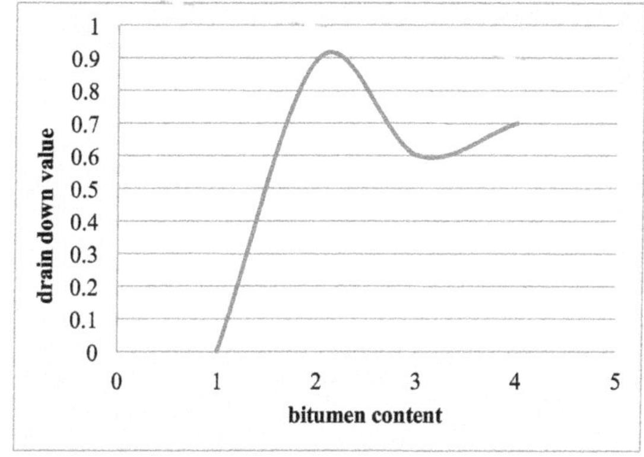

Marshall Stability

It is observed from graphs that with increase in bitumen concentration the Marshall stability value increases up to certain bitumen content and there after it decreases. That particular bitumen content is called as optimum binder content (OBC). In present study OBC for conventional SMA, BC, and DBM mixes are found as 5.5%, 6%, and 6.5% and similarly OBC are found as 7% for modified SMA, BC and DBM mixes with plastic at different concentration. From the graphs it can be observed that with addition of plastic stability value also increases up to certain limits and further addition decreases the stability. This may be due to excess amount of plastic which is not able to mix in asphalt properly. That plastic concentration in mix is called optimum plastic content (OPC) which is found as 1% for SMA and DBM and 1.5% for BC mixes

Table 8 Marshal Stability Test (Figure 2)

Properties	Bitumen content (by weight of aggregate)			
	5.5	6	6.5	7
Without plastic	688	1393.646	987.901	476.30
With plastic	1052	1258	1257	1528

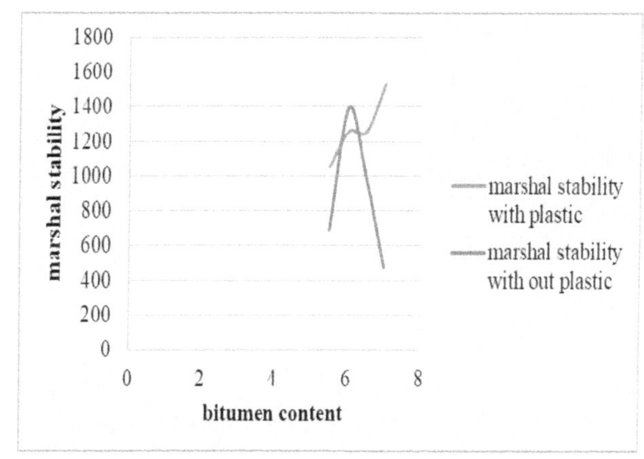

Flow value

It is observed from graphs that with increase in binder content flow value increases but by addition of plastic flow value decreases than that of conventional mixes, again further addition of plastic after OPC the flow value stars to increase.

Table 9 Flow Test (Figure 3)

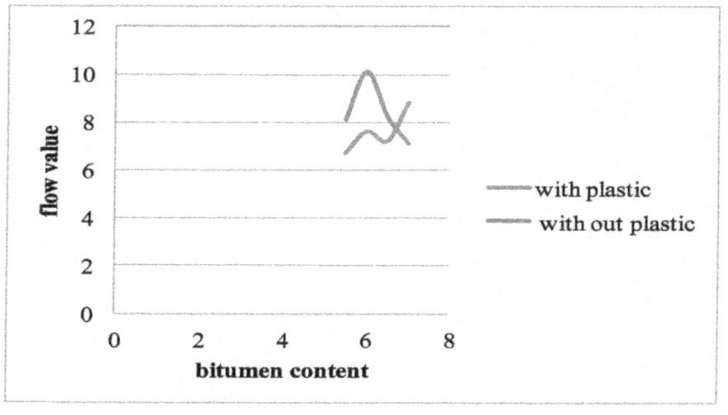

Properties	Bitumen content (by weight of aggregate)			
	5.5	6	6.5	7
Without plastic	8.10	10.10	8.20	7.10
With plastic	6.7	7.6	7.2	8.8

Voids in Total Mix (Vv) Voids in total mix are the volume of small pockets of air between the coated aggregate particles throughout a compacted mix, expressed as a percentage of bulk volume of compacted mix.

Table 10 Bitumen Content (Figure 2)

Properties	Bitumen content (by weight of aggregate)			
	5.5	6	6.5	7
Without plastic	0.745	0.766	1.65	1.187
With plastic	2.75	2.29	2.10	1.808

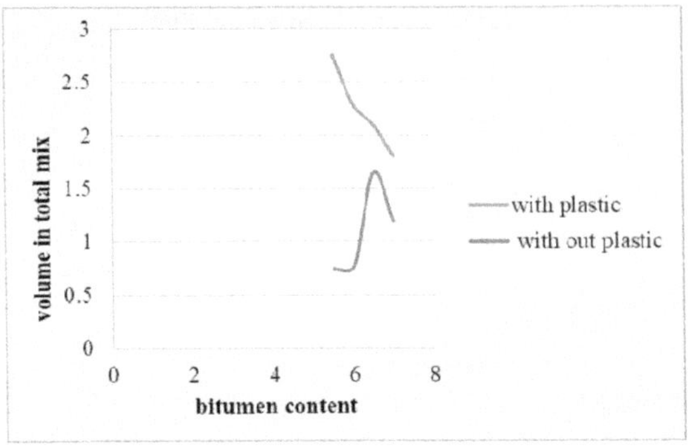

Void in mineral aggregate (VMA)

It is observed that first VMA decreases and then it increases at sharp rate with increase in bitumen concentration in mixes. Variation of VMA values with different binder contents and with different plastic contents are shown in graphs below. From the graphs it is observed that with addition of plastic to mix the VMA values increases than that of conventional mixes.

Table 11 Void in Mineral Aggregate (Figure 3)

Properties	**Bitumen content (by weight of aggregate)**			
	5.5	**6**	**6.5**	**7**
Without plastic	7.41	8.669	17.733	14.739
With lastic	15.01	15.66	16.19	17.303

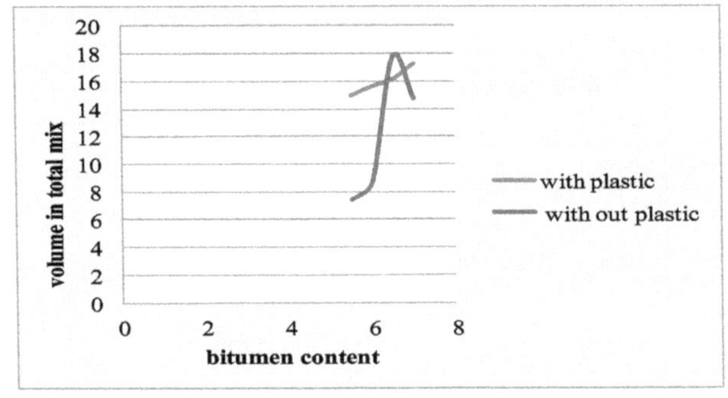

Void filled with bitumen (VFB)

It is observed that VFB values of different mixes increase at sharp rate with increase in bitumen concentration. From these graphs it is observed that with addition of plastic to mixes with fly ash and slag the VFB increases than that of both conventional mixes and mix with fly ash and slag without plastic.

Table 12 Void filled in Bitumen (Figure 4)

Properties	**Bitumen content (by weight of aggregate)**			
	5.5	6	6.5	7
Without plastic	89%	91%	90%	92%
With plastic	81%	85%	87%	89%

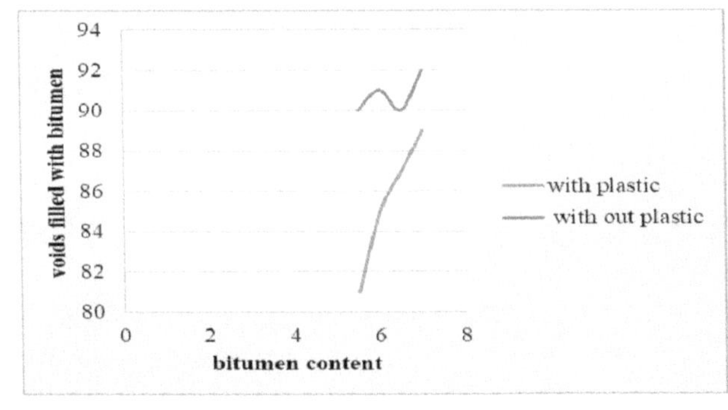

The VCA of the mixture

Table 13 Bitumen Content (Figure 5)

Properties	Bitumen content (by weight of aggregate)			
	5.5	**6**	**6.5**	**7**
Without plastic	46.913	44.69	44.01	40.79
With plastic	34.617	34.64	33.67	35.10

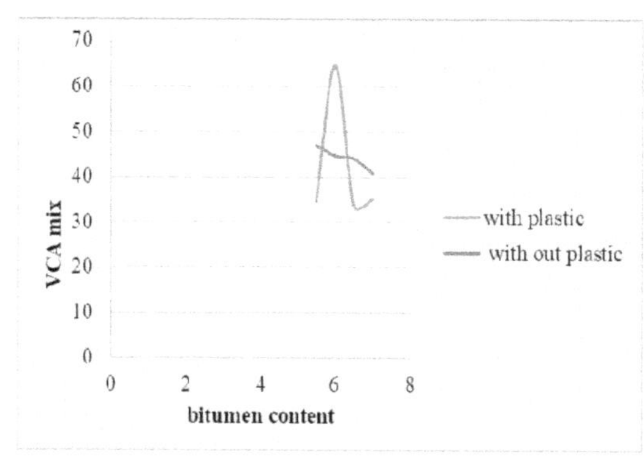

Table 14 Values of testes without plastic

Properties	5.5	6.0	6.5	7.0
Marshal stability	688	1393	987	476
Flow value	8.10	10.10	8.20	7.10
Bulk density	1.865	1.943	1.967	2.08
Volume of voids	2.745	1.766	1.65	1.187
Voids in mineral aggregates	7.41	8.669	17.733	14.789
Voids filled with bitumen	89	91	90	92
Marshal quotient	84.93	137.98	120.47	67.08
VCA MIX	46.43	44.69	44.01	40.29
VCA mix/VCA drc<1	0.46	0.44	0.44	0.40

Table 15 Values of testes with plastic

Properties	5.5	6.0	6.5	7.0
Marshal stability	1052.5	1258	1257	1528
Flow value	6.7	7.6	7.2	8..8
Bulk density	2.297	2.296	2.33	2.28
Volume of voids	2.75	2.29	2.10	1.808
Voids in mineral aggregates	15.01	15.66	16.19	17.303
Voids filled with bitumen	81	85	87	89
Marshal quotient	157.08	165.52	174.5	173.6
VCA MIX	34.617	34.64	33.67	35.10
VCA mix/VCA drc<1	0.34	0.34	0.33	0.35

Conclusions

In this investigation, three sorts of blends for example SMA, DBM and BC are set up with VG30 grade bitumen utilized as a fastener. The impact of expansion of waste plastic in type of locally accessible counterfeit milk with brand OMFED parcels in the bituminous blends has been contemplated by differing fixations plastic from 0% to 2.5% at an addition of 1.0%. Using Marshall Method of blend structure the ideal bitumen content (OBC) and ideal plastic substance (OPC) have been resolved for various sorts of blends. It has been seen that expansion of 2% plastic for SMA and DBM blends and 1.5% plastice for BC blends results in ideal Marshall Properties where stone residue is utilized as filler. Be that as it may, when little division of fine totals are supplanted by granulated impact heater slag and filler is supplanted by fly fiery debris, ideal Marshall Properties for a wide range of blends result with just 1.0% plastic expansion. The OBCs in the event of changed SMA, BC and DBM blends by utilizing stone residue as filler are found 4% and OBCs if there should arise an occurrence of adjusted (I) SMA, and (ii) BC, and DBM by utilizing fly fiery remains and slag are observed to be 5% and 4% individually.

1. Using a similar Marshall examples arranged at their OPCs and OBCs by utilizing both (I) stone residue as filler and (ii) supplanting of stone residue by fly fiery debris and fine total by slag, for test under typical and wet conditions it is seen that the held soundness increments with expansion of plastic in the blends, and BC with plastic outcomes in most elevated held security pursued by DBM with plastic and afterward SMA with plastic. Addition of plastic lessens the channel down impact, however these qualities are not so huge. It might be noticed that the channel down of SMA is marginally more than BC without plastic. In any case, for all blends arranged at their OPC there is no channel down. In general, it is seen that the Indirect Tensile Strength (ITS) esteem diminishes with increment in temperature and for a specific cover, when plastic gets added to the blends the esteem further increments in the two cases. The BC blends with plastic outcome in most noteworthy aberrant elasticity esteems contrasted with SMA, trailed by DBM. It is seen that by expansion of plastic to the blend, the protection from dampness defencelessness of blend additionally increments. BC with plastic outcomes in most astounding elasticity proportion pursued by DBM blends with plastic e and SMA blends with plastic for the two cases. From the above perceptions it is inferred that utilization of waste plastic in type of parcels utilized in milk bundling locally results in improved designing properties of bituminous blends. Henceforth, this examination investigates not just in using most usefully, the waste non-degradable plastics, yet additionally gives an open door in bringing about improved asphalt material in surface courses along these lines making it increasingly sturdy.

Future Scope of the Study

- Many properties of SMA, BC and DBM blends, for example, Marshall Properties, channel down qualities, static elasticity, and static drag attributes have been considered in this examination by utilizing just VG 30 infiltration evaluation bitumen and plastic. Be that as it may, a portion of the properties, for example, weariness properties, protection from rutting, dynamic roundabout elasticity qualities and dynamic killjoy conduct should have been explored.
- In present examination polyethylene is added to them blend in dry blending process. Plastic can likewise be utilized for bitumen alteration by wet blending procedure and correlations made.
- Microstructure of changed bituminous blend ought to be seen by utilizing proper procedure to discover the level of homogeneity.
- Combination of clearing blends shaped with different sorts of plastic squanders which are to a great extent accessible, squanders to supplant regular fine totals and filler a unique kinds of fasteners including altered covers, ought to be attempted to investigate enough extent of finding appropriate materials for clearing blends in case of present requesting circumstances.

References

S. Kumar and S. A. Gaikwad, Municipal Solid Waste Management in Indian Urban Cenetrs: An approach for Betterment", in K. R. Gupta (Ed): Urban Development Debates in the New Millennium, Atlantic Publishers and Distributors, New Delhi (2004) pp. 100-111.

Manual on Municipal Solid Waste Management, Table 3, 6 (2000).

Priya Narayan, Analyzing Plastic Waste Management in India: Case study of Polybags and PET bottles, IIIEE, Lund University, Sweden (2001) pp. 24-25.

CPCB report on Assesment of Plastic Waste and its Management at Airport and Railway Station in Delhi, December (2009) p.8

National Plastic Waste Management Task Force (1997).

Plastics for Environment and Sustainable Development, ICPE, 8(1) (2007).

S.S. Verma, Roads from plastic waste, The Indian Concrete Journal,Nov.2008,43-44

S.Rajasekaran, Dr. R. Vasudevan2, Dr.SamuvelPaulraj, Reuse of Waste Plastics Coated AggregatesBitumen Mix Composite For Road Application – Green Method, American Journal of Engineering Research,2013, Volume-02, Issue-11, pp-01-13 |

Reuse of Waste Plastics Coated Aggregates-Bitumen Mix Composite For Road Application – Green Method S.Rajasekaran1*, Dr. R. Vasudevan2, Dr.Samuvel Paulraj3

Utilization of Plastic Waste in Construction of Roads 1. vatsalpatel 2.snehal popli 3.Drashti Bhatt

Mrs.VidulaSwami,AbhijitJirge,Karanpatil,Suhaspatil,Karansalokhe, Use of waste plastic in construction of bituminous roads,International Journal of Engineering Science and Technology,Vol.4,No.5,2012,2351-235 |

utilization of waste plastic in asphalting of roads amit gawande*, g. s. zamre, v. c. renge, g. r. bharsakale and saurabh tayde reuse of plastic waste in road construction: a review by amol s. bale

Indicative Operational Guidelines on Construction Polymer – Bitumen road" – Central Pollution Control Board, Ministry of Environment and Forests. Probes/101/2005- 2006

Study on Comparison of Properties of Recycled Coarse Aggregate in Concrete with the Conventional Concrete and as A Highway Material

Akula Prakash[1], V. Rajendra[2], M.Venkatesh[3], B. Nikhil[4] and G.Sushanth[5]

[1] Assistant Professor, [2,3,4,5] Students, Department of Civil Engineering,

Gokaraju Rangaraju Institute of Engineering and Technology (Autonomous), Hyderabad

ABSTRACT

This paper for the most part features the examination on functionality and compressive quality of concrete in which reused coarse total is utilized as a halfway substitution of coarse total separately in M45 grade concrete with 1% of super plasticizer is utilized by the weight of the cementitious material .the fluctuating rates received for coarse total are 0%, 20%, 50%, 60%, 80%, 90%, and 100% individually with water to bond proportion of 0.38. The trial thinks about are led on usefulness and compressive quality of new concrete after 7, 14, and 28days relieving period. Solid blends were delivered, tried furthermore, contrasted and the ordinary cement. The examination demonstrated that half substitution of coarse total gives the ideal compressive quality. Additionally the quality accomplished after 28 days restoring period for both customary concrete and cement with reused coarse total had all the earmarks of being practically equivalent .However the investigation demonstrated that the quality of concrete with reused coarse total following 28 days relieving period at 40% substitution of coarse total is considerably 12.6% more than the traditional cement. The properties of total for asphalt development like explicit gravity, compressive test, sway test ,water assimilation test and scraped area test were directed on both characteristic coarse total and reused coarse total and the reused coarse total has more fragile properties than common coarse total.

Keywords: Demolition Waste, Workability, Recycle Aggregate, Compressive Strength, Conventional Concrete.

Introduction

Concrete is a composite material comprise of predominantly water, total, and bond. The physical properties wanted for the completed material can be achieved by adding added substances and fortifications to the solid blend. A strong mass that can be effectively formed into wanted shape can be framed by blending these fixings in specific extents. Over the time, a hard network framed by bond ties the remainder of the fixings together into a solitary hard (unbending) tough material with numerous utilizations, for example, structures, asphalts and so forth., The innovation of utilizing concrete was embraced before on vast scale by the antiquated Romans, and the real piece of solid innovation was exceptionally utilized in the Roman Empire. The coliseum in Rome was constructed generally of cement and the arch of the pantheon is the World's biggest unreinforced solid structure. After the breakdown of Roman Empire in the mid-eighteenth century, the innovation was re-spearheaded as the utilization of cement has turned out to be uncommon. Today, the generally utilized man made material is concrete as far as tonnage.

Benefits

- Keeping concrete debris out of landfills saves landfill space.
- Using recycled concrete can conserve natural resources by reducing the need for gravel mining, water, coal, oil and gas.
- Using recycled concrete as the base material for roadways reduces the pollution involved in trucking material.
- Recycling concrete can create more employment opportunities.
- Recycling concrete drag down the cost for buying raw materials and transporting the waste to landfill sites.
- Recycling one ton of cement could save 1,360 gallons water, 900 kg of CO_2

Advantages of Recycled Aggregate:

- Reduces the amount of virgin aggregates to be created, hence less evacuation of natural resources.
- While being crushed into smaller particles a large amount of carbon dioxide is absorbed. This reduces the amount of CO2 is the atmosphere.
- Cost saving – few research studies have shown a significant reduction in construction costs if RAC is used.
- Conserves landfill space, reduces the need for new landfills and hence saving more costs.
- Creates more employment opportunities is recycling industry.

Disadvantages of Recycled Aggregate:

- Downgrading of quality of concrete.
- Increase in water absorption capacity ranging from 3% to 9%
- Decrease in compressive strength of concrete (10-30%)
- Reduces workability of concrete.
- Lack of specifications and guidelines.
- Less durability of RAC, however few papers have shown an improvement in the durability by mixing it with special materials like fly ash.

Applications of Recycled Aggregate

- Can be used for constructing gutters, pavements etc.
- Large pieces of crushed aggregate can be used for building revetments which in turn is very useful in controlling soil erosion.
- Recycled concrete rubbles can be used as coarse aggregate in concrete.
- Production of RAC also results in generation of many by-products having many uses such as a ground improvement material, a concrete addition, an asphalt filler etc.

Literature Review

As per Asif Husain, and Majid MatouqAssas (2012), In recent years demolished concrete waste handling and management is the new primary challenging issue faced by the countries all over the world. It is very challenging and hectic problem that has to be tackled in an indigenous manner, it is desirable to completely recycle demolished concrete waste in order to protect natural resources and reduce environmental pollution. In this research paper an experimental study is carried out to investigate the feasibility and recycling of demolished waste concrete for new construction. The present investigation to be focused on recycling demolished waste materials in order to reduce construction cost and resolving housing problems faced by the low income communities of the world. The crushed demolished concrete wastes is segregated by sieving to obtain required sizes of aggregate, several tests were conducted to determine the aggregate properties before recycling itinto new concrete. This research shows that the recycled aggregate that are obtained from site make good quality concrete. The compressive strength test results of partial replacement and full recycled aggregate concrete and are found to be higher than the compressive strength of normal concrete with new aggregate.

Prabhatkumar, Abhishek Kumar, Mohd. AfaqueKhan (2012):The recycling of concrete aggregate has been accepted to preserve natural aggregate for other important use. RCA (Recycle concrete aggregate) follow 3R i.e. Reduce, Reuse, Recycle. In many developed country it is used as a substitute of natural aggregate. Also many practical experiments tell that natural aggregate can be replaced by recycle aggregate and can be used for construction purposes. This paper deals with the review of existing literature work for understanding thoroughly about RCA.

MirjanaMalteseVlastimirRadonjanin and SnezanaMarinkovic (2012):A comparative analysis of the experimental results of the properties of fresh and hardened concrete with different replacement ratios of natural with recycled coarse aggregate is presented in the paper. Recycled aggregate was made by crushing the waste concrete of laboratory test cubes and precast concrete columns. Three types of concrete mixtures were tested: concrete made entirely with natural aggregate (NAC) as a control concrete and two types of concrete made with natural fine and recycled coarse aggregate (50% and 100% replacement of coarse recycled aggregate). Ninety-nine specimens were made for the testing of the basic properties of hardened concrete. Load testing of reinforced concrete beams made of the investigated concrete types is also presented in the paper. Regardless of the replacement ratio, recycled aggregate concrete (RAC) had a satisfactory performance, which did not differ significantly from the performance of control concrete in this experimental research. However, for this to be fulfilled, it is necessary to use quality recycled concrete coarse aggregate and to follow the specific rules for design and production of this new concrete type.

Mr.Tushar R Sonawane ,Prof.Dr.sunil , S.Pimplikar (2010) : Use of recycled aggregate in concrete can be useful for environmental protection . Recycled aggregates are the materials for the future. The application of recycled aggregate has been started in a large number of construction projects of many countries. Many countries are giving infrastructural laws relaxation for increasing the use of recycled aggregate. The paper reports the basic properties of recycled fine aggregate and recycled coarse aggregate and also compares these properties with natural aggregates. Basic changes in all aggregate properties are determined and their effects on concreting works are discussed at length. Similarly the properties of recycled aggregate concrete are also determined

Materials and Properties

Table 1 Properties of cement

S.NO	Properties	Test results	IS: 169-1989
1.	Normal consistency	0.32	
2.	Initial setting time	50min	Minimum of 30min
3.	Final setting time	420min	Maximum of 600min
4.	Specific gravity	3.1	

Table 2 Properties of Fine Aggregate

S.No	Description Test	Result
1	Sand zone	Zone- II
2	Specific gravity	2.56

Table 3 Properties of Coarse Aggregate

S.No	Description	Test Results
1	Nominal size used for concrete	20mm
2	Specific gravity	2.7
3	Sieve analysis	20mm

Table 4 Properties of recycled coarse aggregate

S.No	Description	Test Results
1	Specific gravity of RCA	2.35

Concrete Mix Design (AS PER: 10262-2009)

1. Material properties were tested as per IS codes.
2. Mix design for concrete proportion has been developed as per IS 10262-1982.
3. Concrete cube specimens were casted .Cured as per IS procedure.
4. The properties of fresh concrete was tested as per IS 1199-1959.
5. The characteristic strength of hardened concrete specimen was tested as per
6. IS 456 – 2000.
7. Optimum strength for optimum replacement was determined.
8. Results of conventional concrete were compared with that of partial replacement concrete.

Materials Used

Cement	Fineness Consistency Specific gravity Initial setting time Final setting time	=8% =32% =3.1 =30min =600min
Recycled coarse aggregate	Specific gravity	=2.35
Fine aggregate	Specific gravity Fineness modulus	=2.56 =2.16
Coarse aggregate	Specific gravity	=2.7

Mix Design (As per IS10262:2009)

Design Stipulations:

1.	Grade of concrete (designation)	=	M45
2.	Type of cement	=	OPC 53 grade
3.	Minimum nominal size of aggregate	=	20 mm
4.	Workability	=	100 mm (slump)
5.	Exposure condition	=	Severe
6.	Method of concrete placing	=	Normal
7.	Degree of supervision	=	Good
8.	Specific gravity of cement	=	3.10
9.	Specific gravity of coarse aggregate	=	2.70
10.	Specific gravity of fine aggregate	=	2.56
11.	Specific gravity of recycled coarse aggregate	=	2.35
12.	Sieve analysis	=	Zone 2 (IS 383 – 1970)

Design:

At 0% replacement of coarse aggregate with recycled coarse aggregate

Step 1: *Target mean strength*

Fck	$= f_{ck} + 1.65s$	From Table 1 of IS 10262 - 2009
	= 45 + 1.65 x 5	standard deviation, s = 5
Fck	= 53.25N/mm^2	

Step 2: *Selection of water cement ratio*

From Table 5 of IS 456 – 2000

For M45 concrete, maximum w/c ratio = 0.5

Based on experience adopt water cement ratio = 0 .38

Step 3: *Selection of water content*(From Table 2, of IS 10262 – 2009)

Maximum water content for 20 mm aggregate = 186 kg / m^3

(For 25 to 50 mmslump) Increase 3% of water content for every 25 mm slump range.

To attain max of 100mm slump range = 6% increase in water content Estimated water content for 100mm slump = 186 + (6/100) (186)= 197.16 kg / m^3

Step 4: *Calculation of cement content*

Based on super plasticizer reduce W/C upto 20% and above

25% is reduced	=	197 x 0.75 = 147.87
Water content	=	0.38
Cement content	=	147.87 / 0.38= 389.13 Kg / m^3

From Table 5, of IS 456 – 2000

Minimum Cement Content for 'severe' exposure condition = 320 kg / m^3 Above calculate cement content value is > 320 kg / m^3. HENCE OK

Step 5: *Volume of coarse aggregate & fine aggregate content*

From table 3 of IS 10262 – 2009, Volume of coarse aggregate corresponding to 20 mm size aggregate and fine aggregate (Zone II).

For w/c ratio of 0.5 the volume of coarse aggregate per unit volume of total aggregate is 0.48. Therefore, in the present case w/c ratio is 0.5, the corrected proportion of volume of coarse aggregate is 0.47(w/c adjustment) .As thewater cement ratio is increased by 0.05 , the proportion of volume of coarse aggregate is decreased by 0.01(at the rate of -/+ 0.01 for every ± 0.05 change in water cement ratio).

Volume of coarse aggregate is 0.646

For pump able concrete these value should be reduced by 10%

Volume of coarse aggregate = 0.646x0.9=0.5814

Volume of fine aggregate content = 1 – 0.5814 = 0.4186

Step 6: *Mix calculations*

Mix calculations percent volume of concrete shall as follows,

(a) Volume of concrete = 1m^3

(b) Volume of Cement = (mass of cement/sp.gravityorf water)(1/1000)

For water to cement ratio 0.38 = (389.13/3.1)(1/1000)=0.1255m^3

(c) volume of water = (mass of water/sp.gravity of water)(1/1000)

= (147.87/1)(1/1000) =0.1478m^3

(d) Volume of all in aggregate = (a – (b + c))= 1 – (0.1255+ 0.1478)= 0.6914 m^3

(e) Mass of coarse aggregate = (d x volume of coarse aggregate xsp.gravity of coarse aggregate x 1000)

= 0.6914x 0.5814 x 2.7 x 1000 = 1085.34 kg

(f) Mass of fine aggregate = (d x volume of fine aggregate x sp. gravity of fine aggregate x 1000)

= 0.6914x0.4186x2.56x1000

= 740.915 kg

Table 5 Design mix proportions of natural coarse aggregate

Water Cement Ratio (kg/m^3)	Cement (kg/m^3)	Fine Aggregate (kg/m^3)	Coarse Aggregate (kg/m^3)
147.86	389.13	740.915	1085.34
0.38	1	1.9	2.78

Similarly calculated for 20%, 40%, 50%, 60%, 90%and 100% replacements.

Table 6 Mix Proportion for the different Mixes

Mix	Replacement (%) (CA with RCA)	Mix proportion (C:FA:CA) (CA = CA+RCA)	Cement Content (kg/m^3)	W/C Ratio
M0	0	1 : 1.9 : 2.78	394	0.38
M1	20	1 : 1.9 : 2.76	394	0.38
M2	40	1 : 1.9 : 2.73	394	0.38
M3	50	1 : 1.9 : 2.71	394	0.38
M4	60	1 : 1.9 : 2.7	394	0.38
M5	80	1 : 1.9 : 2.64	394	0.38
M6	90	1 : 1.9 : 2.62	394	0.38
M7	100	1 : 1.9 : 2.6	394	0.38

Test Results

Table 7 Workability of RAC in mm (Figure 1)

% Replacement of RCA	Slump in mm
0	98
20	97
40	97
50	95
60	80
80	75
90	74
100	65

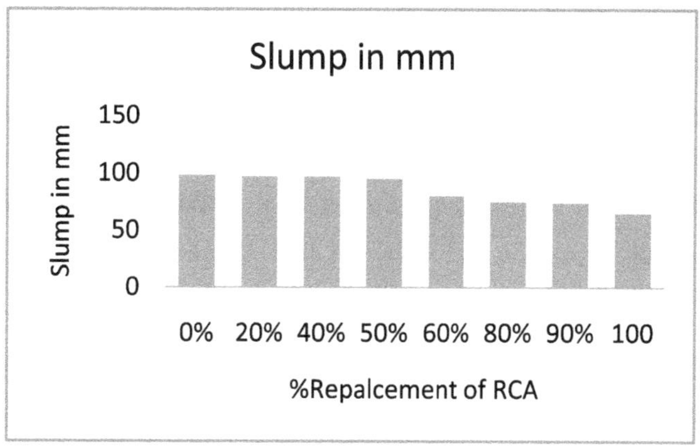

The above figure shows the workability of concrete with varying percentages of recycled coarse aggregate respectively .it can be inferred that workability of concrete is initially 98 mm for conventional concrete. However workability decreased with increase in percentage replacement of recycled coarse aggregate with natural coarse aggregate this can be attributed to the fact that recycled coarse aggregate contains cement particles which increases the water absorption ratio of concrete which consumes more water. Hence the water utility in the concrete increases workability decreases. However the workability decreased after 50% replacement of recycled coarse aggregate with natural coarse aggregate. As the recycled coarse aggregate has more water absorption ratio it consumes more water there by decrease in workability is found

Compressive strength: Figure shows the variation of compressive strength of concrete with varying percentages of recycled coarse aggregate with natural coarse aggregate respectively. The compressive strength of concrete cubes are tested at 7days,14days and 28days of curing period

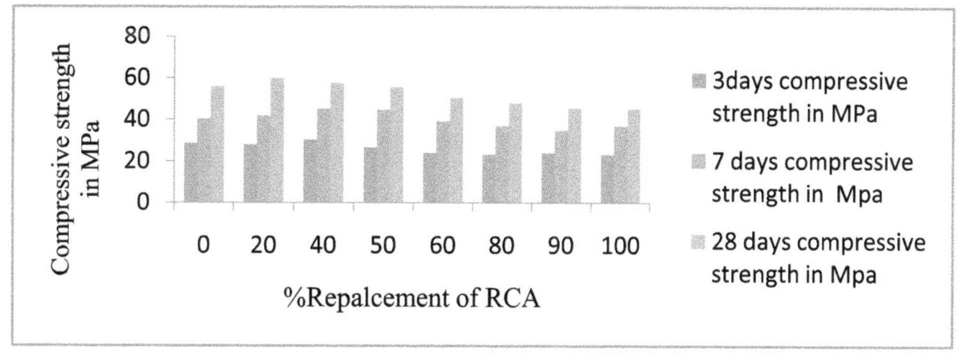

Figure 2 Plot of Percentage of RCA v/s Compressive strength

Table 8 Compressive strength in MPa

%RCA in CA	7 days compressive Strength in MPa	14 days compressive strength in MPa	28 days compressive strength in MPa
0	28.67	40.33	56
20	28	42	57.67
40	30.33	45.33	60
50	26.67	44.67	55.67
60	24	39.33	50.33
80	23.33	37	48
90	24	35	45.67
100	23.33	37	45.33

From the Figure2 It can be observed that compressive strength increased with increase in percentage of replacement of natural coarse aggregate with recycled coarse aggregate. The gain in strength is observed up to 50% partial replacement of recycled coarse aggregate with natural coarse aggregate. it can also inferred that the maximum strength achieved at 40% partial replacement of natural coarse aggregate with recycled coarse aggregate the compressive strength of non-conventional concrete i.e concrete with recycled coarse aggregate found to be higher than the conventional concrete at the end of 28 days curing period upto 40% replacement From this study it can be inferred that recycled coarse aggregate can be used uptoa optimum percentage of 50% as a partial replacement of coarse aggregate.

Road material test results: Figure shows the variation of Specific gravity, Impact Value, Crushing Value, Water

Absorption, Abrasion Value of recycled coarse aggregate with natural coarse aggregate

Respectively

Table 9 Coarse Aggregate Properties

S NO	TEST	NCA	RCA
1	Specific gravity	2.7	2.35
2	Impact value	43%	38.73%
3	Crushing value	47	40
4	Water absorption	0.5%	1.35%
5	Abrasion value	30	45

The specific gravity of aggregates normally used in road construction ranges from about 2.5 to 3.0 with an average of about 2.68. Though high specific gravity is considered as an indication of high strength.From the above experiment, it is found that the specific gravity of RCA is smaller than that of normal aggregate. Hence, RCA can be said to have less density than normal aggregate and hence RCA is lighter.

From the above results, it is found that RCA contains more water than that of conventional aggregates because RCA has a higher amount of adhered mortar and thus absorbs more water than normal aggregates due to larger pore sizes and hence, there is a need to encounter for water absorption. Due to this, RCA will absorb the water during mixing of concrete and this will lead to a bad mixture as there will be a lack of water and thus there will be a need to add more and more water. recycled coarse aggregate has less impact value and abrasion value when compared with natural coarse aggregate because of recycled coarse aggregate has high water absorption and contains adhered mortar particles

Conclusions

Based on the above study, the following conclusions were drawn

The present study indicates that workability of the concrete decreased with increase in percentage of recycled coarse aggregate. Recycled coarse aggregate has rough texture and adhered cement mortar increases the bond strength in concrete and it increases the compressive strength of concrete, with the replacement of recycled coarse aggregate in traditional concrete. The compressive strength of concrete after 28 days curing period is equals to the 28 days strength of non-conventional concrete at 50% replacement. From the results it was obtained that 50% replacement is the optimum percentage of replacement. Recycled coarse aggregate has weaker properties than natural coarse aggregate and RCA is not used for road construction as their properties are not within standards. Workability of concrete decreases by percentage increase in recycled coarse aggregate .It can be compensated by fly ash .Early strength of concrete decreases by fly ash and it can be compensated by quarry dust

References

Asif Husain , and Majid MatouqAssas (2012) Building Materials from Waste. Materials Transactions 44: 1255-1258.

MirjanaMalesevVlastimirRadonjanin and SnezanaMarinkovic (2012)Recycling and Reuse of Building Materials. National Pollution Prevention Center for Higher Education.

MirjanaMalesevVlastimirRadonjanin and SnezanaMarinkovic (2012)Utilization of solid Wastes in Construction Materials. Int J Physics 5: 1952-1963

Mr. Tushar R sonawane ,prof.Dr.sunil , S.pimplikar (2010)Experimental Study of Partial Replacement of Fine Aggregate with Waste Material from China Clay Industries. Int J ComputEng Res 2.

Prabhatkumar, Abhishek Kumar, Mohd. AfaqueKhan(2012)Application of construction and demolition waste. IJITEE 46

Riddhish A. Shah ,JayeshbhaiR.Pitroda (2009) Recycling of construction and demolition wastes. The master builder

Text book of "Concrete Technology" M.S.SHETTY , S.CHAND Publications.

IS 456-2000 Indian Standard code book for RCC design guidelines.

IS 10262(2009): Indian Standard code book for Concrete mix design proportioning

Reducing Construction and Demolition Waste through Reuse and Recycling (Replacement of Fine Aggregate by Demolition Waste Concrete)

Akula Prakash[1], A.Varsha[2], Ch. Shailaja[3], A. Shivaleela[4] and Ch. Swapna[5]

[1]Assistant Professor and [2,3,4,5] Students

Department of Civil Engineering,

Gokaraju Rangaraju Institute of Engineering and Technology (Autonomous), Hyderabad.

akulprakash93@gmail.com

ABSTRACT

Presently a-days the best emergency looked by the construction industry is the accessibility of sand. As the burrowing of waterway sand decimates the stream bed and causes threat for individuals utilizing the stream, burrowing of waterway sand has been made illicit in many streams. So getting stream sand is extremely costly these days as its accessibility is restricted. So more significance is presently given these days for substitution of waterways and as fine total. In this task separated of examining the functionality supplant sand with pounded utilized (pulverized) concrete. The solid made with this total demonstrated nearly. The old cement is squashed and is sieved for impeccable size and utilized. Solid shape throwing with various extents of bond, coarse total and fine total and compressive quality of cement following 7, 14 and 28 days were tried with the structure blend M20 grade concrete has been worked out according to IS 10262-2009. Fine total supplanted by 0%, 18%, 20%, 22%, 40%, and 100% with the wrecked cement by the heaviness of bond. So by utilizing C&D squander concrete as a substitution material the expense of the structure is diminished and the quality can be obtained. The reusing proportion of C&D concrete in the year 2000 was 70%. This 70% of waste is for the most part utilized for landfills and as a subgrade in street development. A similar quality of cement with characteristic sand. This isn't just a lot less expensive than stream sand yet additionally diminishes the transfer of development squanders, which naturalist's state debases the land. These outcomes give a reasonable end that pulverized solid waste can be utilized as a decent substitute for fine total with higher usefulness and qualities at 18% supplanting with ordinary cement.

Keywords: Demolition Waste, Workability, Recycle Aggregate, Compressive Strength, Conventional Concrete.

Introduction

As the time proceeds onward, new creations are appearing. In the comparative manner the world is attempting up to improve the properties of the solid however much as could be expected with low conceivable expense. Concrete is a collection of bond, total and water. The blend of the materials results in a substance response called hydration and an adjustment in the blend from plastic to a strong state happens over some stretch of time. The expense of cement can be diminished by decreasing expense of constituent materials. Cost decrease can likewise be accomplished by utilizing locally accessible elective material, rather than traditional materials. Cement can be casted in any shape. Since it is a plastic material in new state, different shapes and sizes of structures or formworks are utilized to give various shapes, for example, rectangular, roundabout and so on. Different basic individuals, for example, bars, chunks, footings, sections, lintels and so forth are built with cement.

A few Research Works have been done in the past to examine the likelihood of using development destruction squander materials. Different development devastation squander materials like solid, glass, plastics, blocks, steel, clean product, timber. As the reused fine total from the pulverized solid waste are utilized as incomplete substitution of fine total and evaluated the quality parameters and contrasted the benefit rates after substitution and waste materials. The overall utilization of fine total in solid creation is high, and a few creating nations have experienced challenges in gathering the supply of regular fine total so as to fulfil the expanding needs of infrastructural improvement as of late. To conquer the pressure and interest for waterway fine total, analysts and experts in the development businesses have recognized some elective materials, for example, reused fine total, fly fiery debris, slag, limestone powder and siliceous stone powder. In India endeavours have been made to supplant stream sand with reused fine total.

Advantages of Recycled Demolition Concrete Waste

- Reduces the measure of virgin totals to be made, henceforth less clearing of common assets.

- While being smashed into littler particles a lot of carbon dioxide is assimilated. This decreases the measure of CO2 the climate.
- Cost sparing – few research ponders have appeared noteworthy decrease in development costs if annihilation solid waste is utilized.
- Conserves landfill space, diminishes the requirement for new landfills and subsequently sparing more expenses.
- Creates greater work openings in reusing industry.

Disadvantages of Recycled Demolition Concrete Waste

- Downgrading of nature of cement.
- Increase in waster absorption capacity.
- Decrease in compressive strength of cement.
- Reduces workability of concrete.
- Lack of specifications and guidelines.
- Less strength of reused fine total, anyway few papers have demonstrated an improvement in the sturdiness by blending it with extraordinary materials like flyash, quarry dust.

Objectives of the Present Study

(a) To determine the optimum value of replacement of demolition concrete waste in conventional concrete in place of fine aggregate in order to get the estimated workability.
(b) To determine the compressive strength of concrete at optimum replacement of demolition concrete waste in conventional concrete.
(c) Comparative study of workability and compressive strength of conventional concrete with concrete replaced with demolition concrete waste.

Literature Review

As per Mrs. Roopa G Sindigi, Mr. Devaraja, Mr.Paramesh G(2018), Most of the solid waste production is used as landfill. Out of which Construction & Demolition (C&D) waste constitutes a major portion. In present study, utilization of construction and demolition waste in M20 grade concrete is investigated. The fine aggregate in the concrete is replaced with 25%, 50% and 100% recycled aggregate (crushed concrete). The fresh properties like the slump change for these mixes are analyzed. To check the change in compressive strength, split tensile strength and flexure strength, the specimens were tested at 3, 7 and 28 days. Experimental program, the strength variation of M20 grade concrete when replaced with recycled aggregates in various proportions is documented. The rate of gain of 7days and 28 days compressive strength had increased. The split tensile strength shows alternate increase and decrease in early strength for various percentages of replacement, whereas the 7days strength 50%and 100% replacement of fine aggregate with crushed concrete has shown increment in strength than that of control concrete. The flexural strength shows considerable increment in the strength for 1day and 7days but, for 28days a decrease in strength is observed for 25% replacement concert mix. And concluded that observed possibility to replace fine aggregates in concrete to percentage up to 50% to 100% for lowergrade concrete.

J. Vengadesh Marshall Raman and M. Sriram (2017) were investigated Study on Replacement Level of Concrete Waste as Fine Aggregate in Concrete. In this experimental study to replace the fine aggregate by recycled aggregate in different ratio such as 0%, 10%, 20%, 30%, 40% and 50%. Recycled aggregate concrete is found to be 42% greater water absorption than that of natural aggregate. Beyond 30% replacement levels of concrete containing recycled fine aggregate shows 20-40% lower compressive strength is developed at the ages of 7, 28 and 56 days. Both tensile splitting and flexural strength are slightly decreased with the increase of the replacement ratio. The reduction in strength is 15% and 20% when compared to the reference concrete.

K. Radhika and A. Bramini (2017) were conducted an experimental study on the feasibility and suitability of the use of demolition concrete waste also research made on the partial replacement with M_{20} grade concrete the construction and demolition waste on the different samples like M_1, M2, M3, M4, and M5 and the ratios of natural fine aggregate was replaced with the recycled fine aggregate are 100:0%, As Control 80:20%, 50:50%, 20:80% and 0:100% are made and cubes of partially replaced concrete of size 150mm*150mm were casted, cured and then tested for the compressive strength. In this study the properties of concrete, consisting of fine aggregate partially replaced with reucrete fine aggregate, like water absorption, specific gravity, workability etc were determined. The study concluded that the replacement of fine aggregate with demolition concrete waste led to an increase of compressive strength of concrete and were the recycled aggregate the percentage of fines present is more than that of then natural fine aggregate and recycled fine aggregate does not affect the fresh properties of concrete and finally concluded that the recycled fine aggregate has the similar properties to that of natural fine aggregate and compressive strength and tensile strength of concrete increases in the replacement of fine aggregate only up to some percentage.

Mrunalini Deshmukh (2017)were carried out an experimental study on the replacement of sand with demolition concrete waste in the conventional concrete. In the tenure of the studies, the experimentally determined the properties of sand and recruete sand. They also experimentally determined the compressive strength of normal concrete and compared it with the results of compressive strength of concrete. It was concluded that the partial replacement of sand with recruete sand in the 20% of recruete sand and 80% of natural sand for M_{25} grade of concrete was the most effective and showed an increase in the compressive strength. It was also concluded that the concrete cubes made out of combination of crushed sand and recruete sand and also the mechanical properties of recruete sand depend on the source of its raw material hence selection of a quality of demolition waste is very important for obtaining quality fine aggregate and also the environmental impact can be reduced through design. It was finally concluded that the recruete sand can be effectively use 20% replacement of sand in conventional concrete.

Materials and Methodology

Table 1 Normal Consistency of Cement

% of water added	Initial reading	Final reading	Depth of penetration(mm)
26	42	27	15
28	42	24	18
30	42	15	27
32	42	11	31
34	42	8	34

Table 2 Initial Setting Time of Cement

Time taken	Initial reading	Final reading	Depth of penetration (mm)
5	42	34	8
10	42	32	10
15	42	29	13
20	42	24	18
25	42	21	21
30	42	18	24
35	42	17	25
40	42	13	29
45	42	12	30
50	42	9	33

Table 3 Properties of Cement

Sl.No	Test	Result	Requirement as per IS code	IS code number
1	Normal consistency	34%	26-33%	IS 4031 (part-4)
2	Specific gravity	3.15	3-3.2	IS 2720 (part-3)
3	Fineness	6.5%	10%	IS 4031 (part-2)
4	Initial setting time	50 minutes	30-60 minutes	IS 4031 (part-5)
5	Final setting time	1 mm (600 minutes)	600 minutes	IS 4031 (part-5)
6	Compressive strength	25.8N/mm^2(3days) 38N/mm^2(7 days) 54N/mm^2(28 days)	27.8 N/mm^2 37.8N/mm^2 53.8 N/mm^2	IS 4031 (part-7)

Table 4 Results of Fineness modulus of Fine Aggregate

Sieve size in mm	Weight of sand retained (gm)	Cumulative weight of sand retained (gm)	Cumulative % of sand retained
10	0	0	-
4.75	10	10	2
2.36	50	60	12
1.18	70	130	26
0.60	90	220	44
0.30	160	380	76
0.15	80	460	92
Pan	40	500	100
Total	500	1760	352

Fineness modulus = (352/100)

= 3.52

Locally available river sand confirming to **zone II** was used for the project work.

Table 5 Results of Physical Properties of Fine Aggregates

Sl. No	Test	Result	Is code
1	Specific gravity	2.7	IS 2386 (Part-3) -1963
2	Water absorption	1%	IS: 383–1970
3	Fineness modulus	3.52	IS 2386 (Part-1) -1963

Table 6 Results of Fineness Modulus of Coarse Aggregates

Sieve size in mm	Weight of aggregate retained (gm)	Cumulative weight retained (gm)	Cumulative % of weight retained
40	250	250	25
20	450	700	70
10	200	900	90
6.36	75	975	97.5
4.75	25	1000	100
TOTAL	1000		382.5

Fineness modulus = (382.5/100)

= 3.825

Table 7 Results of Physical properties of Coarse Aggregates

S. no	Test	Results	IS code
1	Specific gravity test	2.8(for 20mm aggregates)	IS 2386(part 3)
2	Fineness modulus (%)	3.825	IS 2386 (Part I) –1963
3	Maximum size	4.75m	

Table 8 Properties of Recycled Fine aggregates

S. No	Particulars	Results
1	Type	Building waste
2	Specific gravity	2.45
3	Water absorption	4%
4	Fineness modulus	3.06
5	Grading	Zone II
6	Bulk Density(kg/m^3)	1850

Mix Design Procedure

Characteristic compressive strength (That is, below which only a specified proportion of test results are allowed to fall) of concrete at 28 days f_{ck} (Grade designation)

1. Type of cement (53 grade)
2. Maximum nominal size of aggregate(20mm)
3. Minimum cement content- 320 kg/m^3
4. Limitations on the W/C ratio and the min. cement content to ensure (0.45 as per IS 456:2000)
5. Standard deviation of compressive strength of concrete (5 N/mm^2 as per IS:10262-2009)
6. Workability (slump of 50-100mm)
7. Degree of quality control
8. Exposure conditions (severe as per IS 456:2000)
9. Specific gravity of cement, fine and coarse aggregates (3.15,2.7,2.8)

Procedure

1. TARGET STRENGTH FOR MIX PROPORTIONING
   - Determine the mean target strength ft from the specified characteristic compressive strength at 28-day f_{ck} and the level of quality control. $f_t = f_{ck} + 1.65S$ Where S is the standard deviation obtained from the Table of approximate contents given after the design mix.
   - $f_t = f_{ck} + 1.65$ S , where, S = 5 N/mm^2(from table1 of IS 10262-2009)
     $f_t = 20+1.62(5) = 28.1$ N/mm^2
2. SELECTION OF WATER-CEMENT RATIO
   - The free water cement ratio required for the target mean strength is 0.50 (From the Table 5 of IS 456-2000).
3. SELECTION OF WATER CONTENT
   - For 20mm aggregates the required water content is 186 litres (25 to 50 mm slump for every increase of 25mm slump 3% more is added From the Table-2 of IS 10262-2009).
   - Hence total weight = 186+ 6% (186) =197.16litres.
4. CALCULATION OF CEMENT CONTENT
   - From the steps 2 & 3, cement content is calculated.
   - Water content = 197.16 × 0.80 = 157.728
   - Cement content = 157.728/0.4 =394.32 kg/cm^2> 320kg/cm^2
5. PROPORTION OF VOLUME OF COARSE AGGREGATE AND FINE AGGREGATE
   - From Table-3 of IS 10262-2009, the volume of coarse aggregate per unit volume corresponding to size of fine aggregate (zone 2) for water cement ratio 0.5 is 0.62. For every +/- 0.05 change in w/c ratio there is -/+ 0.01 change in volume of coarse aggregate per unit volume corresponding to fine aggregate.
   - Therefore 0.40 = 0.64.
   - Volume of fine aggregate = 0.36.

Mix Design Calculations (As Per Is 10262:2009)

a) Volume of the concrete = 1 m^3

b) Volume of cement = (mass of the cement/specific gravity of the cement) x (1/1000)

$$cement = \frac{wt.\,of\ cement}{density} = \frac{320}{3150} = 0.101m^3$$

c) Volume of water = (mass of water/specific gravity of water) x (1/1000)

$$\text{Water} = \frac{165}{1000} = 0.165\text{m}^3$$

d) Volume of Air = $\frac{2}{100}$ = 0.02

Total Volume = [a - (b + c+ d)] = [1- (0.101+0.165+0.02)] = 0.714 m^3

e) Mass of coarse aggregate = Total Volume × Volume of coarse aggregate × Specific gravity of coarse aggregate × 1000

For 20mm = 0.714 × 2.65 × 36 % × 1000 = 681.156kgs

10 mm = 0.714 × 2.61 × 24% × 1000 = 447.24kgs

Total = 1128.39kgs

f) Mass of recycled fine aggregate = Total Volume × Volume of Recycled fine Aggregate × Specific gravity of Recycled Fine aggregate × 1000

For R.F.A = 0.714 × 2.45 × 40% × 1000 = 728.28kgs

For $\mathbf{M_{20}}$: Cement - 320kgs

Coarse aggregate - 1128.39kgs

Recycled fine aggregate - 728.28kgs

Water - 182ml

Table 9 Mix proportion

Cement	Fine aggregate	Coarse aggregate	Water
320	728.28	1128.39	182
1	2.27	3.52	0.5

Mould volume - 0.15 × 0.15 × 0.15 mm = 0.003375m^3

For 6 cubes – 0.003375 x 6 = 0.020m^3

Required quantities:

Cement – 320 × 0.020 = 6.4 kg/m^3

Coarse aggregate (20mm)-682 × 0.020 = 13.64kg/m^3

(10mm) – 448 × 0.020 = 8.96kg/m^3

Recycled fine aggregate- 729 × 0.020 = 14.58kg/m^3

Water – 182 × 0.020 = 3.3kg/m^3

Table 10 Quantity of materials for different % of Recycled fine aggregate

Sl.no	Cement (kg/m3)	Recycled fine aggregate (%)	Recycled fine aggregate in weight (kg)	Fine aggregate (kg/m3)	Coarse aggregate (kg/m3)	Water (kg/m3)	w/c
1	320	0	0	728.28	1128.39	182	0.5
2	320	18	131.904	594.18	1128.39	182	0.5
3	320	20	145.656	582.624	1128.39	182	0.5
4	320	22	160.221	568.059	1128.39	182	0.5
5	320	40	291.312	436.968	1128.39	182	0.5
6	320	100	728.28	0	1128.39	182	0.5

Results and Discussions

The results obtained from the tests carried out on 7, 14 and 28 days of curing were compared with the conventional concrete to determine the compressive strength. Based on the results, the strength increases up to 18% of replacement of fine aggregate with recycled fine aggregate. There was slight decrease in workability and strength when the replacement exceeds 18%.

Workability Test

The slump cone test is conducted for the reference concrete and also with the addition of the recycled fine aggregate.

Table 11 Results on Workability test results by Slump Cone test

% of replacement of recycled fine aggregate	Slump observed (mm)
0	90
18	80
20	65
22	62
40	50
100	45

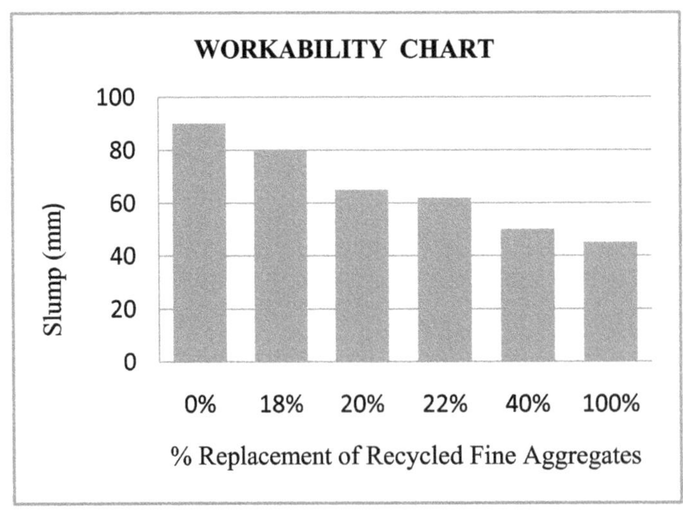

Figure 1

Compressive Strength Test Results

The casted cubes of M20 grade concrete are cured, exposed to temperatures and tested for its Compressive strength at 7, 14 and 28 days after cooling to room temperature.

Table 12 Results of Compressive strength

% of replacement	7days strength (N/mm^2)	14 days strength (N/mm^2)	28 days strength (N/mm^2)
0	13.52	18.67	33.80
18	12.21	14.42	29.40
20	11.65	14.03	25.39
22	10.75	13.58	19.84
40	9.90	12.46	16.42
100	8.86	10.36	15.42

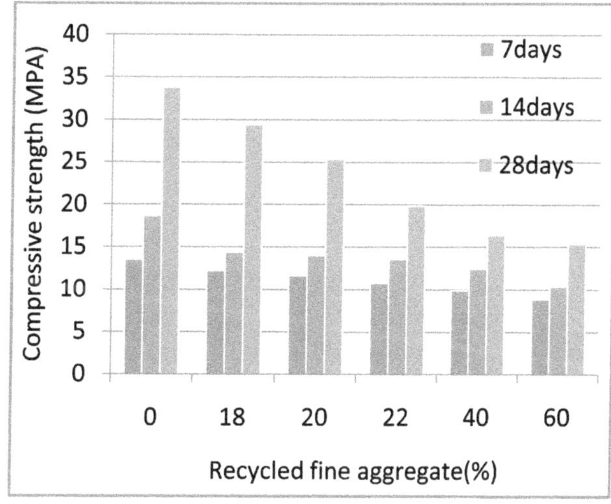

Figure 2

The compressive strength of various concrete mixtures was schematically represented in Fig 2 From the observed results it is found that the compressive strength up to 18% of fine aggregate replaced by recycled aggregate give strength closer to the strength of reference concrete and strength reduction of 25.39% for recycled concrete mix .This reduction is due to insufficient hydration and a weak interface-zone formed between different components of the concrete matrix owing to a large amount of old cement paste on the surface of recycled aggregates, which can be the cause of a poor development of the compressive strength of concrete. In addition, an inconsistent surface of recycled fine aggregate would produce numerous microcracks between aggregates and cement paste, which would reduce concrete compressive strength.

Conclusion

On the basis of results obtained, following conclusions can be drawn:

Based on experimental investigation, it is found Recycled fine aggregate concrete may be an alternative to the conventional concrete. Water required producing the same workability increases with the increase in the percentage of demolished waste. Optimum replacement level of fine aggregate with recycled aggregate is 18%. The RFA does affect the fresh properties of the concrete. The physical properties of Recycled fine aggregate satisfy the requirements of the fine aggregate. Usage of Recycled fine aggregate concrete reduces the cost of concrete because it is a waste material and it also reduces the problems of disposal and prove to be environment friendly thus paving way for greener concrete. The compressive strength is increased upto 18 percentage of the replacement. The ideal percentage of Recycled fine aggregate as partial replacement of sand is 18%. It is found that there is enough workability with the w/c ratio provided at 18% replacement. If the replacement of these and with Recycled fine aggregate increases then the workability of the concrete decreases due to absorption of the water by Recycled fine aggregate. The replacement of the sand with Recycled fine aggregate shows an improved compressive strength of the concrete up to 18% replacement and if it is increased beyond 18% compressive strength gets reduced due to the increased fines and changes in the particle gradation. The properties of the recycled fine aggregate are very similar to the natural aggregates. Workability of concrete mix decreases with increase in Recycled fine aggregate content.

References

Mrs. Roopa G Sindigi, Mr. Devaraja, Mr.Paramesh G "Study On Concrete Replaced With Crushed Concrete Fine Aggregate"*International Research Journal of Engineering and Technology (IRJET) e-ISSN: 2395-0056 Volume: 05 Issue: 03 | Mar-2018 p-ISSN: 2395-0072 © 2018, IRJET Page 1212*

J. Vengadesh Marshall Raman and M. Sriram "Study on Replacement Level of Concrete Waste as Fine Aggregate in Concrete" *Volume 5 Issue II, February 2017 ISSN: 2321-9653 IC Value: 45.98*

K. Radhika and A. Bramini "Construction and demolition waste as a replacement of fine aggregate in concrete" *Volume 6, Issue 6, June 2017, ISSN: 2278 -7798.*

Mrunalini Deshmukh REUCRETE: "Replacement of Fine Aggregate by Demolished Waste Concrete" *International Conference On Emanations in Modern Technology and Engineering (ICEMTE-2017) ISSN: 2321-8169 Volume: 5 Issue: 3 46 - 52*

Lewis Evangelista & J. de Brito "Concrete with fine recycled aggregates: a review" *European Journal of Environmental and Civil Engineering, 2014 Vol. 18, No. 2, 129–172*

Asif Husain, and Majid Matouq Assas"Utilization of Demolished Concrete Waste for New Construction" *International Science Index, Civil and Environmental Engineering Vol:7, No:1, 2013*

Md. Safiuddin "use of recycled concrete aggregate in concrete: a review" *journal of civil engineering and management 19(6):796-810 · December 2013*

Mohd. Monish, Vikas Srivastava, V.C.Agarwal and Rakesh kumar "Utilization of demolished waste as fine aggregate in Concrete" *J. Acad. Indus. Res. Vol. 1(7) December 2012*

J.M. Khatib "Properties of concrete incorporating fine recycled aggregate "*Centre for Infrastructure Management, MERI, Sheffield Hallam University, Howard Street, Sheffield S1 1WB, UK* Received 9 July 2003; accepted 9 June 2004

Sim and Park "Compressive strength and resistance to chloride ion penetration and carbonation of recycled aggregate concrete with varying amount of fly ash and fine recycled aggregate" *in Waste Management 31(11):2352-60 · July 2011*

Hansen, T.C. and Marga 1992 "Strength of recycled concrete made from coarse and fine recycled concrete aggregate" *Ibid .135: 605-612*

M.S. Shetty, Concrete Technology Theory and Practice, 5th edition, S.Chand & Co. Ltd., New Delhi.

IS: 456-2000 — Code of practice for plain and reinforced concrete. Indian Standard Institute, New Delhi.

IS: 6461-1973— Properties of concrete (part VIII). Indian Standard Institute, New Delhi.

IS: 8112-1989 — 43 Grade ordinary Portland cement. Indian Standard Institute, New Delhi.

IS: 383-1970 —Specification For Coarse and Fine Aggregates from Natural Sources for Concrete.

IS: 516-1959 —Methods of Testing for Strength of Concrete.

Land Use Land Cover Analysis of West Zone of Hyderabad Using Remote Sensing and GIS Technologies

M. Ramya[1], M.V.S.S Giridhar[2] and Shyama Mohan[3]
[1]M.Tech Student, [2]Professor, [3]Ph.D Scholar, Centre for Water Resources, IST, JNTUH

ABSTRACT

Land has considered to be a basic component and is regarded as a substantial natural resources currently available. Land use and land cover is the central to debate of sustainable development because it is considered to be one of the key driving forces of change in global environment. Land use land cover has brought serious losses of open land ,vegetation and water bodies at the same time uncontrolled expansion of urban land. It is also responsible for various implications such as pollution, climate change, Battle for land, socio-economic impacts and societal disparities. The present study mainly focuses on how land use land cover transformations occurring at west zone of hyderabad in the span of 1991 to 2018. Geographical changes in land use land cover have been identified by using remote sensing technology. For that Landsat images were acquired from USGS Earth Explorer. Image pre- processing and classification technologies were employed and accuracy assessment has performed to validate the output. The obtained accuracy is between 85 to 100 percent for the all classes. Transformation studies shows the open land has been reduced by 36.07% and urban land has increased by 120.85% , vegetation has decreased by 22.55% and water bodies diminished by 20.82 %. These information is useful to the urban planners and local authorities for sustainable development and betterment of future.

Key words: Land use Land cover, GIS, West Zone, Hyderabad, climate change, urbanization

Introduction

In this context, anthropogenic activity and its concomitant land use and land cover (LULC) changes have become an inevitable issue for the present time and accentuating the risks of environmental degradation around the globe. In an urban environment natural and human-induced environmental changes are of concern today because of deterioration of environment and human health. The study of land use/land cover (LU/LC) changes is very important to have proper planning and utilization of natural resources and their management. Traditional methods for gathering demographic data, censuses, and analysis of environmental samples are not adequate for multi complex environmental studies, since many problems often presented in environmental issues and great complexity of handling the multidisciplinary data set; we require new technologies like satellite remote sensing and Geographical Information Systems (GISs). These technologies provide data to study and monitor the dynamics of natural resources for environmental management.

Remote sensing has become an important tool applicable to developing and understanding the global, physical processes affecting the earth. Recent development in the use of satellite data is to take advantage of increasing amounts of geographical data available in conjunction with GIS to assist in interpretation. GIS is an integrated system of computer hardware and software capable of capturing, storing, retrieving, manipulating, analyzing, and displaying geographically referenced (spatial) information for the purpose of aiding development-oriented management and decision-making processes. Remote sensing and GIS have covered wide range of applications in the fields of agriculture, environments, and integrated eco-environment assessment. Several researchers have focused on LU/LC studies because of their adverse effects on ecology of the area and vegetation.

The concept of LULC change is said to be of a multidisciplinary in nature. when looked at significantly, the term LULC can be divided into Two: land use and land cover. According to various books and authors, land cover can be defined as the type of features which occur on the earth's surface naturally such as vegetation, water bodies etc. On the other hand, land use can be defined as the human activities which take place on the earth including forests, agriculture etc. However, a common factor links these two concepts together is human impact on the earth surface. The main objective of this paper is to detect and quantify the LU/LC in an urban area, west zone of hyderabad in the span of 1991 to 2018.

Study Area

The Hyderabad city situated in the Deccan Plateau and the capital of India occupying 650 square kilometers consist of distinct physical identity categorized by enormous rock formations and water bodies etc. It has the population of about 6.90 million and a metropolitan population of about 7.74 million. It is the fourth most populous city and sixth populous city in the urban agglomeration in India.

The west zone of GHMC i.e study area is spread over 174 square km with a population 1 million (as per 2011 census) and projected as 17° 32’ 18.28” N and 78° 14’20.70”E. The west zone area consist of 4 circles that are 11,12,13,14 .

GHMC has categorized hyderabad into five zones: East zone(L.B nagar), west zone(serilingampally), north zone(Kukatpally), central zone (khairathabad), south zone (charminar)

The areas included in this west zone are serilingampally (circle11), chandanagar (circle 12), patancheru and Ramachandrapuram (circle 13), moosapet (circle 14A),Kukatpally (circle 14B)

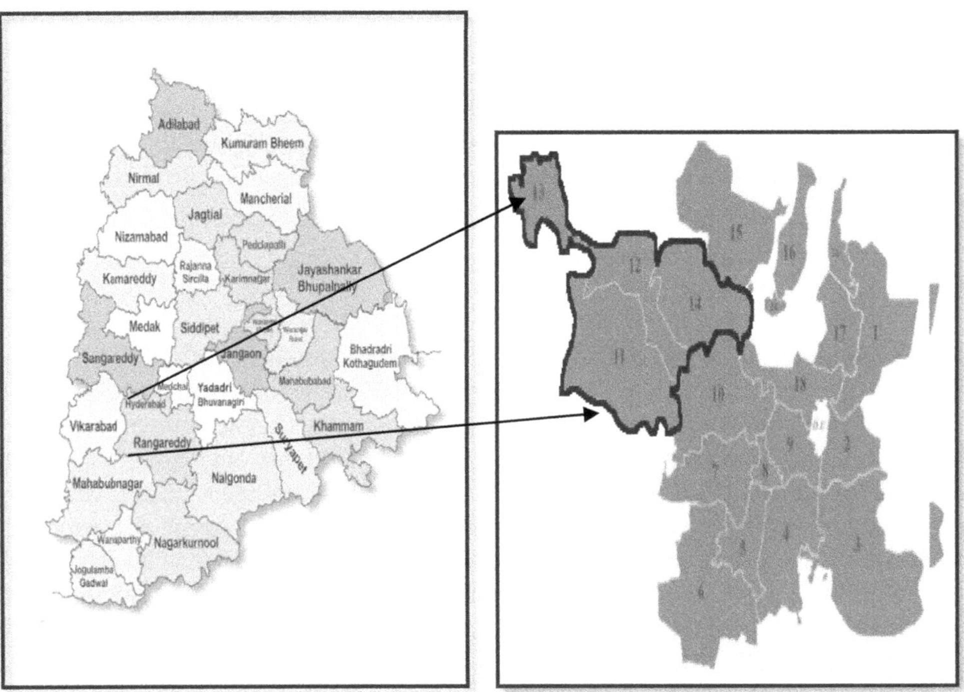

Figure 1 Location of study area

Data and Methodology

Data- The present study involves collecting the Toposheets from the Survey of India and the city map from the Greater Hyderabad municipal corporation and relevant authorities. For the present purpose E44M6& E44M7 Toposheets are requires as the west zone of Hyderabad comes under above maps.

The data products are collected from USGS Earth Explorer contains spatial resolution of 30m, which are used for the LULC Change Detection Analysis shown in the Table 1.

Table 1 Details of Landsat Data Collected

No	Date of image	Satellite/sensor	Reference sytem/pah/row
1	27-01-1991	Landsat5/TM	WRS-2/144/48
2	12-01-2001	Landsat7/ETM+	WRS-2/144/48
3	18-01-2011	Landsat7/ETM+	WRS-2/144/48
4	29-01-2018	Landsat7/ETM+	WRS-2/144/48

Methodology

In the present study we have taken two types of data. These are topographic map and remote sensing data. The topographic map (1:50,000 scale) is obtained from the Survey of India, Hyderabad, which was surveyed and prepared in 1976; it is converted to digital mode using scanning. The topographic map is georeferenced with longitude and latitudes using the ERDAS software and spatial analyst tools and demarcated the boundary of study area. This data is used for reference purpose.

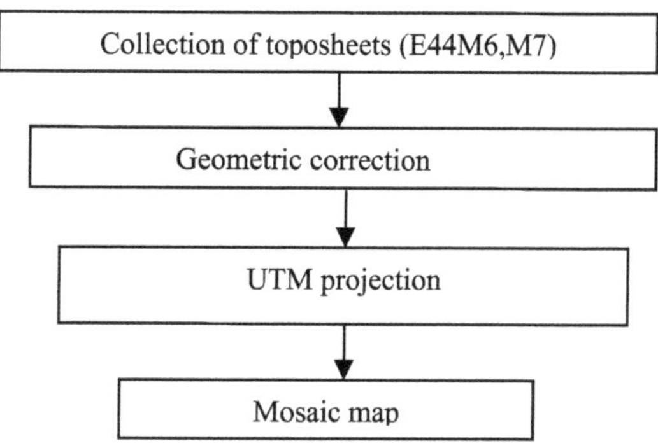

Figure 2 Flow chart of Toposheet processing

The data obtained was stacked and area of interest (Study area) is extracted then a supervised signature extraction with the maximum likelihood algorithm was employed to classify the remote sensing data. The satellite data was enhanced before classification using histogram equalization in ERDAS Imagine 8.7 to improve the image quality and to achieve better classification accuracy.

Image classification is regrouping all image pixels into a number of classes which are defined by the user. Supervised classification uses maximum likelihood classifier algorithm. This algorithm is directed by calculating the likelihood that specified pixel belongs to a set of classes which are predefined, subsequently the algorithm continues by allocating every pixel to the specific class to which the probability is highest. Four classes were identified in this classification which are water bodies, vegetation, urban land, open land.

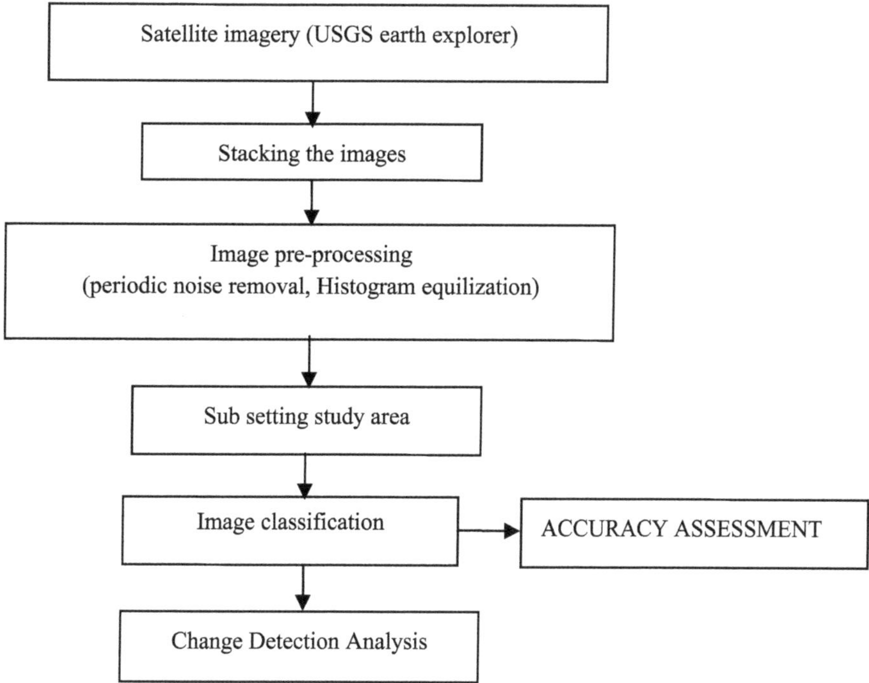

Figure 3 Flow Chart of Methodology

Results and Discussions

The classified images produced after preprocessing and supervised classification showing the Hyderabad city west zone region land use land cover are specified in the following figures. These classified images showing the information about geomorphology i.e physical features of the earth of study area. The blue colour indicates the waterbodies, green colour represents the vegetation land, red colour urban land and cream colour shows the open land.

Land Use Land Cover Images Before Classification

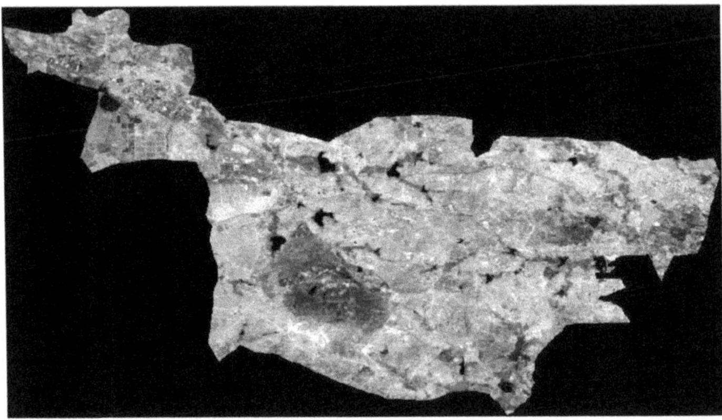

Figure 4 Before Classification Image of Study Area in 1991

Figure 5 Before Classification Image of Study Area in 2001

Figure 6 Before Classification Image of Study Area in 2011

Figure 7 Before Classification Image of Study Area in 2018

Land Use Land Cover Images After Classification

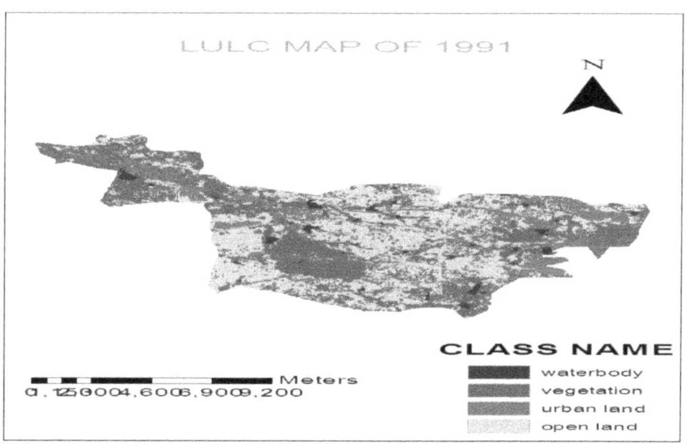

Figure 8 After Classification LULC Map of the study area in 1991

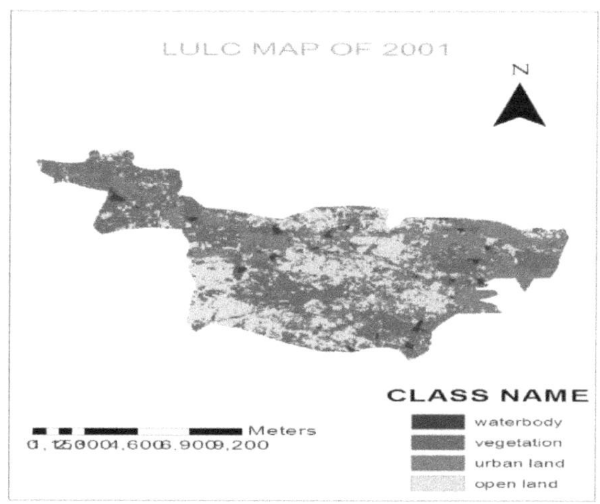

Figure 9 After Classification LULC Map of the study area in 2001

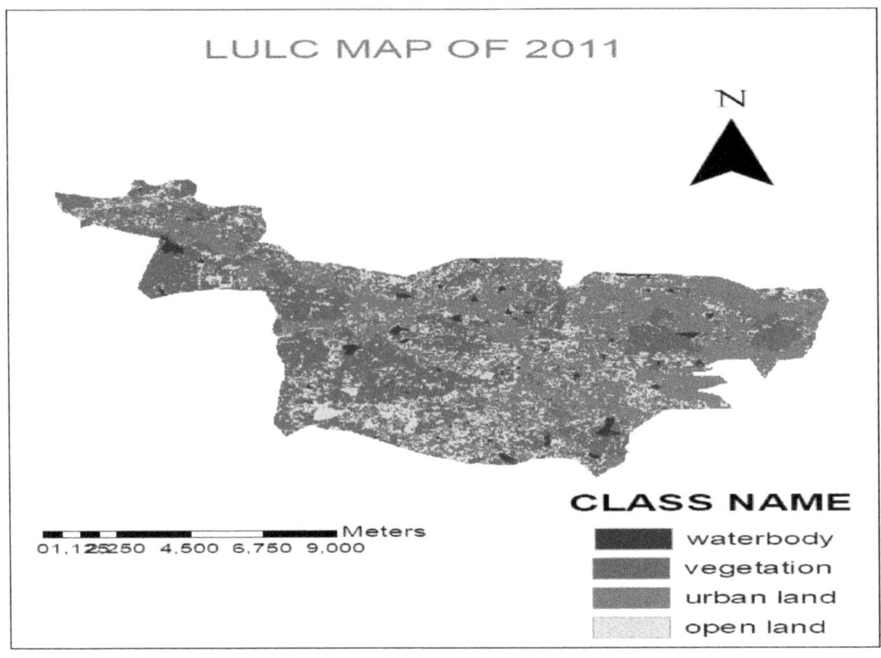

Figure 10 After Classification LULC Map of the study area in 2011

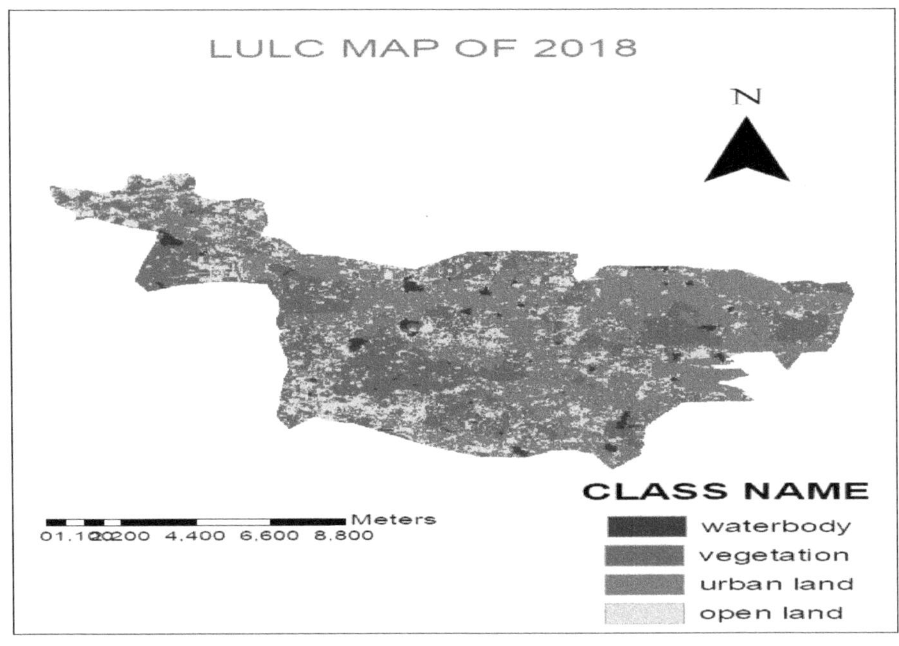

Figure 11 After Classification LULC Map of the study area in 2018

Accuracy Assessment Results

It is used particularly to express the degree to which a classification is considered as correct. Assessment of remotely sensed data is of ultimate important. There are numerous causes behind undertaking an accuracy assessment, as, "to arrange an overall measure of the map quality and an attempt to understanding of errors".

Table 2 Accuracy Assessment Report For the Years of 1991 & 2001

	1991		2001	
CLASS	**PRODUCER'S ACCURACY**	**USER'S ACCURACY**	**PRODUCER'S ACCURACY**	**USER'S ACCURACY**
Water bodies	100.00%	100.00%	100.00%	100.00%
Vegetation	87.50%	100.00%	86.84%	100.00%
Urban land	100.00%	93.33%	93.75%	88.24%
Open land	100.00%	91.67%	100.00%	81.82%
Overall Accuracy	95.16%		91.78%	
kappa statistics	0.928		0.871	

Table 3 Accuracy Assessment Report For the Years of 2011 & 2018

	2011		2018	
CLASS	**PRODUCER'S ACCURACY**	**USER'S ACCURACY**	**PRODUCER'S ACCURACY**	**USER'S ACCURACY**
Water bodies	100%	100%	100.00%	100.00%
Vegetation	100%	98.33%	86.67%	100.00%
Urban land	97.87%	100%	100.00%	96.67%
Open land	95.45%	9545%	91.67%	84.62%
Overall Accuracy	98.45%		94.83%	
kappa statistics	0.975		0.918	

CHANGE DETECTION ANALYSIS RESULTS

The urban change analysis presented in this paper was based on the statistics extracted from the land use and land cover maps of the west zone area. The results of change detection analysis are presented in the Table.4.

Table 4 Change Detection Analysis Result

CLASS	1991	2001	2011	2018
WATER	379.17	281.88	335.97	300.19
VEGETATION	7727.32	7143.48	7170.57	5984.71
URBAN LAND	3289.59	4821.66	6432.84	7265.28
OPEN LAND	5973.74	5122.8	3430.44	3818.92

Conclusion

In order to examine the change in LULC which has taken place in the west zone area of Hyderabad, Landsat images were acquired, classified and finally analyzed. Attention towards monitoring and observing patterns, trends of different types of features, land cover in specific area becoming necessary so LULC transformation study became important parameter. Particularly with changes in LULC occuring at such unprecedented and rapid rate, remote sensing can provide details and overall alteration of one land cover type to other. The water body class between1991-2001 has experienced decrease of 25.65% and the transformation between 2001-2011 was 281.88 ha to 335.57ha and from 2011-2018 was minor i.e from 335.57 ha to 300.19 (-10.63%). The output generated from this study within the time period from 1991 to 2018, indicates that vegetation land between 1991-2001 decreased by 7.55% and between 2001-2011 change in area obtained as 0.37%.It is due to the several factors like high rainfall average and programs which were initiated by HMDA in association with Royal Netherlands department under the GHEP. Overall change in vegetation land from 1991 to 2018 obtained as -22.55%.

The other important class built-up land was experienced major transformation between 1991 -2001 as in the same decade open land has decreased by 14.24% and the land absorption rate also found high in the same decade.

From these analysis it is clear that most of the land transformation was at the cost of conversion of open and vegetation land to the built-up area. Overall, the main driving forces behind west zone LULC transformation especially green cover loss was due to increase in vast number of employment opportunities which influences the individuals to live in close proximity to their place of work as result demand for the land increases and loss of open land, vegetation and water bodies increases. Other causes are easy accessibility, people's residential preferences, institutional establishments and industrial, commercial developments.

References

Rama Chandra Prasad, K. S.Rajan,VijayaBhole and C.B.S.Dutt "IS RAPID URBANIZATION LEADING TO LOSS OF WATER BODIES?" Journal of Spatial Science (March 2009) Vol II (2): 43-52

C.H.Mohammed Koya, "A Case Study on Impact of Urban Population Areas: Hyderabad City" International Journal of Innovative Research in Science, Engineering and Technology;Volume 6, Special Issue 4, March 2017

S.N.Mohapatra, Padmini Pani and Monika Sharma "Rapid Urban Expansion and Its Implications on Geomorphology: A Remote Sensing and GIS Based Study" Hindawi Publishing Corporation Geography Journal Volume 2014, Article ID 361459, 10 pages

Ibrahim RizkHegazy ,MosbehRashedKaloop, "Monitoring urban growth and land use change detection with GIS and remote sensing techniques in Daqahlia governorate Egypt" International journal of sustainable built environment (2015) 4,117-124

Bhagawatrimal, "Urban growth and land use/land cover change of pokhara sub-metropolitan city, Nepal" Journal of theoritocal and applied information technology volume.26 NO.2

A.A.Belal, F.S. Moghanm, "Detecting urban growth using remotesensing and GIS techniques in AI GHARBIYA governorate,Egypt" The Egyptian journal of remote sensing and space sciences, volume14, issue 2, December 2011,pages 73-79

ChimiChimi, Jigme Tenzin, TsheringCheki, "Assessment of Land Use/Cover Change and Urban Expansion Using Remote Sensing and GIS: A Case Study in Phuentsholing Municipality, Chukha, Bhutan" International Journal of Energy and Environmental Science 2017; 2(6): 127-135

YanitMekonnen ,S.K. Ghosh, "Monitoring Urban Growth and Land Use/ Cover Changes using Remote Sensing and Geospatial Techniques in Adama, Ethiopia" IRACST – Engineering Science and Technology: An International Journal (ESTIJ), ISSN: 2250-3498 Vol.7, No.2, Mar-April 2017

Ajay Singh Tomar, U.C. Singh, "Geomorphological Mapping Using Remote Sensing and GIS A Tool for Land Use Planning Around Shivpuri City, M.P., India" IOSR Journal of Computer Engineering (IOSRJCE) ISSN: 2278-0661 Volume 5, Issue 1 (Sep-Oct. 2012), PP 28-30.

C. Weber and A. Puissant, "Urbanization Pressure and Modeling of Urban Growth: Example of the Tunis Met-ropolitan Area," Remote Sensing of Environment, Vol. 86, 2003, pp. 341-352. B. L. Turner, "Local Faces, Global Flows: The Role of Land Use and Land Cover in Global Environmental Change," Land Degradation and Development, Vol. 5,1994, pp. 71-78.

Simulation of Water Distribution System by using GIS Pipe Software

D. Ajay Kumar[1], M.V.S.S Giridhar[2] and Shyama Mohan[3]

[1]M.Tech Student, [2]Professor, [3]Ph.D Scholar, Centre for Water Resources, IST, JNTUH

ABSTRACT

Water distribution is the most essential task after treatment of water. In this paper the classical approach for solving the problems of distribution network is by using specialized software GISpipe. GISpipe is a public domain, water distribution system modeling software includes Geospatial technologies. The purpose of this study is to assess the performance of Ranmnaresh Nagar water distribution system using hydraulic simulation software and also to address any improvements required to existing infrastructure in order to improve quantity and quality of water distributed to the consumers. The main objective of the present study is to design the water distribution network using GISpipe software by keeping in view of future demand of the people of Ramnaresh nagar.

The forecasted population for the design period of distribution network is calculated as 33,880 using Geometrical increase method. The results are obtained in EPANET software by using Hazen-William's formula. They are, minimum velocity is 0.01 m/s & maximum velocity is 0.37 m/s, the minimum pressure head is 9.66 m & maximum pressure head is 19.74 m and the minimum head loss is 0 m & maximum head loss is 1 m. The obtained results are within the range as per HMWS & SB standards.

Keywords: Water distribution, Ramnaresh Nagar, HMWS & SB standards, Hazen-William's formula

Introduction

GISpipe is software that integrates GIS, EPANET, and SWMM, and it can be used easily for designing or operating waterworks and sewerage. More than 80% of all information used by a water utility is geographically referenced. While Geospatial applications for water distribution systems are not new, getting beyond the basic inventory and mapping functions is often challenging. Unless a water system GIS is taken to the operational level, it's just a pretty map. That is why the GIS emphasis is now shifting from computerized mapping to enterprise-wide mission critical applications. GIS applications make the routine business functions easier to perform which improves productivity. GIS applications have the potential to prepare our water distribution systems for the operational challenges of the twenty-first century. The case studies presented in the chapter illustrate that many water utilities are successfully using the GIS applications for modeling and operation of their water distribution systems. GISpipe has wide applicability for water distribution system studies. Representation and analysis of water-related phenomena by GIS facilitates their management. By using GIS pipe from mapping, modeling, facilities management, and work order management, a drinking water distribution system manager can develop a detailed capital improvement program or operations and maintenance plan.GIS pipe consists of EPANET (Environmental Protection Agency Network) and SWMM (Storm Water Management Model).

EPANET is a computer programming that performs extended period simulation of hydraulic and water quality behavior within pressurized pipe networks. A network consists of pipes, nodes, pumps, valves and storage tanks or reservoirs. EPANET tracks the flow of water in each pipe, the pressure at each node and the concentration of chemicals throughout the network. SWMM (EPA's Storm Water Management Model) is used throughout the world for planning, analysis, and design related to storm water runoff, combined and sanitary sewers, and other drainage systems. It can be used to evaluate gray infrastructure storm water control strategies, such as pipes and storm drains, and is a useful tool for creating cost-effective green/gray hybrid storm water control solutions. SWMM was developed to help support local, state, and national storm water management objectives to reduce runoff through infiltration and retention, and help to reduce discharge.

Study Area

Ramnaresh nagar is located northwestern part of Hyderabad, Medchal-Malkajgiri. The piped water supply system for the Ramnaresh nagar was started in 1993. The main language spoken here is Telugu. It is one of the busiest business hubs in Hyderabad famous for its clothing and eateries. Ramnaresh nagar used to be an Industrial corridor in the northwestern part of Hyderabad. Its population began to grow after the early 1990s, with many people migrating from Andhra Pradesh and settling in and around Kukatpally. It is also home to the large chuck of IT goers due to its proximity with Hitech city. There are many small scale industries based in Kukatpally - Sanathnagar belt. The nearest airport is Shamshabad, and nearest metro station is "Miyapur metro railway station". Latitude and longitude coordinates are of 17°29'24" to 17°30'15" North and 78°62'84" to 78°63'58" East respectively as shown in the below figure 1, and the GIS based network model.

GIS pipe has the following features

- **Provides an intuitive GUI:**

  Using the program easily and intuitively reduces time and cost, enabling a more efficient job.

- **Fast and precise:**

  Create a pipe network plan through verification and calibration of vast amounts of GIS facility data faster and more accurately than any existing program.

- **Convenient:**

  Convenient editing functions such as undo, redo, copying & pasting, dividing pipe line, point merge, and result can be intuitively checked after network analysis.

- **One GISpipe is all you can do:**

  No programs like AutoCAD or ArcView are needed and you can do everything with one GISpipe

Basic Usage

If you work according to the following procedure, you can see the results of analysis.

- Import Shape file pipes and meter layers.
- Import contoured or elevation ESRI Shape File.
- Check Waterworks in edit tab.
- Click create grid in management tab.
- Click EPANET in management tab.
- Select layer created layer.
- Input Water Meta demand to excel file in Demand Allocation in pipe network analysis tab.
- Run simulation.
- Confirm result.

Function Description

- **Main Menu Area**

  It consists of management, editing, tools, EPANET, SWMM, Help. Depending on the choice and type of layer, the main number is either yellow or disabled. The types of layers are divided into GIS, water supply and water installations, EPANET, SWMM, Background and raster layers.

- **Layer Area**

  An area that controls a layer that performs functions such as switching layers on and off or exporting layers.

- **Map Area**

  This area operates map such as selection of zoom movements.

- **Extended Period Simulation Area**

  This is an area where you can play analysis results or select time according to time changes.

- **Status bar Area**

  This area shows the current coordinates, scale, expiration date, and so on.
- Pipe: A function to add pipe.
  - If a pipe is open on the map screen, it is displayed as shown below.

  - If a pipe is closed on the map screen, it is displayed as shown below.

- Pump: A function to add pump.
  - If a pump is open on the map screen, it is displayed as shown below.

  - If a pump is closed on the map screen, it is displayed as shown below.

- Valve: A function to add valve.
  - If a valve is open on the map screen, it is displayed as shown below.

  - If a valve is closed on the map screen, it is displayed as shown below.

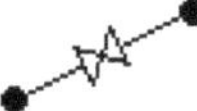

- Junction: The ability to add junction.
- Reservoir: The ability to add reservoir.
- Tank: The ability to add tank.

  Edit Vertices: The ability to edit a green dot as shown in the figure below. You can change the shape by moving the green dot. Clicking the green dot removes the clicked vertex and clicking between the solid lines adds the vertex.

Figure 1 Edit Vertices

- Pipe separation: As shown in the figure below, the first picture shows the pipe before splitting and the second picture after separating. Click the edge point in the fill diagram and click the pipe separation box to detach the pipe.

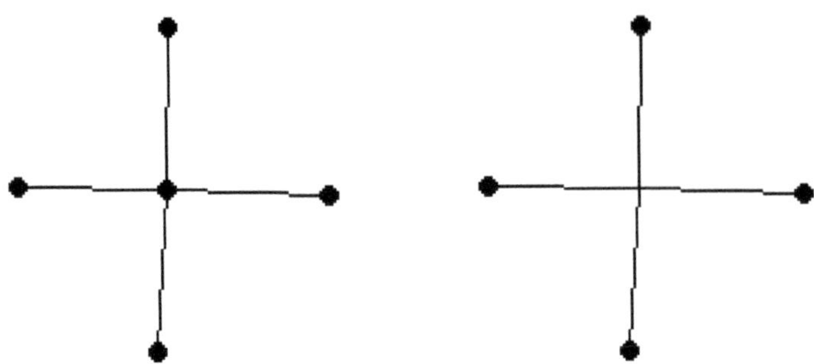

Figure 2 Pipe Separation

- Junction removal function
- Select the J2 junction in the Fig 3.
- When you drag a J2 node to J3 point, J2 node is deleted P2 link and P1 link is connected with J3 node.

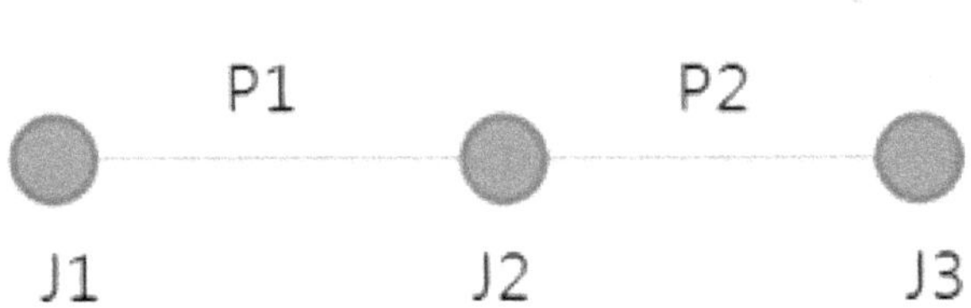

- The J2 node is deleted as shown in the Fig 3.

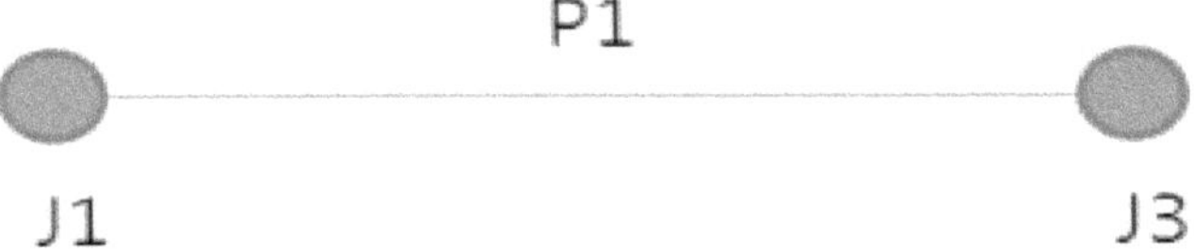

Figure 3 Junction removal function

- Node merge function
- Select the J3 node in the figure below.
- Dragging and dropping the J3 node to the J5 node deletes the J3 node and links the P2 link to the J5 node.

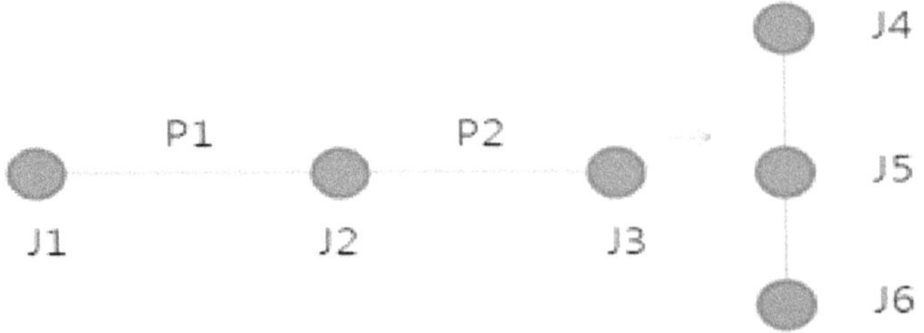

Figure 4 Node merge function

Methodology

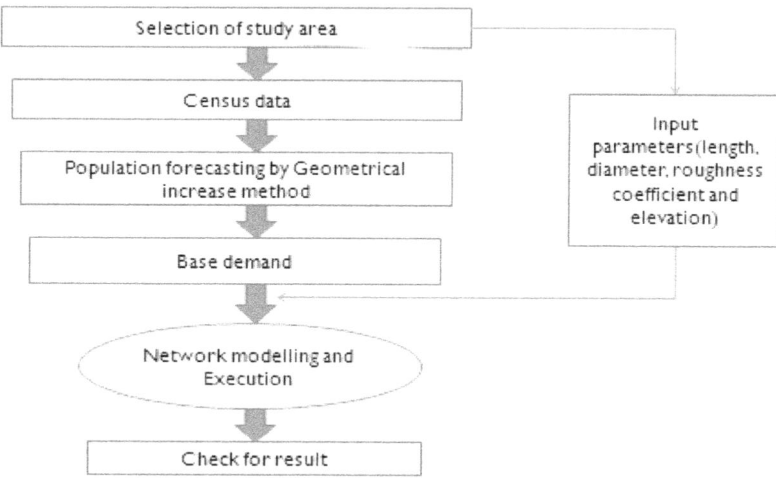

Figure 5 Flow chart of GISPipe software

The input parameters are diameter, start, end nodes, roughness coefficient and length. Computed outputs are includes flow rate, velocity, head loss, Darcy-weisbach friction factor, average reaction rate and average water quality. The hydraulic head loss observed a pipe due to friction with the pipe walls can be computed using one from the below three different formulae.

- Hazen-Williams formula
- Darcy-Weisbach formula
- Chezy-Manning formula

Table 1 Pipe head loss formulae for full flow

Formula	Resistance Coefficient (A)	Flow Exponent (B)
Hazen-Williams	$4.727\ C^{-1.852}\ d^{-4.871}\ L$	1.852
Darcy-Weisbach	$0.0252\ f(\varepsilon,d,q)d^{-5}L$	2
Chezy-Manning	$4.66\ n^2\ d^{-5.33}\ L$	2

where C = Hazen-Williams roughness coefficient shown in below table 4.4.
ε = Darcy-Weisbach roughness coefficient (ft)
f = friction factor (dependent on ε, d, and q)
n = Manning roughness coefficient
d = pipe diameter (ft)
L = pipe length (ft)
q = flow rate (cfs)

The most widely used head loss formula in the United States is Hazen-Williams formula. But it can only be used for water and it was originally developed for turbulent flow. The Darcy-Weisbach formula is the most theoretically used for all liquids. It applies over all flow regimes and to all liquids. The Chezy-Manning formula is generally used for open channel flow.

The head loss formula in between start and end point of the pipe line network is

$$h_l = Aq^B$$

Where A = Resistance Coefficient

B = Flow Exponent are shown in the above table 4.3

Table 2 Roughness coefficient for new pipe.

Material	Hazen-Williams C (unit less)	Darcy-Weisbach e (feet x 10^{-3})	Manning's n (unit less)
Cast Iron	130 – 14	0.85	0.012 - 0.015
Concrete or Concrete Lined	120 – 140	1.0 – 10	0.012 - 0.017
Galvanized Iron	120	0.5	0.015 - 0.017
Plastic	140 – 150	0.005	0.011 - 0.015
Steel	140 – 150	0.15	0.015 - 0.017
Vitrified Clay	110	-	0.013 - 0.015

Geometric Increase Method:

Table 3 Geometric Increase Method

Year	Population	Per decade percentage increase in population
1981	3816	$\frac{5157-3816}{3816} \times 100 = 35.14\%$
1991	5157	$\frac{7171-5157}{5157} \times 100 = 39.05\%$
2001	7171	$\frac{10150-7171}{7171} \times 100 = 41.54\%$
2011	10150	

The population forecasting using Geometric increase method and calculated in above table 5.1. Using the relation,

$$Pn = P0\ [1 + \frac{r}{100}]^n$$

Where,

P0 = latest known population

Pn = Population after 'n' decades

r = Per decade percentage growth rate

n = No. of decades

$r = (35.14 \times 39.14 \times 41.54)^{1/3} = 38.51$

$$P2018 = 10150^* \ [1 + \frac{38.51}{100}]^{0.7} = 12750$$

$$P2048 = 12750^* \ [1 + \frac{38.51}{100}]^{3} = 33880$$

Execution of the software

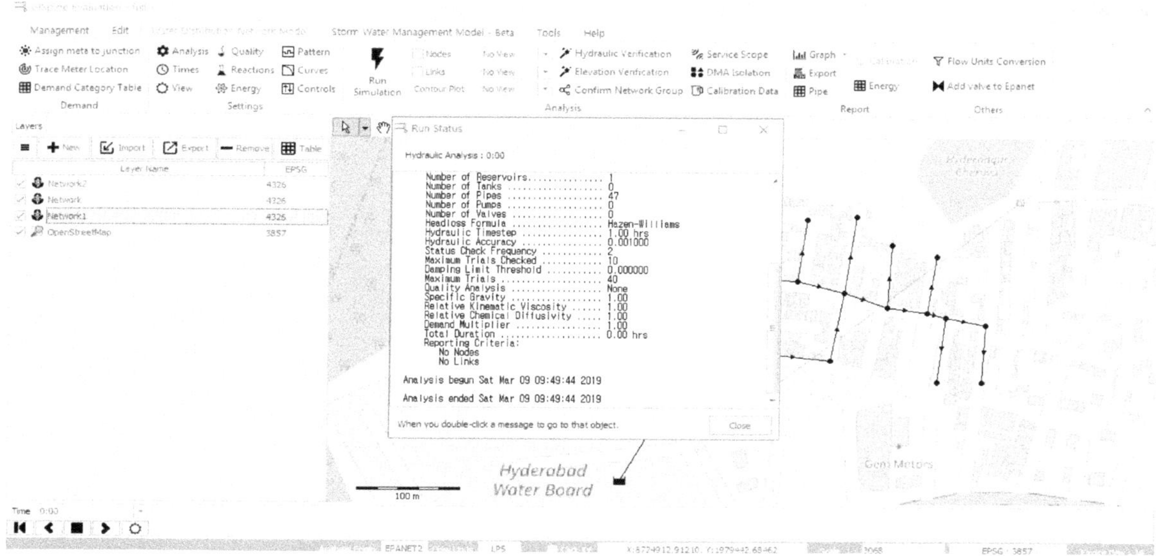

Figure 6 Status of the network

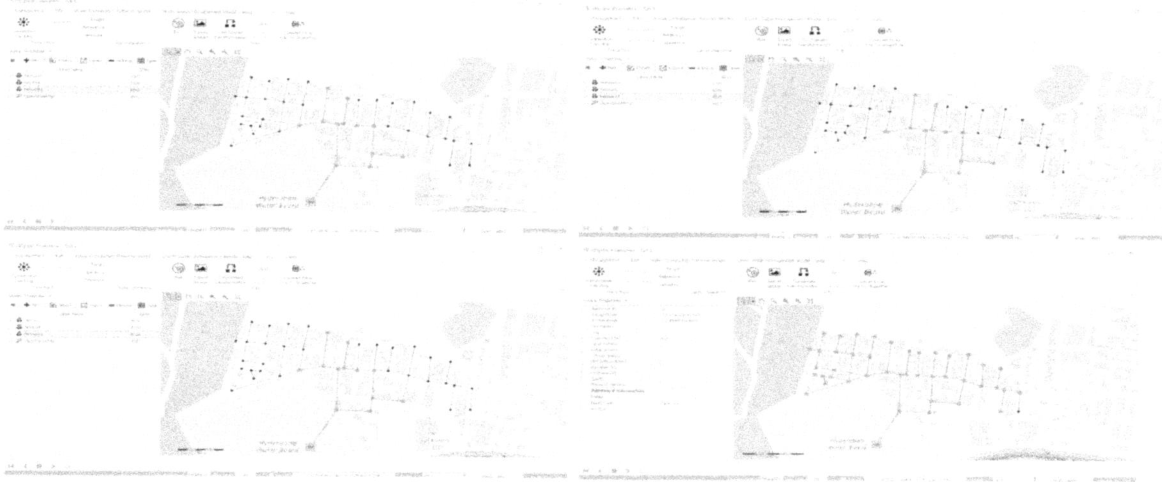

Figure 7 Connection Tracking

Results and analysis

Table 4 Node Results obtained from by GIS PIPE Software

Node ID	Elevation(m)	Base Demand (LPS)	Out Flow (LPS)	Head ()	Pressure (m)
0	620	-	11.02	620	0
44	607	0.36	0.36	619.79	12.79
39	603	0.24	0.24	619.89	16.89
5	603	0.48	0.48	619.81	16.81
13	604	0.12	0.12	619.74	15.74
4	603	0.2	0.2	619.83	16.83
24	606	0.36	0.36	619.79	13.79
42	608	0.24	0.24	619.77	11.77
18	603	0.36	0.36	619.84	16.84
26	607	0.36	0.36	619.78	12.78

Node ID	Elevation(m)	Base Demand (LPS)	Out Flow (LPS)	Head ()	Pressure (m)
38	609	0.36	0.36	619.79	10.79
32	604	0.12	0.12	619.78	15.78
50	602	0.12	0.12	619.85	17.85
9	603	0.36	0.36	619.74	16.74
19	606	0.3	0.3	619.85	13.85
35	605	0.12	0.12	619.78	14.78
8	603	0.36	0.36	619.77	16.77
33	605	0.12	0.12	619.78	14.78
28	608	0.36	0.36	619.77	11.77
36	610	0.36	0.36	619.79	9.79
1	603	0.2	0.2	619.88	16.88
29	608	0.24	0.24	619.77	11.77
21	607	0.36	0.36	619.85	12.85
25	608	0.12	0.12	619.79	11.79
34	605	0.48	0.48	619.78	14.78
40	603	0.36	0.36	619.83	16.83
7	602	0.12	0.12	619.85	17.85
20	607	0.36	0.36	619.85	12.85
31	604	0.36	0.36	619.78	15.78
37	605	0.24	0.24	619.79	14.79
11	603	0.12	0.12	619.74	16.74
15	601	0.12	0.12	619.77	18.77
16	600	0.12	0.12	619.74	19.74
27	607	0.24	0.24	619.78	12.78
30	603	0.12	0.12	619.77	16.77
3	602	0.24	0.24	619.86	17.86
23	605	0.24	0.24	619.79	14.79
6	607	0.36	0.36	619.85	12.85
10	602	0.36	0.36	619.74	17.74
12	602	0.24	0.24	619.74	17.74
14	601	0.12	0.12	619.81	18.81
22	606	0.36	0.36	619.84	13.84

Table 5 Links Results obtained from by GIS PIPE Software

Link ID	Diameter(mm)	Length (m)	Roughness	Flow (LPS)	Velocity (m/s)	Unit Headloss (m/km)
70	150	31.51	120	0.12	0.01	0
66	100	53.34	120	0.12	0.02	0.01
76	150	47.47	120	0.63	0.04	0.02
83	150	71.99	100	0.12	0.01	0
62	100	43.72	120	0.81	0.1	0.21
5	100	66.34	120	0.24	0.03	0.02
58	100	75.6	120	0.36	0.05	0.05
56	100	97.06	120	0.36	0.05	0.05
59	150	68.41	120	0.12	0.01	0
65	100	58.7	120	0.12	0.02	0.01
90	100	84.89	120	0.12	0.02	0.01
4	100	60	120	1.1	0.14	0.37
68	150	55.82	120	0.39	0.02	0.01
72	200	35.5	120	5.52	0.18	0.35

Link ID	Diameter(mm)	Length (m)	Roughness	Flow (LPS)	Velocity (m/s)	Unit Headloss (m/km)
45	200	110	120	2.56	0.08	0.06
92	150	63.44	100	0.12	0.01	0
18	150	117.83	120	0.12	0.01	0
87	100	47.17	120	1.8	0.23	0.92
54	100	74.57	120	0.12	0.02	0.01
84	150	62.82	100	0.12	0.01	0
9	200	78.3	120	4.3	0.14	0.16
8	200	122.66	120	11.02	0.35	0.9
71	200	13.22	100	0.6	0.02	0.01
61	100	45.13	120	0.34	0.04	0.04
63	100	43.36	120	0.21	0.03	0.02
86	100	41.15	100	1.32	0.17	0.73
2	200	150	120	4.1	0.13	0.14
78	100	21.43	120	0.72	0.09	0.17
77	100	21.8	120	0.96	0.12	0.29
89	200	33.52	100	1.54	0.05	0.03
7	100	68.63	120	0.24	0.03	0.02
60	150	59.39	120	0.94	0.05	0.04
11	200	120.55	120	6.48	0.21	0.34
91	100	69.16	100	0.12	0.02	0.01
79	100	23.28	120	0.24	0.03	0.02
75	150	86.49	120	1.95	0.11	0.15
3	100	40	120	1.3	0.17	0.51
73	150	55	120	4.8	0.27	0.79
69	150	38.3	120	0.3	0.02	0
74	150	0	120	1.55	0.09	0.1
55	100	74.48	120	0.12	0.02	0.01
10	150	86.75	120	1.07	0.06	0.05
51	100	25.25	120	0.12	0.02	0.01
12	150	30	120	0.36	0.02	0.01
82	100	24.57	120	0.12	0.02	0.01
88	100	49.47	100	1.3	0.17	0.71
85	150	19.79	100	0.84	0.05	0.04

Conclusion

The salient features of the entire study presented in the paper, after getting the results of GISpipe(EPANET) software, it is concluded that pressure head & demand at the node and also velocity, discharge and head loss at the link results have come within the acceptable range as per HMWS & SB standards.

References

Annelies De Corte and Kenneth Sorensen(2013) “Optimization of gravity fed water distribution network design”, European journal of operational research 228(2013).

Shie-Yui Liong and Md Atiquzzaman(2004) “Optimal design of water distribution network using shuffild complex evoluation”, Journel of institution of engineers Singapore volume 44.

E. Pacchin, S Alivisi and M Franchini(2016) “computed based software for water distribution network design” XVIII international conference on water distribution system analysis 2016.

S.Chandramouli and P.Malleswararao S Chandramouli (2011) “reliability based optimal design of a water distribution network for municipal water supply”, et al International journal of engineering and technology vol3(1).

Darshan Mehta, Krunal Lakhani, Divya Patel and Govind Patel. "study of water distribution network using EPANET", IJARESM (International journal if advanced research in engineering, science and management)

Manual of GISPIPE <https://en.gispipe.com/manual/en/>

Megan L. Abbot(2012), "optimization of redundancy in branched water distribution systems", A report submitted in partial fulfillment requirements for the degree of master of science in environmental engineering in Michigan technological University 2012

" Harsh Srivastava, Anupam Singhal (2018), "design and analysis of optimized water distribution network at bits pilani, pilani campus proceeding of the 3rd World Congress on civil, structural and Environmental Engineering (CSEE'18) paper no. ICESDP 117 DOI: 10.11159/icesdp18.117)

WEBGIS Based Land Information System: Case Study of Cheedikada Mandal, Visakhapatnam District, Andhra Pradesh

D. Seetarama Murty[1], M. Ramya[2], M.V.S.S Giridhar[3] and Shyama Mohan[4]

[1&2]M.Tech Student, [3]Professor, [4]Ph.D Scholar, Centre for Water Resources, IST, JNTUH

ABSTRACT

Cadastral maps are generally regarded as an essential part of the land management infrastructure in most countries yet there is often misunderstanding about their characteristics and role. Due to the vast range of different cadastral systems and resulting cadastral maps, it is very difficult to describe a "typical" cadastral map. It is the authors' view that it is also very difficult, if not impossible, to understand the characteristics and functions of a cadastral map without understanding the respective cadastral system. As a result these project activities to examine the characteristics and functions of cadastral maps by examining the WebGIS based cadastral mapping systems in Cheedikada Mandal, Visakhapatnam District Andhra Pradesh.

It is, in this context, the cadastral maps prepared long ago are still being used and serve the basic purposes at village level. Incidentally they continue to form the legal basis for many situations. On the other hand, the space technologies developed in our own country and elsewhere are identically much more suitable for preparing base maps with high resolution and accuracy. In this context, major programme like (1) Digital India, (2) Make in India, (3) Saansad Adarsh Gram Yojana envisaged and initiated by Government of India has reiterated and impressed the need for the generation of digital maps at village with reasonable accuracy. This study discusses the different characteristics of cadastral maps which have been designed for different users or functions. In particular the project concentrates on the issues concerned with developing digital multi-purpose cadastral maps. The major conclusions from the project are that the creation and maintenance of multi-purpose digital cadastral maps is a difficult and complex task. This complexity arises to a large degree because the characteristics of a cadastral map designed to serve traditional land markets or land registration purposes are quite different from the characteristics of a modern multi-purpose cadastral map.

Keyword: Cadastral data, Village boundary, Toposheet, satellite Imagery and qgis2web.

Introduction

When comparing the software technology used in the surveying industry with that of GIS, there are many commonalities; for example, both systems are dependent on coordinate information, both rely on mapping information about the earth, and both display their data in a pictorial form. Surveyors are well recognized as the earliest contributors to society's spatial information. The ancient cartographers were surveyors, often cadastral surveyors (Brock, 2005) with the advent of GIS technology in the 1960's, many new kinds of specialized jobs evolved, including the GIS cartographer. Through time the roles of the surveyor versus the GIS cartographer have diverged; this divergence can be attributed to the technology of digital maps. Though different technology paths have been traversed, it is ironic that new technology is now bringing the paths of cartographer and surveyor back together.

Study Area

Cheedikada is a Mandal in Visakhapatnam District of Andhra Pradesh State, India. Cheedikada Mandal Head Quarters is Cheedikada town. It belongs to Andhra region. It is located 49 KM towards west from District head quarters Vishakhapatnam. Cheedikada Mandal is bounded by Devarapalle Mandal towards East, Chodavaram Mandal towards South, Butchayyapeta Mandal towards South, Madugula Mandal towards west Anakapalle City, Narsipatnam City, Visakhapatnam City, Bheemunipatnam City are the nearby Cities to Cheedikada. Cheedikada consist of 47 Villages and 23 Panchayats, Vellanki is the smallest Village and Cheedikada is the biggest Village. It is in the 30m elevation (altitude). Simhachalam, Visakhapatnam (Vizag), Ananthagiri, Araku Valley, Vizayanagaram are the nearby Important tourist destinations.

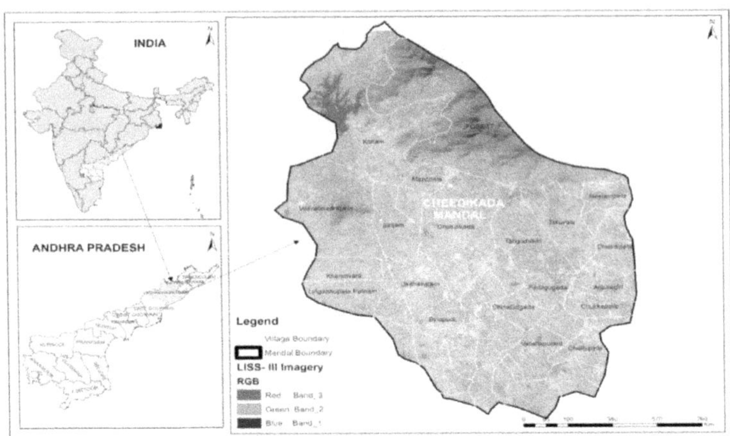

Figure 1 Study Area Location Map

Data used:

- LISS -III Satellite Imagery (23.5 resolution)
- TOPOSHEET 65K/13 (1:50,000 scale)
- CADASTRIAL MAPS

Methodology

The methodology involves the step by step approach as shown below:

- The first step was to acquire the required data such as cadastral maps, Toposheet and satellite imagery of Cheedikada Mandal.
- Acquiring boundary of the Cheedikada Mandal by geo-referencing the cadastral maps using Toposheet 65K/13.
- The next step was to upload the land survey information regarding each parcel. The data obtained from the geo-referenced cadastral maps is used for this process.
- Extract land use land cover information using the satellite imagery.
- The final step was to prepare the location specific GIS data such as land survey information and Land use pattern then export to web application by using qgis2web plugin.

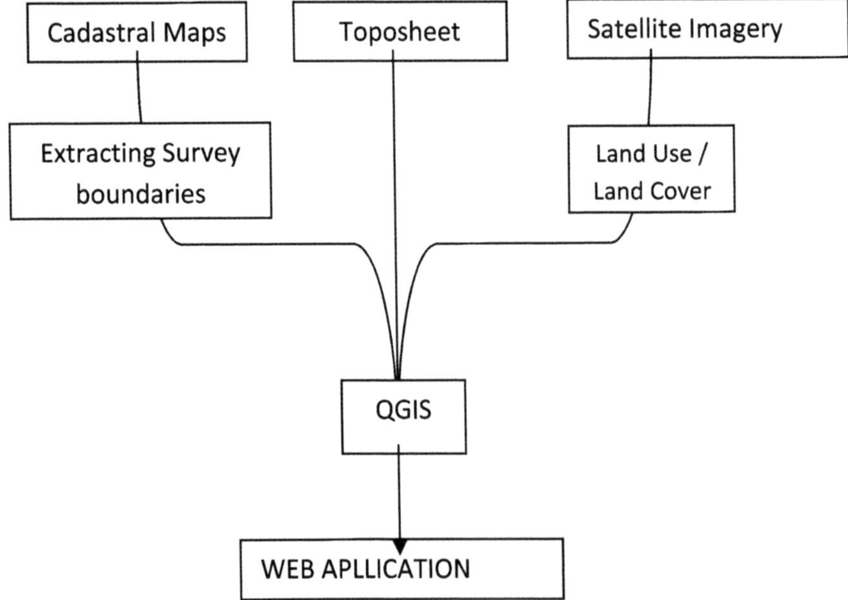

Figure 2 Methodology Flow Chart

Scope of the Project

The preparation of land information system development plans requires location specific information which is up-to date, accurate and relevant to bridge the critical gaps of development through participatory planning process. In the process of preparation Road agricultural activates, it has been realized that the scheme as of now has no location specific digital database. The absence of such data has resulted delay in the preparation of detailed plan at grass root level. Under such considerations and with a view to equip the scheme to make plan realistically, the study is made so that there is generation of web enabled multi-layer GIS database in the Cheedikada madal.

In the present project, prime parameters of land parcels of Cheedikada mandal, was taken as a model to demonstrate the GIS based web information system using free open source software, QGIS 2.14 version. The purpose of this project is to monitor and manage land parcel at divisional level for effective functioning of the all the departments.

A GIS based will help the government in planning, implementation and monitoring of various projects for development in different fields at much faster rate which in turn will make the state technologically more developed.

Preparation of Cadastral Level Base Map

Cadastral Mapping

Land is the most important assets of any country. The Land also contains all other valuable assets of the country such as forests, rivers, Minerals, oil, agricultural resources such as crops, orchards, industrial resources such as hydroelectric power, factories, markets, etc. Cadastral surveying is the branch of surveying which is concerned with the survey and demarcation of land for the purpose of defining parcels of land for registration, and creation of land records in a land registry. The record consisting of Record of ownership describing the nature of rights or interest in the respective parcels map of land parcels linked to other records.

The primary objective of cadastral mapping is to maintain Ground to Record-Record to ground- truthfulness at all times. Only then that it can be used for legal, administrative, planning and developmental purpose. There are several ways of presenting cadastral information. The most common and popular one is through maps. Generally in India, cadastral information is stored as written records containing lists of parcels, areas, landholders, assessed values and other items regarding the land.

The Cadastre (Cadastral Information) is classified into Graphic and Numerical Cadastre. The Graphic Cadastre system contains the land parcel data in the form of graphic record, from which area & other dimensional measurements has to be extracted. The Numerical Cadastre contains each individual land holding and is physically measured and parcels are recorded in the FMB. This system is also known as FMB (Field Measurement Book).

In developed countries as well as in India, cadastral survey uses the following three methods namely

(i) Ground based survey using Total station and GPS

(ii) Aerial photographs

(iii) High resolution satellite images.

Base Map Preparation for Village Information System

The Village Information System requires base map showing parcel boundaries at a scale of about 1:3960. Conventional base maps prepared from Survey of India Topographic Sheets at 1:50,000 scale may not be useful for this case. Hence a methodology was formulated considering the availability of resources and limitations of various partner institutions of the network project. The methodology for base map preparation at 1:3960 scale for VIS project is detailed below.

Part I (Cadastral map)

- Procure village maps (Cadastral map) from authorised agencies like Survey and Land Records Department of the State Government preferably at 1:5000 scale or whichever is available.
- Scan the cadastral map in JPEG or TIFF format at 300 dpi resolution or higher.

- Digitise the cadastral map using any GIS package and extract the features like village parcel/survey polygons, water bodies, road/rail network, water and soil information, drainage, settlements, plantation etc.
- Through field survey identify Ground Control Points (through GPS/DGPS).
- Georeference the cadastral layers using GCP's.
- Set datum and projection for these vector layers i.e., WGS 84 datum and UTM projection.
- Using geospatial adjustment tool available in GIS packages, spatially adjust the digitised cadastral layers to fit with corresponding features seen on High Resolution Satellite Imagery (georeferenced and orthorectified) to improve positional accuracy.
- Attach survey numbers to each of these polygons

Part II (High Resolution Satellite Imagery)

- Procure High Resolution Satellite imageries (HRSI) (geo referenced & ortho-rectified) either from NRSC.
- Further ortho-rectification can be taken up using local ground control points to get a fairly good horizontal accuracy in distances.
- Ground control points identified and collected earlier can be used for ortho-rectification of HRSI.
- Extract the features like village parcel/survey polygons, water bodies, road/rail network and other physical features from Satellite Imagery by keeping scale at 1:500 in GIS software while digitising.
- The cadastral maps are to be used as reference while extracting information from HRSI.
- To assist digitization using HRSI, we may physically shift the cadastral vector layer suitably over the imagery to clearly identify/demarcate the parcels and sub-parcels/Hissa is also extracted through HRSI.

Part III (Field verification)

- Field Verification and Validation of parcel boundaries to identify and update the left over (parcels which could not be identified from satellite imagery) and unidentified parcels.
- Identification of the land parcels on the ground also fixes its location on the ground, its boundary polygon and its place in the village with reference to a spatial standard reference point already established and accepted by the Government.
- Edit/mark the survey boundary with existing/ new PIN and clear adjacency.
- Field measurement of 5% of the parcels using Electronic Total Station for quality assessment of parcel mapping and to check the area with respect to records available with revenue department.
- Linking of the attribute information available with the departments to digital vector layers i.e., for both survey and hissa polygons.
- Preparation of Thematic Maps like Land use/ Land cover using base map

Part IV (Unique identification)

- Unique identification code for parcel/hissa – Census revenue village code +c+ parcel id (three digit)+ hissa id (alphanumeric - 5digit)
    - The unidentified parcels will be identified by number starting from 500.
    - Parcel id – 3 digit: e.g., if parcel id is '1' record it as 001.
    - Hissaid – 5 digit: e.g., if hissa id is '1/a' record it as 001/a.
- Unique identification code for habitation - Census revenue village code +h+ plot id (three digit)
- Unique identification code for assets - Census revenue village code +a+ asset id (alpha numeric – 5 digit)

The following information is mostly available with all state SSLR departments and can be linked to features.

- Parcels (revenue, non-revenue, water bodies, forest)
- Survey Number and Hissa Number of Land.
- Total Land under the Pahani. Land Revenue details.
- Land Owner's name with Extents and Khatha Number.
- Classification of the Soil. Number of Trees.
- Source of irrigation (rain fed/bore well/canal etc.)

- Area irrigated. Utilisation of land under various categories.
- Details of Crops grown season-wise, Details of Mixed Crops.

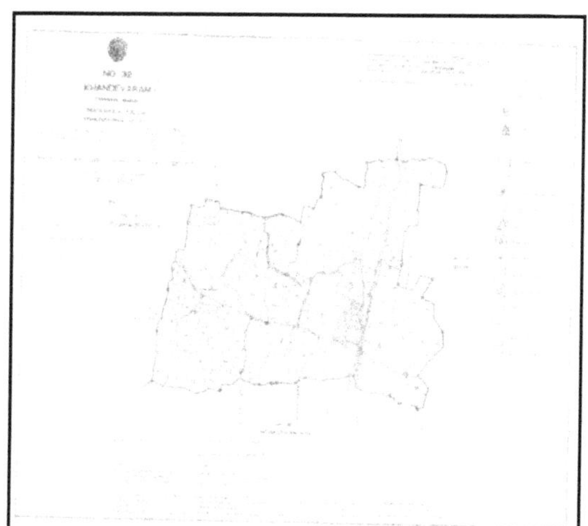

Figure 3 Cadastral map of khandevaram village

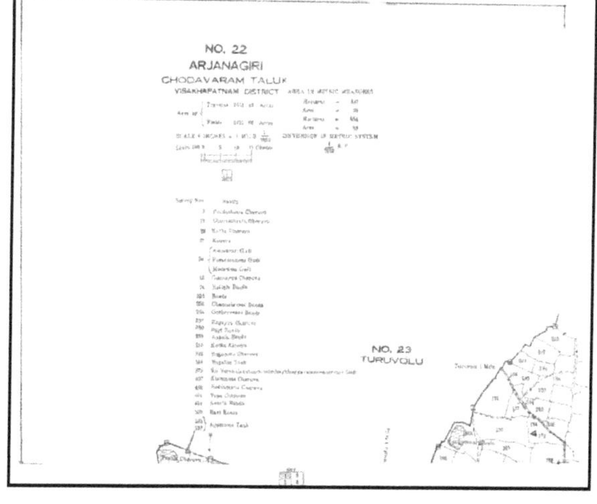

Figure 4 Cadastral map of Arjanagiri village

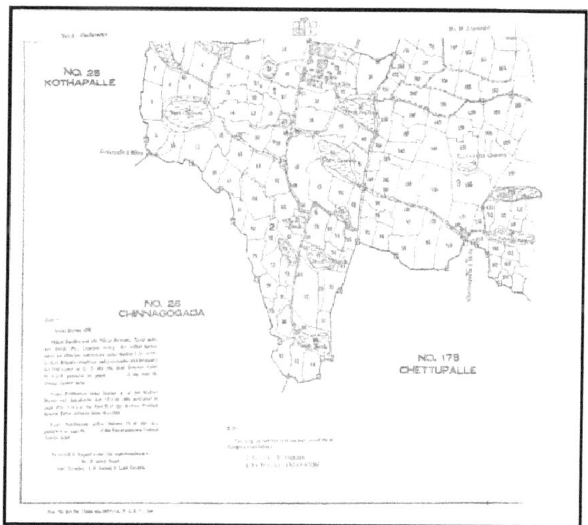

Figure 5 Cadastral map of Kothapalle village

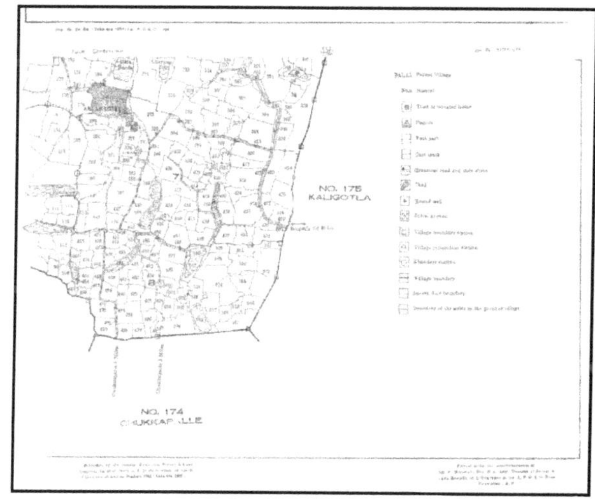

Figure 6 Cadastral map of Kaligotla village

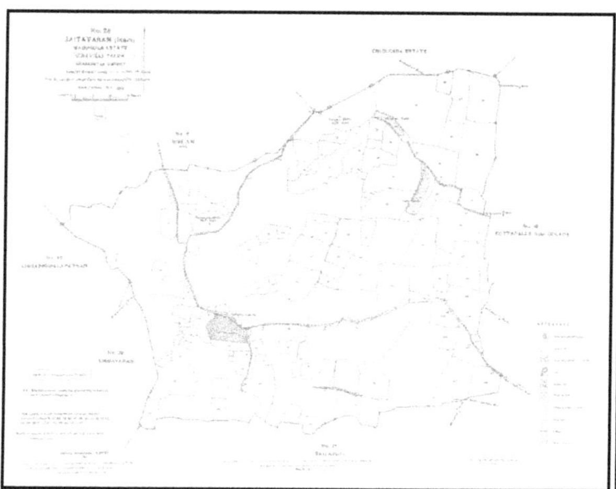

Figure 7 Cadastral map of jaitavaram village

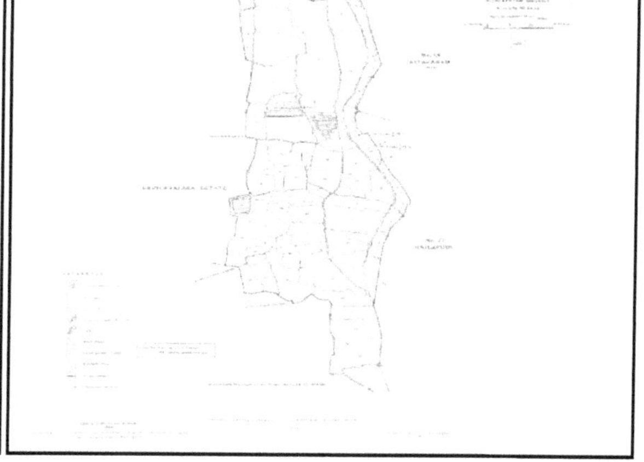

Figure 8 Cadastral map of Manchala village

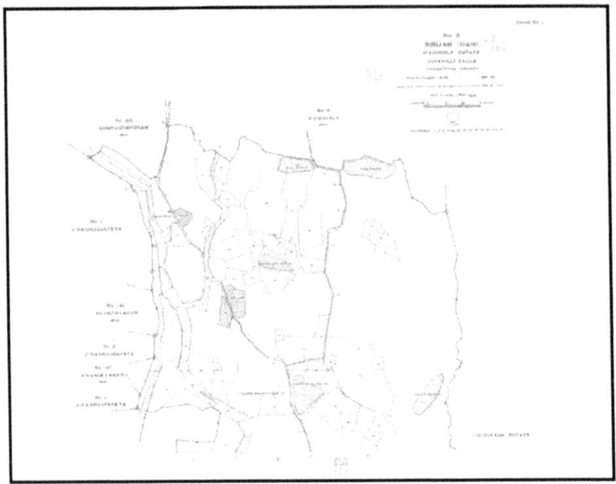

Figure 9 Cadastral map of Sirijam village

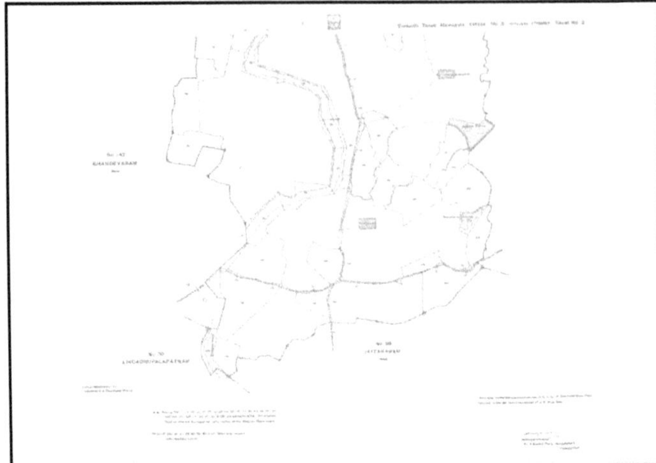

Figure 10 Cadastral map of khandevaram village

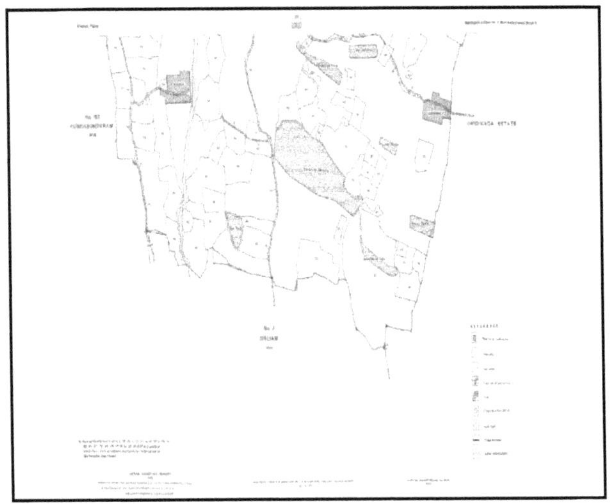

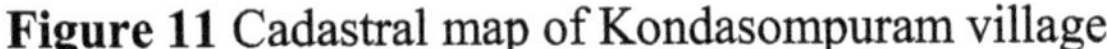

Figure 11 Cadastral map of Kondasompuram village

Figure 12 Cadastral map of Konam village

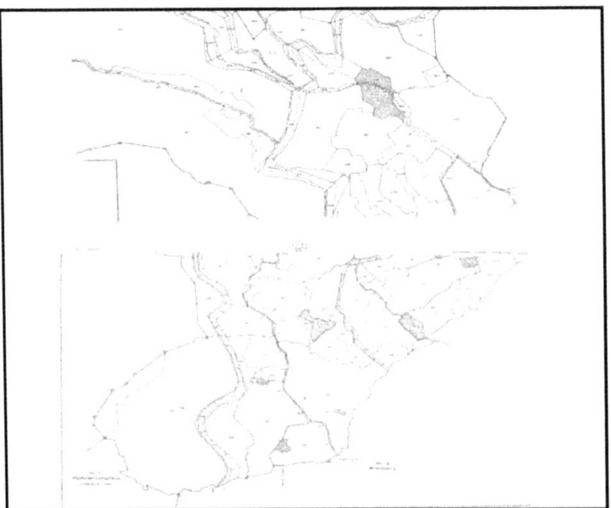

Figure 13 Cadastral map of Neelampeta village

Village Maps

A village map consists of land parcels which are numbers sequentially. A land parcel is an extent of land and shown in the Government record associated with a set of ownership rights officially recognized as property. The physical boundary of the parcel is contiguous and is defined uniquely in the official record. The said boundary is depicted in the form of two-dimensional map known as village map prepared by methods of plane tabling and chain and tape measurement. The village map shows individual parcels of all free hold, public and government land, encroachment and adverse possessions. The scale of the village maps varies in different states but the most common being 1: 4000. The land parcel boundary is quite old which calls for fresh mapping / updating. Also for developmental purposes it is important to associate present natural resources information with the land parcels. The availability of high-resolution data from satellite based platform has opened up new possibilities for land parcel mapping and updating with unique opportunity of natural resources appraisal and its change analysis. The cadastral maps for villages are prepared and maintained by Land Records Department. Most of the village maps have either no projection or the Cassini projection. As these maps are not geo-referenced, hence, they cannot directly be overlaid on any other resources map prepared using remote sensing data or other method.

Web Based Application Development

QGIS is a open source software by using we can develop WebGIS application that helps take the geographic information and make it available to others. QGIS software to create GIS resources, which are the maps, geo-databases, and other tools need for storing and using geographic information. QGIS gives the power to take the GIS resources on computer and make them available to a wider group of users throughout a network of computers. In QGIS web application, publish a GIS resource to others is through a service, with web reference. QGIS open source software gives the ability to create, manage, and distribute GIS services over the Web to support desktop, mobile and Web mapping applications. Qgis2web pluging for web application simplifies access to GIS services for GIS professionals, mobile workers, as well as knowledge workers without any GIS experience. We stay in control of our content through centralized management of spatial data, including imagery. In addition, Qgis2web pluging application provides with a scalable GIS server platform that can be deployed on a single machine to support small workgroups, or it can distributed across multiple servers for supporting enterprise applications. We can also deploy web application on Cloud infrastructure. A service is a representation of a GIS resource that a web page is making available to other computers on a network. This network can be a local one, such as company's computer system, or it can be a broader network, such as the Internet. The computers on the network that access your service are called clients. When we use Qgis2web pluging to publish a service, we are giving clients access to a GIS resource.

Results and Conclusions

The resultant web application can be access using URL in web Browser.

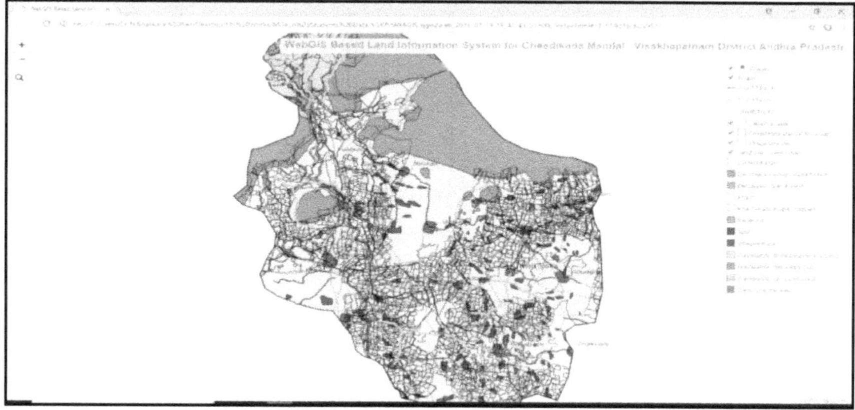

Figure 14 Study area all the spatial information

This application is having two services i.e. are all Thematic Layers and OSM standard maps.

In Thematic service is consisting of existing locations of Transmission mains Roads, Land use Land cover, village boundary, Cadastral Boundary and village locations. This information about existing land survey numbers locations of cadastral data useful to get information Transmission of the survey numbers and getting information for agricultural information system useful for persons who are working in different sectoral departments and even common people. The proposed locations of infrastructure are help to village administration for quick and easy planning under various schemes. The Thematic layers give all information regarding cadastral survey details and all other facilities available in study area.

The thematic layers service contains different thematic layers like locations and roads. This information is very helpful to concerned departments like water resource department, Irrigation, Road Transportation, Health Department and rural water Supply projects for optimum planning.

The OSM image is an open service, which is incorporated in this application. Which allows the user can visualize image at larger scale. In this web application, added the functionalities like Zoom in and Zoom out, identity. The screenshots of the above mentioned functionalities are given below.

Figure 15 The village boundary information

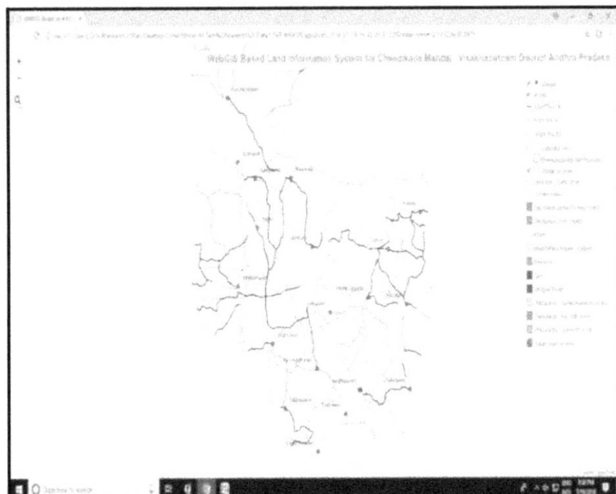

Figure 16 The village boundary and Road information

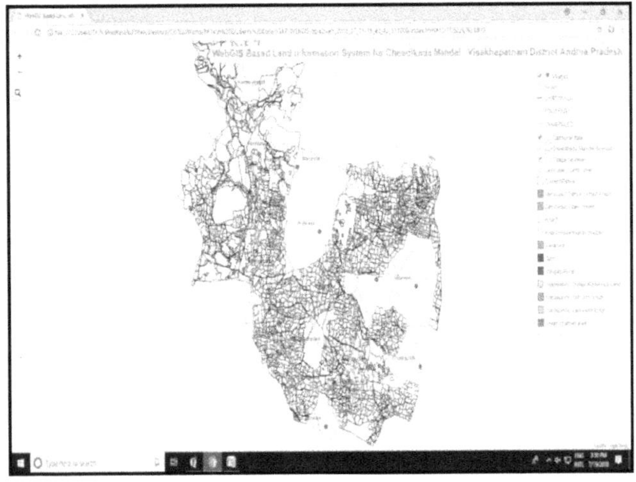

Figure 17 The cadastral survey boundary information

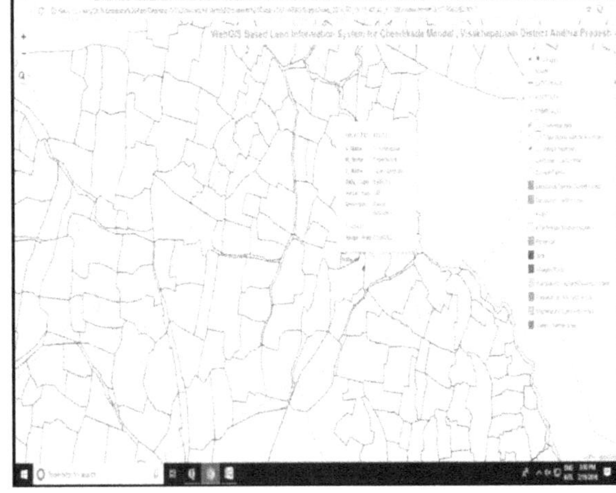

Figure 18 Cadastral survey boundary information with attribute

Figure 19 LULC information

Figure 20 LULC information with attribute

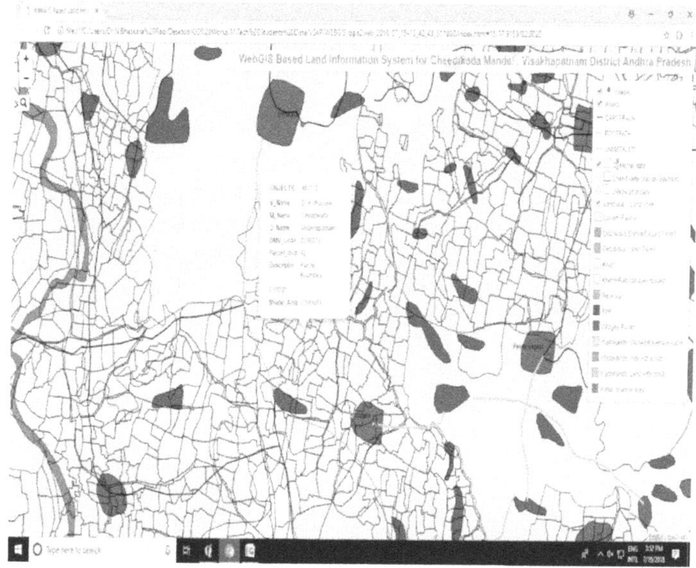

Figure 21 LULC and cadastral information

Conclusion

The convergence of Geographic Information Systems (GIS) and online technology by using open source free software QGIS has led to a proliferation of easily accessible yet surprisingly powerful web-based mapping applications. These tools can help planners visualize, analyze, and share information on a wide range of topics. While these capabilities were previously reserved for highly skilled GIS technicians, today, any practitioner with access to the internet can easily produce attractive, informative maps on a wide range issues related to housing, transportation, the environment, and more. This document provides a practical overview of useful web-based GIS applications. This resource is tailored to the needs of planners and those in planning-related professions. Planners must be able to convey where and to what degree their plans will affect the built environment and conduct analysis on spatial information – tasks intrinsically suited to the use of GIS and mapping tools.

The integration of information derived from QGIS technologies with other data sets, both spatial and non-spatial formats, provides tremendous potential for characterization and analysis of earth surface resources. Such derived information needs to be understood carefully in relation to the socio-economic situation for planning and development for sustainable growth employing GIS tool. The present project emphasizes the power of GIS technology which will help the decision makers at block level to better understand and evaluate spatial data by creating graphic displays using information stored in the database. A GIS based planning will help the government in planning, implementation and monitoring of various projects for development in different fields at much faster rate which in turn will make the state technologically more developed.

This analysis has confirmed the need for spatial planning, which can achieve the desired results of economic, social interaction and overall development of a region. This study involves a methodology for rank ordering the various settlements, so that, priorities for linking the settlements can be worked out. This GIS database will be customized to meet the requirements of stake holder departments/concerned in providing the digital resource databases and any other requirements of the block, based on their problems/ priority/ developmental programmes at Village/ Block/ Taluka level.

References

B.A.U.I.Kumara, 2008 Application of participatory GIS for Rural Community Devlopment and local level Spatial Planning System

Bahree, M., 2006, Map Mania, Forbes Magazine.

Biazin, B.; Sterk, G.; Temesgen, M.; Abdulkedir, A.; Stroosnijder, L. Rainwater harvesting and management.

Boers, T.M, Ben-Asher, J. A review of rainwater harvesting. Agric. Water Management, 5,145–158.

Borowitzka LJ, Post FJ, Borowitzka MA, Mackay B, Moulton T (1983) The protozoa of a Western Australian hypersaline lagoon. Hydrobiologia 105:95–113.

Brock, J. F., 2005, Four Surveyors of the Gods: In the XVIII Dynasty of Egypt – New Kingdom c. 1400 B.C., From Pharaohs to Geoinformatics, FIG Working Week 2005 and GSDI-8, Cairo.

Buckner, R. B., 1983, Surveying Measurements, and their Analysis.

Chatterjee A, Roy K, Ghosh UC, Pramanik T, Kabi SP, Biswas K (2008) Fluoride in water in parts of Raniganj coalfield, West Bengal. Curr Sci 94(3):309

Cook RN (1976) Land information systems. In: Unpublished paper prepared for the U.S. Department of Agriculture with assistance from a special committee of the North American Institute of Modernization of Land Data Systems

Craig BA (2003) Cadastral survey accuracy standards. Surv Land Inf Sci 63(2):87–106

Hydrological Viability Analysis for Effective Utilization of Palli Tank for Irrigation in Anandapuram Mandal, Visakhapatnam District

VVS Ramakrishna[1], M. Ramya[2], M.V.S.S Giridhar[3] and Shyama Mohan[4]
[1&2]M.Tech Student, [3]Professor, [4]Ph.D Scholar, Centre for Water Resources, IST, JNTUH

ABSTRACT

Most of the minor irrigation tanks are interconnected cascades, which allows run-off and return flow of the upstream command area to the downstream tank. This facilitates reuse of water in the command area and increases available water for irrigation. The nature has can use of water transportation to irrigation tanks, into the storage results in of storage capacity which in turn of tank performance. Rehabilitation projects have been undertaken to brighten more agriculture land under from on assured irrigation system. The prediction of water availability in a tank is in important process for sustainable use of water resources in a tank cascade system. The land use/land cover map is prepared using IRS, LISS-IV (5.8m) Resolution satellite imagery reveals water usage and demand for different development activity. The problems in drainage courses and tank catchment boundaries identified using SOI 1: 50,000 scale toposheet and spatial discrepancy have been adjusted with the satellite image. This study identified issues of earthen channel, catchments and head works, area the major problems are to be restored. Remedial measures suggested improving the tank cascade system.

Key Words: Irrigation, land use land cover map, catchment, satellite imagery

Introduction

The term Remote Sensing is applied to the study of earth's features from images taken from space using satellites, or from nearer the earth using aircrafts. The technique of remote sensing has picked up in the past half a decade, largely due to the availability of digital computers, improved communication systems, digital imaging techniques and space technology. Remotely sensed data can be said to have its origin in photography, where the information about a target area is interpreted from photographs. Later this technique was extended to aero plane - borne cameras giving rise to the science of aerial photography. This technique is still used, but largely the signal cameras have been replaced by Laser operated ones where the reflectance of a Laser beam projected from the bottom of the aircraft is sensed by electronic sensors. In this chapter we shall discuss remote sensing using satellite as India has strived ahead in this field and made good use of satellite images. The satellite launching program of our country is one of the most ambitious in the world, and is still continuing to be so in the future as well. Amongst other fields, the Water Resources Engineers have benefited greatly by using satellite imaging techniques, some applications of which have been highlighted in this chapter. The other topic that is discussed in this lesson is the Geographic Information System (GIS) that has wide applications in planning any spatially distributed projects. Fundamentally, a GIS is a map in an electronic form, representing any type of spatial features. Additionally, properties or attributes may be attached to the spatial features. Apart from its spatial data analysis capabilities, it provides an interface to remotely sensed images and field surveyed data. This technique has specifically benefited the Water Resources Engineers, which has been discussed in some detail. This project is proposed for improvement of independent containing 1 tank of which Palli tank is last in the order. Hence it is named as Palli Tank independent. The tank is covered in Anandapuram Mandal of Visakhapatnam district and situated on North-West of Kasulavada Village.

Study Area

The study area covers Kasulavada village, Anandapuram Mandal, Visakhapatnam district Andhra Pradesh state. The catchment boundary is taken from SOI Map format and Geo-rectified to study Drainage and tanks all along the catchment. The tank catchment boundary area covered in 65O 5 SOI topo-sheets of 1: 50,000 scale. Palli Tank lies in 83° 19' 44" East Longitude and 17° 56' 41" North Latitude and capacity of the tank is about 1.77 Mcft and provides irrigation facilities to an Ayacut of 105 Ac. It comes under Gosthani basin. Influencing rain gauge station is Padmanabham. It is at 6.62 Km from Padmanabham station. The Total catchment area, the tank was across the Local Vagu of Gosthani minor basin of Gosthani major basin.

Palli Tank is the only one study tank in the cascade. It is considered as independent tank. There are three intercepted tanks in the cascade which do not have any ayacut i.e., small kuntas.

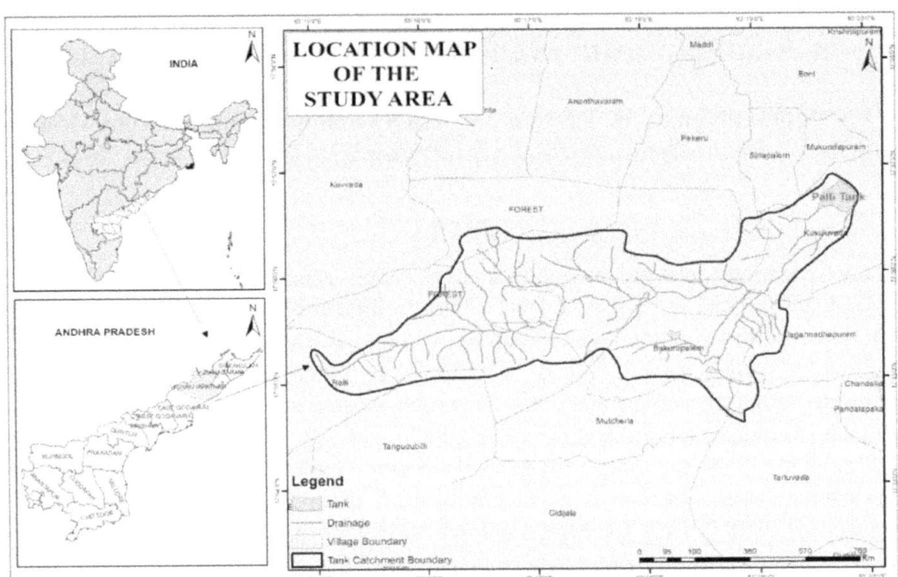

Figure 1 Location map of the study area

The proposed independent tank is situated at about 1.00 km towards North-West of Kasulavada (V), Anandapuram (M) in Visakhapatnam District.

The Global Coordinates are as follows:

Unique code of the tank	:	0302325005010501
Latitude	:	17° 56′41.00″/1985615.00 m N
Longitude	:	83° 19′44.00″/746674.00 m E
Zone	:	44PQ
SI Sheet No	:	65-O/5

The tank was across the Local Vagu of Gosthani minor basin of Gosthani major basin.

Data and Methodology

Datasets used

The Topo sheet 65O5 are collected from the Survey of India and is digitized and then the study area is delineated using the QGIS software. Further, the tanks which are connected in a cascade manner are delineated.

a. Tank information from WRD
b. Survey of India Toposheets (1:50,000)
c. Latest satellite data (LISS IV) of swath 5.8m
d. Thematic maps like soils, land use/land cover, drainage, catchment boundaries.
e. The non-spatial data like command area, beneficiaries, meteorological data (DES/IMD) information and other resource attributes.

Softwares used

The following software's are used in this study

a. Google Satellite Image
b. ArcGIS 10.3
c. Google Earth

Methodology

The methodology to achieve the goals and objectives of the present study is shown in this Fig. 2 which shows the flow chart for the model development. The various steps involved in the model development are as follows

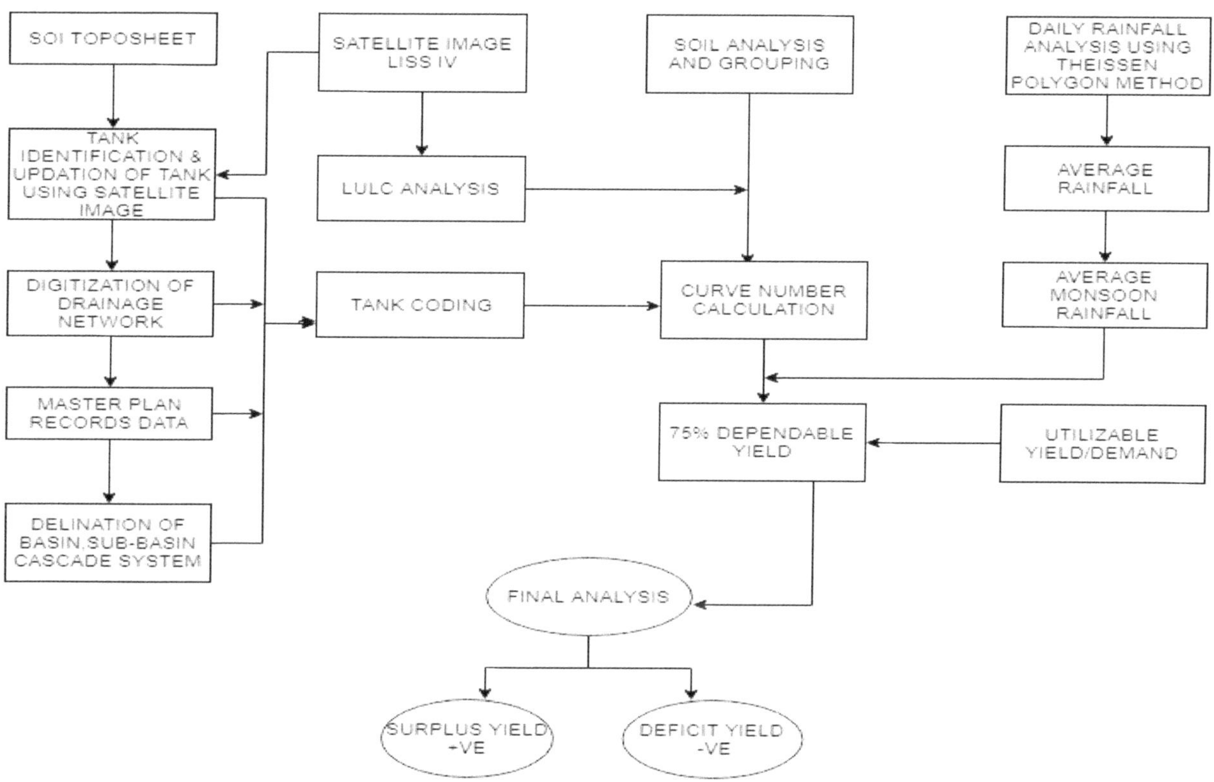

Figure 2 Flowchart of Methodology

Description of Palli Tank

	Ayacut in hectares	Ayacut in acres
Palli Tank	42.50	105

Physical Features

Main Components of Tank

Palli Tank

Earth Bund

a. Length of the bund = 780 m

b. FTL = 36.200 m

c. MWL = 36.800 m

d. TBL = 38.300 m

e. Slopes U/s & D/s = 1.5:1 / 2:1

Proposed 2:1 on U/s

f. Top width of bund = 3.00 m

g. Max. Height at U/s bund = 4.00 m

Sluice

a. No. of Sluices = 2 No

b. Sill level of Sluice – I at chainage = 34.900 m 0.005

c. Sill level of Sluice – II at Chainage = 34.300 m 0.565

d. Size of vent (Pipe Dia / Barrel) / Type

		Size	**Type**
I	=	0.300 m	Pipe
II	=	0.300 m	Pipe

Surplus Weir

a. No. of Weirs = 1 No

b. Length of Body wall for weir = 26.40 m

c. FTL = 36.200 m

d. MWL = 36.800 m

e. Type of weir = B.C. Weir

(Proposed Type of Weir) = H.C. Weir

f. Abutment top level = 38.300 m

g. Length of D/s Apron for weir = 3.400 m

h. Head of flow = 0.600 m

i. Discharge dispose for weir = 27.70 cumecs

Irrigation Channel

a. No. of Irrigation channels = 2 No

b. Length of main canal under Sluice – I = 0.700 km

Bed Width = 0.400 m

FSD = 0.200 m

c. Length of main canal under Sluice – II = 1.200 km

Bed Width = 0.500 m

FSD = 0.200 m

Supply Channel

a. No of Supply Channels = 1 No

b. Length of supply channel from Jaggam Vagu to tank = 1.600 km

c. Bed width = 1.300 m

d. FSD = 0.550 m

Surplus Course

a. No. of Surplus course = 1 No

b. Length of Palli Tank Surplus Course = 0.250 km

Bed Width = 10.000 m

FSD = 1.300 m

Hydrology

The main source for getting of water to Palli Tank is through the monsoon run off from its total catchment area of 1.52sq.km. The catchment area covered with hilly area and semi plain area. Considering the overall physiography of the catchment on the basis of experience it can be classified as average catchment according to Strange's table.

Table 1 Catchment Area details

1	Free Catchment area	=	1.52	Sq.km
2	Gross Catchment area	=	**1.52**	Sq.km

Scs-Curve Number Method

Surface Runoff Estimation

Rainfall - Runoff modelling is an essential part in water resources planning and management. The soil conservation service curve number (SCS-CN) is a simulation model that analyzes runoff volumes from the rainfall. It is one of the efficient methods to estimate direct runoff volume in ungauged catchments (Hawkins, 1993; McCuen, 2002; Michel et al., 2005; Ponce and Hawkins, 1996). It uses the curve number (CN) to determine the runoff volumes. Curve number varies with the terrain conditions. The SCS-CN method uses Land use / Land cover, Soil information and antecedent soil moisture conditions of the catchment. The modifications suggested by the Ministry of Agriculture, Govt. of India, (Hand book of Hydrology, 1972), to suit Indian conditions are included in the study. A brief account of inputs and formulae of the SCS method adopted for the present study is given below.

a. Rain gauges in the neighborhood of the tank catchment are taken and weighted average rainfall of catchment arrived using Thiessen polygons.
b. Hydrological soil groups map is prepared based on soil classification map.
c. Integrating Hydrological soil groups and Land use / Land cover information, a weighted Curve number - CN (II) for each tank catchment is arrived based on cure number table.

Calculation of Cn Value

Weighted Average Curve Number (II) = Total CN / Total Area

Area (sq km)	Land Use / Land Cover	Soil Type	HSG	CN (CN-II)	TCN	WCN
0.03	Built Up-Rural-Rural	Gravelly clayey	C	91	2.78	91
0.04	Built Up-Mining / Quarry-Quarry	Gravelly clayey	C	91	3.36	91
0.10	Agricultural Land-Cropland-Kharif	Gravelly clayey	C	81	8.36	81
0.05	Agricultural Land-Cropped in two Seasons	Gravelly clayey	C	95	4.81	95
0.06	Agricultural Land-Cropped in two Seasons	Gravelly clayey	C	95	5.88	95
0.01	Agricultural Land-Cropped in two Seasons	Gravelly clayey	C	95	0.90	95
0.05	Agricultural Land-Cropped in two Seasons	Gravelly clayey	C	95	4.69	95
0.26	Agricultural Land-Cropped in two Seasons	Gravelly clayey	C	95	24.52	95
0.08	Waterbodies-Reservoir/Tank-Seasonal	Gravelly clayey	C	88	6.86	88
0.19	Agricultural Land-Fallow land	Gravelly clayey	C	76	14.12	76
0.08	Agricultural Land-Fallow Land	Gravelly clayey	C	76	6.13	76
0.07	Agricultural Land-Fallow land	Gravelly clayey	C	76	5.51	76
0.03	Built Up-Mining / Quarry-Quarry	Gravelly clayey	C	91	3.17	91
0.04	Built Up-Mining / Quarry-Quarry	loamy skeletal	C	91	3.32	91
0.24	Agricultural Land-Cropped in two Seasons	fine loamy	C	95	22.89	95

Area (sq km)	Land Use / Land Cover	Soil Type	HSG	CN (CN-II)	TCN	WCN
0.01	Agricultural Land-Cropped in two Seasons	fine loamy	C	95	0.89	95
0.00	Agricultural Land-Cropped in two Seasons	fine loamy	C	95	0.00	95
0.00	Agricultural Land-Cropped in two Seasons	fine loamy	C	95	0.24	95
0.03	Wastelands-Scrub land-Open	loamy skeletal	C	79	2.40	79
0.03	Wastelands-Scrub land-Open	fine loamy	C	79	2.15	79
0.00	Waterbodies-Reservoir/Tank-Seasonal	fine loamy	C	88	0.08	88
0.02	Waterbodies-Reservoir/Tank-Seasonal	fine loamy	C	88	2.06	88
0.02	Agricultural Land-Fallow land-Fallow land	fine loamy	C	76	1.49	76
0.05	Agricultural Land-Fallow land-Fallow land	fine loamy	C	76	3.68	76
0.12	Agricultural Land-Fallow land	fine loamy	C	76	9.19	76
0.00	Built Up-Mining / Quarry-Quarry	fine loamy	C	91	0.13	91
0.00	Built Up-Mining / Quarry-Quarry	fine loamy	C	91	0.04	91
0.04	Built Up-Mining / Quarry-Quarry	loamy skeletal	D	93	3.28	93

Estimation of Runoff using Curve Number Method

The Curve Number method is used for estimating the direct surface runoff volume using the recorded rainfall data and weighted curve number of the Palli tank catchment. SCS-curve number method takes into account the parameters characterizing a farm such as land use, soil cover, rain fall condition and wetness for estimating yield from the area. These parameters are used to calculate the curve number for Palli tank catchment as shown in pervious figure. The curve number indicates the runoff potential of a complex storm during the particular period.

After calculate the curve number value, the maximum potential retention (S) is calculated. By considering rain fall condition of previous 25 years rainfall, maximum potential retention and rainfall data the yearly runoff was estimated. The runoff for different sites of palli tank catchment was calculated and the total rainfall, calculated runoff and the percentage runoff for the year 2011 are presented in Table 3.

To check the performance of curve number method, the calculated daily runoff and the observed daily runoff data were plotted using MS Excel to develop the relationship between them. Different models were used to find the best correlation coefficient of the data. The coefficient of determination between observed and calculated daily runoff for different.

75% Dependable Yield Calculation

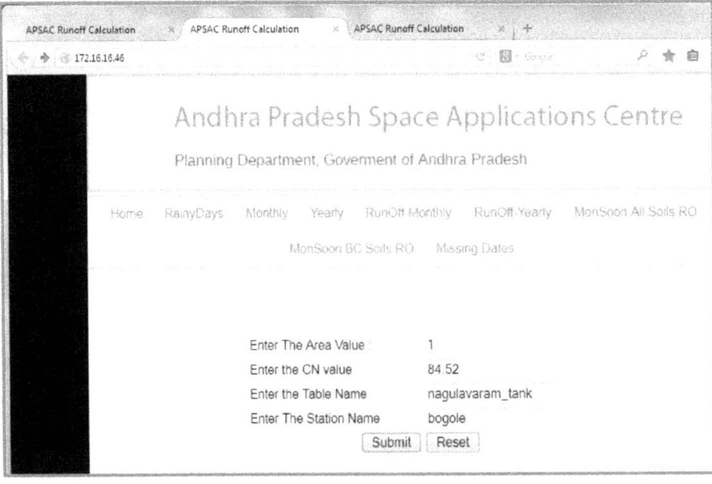

Figure 3 APSAC Runoff Calculation

Home RainyDays Monthly Yearly RunOff-Monthly RunOff-Yearly MonSoon All Soils RO MonSoon BC Soils RO Missing Dates

Enter Table Name Select Order Submit Cancel

The Details of the Table : **nagulavaram_tank** CN Value : **84.52**

SNO	YEAR	JUNE	JULY	AUGUST	SEPTEMBER	OCTOBER	NOVEMBER	DECEMBER	RAINFALL_SUM	ALL_SOILS_RUNOFFSUM
1	1991	139.12	0.00	0.00	47.44	256.65	362.42	45.92	1516	851.55
2	2007	55.05	39.41	21.60	0.00	652.43	23.53	0.00	1288	792.02
3	1996	106.58	0.00	0.00	0.00	387.34	75.91	140.46	1359	710.29
4	1994	0.00	72.17	0.00	0.00	211.22	369.11	54.90	1361	707.40
5	2015	0.00	0.00	0.00	39.78	5.33	502.81	106.04	1264	653.96
6	2005	0.00	0.00	1.19	0.00	518.24	97.27	7.09	1157	623.79
7	2001	0.00	0.00	0.00	34.02	221.89	253.22	69.29	1228	578.42
8	2006	0.00	0.00	0.00	51.70	166.05	317.48	0.00	1089	535.23
9	1997	0.00	22.20	0.00	52.49	25.40	146.47	273.91	1365	520.47
10	1990	2.48	5.43	42.58	37.84	226.65	176.58	0.00	1152	491.56
11	2008	0.00	17.87	0.00	0.00	222.01	245.74	0.03	1044	485.65
12	2010	45.82	10.79	33.04	0.00	76.62	262.48	53.76	1090	482.51
13	2002	0.00	0.00	0.00	0.00	278.10	167.55	0.00	856	445.65
14	2009	9.35	0.00	45.05	6.60	0.00	361.17	16.70	931	438.87
15	2000	84.08	87.58	162.86	0.00	22.99	44.81	24.80	930	427.12
16	1993	0.00	0.00	12.55	0.00	126.21	66.69	210.59	931	416.04
17	2011	0.00	0.00	5.33	6.04	254.15	147.42	2.05	953	414.99
18	1992	0.00	0.00	196.79	13.07	35.66	160.32	0.00	951	405.84
19	1989	0.00	56.93	0.00	19.80	0.00	215.43	97.82	785	389.98
20	2013	0.00	29.96	27.04	0.00	181.51	139.22	0.00	778	377.73
21	2003	0.00	17.24	42.94	42.19	138.83	5.63	81.56	848	328.39
22	2012	0.00	0.00	8.23	0.00	84.80	212.24	0.00	724	305.27
23	1998	0.00	0.00	0.00	59.42	176.97	58.64	0.00	844	295.03
24	2014	0.00	0.00	0.00	16.02	60.28	81.28	75.75	607	233.33
25	1995	0.00	36.07	16.26	1.98	30.68	58.43	0.00	670	143.42
26	1999	0.00	7.09	33.33	0.00	77.88	7.55	0.00	615	125.85
27	2004	0.00	0.00	0.00	86.69	0.00	0.00	0.00	300	86.69
28	1988	0.00	0.00	0.00	85.86	0.00	0.00	0.00	198	85.86

Figure 4 Monthly Runoff Data

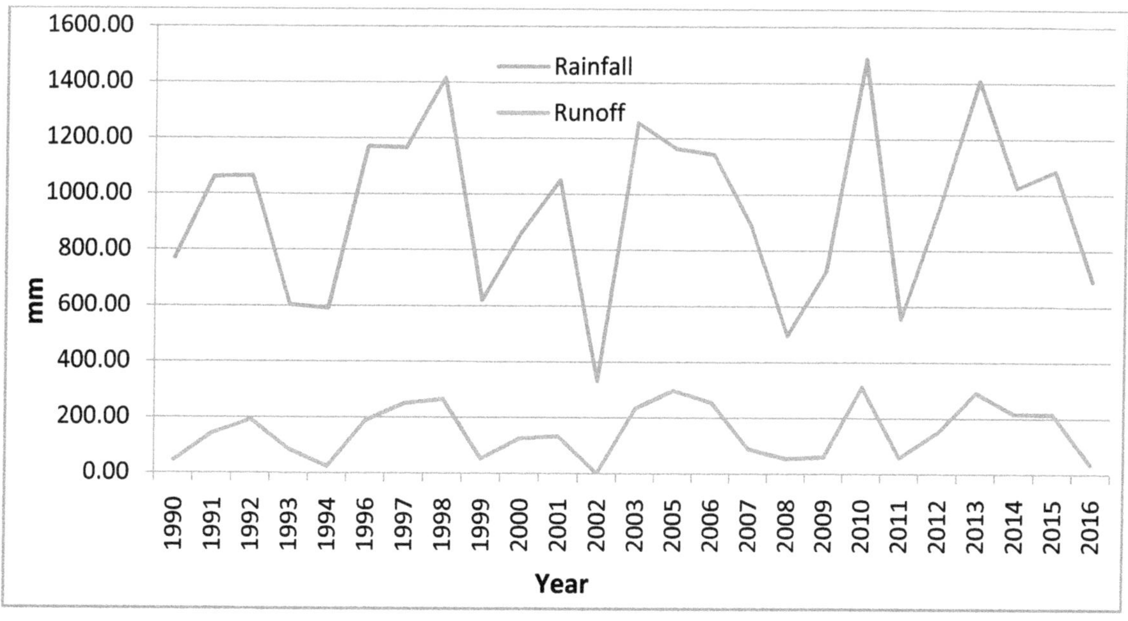

Figure 5 Palli Tank Monsoon rainfall and runoff (mm)

Table 2 Palli Tank 75% yield Estimation (Monsoon)

S.No	Year	Rainfall	Runoff
1	2010	1487	312.40
2	2005	1164	297.21
3	2013	1409	290.93
4	1998	1416	264.51
5	2006	1143	253.79
6	1997	1166	249.37
7	2003	1255	233.01
8	2014	1025	216.28
9	2015	1084	215.68
10	1992	1065	191.32
11	1996	1171	188.68
12	2012	959	148.26
13	1991	1062	143.39
14	2001	1051	132.90
15	2000	861	125.26
16	2007	889	88.69
17	1993	603	85.09
18	2009	726	60.30
19	2011	555	57.54
20	2008	495	55.06
21	1999	620	52.44
22	1990	770	45.29
23	2016	691	34.80
24	1994	591	23.76
25	2002	331	0.90

75% yield in mm - 57.54
75% yield in mcft - 31.78

Results and Discussions

Analysis

Land use / land cover classes and type of soils were extracted using the latest satellite data. The cascade wise land use and soils were merged using Arc map (GIS) software. Total curve number was assigned using the unique land use and hydrological soil group. The ratio of total curve no and its corresponding area gives the weighted curve number (WCN). This generated weighted curve number is a factor for runoff estimation. 75% dependable yield was estimated for each tank in the cascade using SCS Curve method. The tank wise details are as follows.

Table 3 Tank wise details

Mandal	Name of Tank	Registered Ayacut in Acres	Catchment Area (Sq.km)	Basin Name	Tank Capacity (McFt)	75% Dependable Yield (Mc.Ft)	Demand (0.15 Mcft per acre)	Balance Yield (McFt)	Status
Anandapuram	**Palli Tank**	105	15.64	Gostani	1.77	31.78	15.78	16.03	Viable

The objective of the study is to map the land use/land cover of the focused area, to have detailed information on the spatial variation from Remote Sensing Satellite Imagery, to identify the problems in the drainage course of the tank cascade system through GPS survey and to suggest suitable analysis to measure the data base.

The Minor irrigation tank system components consist of drainage course, catchments, tank bund, sluice and surplus weir locations. The problems identified are excessive weed growth-choking of sections of canal, thereby reducing the velocity of flow and causing the deposition of sediment, silt accumulation, aquatic weed growth in water submerged areas is a problem because some of this growth eventually reaches the canals,

Total curve number was assigned using the unique land use and hydrological soil group. The ratio of total curve no and its corresponding area gives the weighted curve number (WCN). This generated weighted curve number is a factor for runoff estimation. Demand calculated 10 acres per one Mcft of water. 75% dependable yield was estimated for each tank in the cascade using SCS Curve method. The tank wise details are as follows.

Conclusion

The need to Registered Ayacut area water demand and improve irrigation systems results from deferred maintenance, increased scarcity of the water resources, and a disproportion between the technological state of agricultural production and rising environmental concerns. Changing institutions and management within irrigation systems are now viewed as a catchment area and tank improvement process. This study is conducted towards hydrological viability of minor irrigation tank system. It was found that most of the irrigation tanks do not receive the surplus from the upstream tanks. Through Rainfall, Runoff and GIS tracking over the drainage course and tank system the problems were identified and suitable remedial measures were suggested for the tank cascade for augmenting more water. For optimum utilization of financial resources, it is suggested to receive and restore only those tanks that are hydrological viable. Hence it is recommended that the rehabilitation of irrigation tank system is very essential and GIS and GPS are the most effective tools.

Rainfall runoff simulation is important for water resources planning and management. This study is undertaken to estimate the surface runoff and flow volume in upstream of two water harvesting structures on Palli catchment using SCS Curve Number method along with RS and GIS. Estimation of curve number was carried out with extracted geo-morphological data such as land use, soil type, drainage, etc under the environment of RS and GIS. In this study it is found that combination of curve number method with environment RS and GIS can improve the model performance significantly. In this study also found that GIS is the efficient tool for preparation of thematic maps as input for SCS curve number method. It is to be noted that SCS-CN method was formulated for humid catchments, yet it is found to be a suitable method for semi-arid regions of Gujarat. Overall it is found in this study that curve number method along with environment of RS and GIS techniques is very efficient and accurate for modelling rainfall-runoff and to estimate total surface runoff.

References

Anbumozhi. .Matsumots. K, Yamaji.E., "Towards improved performance of irrigation tanks in semi-arid regions of India: Modernization opportunities and challenges" , Irrigation and Drainage System Vol. 15,pp 293-309. 2001.

John bosco. V.J. "Monitoring and evaluation of tank modernization works panapakkam tank" M.E. thesis CWR Anna university. 1987.

Madduma Bandara, C. M. Catchment eco-systems and village tank in the dry zone of Sri Lanka. In Strategies for River Basin Management. Pp 265-277. In: Lundqvist. J., Lohm U. and Falkernmank, W. (ED), Reidel Publishing Company. 1985.

Palmyra Centre for ecological land use, water management and rural development, Aurovile, "Rehabilitation of Integrated tank management systems. In the kaliveli watershed Villupuram district, Tamil Nadu", Final report 1999 – 2006.

Panabokke, C.R. The small tank cascade systems of the Rajarata, their settings, distribution pattern and hydrography. Mahaweli Authority – International Irrigation Water Management Institute, Srilanka. 1999.

Musgrave, G.W. 1955. How much of the rain enters the soil? In Water: U.S. Department of Agriculture. Yearbook. Washington, DC. pp. 151–159.

Nielsen, R.D., and A.T. Hjelmfelt. 1998. Hydrologic soil group assessment. Water Resources Engineering 98. In Abt, Young-Pezeshk, and Watson (eds.), Proc. of Internat. Water Resources Eng. Conf., Am. Soc. Civil Engr: pp. 1297–1302.

Rawls, W.J., and D.L. Brakensiek. 1983. A procedure to predict Green-Ampt infiltration parameters. In Advances in infiltration. Proc. of the National Conference on Advances in Infiltration. Chicago, IL.

U.S. Department of Agriculture, Natural Resources Conservation Service. 1993. Soil Survey Manual. Agricultural Handbook No. 18, chapter 3. U.S. Government Printing Office, Washington, DC.

U.S. Department of Agriculture, Natural Resources Conservation Service. 1993. National Engineering Handbook, title 210–VI. Part 630, chapters 9 and 10. Washington, DC.

Study of Soil and Water Retention Characteristics using Pedotransfer Function

V. P. Pandagale[1*], G. U. Satpute[2], Snehal Dongardive[3] and R. S. Patode[4]

[1*]*Junior Res. Asstt, AICRP for Dryland Agriculture*

[2] *Professor, Department of Soil and Water Conservation*

[3]*Junior Res. Fellow, Department of UCES & EE*

[4]*Senior Scientist, SWCE, AICRP for Dryland Agriculture*

Dr. PDKV, Akola (M.S) India.

ABSTRACT

Soil hydraulic properties are essential for land use planning, water resource management and the development of water harvesting structures. Assessment of soil water regimes is an important step in water resource management in agriculture. *In-situ* or laboratory measurements of hydraulic properties are time consuming, labour intensive and expensive. The most preferred way to acquire these data is indirect estimation (pedotransfer function). Pedotransfer function has potential use in soil and water management. The study related to soil and water dynamics needs an input hydraulic property. The geographical area of this watershed is 22,400 ha in Buldhana district of Maharashtra state. Sixty sampling points were marked at a grid of 2 km × 2 km using GPS. Soil samples were collected from each sampling point at a depth ranges from 0-30 cm. Results revealed that neural regression is a better tool for calibrating PTFs, as it resulted in predictions with lower errors. However in testing dataset the results were mixed implying that more data were required for calibrating robust PTFs. The predictions by neural PTFs utilizing SSC as an input and PTFs using SSC, BD and OC as input, both had acceptable error of prediction. It was inferred that a user of PTFs thus could use limited data of textural composition to predict FC, PWP and thereby AWC. It was noted that PTFs to predict PWP were in general better.

Keywords: pedotransfer function, hydraulic properties, neural regression etc.

Introduction

The demand of fresh water all over the world is increasing every year. The depletion because of over exploitation and or due to reduction in recharge and the threatening of this fresh water by contamination is especially due to expansion of Industry to the developing world in addition to local human contamination is intensifying. More than 98% of the available fresh water is groundwater which by far exceeds the volume of surface water (Fetter, 2001). Infiltration and evaporation are the most significant hydrological processes determining soil water storage in a rain-fed ecosystem. The soil hydraulic properties determine the behavior of the soil fluid within the soil system under specified conditions. It is pertinent that soil-water dynamics is studied in detail for managing rainfed as well as irrigated agriculture. The soil water dynamics needs an input of two important soil hydraulic properties – i) saturated hydraulic conductivity and ii) soil water retention characteristics. Infact in-situ or laboratory measurements of hydraulic properties are time consuming, labour intensive and expensive. The most preferred way to acquire these data is indirect estimation. Soil water retention characteristics are needed to describe the availability of soil water to plants and to model movement of water and solutes in unsaturated soils. the difference in water retention corresponding to two matric potentials (available water capacity) was predicted from textural and structural properties. Using particle-size analysis, developed relationships that describe water retention characteristics between 0.33 and 15 bar.

Rosetta is a program that estimates unsaturated hydraulic properties, from surrogate soil data such as soil texture and bulk density. This Pedotransfer functions are transform soil basic data in hydraulic properties. It's potential use is soil and water management.

Materials and Method

This section deals with the material and methods to achieve the objectives.

Methodology

Location, topography and climate of the study area

Area selected for the study is Shirala Nemane watershed is 22,400 ha in Buldhana district of Maharashtra state. It is located between is located between 76019'23.72''E- 76042'55.32''E longitude and 20017'32.48'' N – 20030'23.42'' N.

Collection of the soil samples

Seventy five sampling points were marked at a distance 2km × 2km intervals using Geographical Positioning System (GPS). Soil samples were collected from each sampling point at a depth of 0-30 cm.

Soil analysis

Collected soil samples were then analysed for calculation of physical and chemical soil properties such as bulk density by clod method (1965), particle size distribution by Bouyoucos Hydrometer (1986), moisture retention parameter by pressure plate apparatus (1986) and organic carbon by Walkely and Black method (1967).

Development of Point PTF's

Pedotranfer (PTFs) are generally developed using one of the following methods:

(i) Regression analysis
(ii) Neural networks
(iii) Group method of data handling and
(iv) Classification and regression trees (CART)

In a point-specific approach, water retention at a defined pressure (suction) head is estimated serving specific-point interest for application (Gupta and Larson, 1979; Rawls and Brakensiek, 1982; Campbell, 1990).

General form of the equation is,

$$\theta(h) = C_1 sand + C_2 silt + C_3 clay + C_4 bd + \ldots\ldots\ldots.. + C_i x_i \qquad \ldots.. 2.1$$

Where,

$\theta(h)$ = water retained at pressure h.

C1, C2, & Ci = regression coefficients.

X = variable represents predictor soil property that is readily available or easily measured.

The regression analysis is the oldest method, neural analysis is now most widely accepted due to its ability to develop a relationship between non-linear inequalities. Multiple regression technique are used to calibrate pedo transfer (PTFs) for predicting soil water retention at -33 and -1500 kPa (FC and PWP).

Software Used

Rosetta

Rosetta is able to estimate the Van Genuchten water retention parameters (Van Genuchten, 1980) and saturated hydraulic conductivity (Ks), as well as unsaturated hydraulic conductivity parameters, based on Mualem's (1976) pore-size model (Schaap et al., 2001). The retention function is given by Equation (2.2)

$$\theta\,(h) = \theta_r + (\theta_s - \theta_r)\left[\frac{1}{(1+\alpha h)^n}\right]^{(1-1/n)} \qquad \ldots.. 2.2$$

Where,

θ (h) = volumetric water content (m^3/m^3).

θr = residual water content (m^3/m^3).

θS = saturated water contents (m^3/m^3).

α = inverse of the air entry suction (> 0, in cm-1) and

n = pore-size distribution (> 1).

To render the pedotransfer (PTFs) of Rosetta as widely applicable as possible, a large number of records of soil hydraulic data and corresponding predictive soil properties were obtained from three databases (Schaap and Leij, 1998; Schaap et al., 2001).

A hierarchical approach with limited or more extended sets of predictors was used to estimate the water retention parameters (θr, θS, a, n) and Ks (Schaap et al., 2001). The first model (H1) is a class PTF, consisting of a lookup table that provides parameter averages for each USDA textural class. The second model (H2) uses sand, silt, and clay as inputs (SSC). The third model (H3) includes bulk density as a predictor (SSCBD) in addition to the input variables of the second model. While model H1 is a simple table with average hydraulic parameters for each textural class.

Performance Evaluation of Models

Performance of the pedotransfer (PTFs) will be evaluated based on one to one correspondence between measured and predicted values of Ks. Statistical indices. i.e. root mean square error (RMSE), mean absolute error (MAE), index of agreement (d), and maximum absolute error (ME) are based on squared difference between measured (Mi) and estimated (Ei) value, where 'i' indicates i^{th} value of dataset containing 'n' values.

Result and Discussion

The ranges of measured of soil water retention characteristics data of different soil textural classes are presented in Table 1. From Table 1 it is revealed that there is great variation in soil water retention characteristics of the soil groups.

Table 1 Range of soil water retention characteristics of different soil textural classes

Textural classes	FC (%)	PWP (%)	AWC (%)
Sandy loam	18.0-21.5	8.0-9.75	8.0-10.66
Sandy clay loam	24.0-26.0	13.4-16.66	8.0-12.38
Loam	25.0-27.0	10-17.76	13.0-16.0
Silt loam	25.0-28.76	9.0-18.5	16.0-17.76
Silty clay loam	33.0-35.88	16.0-20.0	16.0-18
Silty clay	31.52-36.0	17.5-22.0	14.0-19.69
Silt	34.0	17.0	17.0
Clay loam	30.0-37.5	15.0-22.5	14.0-20
Clay	38.0-43.88	29.0-31.0	14.0-27.8

PTFs were calibrated and tested using statistical and neural regression. The total data points were 60, of which 44 were used for developing the PTFs and 14 points were used for testing the calibrated PTFs. The input data used in PTFs was as follows

1. SSC
2. SSCBD
3. SSCOC
4. SSCBDOC

Thus four PTFs each were developed to estimate FC, PWP and consequently AWC.

Table 2 Equation for FC

Model	Equation for FC
H1	FC = (0.71)+(0.81)Sand +(1.10) Silt + (1.09) Clay
H2	FC = (-54.63)+(0.65)Sand +(0.95) Silt + (0.93) Clay+(-0.77) Bulk density
H3	FC = (-67.78)+(0.77)Sand +(1.06) Silt + (1.05) Clay+(-0.063) organic carbon
H4	FC = (-0.53)+(0.64)Sand +(0.93) Silt + (0.92) Clay+(-0.75) Bulk density +(0.027) organic carbon

Table 3 Equations for PWP

Model	Equation for PWP
H1	PWP = (58.37)+(-0.49)Sand +(-0.48) Silt + (-0.35) Clay
H2	PWP = (60.63)+(-49.20)Sand +(-0.49) Silt + (-0.35) Clay+(-1.50) Bulk density
H3	PWP = (59.35)+(-0.50)Sand +(-0.49) Silt + (-0.36) Clay+(0.01) organic carbon
H4	PWP = (86.47)+(-0.75)Sand +(-0.74) Silt + (-0.61) Clay+(-1.42) Bulk density + (-0.051) organic carbon

The development of point PTFs and performance of the PTFs were evaluated based on one to one correspondence between measured and predicted values of hierarchical approach with limited or more extended set of predictors was used to estimate the water retention parameters. In present study following parameters were studied.

- Basic soil physical and chemical properties

- Hydrological properties of soils

-Soil moisture retention

- Development and Performance of the PTFs

Table 4 Prediction of AWC using Neural Regression PTFs

Development set					
AWC	RMSE	D	ME	MAE	R2
SSC	0.0402	0.85	0.1097	0.0329	0.57
SSCBD	0.0355	0.91	0.0892	0.0263	0.72
SSSOC	0.0288	0.91	0.0889	0.0214	0.78
SSSBDOC	0.0279	0.93	0.1050	0.0182	0.81
Mean	0.0331	0.90	0.072	0.0247	0.72

The developed relationships were evaluated using statistical indices like root mean square error (RMSE), mean error (ME), mean absolute error (MAE), degree of agreement (d) and coefficient of determination (R2). The measured and estimated values were compared with one to one correspondence. The point PTFs of neural regression has better result predicted than statistical regression. The RMSE in developing PTFs to estimated FC ranges from predicting 0.0212 to 0.0353 m3 m-3 as against 0.0489 to 0.0498 m3 m-3 in statistical regression. The degree of agreement in developing PTFs to estimated FC ranges from 0.86 to 0.96 as against 0.79. ME in predicting FC ranged from 0.0787 to 0.1139 m3 m-3 as against 0.126 to 0.128 m3 m-3. MAE in predicting FC ranged from 0.014-0.028 as against 0.13 – 0.036 and The coefficient of determination in developing PTFs to estimated FC ranges from 0.63 - 0.87 as against 0.46 - 0.47. It's conclude that neural regression was a better tool and all the models (H1,H2,H3,H4) are good in development data set.

Conclusion

Based on the results obtained from the study, following conclusions are drawn.

The neural regression is a better tool for calibrating PTFs, as it resulted in predictions with lower errors. However in testing dataset the results were mixed implying that more data were required for calibrating robust PTFs. The predictions by neural PTFs utilizing SSC as an input and PTFs using SSC, BD and OC as input, both had acceptable error of prediction. It was inferred that a user of PTFs thus could use limited data of textural composition to predict FC, PWP and thereby AWC. It was noted that PTFs to predict PWP were in general better.

References

Batjes, N. H. 1996 Development of world dataset of soil water retention properties using pedotransfer rules. Geoderma 71: 31-52.

Black, C. A. Evans, D. D., Esingoger, L. E., White, J. L., and Clark, F. E. 1965. Physical properties In: Methods of Soil Analysis. Part I and II, Amer. Soc. Agron. Madison, Wisconsin, U.S.A.

Campbell, G. S. and Sho Shizawa, 1990. Prediction of hydraulic properties of soils using particle size distribution and bulk density data. Proceedings of International Workshop in Indirect Methods for Estimating the Hydraulic Properties of Unsaturated Soils. USDA ARS/University of California, Riverside, CA.

Fetter, C. W., 2001. Applied Hydrology (4th edition). New Jersey Upper Saddle River, New Jersey: Prentice Hall.

Gee, G. W. and J. W. Baunder. 1986 Particle – size analysis P. In A. klute (ed.) Methods of soil analysis, Part-1, Physical and mineralogical methods. Second Edition, Agronomy

Gupta, S. C. and Larson W. E. 1979. Estimating soil water retention characteristic from particle size distribution, organic matter percent and bulk density. Water Reourc. Res. 15: 1633-1635.

Mdemu, M.V. and Mulengera, M.K. 2002. Using pedotransfer functions (PTFs) to Estimate Soil Water Retention Characteristics (SWRCs) in the Tropics for Sustainable Soil Water Management: Tanzania Case Study.

Pachepsky, Ya. A., and Rawls W. J. (1999). Accuracy and reliability of pedotransfer functions as affected by grouping soils. Soil Sci. Soc. Am. J. 63: 1748-1756.

Schaap, M. G., Feike J. L., Martinus L., and van Genuchten M. Th. 2001. ROSETTA: a computer program for estimating soil hydraulic parameters with hierarchial pedotransfer functions. J. Hydrology 251: 163-176.

Schaap, M. G., Leij F. L., and van Genuchten M. Th. 1998. Neural network analysis for hierarchical prediction of soil hydraulic properties. Soil Sci. Soc. Am. J. 62: 847-855.

Shirazi, M. A., and Boersma L. 1984. A unifying quantitative analysis of soil texture. Soil Sci. Soc. Am. J. 15: 142-147.

Tietje, O. and Tapkenhinrichs M. (1993). Evaluation of pedo-transfer functions. Soil Sci. Soc. Am. J. 57: 1088–1095.

Van Genuchten, M. Th., 1980. A closed-form Equation for predicting the hydraulic conductivity of unsaturated soils. Soil sci. Soc. Amer. J.44: 892-898.

Williams, R. D., Ahuja L. R., and Naney J. W. 1992. Comparison of methods to estimate soil water characteristics from soil texture, bulk density, and limited data. Soil Science 153(3): 172-184.

Wosten, J. H. M. and van Genuchten M. Th. (1988). Using texture and other soil properties to predict the unsaturated soil hydraulic functions. Soil Sci Soc. Am. J. 52: 1762-1770.

Significance of Public Participation through Crowdsourcing in Planning and Development for a Better Infrastructure

Padmaja G[1], Giridhar MVSS[2], Shiva Chandra V[3] and Akhil G[4]

[1,3]Dept. of Civil Engg., MVSR Engg. College, Hyderabad,Telangana, India

[2]Center for Water Resources, IST, JNTU, Hyderabad,Telangana, India

[4]Student, Department of Computer Science, AGI, Hyderabad,Telangana, India

ABSTRACT

In today's world smart phones and internet play a pivotal role, enable boundless applications which can be used effectively to solve various infrastructure problems faced by mankind. Crowdsourcing is one such application that involves collection of data by "crowds" about the infrastructure system, damages caused, issues relating non-functionality of structures etc. It involves people's participation to gather information and also allows them to incorporate their ideas for improving either the existing facilities or build new ones. The role of participation by public is the key factor, thus crowdsourcing integrated with GIS platform can be a helpful tool in efficient infrastructure planning more so in urban areas.

The aim of this paper is to understand the importance of crowd sourcing in connecting communities and individuals, engaging them in creative activities by capturing their ideas, proposals and innovative solutions to the existing problems. The ability of crowd sourcing to solve problems, as citizens themselves become active participants who collect the data, also give advice to solve the existing problem and get involved in activities there by evaluating its feasibility in attaining solutions in the present scenario.

Keywords: Crowd Sourcing, epicollect, GIS.

Introduction

The exponential population growth, migration of people from rural areas to urban areas poses numerous challenges to the public living in cities. Urbanization provides people with better amenities, job opportunities and the difficulties as well. These include traffic congestion, pollution, scarcity of non-renewable energy resources, management of solid waste, poor hygiene, improper drainage facilities among others. Rapid urbanisation has contributed to drainage systems getting clogged and is no longer capable of taking the pressure of excess water that gets accumulated due to heavy rain, which invariably leads to waterlogging and urban floods. Most of these urban floods are induced by human and are triggered due to improper maintenance of drainage networks, choking caused by plastics, rubble, shrinkage of open spaces by encroachments, which contribute to accumulation of water on roads even for a moderate rainfall.

Crowd sourcing is defined as the use of digital technologies to gather and organise information from a multitude of non-professional individuals to develop innovative solutions to impending problems. It is an approach in which non-expert people, called "crowds", join a project to collect the data by using a simple smart phone on hand. It is the use of digital technologies to gather and organize contributions from citizens in order to improve the urban liveability and aim at a better Infrastructure System. Crowd sourcing processes can be widely used across the globe to increase the community participation and improve the local government decision making processes for better standards of living. The opinion that crowd sourcing can effectively be useful in urban planning projects is widely shared (BRABHAM, 2009; HILGERS, IHL, 2010). Public participation is a key parameter in crowd sourcing and is a good way to engage citizens and facilitate an open dialogue between them and urban planners (BUGS, et al., 2010; ADAMS, 2011).

The study is initiated to familiarise the benefits of crowd sourcing in mapping the conditions of open drains in cities which often lead to urban floods due to their poor maintenance and clogging due to various obstacles. Crowd sourcing would be a good solution for problems requiring large amount of accurate information. EpiCOLLECT5 a crowdsourcing application tool integrated with QGIS, an open GIS platform can be used for addressing most of the drainage blockage problems in cities. The data collection is performed by non-expert operators or crowds on a regular basis to check the status of drains. The data is uploaded on to the web server on a routine basis so that required measures would be directed by officials for the smooth running of the networks. Thus the digital platforms created should empower the local residents by expanding civic engagement in collection of data and initiate mitigation measures that will have tangible effect on the way they live and also provide smart solutions to their own problems by means of their cooperation, creativity and ideas.

Study Area

The State government's consideration about the desilting of nalas by GHMC officials revealed that 70 % of the solid waste blocking nalas and drains in the city is due to plastic waste. Most of the drainage overflow complaints received from the public in peripheral areas was due to plastics, eateries, oils and other wastes there by clogging drains. Lal Bahadur Nagar, well known as L. B. Nagar, is a commercial and residential hub in Hyderabad, Telangana, India is considered for the proposed study.

Figure 1 Study area

EpiCollect5

- *EpiCollect5* was developed by Imperial College London to provide a simple and intuitive method for complete online project creation, data storage and visualisation of data captured using smartphones. It can be used for a broad spectrum of data collection applications and projects by just using a smartphone. We can create specific projects based on the users requirements and can design the data collection form on line by following a simple process at the EpiCollect5 website (www.five.epicollect.net). It creates a project in just a few easy steps and the Data can be collected online / offline. If the project contains location as one of the parameters then it can be viewed on a map.

  EpiCOLLECT5 software consists of two parts

1. A mobile application
   * To collect the Data using a Mobile handset that runs on Android or iPhone Operating System and can be used for running EpiCollect5.
2. A web application
   * For developing or generation of forms
   * For storing the data collected using the epicollect5
   * Freely hosted project websites for mobile data collection projects.

Advantages of epiCOLLECT5

• Simplicity • Fewer errors • Quicker • Easily updated • Data immediately available • Cost effective • Smart phones that are always handy • Flexible • No skill required for data collection • No other equipment needed

Methodology

Form Builder → Field Work → Integrate with QGIS

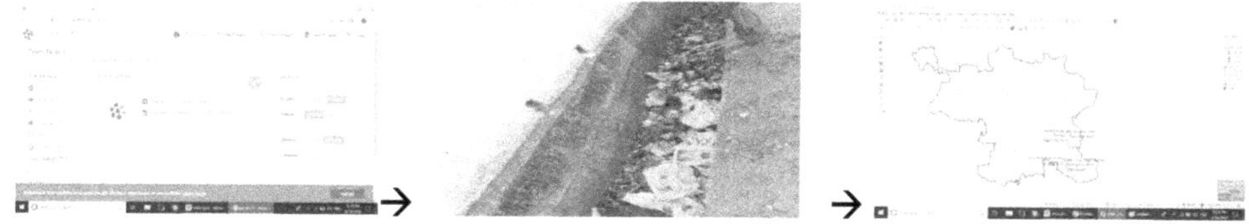

Figure 2 Methodology adopted

1. **Create Project:** This is the first step in which the Name of the project is assigned and a permanent project homepage is created. The 'Create project' link at https://five.epicollect.net/project generates a prompt for entering a new project name – in this case "Open Drains".

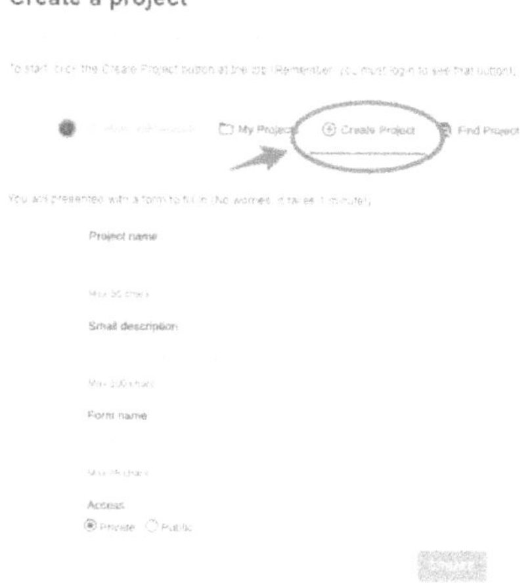

Figure 3 Web Application Interface

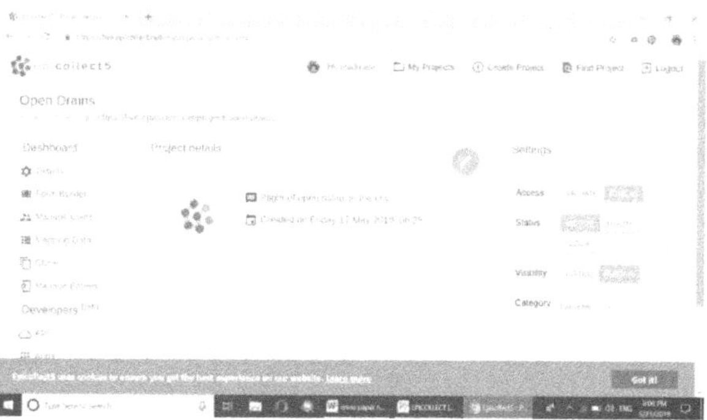

Figure 4 Create Project Interface

2. **Project Design:** In this phase it is required to identify the survey questions based on the requirements of the project initiated by using a 'drag and drop' form builder. A data collection form having detailed questions to be asked during the collection of the data based on the users / projects requirements is now generated. The project owner has to login to five.epicollect.net using a Gmail account and only this Gmail account holder can modify the project form at a later stage. Once logged in, the project owner is presented with a 'drag and drop' form builder, which provides a simple and intuitive way to state questions presented to the user. There is no limit on the number and type of questions that can be specified in a single form. Forms containing different questions can be built easily online and personalised to the requirements of the project. Upon completion of the form, now the project is ready to be loaded onto one or more mobile phones which can be used by various users for data collection. The form builder has three columns layout. 1. Left column: Available inputs2. Middle column : Inputs added to each form based on user requirements, 3. Right column: Currently selected input settings

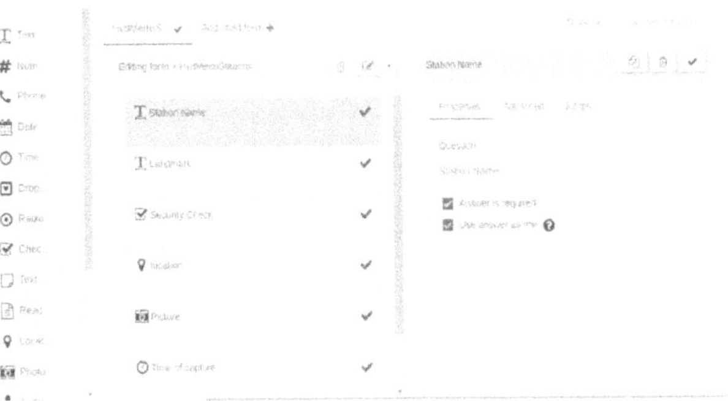

Figure 5 Form Builder Interface

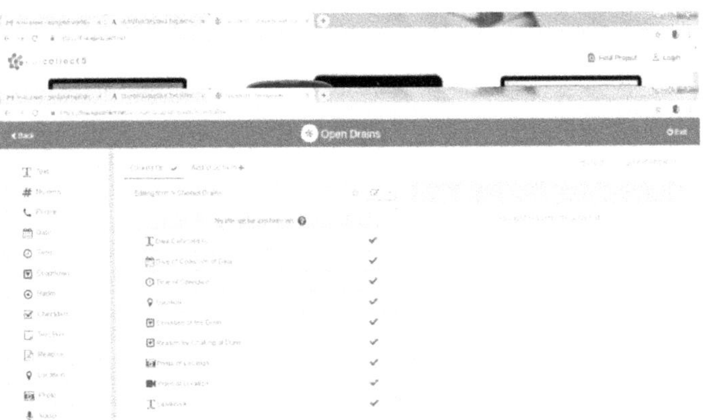

Figure 6 Designing the Form using the form builder in epiCOLLECT5

3. **Collection of data:** The EpiCollect5 mobile app is first installed on each Android or IPhone to be used in the survey. This is the important phase in which various users or crowds can load the project onto their mobile phones and start collecting data. From the mobile app's home screen, entering the project name in the 'Load Project' menu option - "Open Drains". Once loaded 'New entry' is seen on the home screen by clicking which the user can start the data collection. Each new entry is assigned a unique ID and automatically date and time are assigned with location information which can be further be integrated into QGIS for further analysis. A photo or video can be captured using the phone's camera and can be attached to an entry and then the questions defined in the survey form are answered. Each completed entry is stored in the phone's database, allowing many entries to be collected on a single mobile. Many users can collect the data simultaneously using multiple devices and finally upload the data to the project website. Data connectivity is required while initially loading the project into the app and when synchronising the data with the project database.

Entry: d99a5d2b-267c-429c-b46b-8e958f26c5f1

Question	Answer
Data Collected by	Akhil
Date of Collection of Data	19/05/2019
Time of Collection	18:09:00
Location	17.341871 78.553065
Condition of the Drain	Overflowing
Reason for Choking of Drain	Misc
Photo of Location	
Video of Location	
Landmark	dvk

Figure 7 Data collection on the mobile

4. **Visualise:** The data collected from either one or more users during the survey can be viewed via the project homepage. The data can be listed in a tabular view, with columns based on the defined form fields, and all data can be downloaded in a either CSV or XML format for further use and interpretation outside of EpiCollect5.
5. **Integrate with QGIS:** The data collected using epiCOLLECT5 is now downloaded into .csv format and is incorporated into QGIS environment for further analysis. The maps can be generated and further used for analysis during the event of any disasters like urban floods. Flood hazard maps can be developed from the data collected and potential zones which are vulnerable to inundation during heavy rains can be identified.

Applications

The potential applications of EpiCOLLECT5 are as vast as imagination. Virtually any information gathering task can be replicated through a survey platform, and made faster and more efficient. Some of the potential applications of EpiCOLLECT5 are Infrastructure Mapping, Mapping and rating road infrastructure such as ditches and culverts based on their maintenance and efficacy, Location of Solid waste dumping sites in cities, Mapping of potential zones for Rain water harvesting etc.

Results

Figure 8 Data collection using smart phone

Figure 9 Data downloaded from epiCOLLECT5

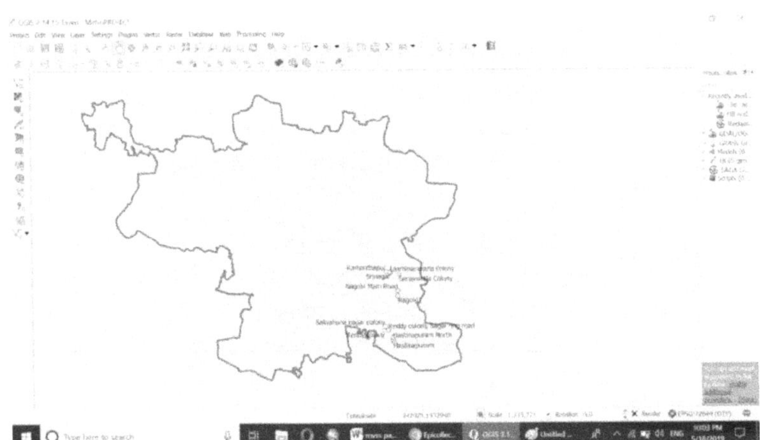

Figure 10 Integration of downloaded data from epiCOLLECT5 in QGIS

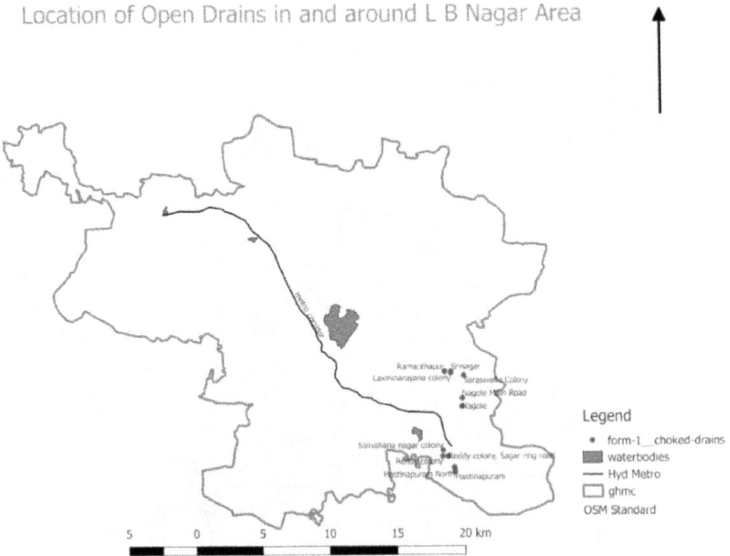

Figure 11 Generation of Map in QGIS

Conclusion

In this assignment, the study area considered was limited to 10 km × 10 km. The scope of the study can be extended to greater areas for better results. There is still a lot of space to be investigated in conditions of assigned region, complexity of urban area, cultural attitudes, type of communities residing in those areas, types of soil and maintenance issues in drains etc. Designing and offering services for citizens can present challenges and critical issues since they impact on their basic needs, interests, and cultural attitudes of groups. Communities play a significant role in all urban issues and can seriously affect both delivery and sustainability of services. Quantification of community participation is not an easy task. People's involvement is essential in addressing issues like clogging of drains which have impact on their living standards requires a great effort by local governmental institutions. Motivation of public is essential and their involvement in gathering information is crucial if municipalities have to engage citizens who would be stake holders and face the consequences. Citizens must be encouraged as "crowds" who provide necessary and useful data and also they must be ensured that their suggestions are taken into account in solving impending problems. Indeed, many issues concerning crowdsourcing in urban areas where huge populations reside need to be explored more in-depth. Therefore the use of crowdsourcing for developing large-scale projects and encouragement of public participation for better standards of living is the need of the hour. In the context of urban innovation, Lucifer's exhortation, contained in Milton's poem, to "Awake, arise, or be forever fall'n" is very apt for the present task on hand, but we should negotiate this inducement within the scope of a collaborative approach, in a renewed engagement aimed at social liveability and social harmony.

References

Alexander Wilson , Mark Tewdwr-Jones , "Let's draw and talk about urban change: Deploying digital technology to encourage citizen participation in urban planning".

Brian Spratt , Derek Huntley Imperial College London, EpiCollect+: Linking smartphones to web applications for complex data collection projects.

C.J. Lukensmeyer and L.H. Torres "Citizen sourcing: Citizen participation in a networked nation", 2008.

Daren C. Brabham "Crowdsourcing the Public Participation Process for Planning Projects".

Epicollect5 Data Collection User Guide, https://epicollect5.gitbooks.io/epicollect5-user-guide

gilberto marzano rezekne academy of technologies, joanna lizut wyższa szkoła pedagogiczna im. janusza korczaka, poland, warsaw, "ˆUrban crowdsourcing: potential and challenges".

Parul Srivastava 1 and Ali Mostafavi 2,*" Challenges and Opportunities of Crowdsourcing and Participatory Planning in Developing Infrastructure Systems of Smart Cities".

Sustainable Management of Water Supply Using Hard Rock Aquifers in Water Scarce South Andaman Island

Amlanjyoti Kar

Central Ground Water Board, CP-6, Sector-V, Salt Lake City, Kolkata.

karamlanjyoti@gmail.com

ABSTRACT

An area of about 200 sq.km in South Andaman and Rutland Island is underlain by hard rocks comprising maficultramafic cumulates, plagio-granite and thick pillow lava sequence of Ophiolite igneous suite. These rocks are characterized by shallow weathering and potential water bearing fractures that occur up to a depth of 100 - 120 m. Extensive fracturing has also facilitated the genesis of potential springs in the Ophiolites in many islands. Discovery of such potential springs by the author in the Rutland Island has enabled the Island administration to formulate water supply scheme for interisland transfer of water to the Port Blair town. Hard rock aquifers in parts of the study area are continuously under development by private bore wells for irrigation, drinking and poultry purposes, since the late nineties. Unscientific use of groundwater through more than 200 bore wells has resulted in overexploitation of this precious natural resource as envisaged during 2004, and the mega earthquake (M = 9.3) of 26.12.04 had made a severe impact on the hard rock aquifer system whose effects are still persisting and will continue further. The earthquake generated new set of fractures in the brittle formation which has facilitated the movement of groundwater to the topographic lows. After the event, in the parts of the study area, the ground water and surface water sources located at higher altitude either dried up or their discharges dwindled. Interestingly there was increase in discharge in the water sources available at lower altitude. New springs generated in the topographic lows. The areas at relatively higher altitude are suffering from acute crisis of drinking and irrigation water after the mega disaster. Intensive studies were carried out to find out the solution with artificial recharge to the aquifers and rejuvenation of drinking water supply and irrigation in the area. Peer surveys have delineated and pinpointed subsurface fractures which can be tapped through water wells and grouted. The paper highlights the studies carried out in the area with a holistic approach and the recommendations to manage the ground water for various requirements in the aftermath of the colossaldisaster.

Keywords: Hard rock aquifer, Over exploitation, Fracture, Mega earthquake (M=9.3), Artificial recharge, Grouting, Drinking and irrigation water.

Introduction

The study area (12^{0}38'30" to 12^{0}28'30"N and 92^{0}37'30"to 92^{0}45'30"E) spans for about 200 sq km and is located at the outskirt of Port Blair town in the South Andaman Island and includes the contiguous Rutland Island. The area falls under the jurisdiction of newly created South Andaman district of the Union Territory of Andaman and Nicobar Islands in the Republic of India (Fig. 1). Andaman and Nicobar group of Islands form an arcuate chain of about 720 km length and 24 km width in the Bay of Bengal. The cordillera is composed of 532 picturesque small to large islands, few of which are important tourist destination. The capital town of the island territory is Port Blair (11^{0}40"N, 92^{0}46"E). The location of the islands is highly interesting in matter of collision tectonics. The Andaman and Nicobar Islands forms the outer arc ridge of the subduction zone, where the Indian plate collides with the Eurasian plate. Besides generating the killer tsunami, the great Indian Ocean earthquake of magnitude 9.3 in the Richter scale on December 26, 2004, caused significant vertical changes in its rupture zone. About 800 km of the rupture is co-seismic deformation along the exposed land and could be observed as uplift/subsidence (Rajendran et al. 2008). During the mega disaster as also in its aftermath many significant changes on land and especially in the water resources of the islands occurred whose case studies are forming invaluable evidences in the literature (Jain 2005; Rao and Chary 2005; Ramachandran et.al 2005; Jade et.al 2005; Kar and Adhikary 2005; Subarya et al. 2006; Thakkar and Goyal 2006; Mondal and Rai 2006; IGRAC 2006; Banerjee et al. 2007; Rajendran et al.2007).

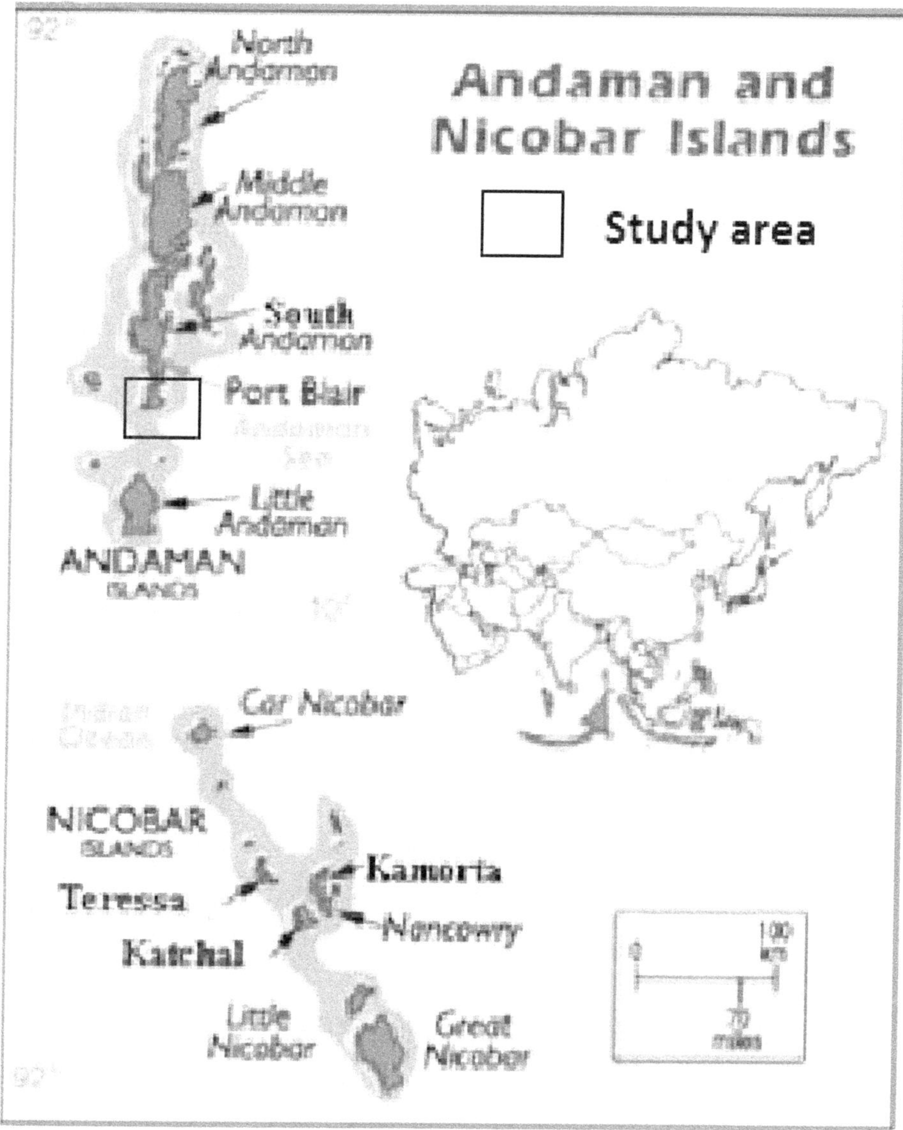

Figure 1 Locationmap

The islands were devastated by both Mega earthquake (M = 9.3) and tsunami. The islands in the Nicobar group as also in the Andaman group up to Rangat ($12^{0}30^{'}25^{''}$N,$92^{0}55^{'}01^{''}$E) in Middle Andaman were subsided while the rest part in Middle and North Andaman were uplifted (Jain 2005; Kar and Adhikary 2005). Tsunami could not damage the study area, because of its topographical height barring the low lying coastal stretches in the south near Chidiyatapu. However, earth quake had severely battered the area which had influenced the water resources to a great extent. Pre- and post- tsunami geological studies of Andaman and Nicobar islands were carried out by various workers of the Geological Survey of India (Vohra et al. 1989; Roy et al. 1988; Bandyopadhyay et al. 2008; Kayanne et al. 2007; Ray and Acharya 2007, Som et al. 2009). Hydrogeological and geophysical studies of the islands were carried out by many workers of the erstwhile ground water wing of the Geological Survey of India and subsequently by the Central Ground Water Board of the Union Ministry of Government of India (Ghose 1972; Adyalkar and Najeeb 1981; Banerjee et al. 1988; Bhattacharya 1997; Kar 2001; Kar 2002a,b,c; Kar 2003; Kar 2004a,b). Ground water resources of the entire Andaman and Nicobar islands as also in the study area have been thoroughly investigated in the post-tsunami period (Kar and Adhikary 2005, Kar et al. 2006 a,b; Kar 2006, Kar and Gawri 2006; Kar and Gawri 2007; Sarkar 2007; Kar and Srivastava 2008; Raja et al. 2009). During 1984-91 deep ground water exploration activities were undertaken in South, Middle and North Andaman including Great Nicobar islands which had unraveled the groundwater potential of the deeper aquifers of the geological formations in the island territory for the first time.

Climate and Rainfall

The overall temperature in the islands remains almost identical throughout the year while the mean maximum and minimum temperature range from 23° to 30°C. Relative humidity remains high in the rainy season ranging from 79 to 89%. Mean humidity varies from 77 to 82%. The islands receive copious rainfall both from northwest and southeast monsoons and the average cumulative rainfall is 3180 mm per annum. The monsoon generally sets in May and continues till November. In the month of December good amount of rainfall occurs in the area. During the last two decade highly erratic behavior of rainfall has been observed (Fig. 2) in the islands. In thirteen years there has been negative departure from the normal rainfall which has disturbed the water supply recurrently, leading owing to the miseries of rural and urban population in the South Andaman Islands. In the post-tsunami period there was positive departure of rainfall from normal in 2005 and 2008, was almost identical in 2006 while a negative departure of 20% was recorded in 2007 (Fig. 2). The rate of evaporation (Fig. 3) in the islands as recorded in Port Blair is on an average of 1400-1500 mm per year because of geographical location of the islands in the Tropical zone. Because of high evaporation rate, the effective rainfall is relatively less which warrants judicious conservation.

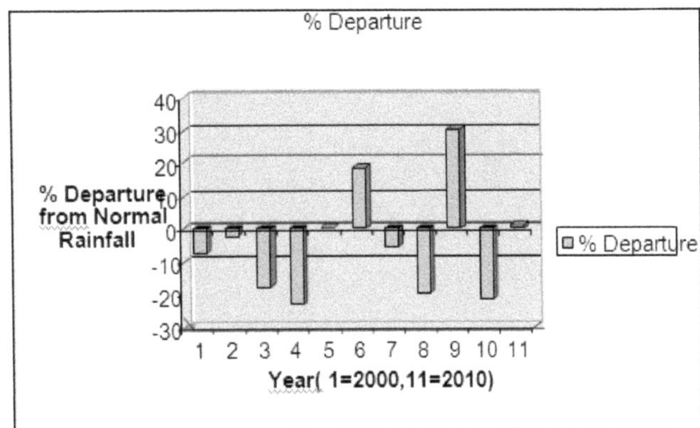

Figure 2 Percent deviation of rainfall from normal

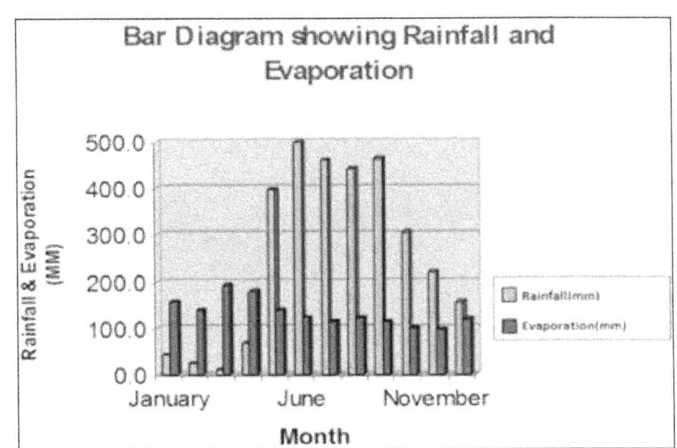

Figure 3 Monthly evaporation verses rainfall

Geomorphology and Drainage

The topography of the study area is highly undulating. The area is characterized by low conical hills and deep slender valleys which occur within the elevation range of 60 to 100 m from mean sea level. Relatively high topographic position saved the area from the invasion of tsunami waves barring the low lying fringe along coast line in the east and in the southern tip near Chidiyatapu and near Bimblitan and Sippighat in the west. Drainage pattern is dendritic and immature and it is fully controlled by the structural weak planes. Small streams, locally called *nalas*, like Burma *Nala*, Kodiaghat*Nala*, Brichganj*Nala*, Bimblitan*Nala*, Barakhari*Nala*, Chain *Nala*etc. drain the area and show meager to significant flow even in dry period.

Geology

Late Cretaceous igneous ophiolite suite of rocks, marine sedimentary classic rocks of Paleocene to Oligocene age, Recent to sub Recent beach sand, mangrove clay, alluvium and coral rags underlie the areas in south Andaman and Rutland islands (Fig. 4). While a lion share of the South Andaman island (more than 80% of the total geographical area) is occupied by the sedimentaries, an area of about 200 sq km, that is, the study area, in South Andaman and Rutland Islands is underlain by hard rocks comprising mafic-ultramafic cumulates, plagio-granite and thick pillow lava sequence of Ophiolite igneous suite. These ophiolitic rocks are exposed in the tract of Brookshabad-Brichganj to Calicut-Maccapahar- Chidiyatapu sector and extended in the west up to Teylerabad-Bimlitan areas. In further west ophiolites crop out at Sippighat-Lalpahar-Chouldari-Badmaspahar areas. Since the islands are located in a tectonic zone, the rocks are highly disturbed and intensely fractured and sheared.

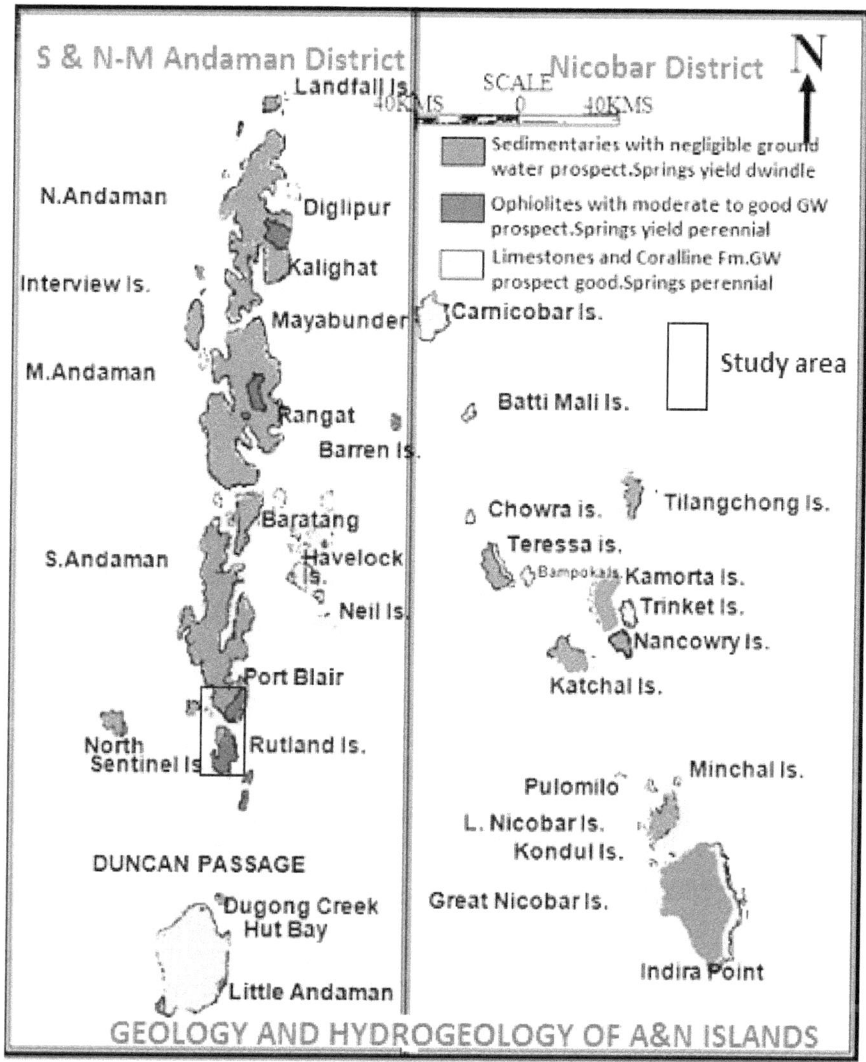

Figure 4 Geology and hydrogeology of Andaman Islands

Hydrogeology

Availability of ground water in hard rock is unubiquitous and if it is in island certainly occurrence of fresh ground water is to be ascertained more cautiously keeping in view of its scientific withdrawal to avoid sea water intrusion. Several published literatures are available on hydrogeological studies in the islands underlain by mafic and ultramafic igneous rocks (Peterson 1972; Peterson 1973; Woodhall 1974; Virgilio et al. 2001). The ophiolite suite of rocks in the study area are characterized by shallow weathering down to a depth of 1 - 12 m and potential water bearing fractures occur up to a depth of 70 - 80 m. The potentiality of the hard rock aquifer in the study area was first discovered through the exploration made by the Central Ground water Board (CGWB) in 1984 at Calicut village ($11^0$36'N $92^0$42'30"E). Ground water occurs under unconfined condition in the weathered mantle while confined to semi-confined conditions exist in the deeper fractured mantle as evident from the occurrence of the auto flowing wells in the area. The weathered and fissured horizons in ophiolite are tapped by dug well and bore wells with moderate yield (Table 1).

Table 1 Hydrological parameters of shallow colluvial-regolithic and deeper fractured aquifers in the hard rock areas of South Andaman

Sl. No	Location	Aquifer	Type of structure	Depth (mbgl)	Dia(m)	SWL (mbgl)	Yield (LPD)	Duration ofpumping (min) during test	Recovery (m)	Sp. Capacity(m3/min)	Transmi ssivity (m2/day
1	Corbyn□sc ove	W.PL	Dug Well	5.42	1.43	3.02	2500	15	0.73m in1225 min	0.0458	19.05
2	Brooksh-abad on the way to Quarry	W.PL	Dug Well	2.68	2.55	1.72	3000	16	0.55m in 1320 min	0.0058	44.46
4	Chidiyatapu	W. PL	Dug Well	5.75	1.46	3.95	6000	50	1.20m in 350min	0.0095	271.6
5	Brichganj	W.PL	Dug Well	4.17	1.88	2.57	2500	15	0.128 m in 438 min	0.0079	23.64
6	Calicut Junction near Mosque	WF. PL	Bore Well	80.2	0.203	1.30	375x103 (500min pumping)	500	4.05m in 650 min	0.090	163.79
7	Beadn-abad near Culvert	Coll. and W.PL	Tube Well	17.0	0.203	3.28	3729x10 2 (500min pumping)	500	4.67m in 484 min	0.131	127.03

Abbreviations used: m - metre, m bgl- metre below ground level, min- minute, Sp. Capacity - Specific capacity, W - Weathered, PL - Pillow lava, Coll.- Colluvium, Dia.- Diameter, LPD -Litres per day

The intermontane valleys are underlain by the porous colluvial materials as also alluvial valley fill materials of various size fractions borne by the streams and carries good amount of base flow even in the summer. Thickness of the valley fill varies from 1 to 18 m below ground level (bgl) as revealed from the detailed hydrogeological studies coupled with electrical resistivity sounding and exploratory drilling carried out in the area in the pre- tsunami period. Hydrogeological studies revealed that the valley fills form potential aquifers at shallow depth which can be intercepted by tube wells and collector wells with subsurface dams. At places where potential fractures are available at shallow depths, dug-cum-bore wells can also be constructed with subsurface dams. Depth to water levels in various geomorphic settings is shown in Table 2. The hydrogeologic situation of the area has thoroughly changed after the mega-earthquake (M = 9.3) on 26.12.04 (Kar and Adhikary 2005; Kar 2006; Kar 2007). Pumping and slug tests were conducted in dug wells and bore wells tapping both shallow and deeper aquifers in the ophiolites to adjudge the aquifer potential and to determine the aquifer parameters. Good number of slug tests were also conducted in the shallow experimental pits to verify the transmissivity of the valley fills for construction of subsurface dams and check dams (especially in areas where good surface flow exist during dry season). Computation of hydrologic parameters were done following the methods of Slichter (Slichter 1906), Hvorslov (Hvorslov 1931) and Cooper (Cooper 1967).The results are summarized in Table 3. Thickness of weathered horizon generally tapped by the dug wells varies in depth from 3 - 6 mbgl.

Table 2 Seasonal fluctuations and water levels in different physiographic settings

Sl. No	Physiogra phy	Depth to water level (m bgl)		Fluctuation (m)	Remarks
		Pre-monsoon	Post-monsoon		
1. 2.	Intermontane valley Hill slope	0.6-3.66 3.16-7.58	0.47-2.69 0.98-5.95	0.31-0.97 1.63-2.18	At places auto flowing conditions were noted. Good springs are developed in ophiolites. The water level situation was changed after the mega (M = 9.3) earthquake.

Besides tapping ground water from wells, the other mode of availability of ground water through natural gravitational flow is the spring water. Spring water forms a dependable source of water supply in the islands since the establishment of these islands by Britishers. These springs are tapped for water supply in the rural areas of South Andaman Island by Andaman Public Works Department (APWD). Number of such topographic and fracture springs is originated in the forest clad hills of South Andaman and Rutland islands. Studies carried out by CGWB (Kar 2002; Kar 2004) have proved that the springs originating in the fractured ophiolites are more potential and sustainable than the springs in sedimentary formations. The entire South Andaman Island including Port Blair urban area suffers from acute drinking water crisis in the years with deficit of rainfall (Fig. 2). During 2002, in the wake of acute drinking water crisis in Port Blair and environs, a reconnaitory study on the yield potential of the springs of Rutland island and their feasibility for interisland transfer to Port Blair city was carried out by the author and a recommendation was put forward to A&N Administration for implementation (Kar 2002). The study revealed that nearly 36,54,722 gallons of water per day was unabatedly flowing to the sea (Table 4). A short term option was immediately implemented by the island administration in 2002 by transporting spring water from the Chain *Nala*by barges. The implementation of the project was deferred in view of its environmental clearance and expenditure sanction. Recent severe water scarcity in the summer of 2007 was instrumental for final approval of the inter-island transfer of spring water scheme from the Rutland Island and the pipeline work isunderway.

Status of ground water development in the pre-tsunami period

After the establishment of the villages in the past in the study area, only dug wells were constructed by the local inhabitants tapping the weathered regolithic aquifers to meet the drinking and domestic water requirements. However, perennial springs were tapped by the Andaman PWD for rural piped water supply.

Table 3 Result of slug tests carried out in the streams flowing through the areas underlain by hard rocks in South Andaman Island, India

Sl. no	Location of the Stream section	Depth of the alluvium/colluvium above the massive/ fractured basement (m)	Average width of the Stream section (m)	Nature of the Stream and flow condition in lean period	Transm-issivity (m^2/day)	Conservation and recharge structures proposed to harvest the flow
1	Beadnabad Near Pump House	3 - 5	10 - 15	P, MF	159.23	CheckDam
2	Near culvert of Maratha Rifle, Brichganj Cant.	3 - 4	12 - 15	P, SF	779.93	SSD, DCB
3	EME Workshop, Brichganj Cant.	2 - 3	5 - 8	P, MF	6.33	SSD, DCB
4	Taylerabad near Quarry	5 - 6	5 - 8	NP, BF	381.97	SSD, DW
5	CSD Depot, Brichganj Cant.	5 - 8	8 - 10	P, SF	112.75	SSD, DCB
6	Spring zone of BadmasPahar	3 -5 m	2 - 3	NP, BF	1.74	SSD, DW
7	Bird Line. Near explosive godown	5 - 8	5 - 8	P, SF	679.27	SSD, CD, SH, DCB
8	LalpaharNalanear pumphouse	3 - 5	5 - 8	P, MF	120.78	SSD, DW
9	Brookshabad on the way to quarry	3 - 5	15 - 20	P, SF	57.81	SSD, DW
10	Near J&K Rifle,Brichganj Cant.	5 - 8	5 - 8	P, SF	81.66	SSD, DCB
11	Near Brichganj Cant. Canteen	5 - 6	8 - 10	P, SF	91.67	SSD, DCB

Abbreviations used: Cant.-Cantonment, P -Perennial, NP-Non-perennial, MF-Meagre flow, SF-Scanty flow, SSD-Subsurface Dam, DCB-Dug-cum-bore well, DW- Dug well

Table 4 Details of springs in Rutland Island

Sl. No	Date of inspection	Name of spring	Location of place of origin of the spring from sea coast (km)	Reading through 900 V-notch plate (cm)	Discharge (gallons per day; GPD)	Electrical conductivity (micro-Siemens/cm)
1	6.2.02	BadaKhadiNala	2.5 5	20 20	460320	270 270
2	9.2.02	KendiNala no-I	2.5 3.5	20 21.5	460320 555200	320 310
3	9.2.02	KendiNala no-II	2.5 3 .0	14 14	189760	350 330
4	9.2.02	KumdaNala	2.5	14	189760	350
5 6 7 8	12.2.02 12.2.02 12.2.02 12.2.02	Beth Nala-I Beth Nala-II Grass NalaNimbooNala	2.5 2.5 2.5 2.5	Dry Dry Dry Dry	Negligible NegligibleN egligibleNeg ligible	290 300 340 320
9	12.2.09	KichedyNala	2 .0	8	45120	350
10	12.2.09	Bamboo Nala	3 .0	13	156800	340
11	11.2.02	Chain Nala	0.75	23	657600	240
12	11.2.02	MeethaNala	1.0	11	104320	280
13	11.2.02	PuranaNala-I	0.5	3	4038	310
14	11.2.02	PuranaNala-II	0.75	12	126400	330
15	11.2.02	PuranaNala-III	0.75	9	63630	320
16	11.2.02	KomioNala	0.5	23.	657600	240
Cumulative discharge from all Springs in Rutland Island (GPD)					32,10,278	
Subsurface discharge as base flow along all spring locations (GPD)					4,44,444	
Total Ground water availability for Inter Island transfer to Port Blair (GPD)					36,54,722	

Large scale exploitation of the hard rock aquifers in parts of the study area located at Brichganj, Kamrajnagar, Calicut, Maccapahar, Teylerabad, Bimbllitan, new Bimblitan was started after the maiden ground water exploration done by the Central Ground Water Board in 1984 at Calicut village and at Beadnabad. Since then the ground water in the fractured ophiolite aquifers in the area was continuously developed through private bore wells and gradually 200 private bore wells were constructed in the area for irrigation, drinking and poultry business. The first irrigation well in the entire islands was constructed in the study area. Thereafter, unscientific use of ground water has led to its over-exploitation as envisaged from the information supplied by APWD. As per the data the static water level (SWL) of the bore well constructed by the Central Ground Water Board (CGWB) was 1.30 m bgl during March-April, 1985. In April 2002 the SWL dropped to 5.34 m bgl while in April 2004 the SWL was recorded as 7.01 m bgl. The borewell yield was 2, 95,000 litres per day (LPD) in April 2000 while in April 2002 it was 95,000 LPD. During the survey in 2004, it was reportedbythelocalvillagersthattheSWLandthedischargeoftheirrespectiveborewell has significantly dropped in the last 2 - 3 years. Based upon the studies carried out by the author, the Andaman and Nicobar Islands Administration was informed by CGWB and it was decided to stop further drilling of borewells in the area and to notify the area under the Central Ground Water Authority. Accordingly for creating public awareness before notification, regular publication in local print media regarding the scientific observation on overexploitation of groundwater in the area and its after-effects was initiated along with telecasting and broadcasting of scientific programs on groundwater development status in the islands. However, before enacting the regulatory measure and initiation of large scale artificial recharge and conservation of ground water, the area was rocked by the mega (M = 9.3) earthquake on 26.12.04 which was also responsible for the killer Indian Ocean tsunami.

Besides exploitation of groundwater through bore wells, dug well and springs, detailed studies were carried out by CGWB (Kar 2001; Kar 2003) on conservation and recharge of perennial subsurface flow (base flow) along stream channel flowing unabatedly towards the sea and also on conservation of surplus monsoon runoff. From the studies, models were prepared for augmentation of drinking and irrigation water in the entire Andaman and Nicobar islands including the study area. Studies revealed that good amount of ground water flows along the stream *(nala)* sections as base flow even in dry season when the islands suffer from severe crisis of drinking and irrigationwater.

While the formation materials in the islands at many places are impervious and unproductive in terms of ground water, the stream sections are underlain by appreciable thickness of porous alluvium/colluviums which can yield good amount of water if the base flow is intercepted through subsurface barrier or subsurface dams. Profuse annual and intermittent rainfall even in dry spell gives ample scope for recharge of the subsurface reservoir in succession if recharging structures like check dams, shafts are constructed. The subsurface perennial flow (base flow) is required to be obstructed by subsurface dams, which can be tapped fully through properly designed dug wells or collector wells. The yield of these wells could be estimated using the following formula derived from modification of Darcy□s law. Q= TIL where Q= Flow along any stream section in m^3 or litres per day, T= Transmissivity in m^2/Day, I= Hydraulic gradient and L= width of the stream section inmetre.

To assess the flow through subsurface aquifers, good number of slug tests was conducted in test pits conducted in the stream beds and transmissivity of the formations were ascertained. Accordingly after finding out the hydraulic gradient and stream width, sectional flow in each study area was determined (Table 5). Tables 6 and 7 shows the recommendations made and the status of implementation.

Similarly, to tap the spring flow in dry season, base flow and optimal conservation of surplus run-off, construction of check dams with intermittent subsurface dam and wells, all along the length of the stream, from hill to sea were advocated by the author (Kar 2003). The recommendation was fully accepted by the island administration for implementation to augment irrigated agriculture. In 2003-04 101 check dams were constructed in the entire Andaman group of islands while 17 check dams were constructed in the study area during the first phase (Table 5). This also marked the beginning of irrigation from surface water in the islands.

In addition to all the above recommendations, for optimal conservation of rainwater in the island water sheds along with recharging of ground water, construction of ponds were advocated. Accordingly ten ponds were constructed in the area.

Table 5 Impact assessment of check dams in the study area during the second week of December□04

Sl.no	Location	Width (m)	Depth of Water (in m)	Volume of water impounded (M^3)	Irrigation potential created (Hectare)	Cropping pattern
1	Beadnabad-I	4.46	0.9	137.68	6.0	Betel vine, ladies finger, pumpkin, ridge gourd, cowpea, Radish
2	Kodiyaghat	4.50	0.55	24.85	5.0	Cowpea, Betel gourd, Ladies finger, Radish, Bean, French bean, Bottle gourd, pumpkin
3	Maccapahar-I	3.8	0.88	26.51	3.0	Drumstick, Betel vine, Areca nut
4	Maccapahar-II	4.0	0.7	34.66	4.0	Coconut, Areca nut, Cowpea, Ladies finger, ginger
5	Maccapahar-III	6.02	1.2	40.47	4.0	Coconut, Areca nut, Cowpea, Ladies finger
6	New Bimblitan-2	2.10	1.0	11.52	2.5	Coconut, Areca nut, Cowpea, Ladies finger
7	Calicut-II	8.5	1.0	119.98	10.0	Coconut, Areca nut

Hydrogeological changes in the post-tsunami period

It is already explained in the preceding paragraph that barring the devastation from the mega earthquake of 26.12.04, the study area was spared by the tsunami invasion because of topographic elevations, except the fringe areas. However, one notable hydrogeological event occurred in Calicut area during the progression of tsunami. High mineralized water with electrical conductivity of 14500 microsiemens per cm had pushed through the fractures and reached uplands (103 m above mean sea level) by dint of the thunderous pressure generated by the propagating tsunami. The water had pungent smell continuing various volatile compounds. The upcoming highly mineralized saline water was detected at Baratang island and it was also reported from Little Andaman island in the uplands during tsunami event (Kar and Adhikary 2005; Kar and Gawri 2006, Kar2007).

The aquifer system in the study area was highly influenced by the mega earthquake (M=9.3) of 26.12.04 and its impact on the brittle hard rock formation i.e. ophiolite suite of rock in the study area was grave. On the other hand the pervasive sedimentary formations in the islands were also highly disturbed during the earthquake. Deep fractures were produced in the comparatively more elastic sedimentaries comprising mainly shale, siltstone and fine grained sandstone. In many places groundwater levels in the highland fell while in low lands it rose. The water in the ponds moved out through the newly developed fissures. But interestingly after a gap of four years these cracks and voids in the sedimentary terrain have been infilled with the weathered clay minerals. The gradual natural revival of the aquifer characteristics is evident through continuous groundwater monitoring in the islands.

Table 6 List of Recommendations of Artificial recharge and base flow conservation structures in hard rock areas of South Andaman during Pre andPost-tsunami

Sl. no	Location	Structures proposed	Remarks
1	Austinabad . Near Mariamma temple	15M long,1m width,3.5m deep SSD,DW one no,5-6m dia,6m depth(Pumping device with pipe lines)	Status- Completed in the Pre tsunami Estimated yield(Pre project) - 40,000 lpd Post project yield (observed)- 50.000LPD
2	Bird Line. Near explosive godown	15M long,1m width,3 - 4 m deep SSD,DCB one no,5-6m dia,6m depth,5 Check dam 3-5 m long as per required engineering design, 3-5 nos concentric shaft 1.5-2m depth1.5m dia in between two check dams, one bore well 4// dia,25-30m deep inside the well (Pumping device with pipe lines) .	Status- Completed in the Pre tsunami Estimated yield(Pre project) - 40,000 lpd Post project yield(observed)- 50.000LPD
3	Brookshab ad On the way to quarry	50M long arcuate,1m width,3.5m deep SSD,DCB one no,5-6m dia,6m depth,2 Check weirs(dam),20m long as per engineering design, 3 bore wells 4// dia,15-20m deep inside the well ((Pumping device with pipe lines)	Status- Status- Completed in the Post tsunami Estimated yield(Pre project) - 40,000 LPD Post project yield(observed)- 50.000LPD
4	Taylerabad near quarry	15M long,1m width,3.5m deep SSD,DW one no,5-6m dia,6m depth(Pumping device with pipe lines)	Status- Sanctioned Estimated yield(Pre project) - 40,000 LPD
5	LalpaharNa la	30 M long,1m width,3.5m deep SSD, DW one no,5-6m dia,6m depth(Pumping device with pipe lines)	Status- Completed in the Pre tsunami Estimated yield(Pre project) - 40,000LPD Post project yield(observed)- 50.000LPD
6	Sippighat near Agriculture Farm	10M long, 1m width, 3.5m deep SSD,DW one no,5-6m dia,6m depth(Pumping device with pipe lines) .	Status- Completed in the Post tsunami. Estimated yield(Pre project) - 60,000 LPD Post project yield(observed)- 80.000LPD
7	New Bimblitan near Karpaswa my temple	10M long,1m width,3.5m deep SSD,DW one no,5-6m dia,6m depth, one check dam 20m width as per appropriate design(Pumping device with pipe lines).	Status- Completed in the Post tsunami Estimated yield(Pre project) – 1,50,000 LPD Post project yield(observed)- 2,00000 LPD

Sl. no	Location	Structures proposed	Remarks
8	Dhanikhari village near Check dam	15M long,1m width,3.5m deep SSD,DW one no,5-6m dia,6m depth(Pumping device with pipe lines)	Status- Completed in the Post tsunami Estimated yield(Pre project) - 70,000 LPD Post project yield(observed)- 80.000LPD
9	Maccapaha r-Burmanala project close to teak plantation	30 M long,1m width,10 m deep SSD,DCB one no,8-10 m dia,8-10 m depth, Check dam 3nos, 15- 20m long as per engineering design, 3 bore wells 4" dia,25-30m deep inside the well (Pumping device with pipe lines) .(Fig-5&6)	Status- Ongoing Estimated yield(Pre project) – 1.0 MLD(Million litre per day)

Abbreviations used: SSD - Subsurface Dam, DCB - Dug- cum -bore well, DW - Dug well, LPD - Litres per day, Dia - Diameter, m - Metre

Table 7 List of Recommendations of construction of Check dam in hard rock areas of South Andaman during Pre and Post-tsunami

Sl. No	Location	Structures proposed	Remarks
1	New Bimblitan village	Cement concrete (C.C) check dam 10 m width with wing wall and appropriate foundation as per engineering design to tap the baseflow, spring discharge and surface water flow.	Status- Completed in the Pre tsunami The structure was highly successful but it dried up at the aftermath of the maga earthquake (M=9.3) on 26.12.04.
2	Calicut village at the upstream of SupariBagicha stream	Cement concrete (C.C) check dam 10 m width with wing wall and appropriate foundation as per engineering design to tap the baseflow, spring discharge and surface water flow.	Status- Completed in the Pre tsunami The structure was highly successful but a crack was developed during the maga earthquake (M=9.3) on 26.12.04.
3	SupariBagicha stream, Calicut village	Cement concrete (C.C) check dam 40M width with wing wall and appropriate foundation as per engineering design to tap the baseflow, spring discharge and surface water flow.	Status- Status- Completed in the Post tsunami.
4	Beadnabad Stream close to the main road and the pump	Cement concrete (C.C) check dam 30M width with wing wall and appropriate foundation as per engineering design to tap the baseflow, spring discharge and	Status- Status- Completed in the Post tsunami.
5	Chidiyatap u Stream For water supply	Cement concrete (C.C) check dam 30M width with wing wall and appropriate foundation as per engineering design to tap the base flow, spring discharge and surface water flow. surface water flow.	Status- Status- Completed in the Post tsunami.

As in the case of sedimentaries, the earthquake movement caused development of new fractures in the brittle ophiolite and groundwater movement occurred towards the topographic lows. After the event, in the higher topographical parts of the study area in the villages of New Bimblitan, Maccapahar, Calicut, Teylerabad, Brichganj, Brookshabad, Burmanala etc. groundwater and surface water sources either dried up or their discharges dwindled. Cracks developed in the check dam at Calicut, bore wells in many cases dried up and auto flowing wells ceased to flow. Water levels in wells fell down alarmingly (Table 8). A pond in Maccapahar village was full of water just before the tsunami and about 25,000 gallons per day of water was supplied from this pond to Arvind International, a local hotel at Port Blair for its daily need. Just after the mega earthquake the pond completely dried up and till very recent time the situation has not improvedmuch.

The area is suffering from severe drinking water crisis, irrigation system has collapsed and agriculture has turned rainfed after the mega disaster. While the ground water situation in higher altitudes deteriorated, discharge of the springs increased in the lower altitudes. New springs also developed in the topographic lows. The discharge of the stream in the upper part near New Bimblitan decreased sharply, the same stream in the lower stretches, near teak plantation area at Maccapahar close to Burma Nala, was found to discharge copiously (75,000 gallons per hour). Similarly, the stream near Beadnabad started discharging heavily (more than 50,000 gallons per hour) even in the lean period. Many private water tanker owners earned easy money through carrying water from the free flowing stream at the former location to the water scarce Port Blair town in the post-tsunami period at least till February, 2009. The springs in the Rutland Island did not show much variation indischarge.

Studies carried out in the Post tsunami period, recommendations and their implementation

After the disaster, there was an adverse effect on the groundwater condition. The drinking water supply and irrigation systems failed in the area. The acute scarcity of water invited detailed hydrogeological studies to find out the root cause of the change and to recommend at least some remedial measures to restore first the drinking water supply followed by irrigation water supply. Consequently, no action was taken to tackle over-exploitation which occurred before the tsunami took place as contemplated in the Notification of the Central Ground Water Authority but intensive hydrogeological and geophysical studies (Table 9) were carried out to find out the solution of the newly developed problems. Detailed surveys (Kar and Gawri 2006; Sarkar 2007) have delineated the existence of subsurface water bearing which can be tapped for water supply and dry fractures which may be grouted. From the hydrogeological investigation and resistivity surveys it is clear that formation of shallow fractures at high altitudes are responsible for flow of ground water from higher altitudes to lower altitudes and loss of surface water from the ponds, check dams, dug and bore wells . This is also corroborated by the fact that in many places the water level has gone down to much deeper level and auto flow condition has stopped due to lowering of the piezometric surface.

Table 8 Fluctuation of water levels in bore wells of Calicut-Maccapahar area, South Andaman before and after theearthquake

Village Name	Name of the owner	District & Sub - Division	Type	Total Depth (m)	Static water level (m)		Affected irrigable area after the earthquake (Hectare)
					Pre-tsunami	Post tsunami	
Maccapahar	S/Sh. Sunderraj	South Andaman	Bore well	45.75	1.5	15.00	0.62
Do	Ramaswamy	Do	do	34.00	3.0	12.00	0.6
Do	BalMurugan	Do	do	54.00	0.6	24.00	0.70
Do	Arumugam	Do	do	40.00	7.6	12.19	0.54
Do	Rajendran	Do	do	33.50	3.0	15.00	0.60
Do	N.Hamza	Do	do	33.50	7.6	12.19	2.10
Do	R.Manoharan	Do	do	30.48	Auto flow	6.00	0.55
Do	R.Veluswamy	Do	do	30.48	1.2	9.14	0.65
Do	R.Ramaswamy	Do	do	33.50	3.0	12.19	0.56
Do	Muthuswamy	Do	do	44.20	3.0	6.00	NA
Do	Ganeshan	Do	do	33.50	3.0	12.19	NA
Do	Arunachalam	Do	do	33.50	Auto flow	6.00	NA
Do	GunaSekharan	Do	do	22.86	Auto flow	7.62	NA
Do	MasnaMuthu	Do	do	33.50	6.0	12.19	NA
Do	S.Muthaiah	Do	do	30.48	3.35	15.00	NA
Teylerabad	Vellakanni	Do	do	30.48	6.0	15.00	NA
Calicut	P.Moideen	Do	do	30.48	3.0	Dried up	NA
Do	do	Do	do	45.73	24.0	Dried up	2.10

Village Name	Name of the owner	District & Sub - Division	Type	Total Depth (m)	Static water level (m)		Affected irrigable area after the earthquake (Hectare)
					Pre-tsunami	Post tsunami	
Do	V.Subramanium	Do	do	30.48	2.43	18.29	NA
Do	K.Ali	Do	do	30.48	2.74	19.80	0.50
Do	M.Mohammed	do	do	42.68	3.00	15.24	0.50

Table 9 Subsurface information from VES in hard rock areas of South Andaman Island

VES No.	Resistivity (Ohm-m)				Thickness (m)			Fracture Zones (m bgl)
	P1	P2	P3	P4	H1	H2	H3	
1.	12.82	39.36	V.H	-	1.76	35.6	-	10-15,25-40,60-100
2	4.70	12.66	V.H	-	0.39	16.86	-	10-20
3	11.44	54.99	-	-	3.06	-	-	25-30,40-60
4	11.19	138.89	-	-	4.60	-	-	30-40,60-70
5	8.08	112.87	-	-	6.54	-	-	-
6	11.83	127.88	-	-	3.66	-	-	90-100
7	115.85	11.93	V.H	-	4.72	50.87	-	40-50,80-100
8	87.07	High	-	-	10.77	-	-	70-80,90-100
9	12.72	V.H	-	-	6.60	-	-	-
10	34.10	98.06	-	-	4.68	-	-	30-40,60-70,80-100
11	11.11	50	V.L	-	7.0	53.0	-	-
12	8.45	276.53	400.00	-	1.66	28.34	-	-
13	8.34	511.22	-	-	4.76	-	-	-
14	20.66	123.03	-	-	4.0	-	-	15-20,60-80
15	77.80	26.35	133.56	-	0.77	9.84	-	40-50,80-90
16	4.0	70.0	20.0	100.0	1.0	15.0	39.0	60-70,80-100
17	15.6	V.H	-	-	7.21	-	-	80-90
18	9.66	293.25	-	-	3.72	-	-	20-30
19	5.63	32.75	V.H	-	0.30	13.64	-	-
20	5.62	36.07	V.H	-	0.30	23.21	-	30-40,80-90
21	50.43	5.20	V.H	-	11.23	8.88	-	20-50,80-90
22	31.9	537.49	-	-	7.71	-	-	-
23	39.04	254.65	-	-	14.65	-	-	-
24	22.0	300.0	40.0	-	6.0	14.0	-	20-30,40-90
25	22.45	48.84	V.H	-	3.47	55.60	-	80-100
26	6.04	51.51	V.H	-	0.75	46.41	-	15-25,70-80,90-100

Grouting of the water loosing fractures at strategic locations will rejuvenate springs in the upstream which will in turn will revitalize ground water in dug wells, bore wells and ponds located in the highland. Grouting is yet to be implemented by the island administration. To meet the acute scarcity of drinking water, construction of two bore wells at Lalmitty area and two wells at Teylerabad were recommended (Kar 2007). The Lalmitty wells (80 m deep, 6// diameter) have been constructed and these are yielding cumulatively 25,000 gallons per day of water. Following the research and development studies during the pre-tsunami on artificial recharge and conservation of water, a major water supply project near the Teak Plantation area has been recommended (Kar 2007) considering the baseflow, surface flow, potential fractured basement aquifer. The project envisages construction of a 8-10 m diameter collector well with 3 - 5 shallow borewells, feeder wells with infiltration gallery, check dams and a sub-surface dam. The well will be capable to supply one million litres of potable drinking water per day (Fig. 5). For conjunctive use of surface water and ground water and their conservation two dam projects have been also recommended one at SuperiBagicha, Calicut village and the other at Beadnabad . It is contemplated that with the completion of this project the water scarcity in the entire southern part of South Andaman will besolved.

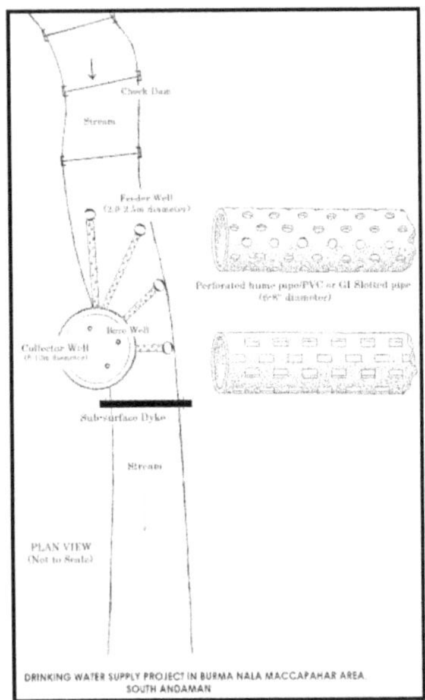

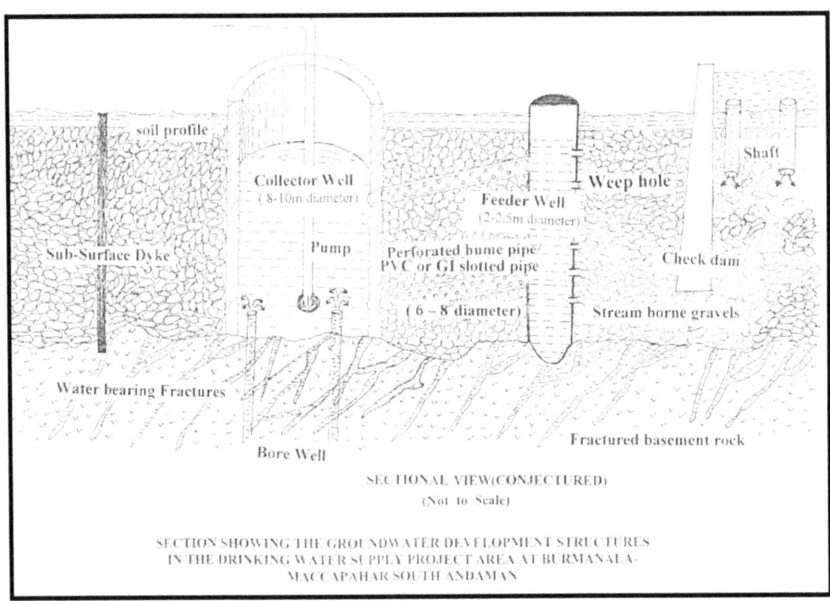

Figure 5 Plan and section of MaccaPahar-Burma Nala water supply project in the post tsunami

The two projects have been approved by the island administration and implemented. The former one is completed and the latter one is under completion. In addition to the new recommendations (Kar 2007), the old recommendations advocated in the pre-tsunami period (Kar 2002, 2004) have been approved by the Andaman and Nicobar Island Administration and are under implementation (Table 6).

In addition to these projects, detailed ground water investigation was also carried to undertake an artificial recharge project in the Defence establishment at Brichganj. The cantonment is located in the comparatively lower topographic locales. After a through study a sum total of fifteen (15) bore wells, six (6) Dug cum bore wells with subsurface dam,seven (7) recharge bore wells with recharge shafts from the roof tops were advocated (Kar 2007). As per the information of Army engineering authority received, six bore wells were constructed immediately to tide over the extreme water scarce situation and all were successful, yielding water to the tune of 30,000 to 50,000 gallons per day. The other recommendations are yet to be implemented.

Conclusion

Detailed hydrogeological studies encompassing an area of 200 sq.km in the South Andaman and Rutland islands were carried out. The area in underlain by ophiolite suite of rocks comprising mafic and ultramafic igneous rocks fractured down to 70 - 80 m from ground level. Regoliths and the fractured hard rock aquifers in the igneous complex are suitable for construction of dugwell, bore well and dug-cum bore well. Besides availability of ground water in the subsurface, perennial springs are also present in the ophiolite which can supply copious amount of water for various uses. Base flow along the streams could be tapped through subsurface dams and properly designed dug well, collector well with or without infiltration gallery, dug-cum-bore wells etc. Furthermore, the subsurface reservoirs could be recharged if check dams are constructed with recharge shafts. Conjunctive use of surface water, ground water and rainwater is possible in the study area through check dams and ponds. The deep aquifers were being continuously pumped by more than 200 bore wells during the pre- tsunami period leading to their over-exploitation as revealed from the survey carried out by the author in 2004. Extensive hydrogeological studies were carried out in the pre-tsunami times by the author for optimum development of ground water, surface water and rainwater though wells, bore wells , dug-cum bore wells, check dams, sub-surface dams, ponds and by harvesting of spring water from Rutland island for inter-island transfer of water to Port Blair town. Many of the recommendations were implemented by the island administration to meet the drinking as also irrigation needs of the studyarea.

Acknowledgments

The author expresses his sincere gratitude to ShriK.B.Biswas, Chairman, CGWB for his permission and enthusiasm to publish the paper in the International Conference. The author is indebted to Dr. D. Saha, Member (SAM), CGWB, for his inspiration to publish the paper. Regional Directors of Eastern Region, Kolkata office of CGWB are acknowledged for providing infrastructural facilities. The author expresses his gratitude to the Lt. Governors, Andaman and Nicobar Islands, Chief Secretaries, Development Commissioners, Secretaries of PWD, A&N Administration, Chief Engineers, APWD, MES and ALHW, Director of Agriculture, A&N Administration, for their constant enthusiasm and needful logistics during the long course of work in the remote islands. Their keen attention to implement the projects had enhanced the working experience and spirit of the author. Dr. S. P. Sinha Ray, Former Member, CGWB, President and organizer of the International Conference is gratefully remembered for his inspiration to submit a paper.

References

IGRAC (2006): Web site of International Groundwater Resources assessment Centre.

Jain, S.K. et al. (2005). Effects of M9 Sumatra earthquake and tsunami of 26 December 2004.Curr, Sci. 88(3), 357-359.

Kar, A. (2001). Hydrogeology of Andaman and Nicobar Islands (Unpub Report of CGWB).

Kar, A. (2002a). Preliminary reports on investigation of springs as a possible water supply source to Port Blair town from Rutland island (Unpub. Report of CGWB).

Kar, A. (2002b). Hydrogeological investigation in Carnicobarisland with special reference to Mus break water site. (Unpub.Report of CGWB).

Kar, A. et al. (2002c). Water supply to rural and urban areas of Andaman and Nicobar islands with special reference to the application of artificial recharge and conservation technique.Nat. Sem. of Ind. Desalination Assoc., Kolkata. Abs Vol.

Kar, A. (2002d) Artificial recharge of ground water for augmentation of water supply and landscape development in Port Blair town and A&N islands. Nat. Sem., LAWMI, Kolkata, Abs.Vol.

Kar, A. (2003). Hydrogeological studies in Port Blair Municipal area to augment the water supply situation through construction of new structures and renovation of old dug wells and ponds. (Unpub.Report of CGWB).

Kar, A. (2004a): Studies on Hydrogeology of South Andaman Islands with special reference to application of artificial recharge and conservation techniques of ground water and inter island transfer of spring water for sustainable water supply to Portblair town and environs- (Trans.M.G.M.I., Kolkata, 100 (1&2),(pp-37-59).

Kar, A. (2004b). Ground water investigation in water scarce Chowraisland, Nicobar, District, to explore the possibilities and augmentation of fresh water source through rain water harvesting. (Unpub.Report of CGWB).

Kar, A. et al. (2004c). Conservation and artificial recharge of Ground Water- an urgent strategic option for water supply to Portblair and Surroundings in Andaman Islands. (Abs. Vol. International Geological Congress, Florence, Italy,)

Kar, A. et al. (2005). Impact of 26thDecember, 2004 tsunami and earth quake on ground water resources in A&N islands: Post tsunami studies by Central Ground Water Board. (Unpub.Report ofCGWB).

Qualitative Analysis of Cations in Surface Water Samples of Pragathi Nagar Cheruvu, Medchal District, Telangana State, India

Hepsi Swarupa Rani Guduri[1] and C. Sarala[2]
[1]Research Scholar, [2]Professor
Centre for Water Resources, IST, Jawaharlal Nehru Technological University, Hyderabad, Telangana, India.
hepsiguduri@gmail.com, sarala2006@gmail.com

ABSTRACT

Water is becoming very scarce due to rapid growth of population and also reduction in frequency of rainfall. The quality and availability of water has become a critical issue to living organisms. Abstract This paper determined cations concentrations, Many instruments were used in this work: flame photometry (K, Na), EDTA titration (Ca, Mg), The study area is Pragathi Nagar lake located in Pragathi Nagar, Hyderabad, Medchal district, Telangana State. Water samples were collected from 13 locations during post monsoon period of the year 2015. The primary objective of this paper is to analyze the surface water quality of Cations like Calcium (Ca2+), Magnesium (Mg2+), Sodium (Na+) and Potassium (K+). All the other parameters values exceeded the permissible limit in the study area.

Keywords: surface water, cations, water quality, pollution, environment, human health.

Introduction

Water is the most valuable resource of the nature. The demand for water in India is steeply increasing and the prime reason is population explosion which was projected as 1.66 billion in 2050. The growing problem of water scarcity is further aggravated with to rapid urbanization. In 2016, 54% of the Indian population was living in urban areas and the urban population is expected to increase to 66 % by 2050. The agriculture development will be more on water intensive cash crops and there will be 80% increase in the demand for water by 2050.It is therefore necessary to address the bottlenecks affecting the water supply and optimize surface and groundwater use in all sectors of economy.

Objectives of the Study

(i) To collect the post monsoon water samples from the thirteen locations in the study area.

(ii) To analyze the cations in the surface water samples for the pre monsoon period.

(iii) To prepare the cation concentration graphs.

Study Area

The present study area Pragathi Nagar is a residential colony in Kukatpally, Hyderabad, Rangareddy District in Telangana State, India. It's a major residential suburb in Kukatpally and situated at a distance of 3 kilometers from Kukatpally on National Highway 9 which leads to the city of Mumbai. It is about 2 and half kilometers from the famous university called Jawaharlal Technological University, Hyderabad. It has an average elevation of 33 metres (108 ft). Pragathinagar Study area which is lies in between 17.53'27'' North Latitude and 78.39,64'' East Longitude.

The surface water quality analysis has been carried out for the thirteen samples collected at Peddanalla Starting, Peddanalla ending, Pedda Banda, Allugu, Allugu1, Allugu2, Pragathi Nagar Bridge starting, Pragathi Nagar Bridge ending, Near fish market beginning, Near fish market ending, Sofit, Sofit1, Sofit2 Heritage Bus stop. In the present research work attempts have been made to detect the anions present in the collected samples. The location map of the study area is shown in figure1.

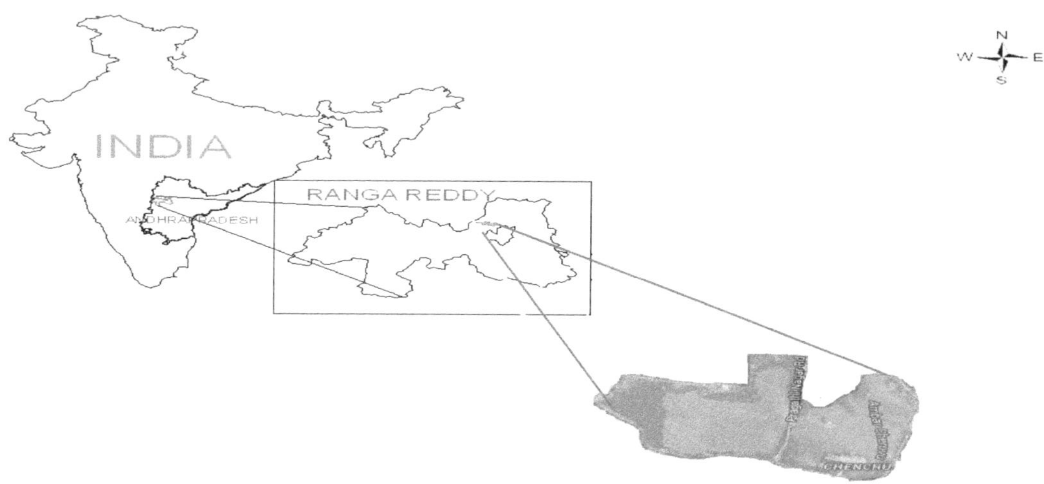

Near fish Market beginning

Near fish Market ending

Sofit 1

Sofit 2

Heritage Bus Stop

Figure 1 Location Map of Study Area

Methodology

In this work, the data used is Survey of India Topographic maps and sample locations taken with the help of Global Positioning Systems (GPS). Water is for drinking, Industrial and Agricultural purposes has increased manifolds but consequently it is observed that the water is polluted and affecting the human health, soil nutrients, live stock, Biomass and Environment in certain areas.

The present study is restricted to thirteen water samples from thirteen locations only. It can be extended to some more samples for more accurate results. The study may be extended to few more seasons or years to know complete surface water characteristics.

Analysis of Water Quality Parameters for Thirteen Locations in the Study Area

The physical parameters analyzed in the study area are P^H, Electrical Conductivity, Calcium, Magnesium, Sodium and Potassium.

Analysis for p^H

Although P^H values upto 6.5 to 8.5 are permissible in potable water. The p^H values varied between 6 to 10.5. P^H values in the study area for post monsoon period is shown in the figure 2.

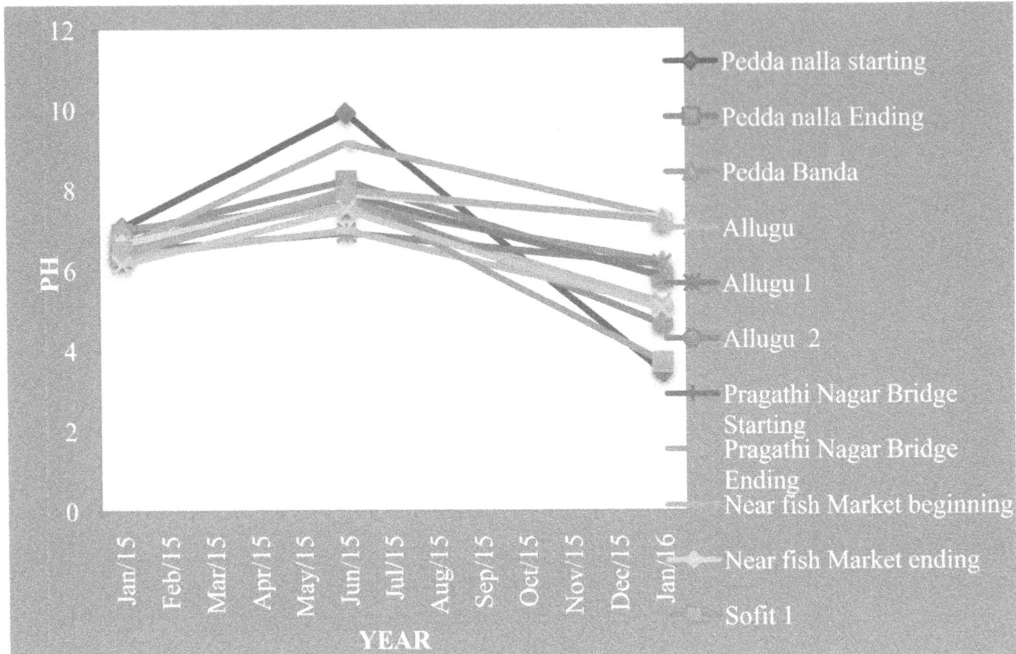

Figure 2 P^HVALUES IN THE STUDY AREA

Analysis for Electrical Conductivity (EC)

The specific electrical conductivity water samples at room temperature is determined by the conductivity meter and expressed in micro simens/cm at 25^{0}C. Rough estimation of dissolved ionic contents of water sample can be made by multiplying specific conductance by an empirical factor which may vary from 1814 to 2228 depending on the soluble components of water and that temperature of measurement. The highest EC values of 2228, 2193, 2132, 2007, 1931, 1904, 1900, 1894, 1893, 1885, 1882, 1880, 1878 umhos/cm are observed at Peddanalla Starting, Peddanalla ending, Pedda Banda, Allugu, Allugu1, Allugu2, Pragathi Nagar Bridge starting, Pragathi Nagar Bridge ending, Near fish market beginning, Near fish market ending, Sofit, Sofit1, Sofit2 Heritage Bus stop. The EC values of the samples ranged between 1601 to 2228 umhos/cm. The EC values of the study area are shown in figure 3.

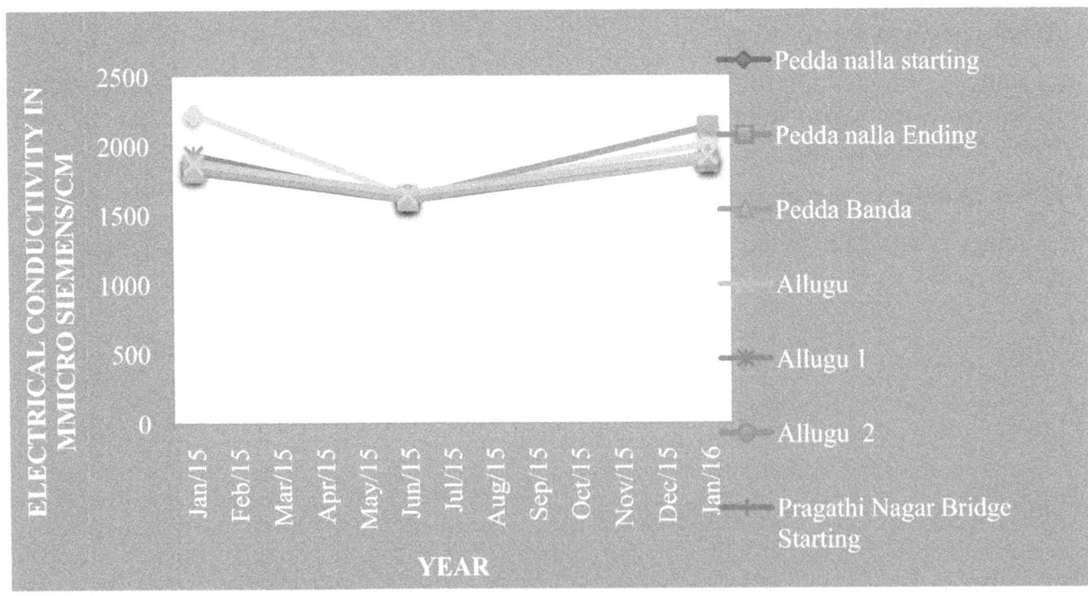

Figure 3 Electrical Conductivity Values In The Study Area

Analysis for Calcium Concentration

Small water supplies using groundwater often encounter significant levels of hardness, but some larger surface water supplies also have the same issue. Calcium concentrations up to and exceeding 100 mg/l are common in natural sources of water, particularly groundwater.Inadequate intakes of calcium have been associated with increased risks of osteoporosis, nephrolithiasis (kidney stones), colorectal cancer, hypertension and stroke, coronary artery disease, insulin resistance and obesity. Most of these disorders have treatments, but not cures. Owing to a lack of compelling evidence for the role of calcium as a contributory element in relation to these diseases, estimates of calcium requirement have been made on the basis of bone health outcomes, with the goal of optimizing bone mineral density. Calcium is unique among nutrients, in that the body's reserve is also functional; increasing bone mass is linearly related to reduction in fracture risk. The total body stores are on the order of 1200 g, with about 99% in bones and teeth.In the study area Magnesium concentration varied between 160-1000mg/l. The Calcium values of water samples are shown in Figure 4.

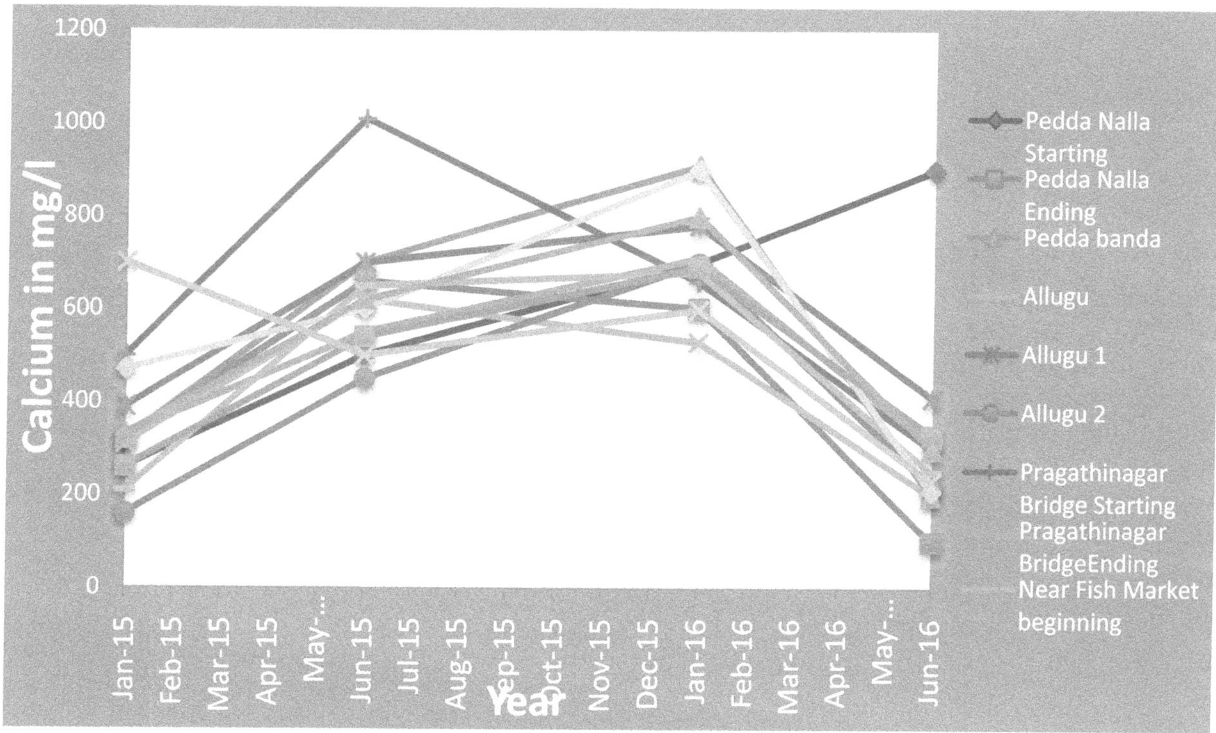

Figure 4 Calcium Values In The Study Area

Analysis for Magnesium Concentration

Magnesium is present in natural groundwater usually at lower concentrations (from negligible to about 50 mg/l and rarely above 100 mg/l), so calcium-based hardness usually predominates (National Research Council, 1977). HARDNESS IN DRINKING-WATER 2 Estimated daily intakes of magnesium from water of about 2.3 mg and 52.1 mg in soft-water and hard-water areas, respectively, have been reported, based on adults drinking 2 litres of water per day (Neri et al., 1985). In the study area Magnesium concentration varied between 310-1700mg/l. Magnesium concentration is shown in the figure 5.

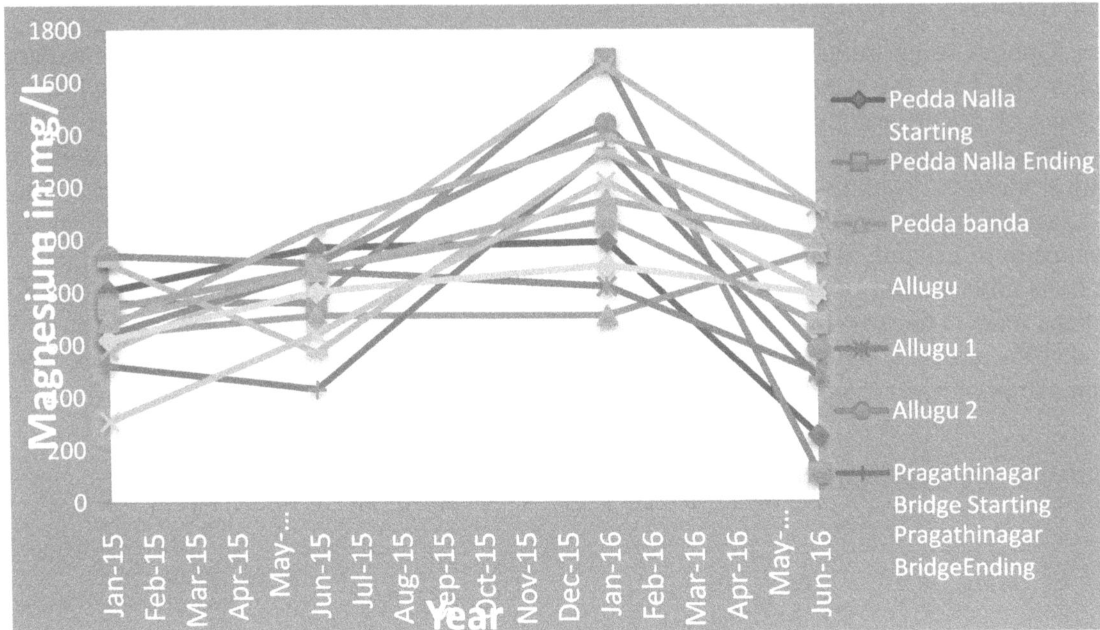

Figure 5 Magnesium Values in the Study Area

Analysis for Sodium Concentration

Sodium ion is ubiquitous in water, owing to the high solubility of its salts and the abundance of sodium-containing mineral deposits. Seawater contains about 30,000 mg of sodium chloride per liter (mg/L). Groundwater typically contains higher concentrations of minerals and salts than surface waters, especially in areas with an abundance of sodium mineral deposits or in areas with sea or estuarine water intrusions (WHO 1979).In the study area Sodiumconcentration varied between 410-450mg/l. All the thirteen samples shows the values exceeded the permissible limit. The sodium concentrations are shown in figure 6.

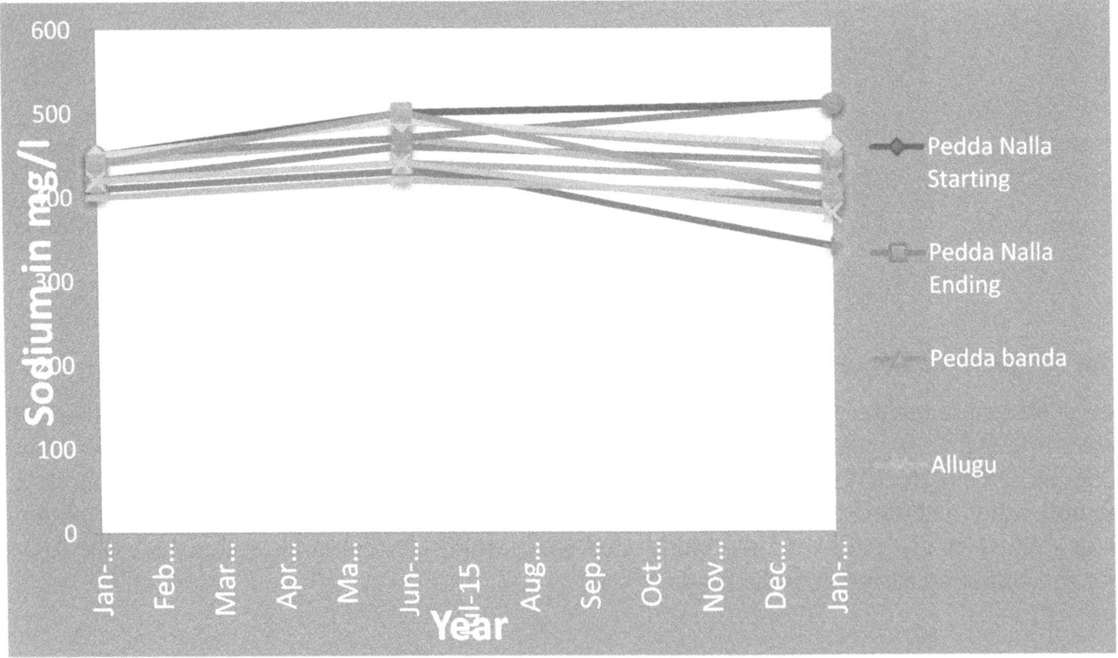

Figure 6 Sodium values in the Study Area

Analysis for Potassium Concentration

In the study area Potassiumconcentration varied between from 0 to 0.5 ppm. Potassium values in drinking water are usually below1.0-8.0 mg/l. Concentration of Potassiumexceeded the permissible limit in the study area and the Potassiumconcentration is shown in figure 7.

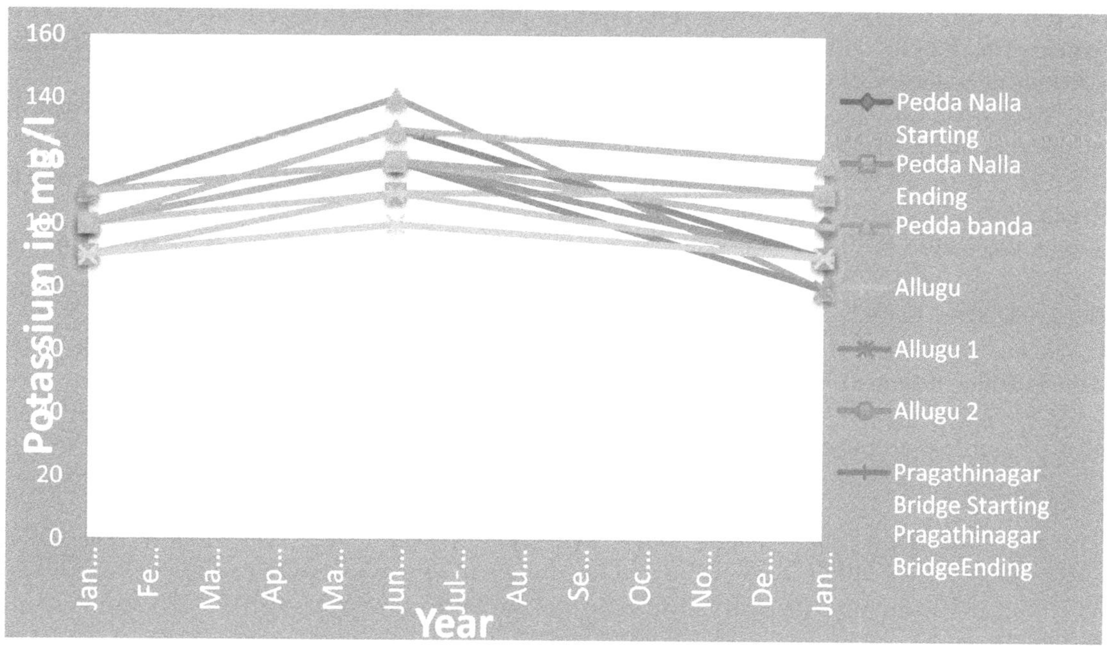

Figure 7 Potassium Values in the Study Area

Table 1 Analysis Results of Water Quality Parameters for Thirteen Locations in the Study Area

S.NO.	Locations of the study Area	pH	EC	Calcium (mg/l)	Magnesium (mg/l)	Sodium (mg/l)	Potassium (mg/l)
1	Peddanalla starting	9.901	1814	1000	1700	450	100
2	Peddanalla Ending	8.212	1812	320	760	420	90
3	Pedda Banda	7.635	1829	320	640	450	110
4	Allugu	7.606	1849	310	580	440	100
5	Allugu 1	6.924	1931	390	640	410	100
6	Allugu 2	7.821	1854	160	940	450	110
7	Pragathi Nagar Bridge Starting	7.9	1904	500	520	420	100
8	Pragathi Nagar Bridge Ending	9.153	1872	330	670	450	110
9	Near fish Market beginning	6.983	1892	210	910	400	90
10	Near fish Market ending	7.89	2228	470	620	450	100
11	Sofit 1	7.829	1866	260	740	440	100
12	Sofit 2	7.986	1840	330	690	420	100
13	Heritage Bus Stop	7.485	1874	700	310	420	90

Conclusion

The PH values in the study area varied between 0.043 – 9.901. The electrical conductivity values varied between 1601 -2259 micro siemens/cm and the highest value is observed at Pedda Nalla Starting. All the other parameters values exceeded the permissible limit in the study area .

References

Nagamani C. Physico-chemical analysis of water samples. International Journal of Scientific and Engineering Research, 2015, 6: 2149.

Gangwar R.K., Khare P, Singh A.P.Assessment of physicochemical properties of water: River Ramganga at Bareilly, U.P.Journal of Chemical and Pharmaceutical Research, 2012. 4(9): 4231-4234.

Bellingham, K. Stevens, Water monitoring systems. 2012: 205-209.

Haruna R, Ejobi F, Kabagambe EK.

The quality of water from protected springs in Katwe and Kisenyi parishes, Kampala city, Uganda. Afr Health Sci, 2005, 5: 14

Garg RK, Saksena DN, Rao RJ. Assesment of physico-chemical water quality of Harsi reservoir, District Gwalior, Madhya Pradesh. Journal of Ecophysiology and Occupational Health, 2006, 6: 33.

Devaraju TM, Venkatesha MG, Singh S. Studies on the physico-chemical parameters of Maddur Lake with reference to suitability for aquaculture. Nat Environment and Pollution Technology, 2005, 4: 287.

Biolife | 2014 | Vol 2 | Issue 3 APHA, Standard methods for the examination of water and wastewater. 21st ed. American. Public Health Association Inc. Washington D.C.: 1135- 1219. 2005.

Chaterjee, C.and Raziuddin. M., 2002. Determination of water quality index (WQI) of a degraded river in Asanol Industrial area.

Raniganj. Burdwan, West Bengal. Nature, Environment and Pollution Technology. 1(2): 181-189.

Bhatt L. R., Lacoul P., Lekhak H. D., Jha P.K., Physicochemical characteristicsand phytoplankton of Taudaha Lake, Kathmandu. Pollution Research1999: 18(14):353-358.

GIS Based Morphometry Analysis: A Case Study of Vaippar River, Thoothukudi, India

Monica. S[1] and Giridhar. M.V.S.S[2]

[1]*M.Sc Department of Geography, Central University of Tamlinadu, Thiruvarur.*

[2]*Professor,Centre for Water Resource, JNTU, Hyderabad.*

ABSTRACT

Geographical information system (GIS) has emerged as a professional tool in demarcation of drainage pattern and ground water potential and its planning. GIS and image processing techniques can be utilized for the identification of morphological characteristics and investigating properties of basin. Morphometric analysis in any hydrological investigation is inevitable in performing the development and management of drainage basin. The development of morphometric techniques was a major advance in the quantitative description of the geometry of the drainage basins and its network which helps in characterizing the drainage network. The geomorphological properties which are important from the hydrological studies point of view include the linear, aerial and relief aspect of the watersheds. The study indicates that analysis of morphometric parameters with the help of geographic information system (GIS) would prove a viable method of characterizing the hydrological response behaviour of the watershed. In the present study a morphometric analysis of Vaippar river basin area has been carried out using geoprocessing techniques in GIS.It is a eighth order drainage basin and is mainly of dentritic pattern and the bifurcation ratio is of 1.89.

Introduction

QGIS (Quantum GIS) is a free and open-source software, they are freely available on the internet which is designed to store, manipulation, analyse, manage and present spatial or geographic data. It allows the user to perform interactive queries, analyse spatial information, edit data in maps, and present the results of all these operations. They offer an integrated environment for raster and vector analysis, image processing and map creation.

Water is known as the liquid for sustenance of life. All living beings are depending on water, without which no life exists on the earth. Earth has plentiful water due to the presence of Hydrological cycle on it, but most of it is unfit for living beings use and consumption. The study of the watershed morphometric analysis provides the beneficial parameters for the assessment of the ground water potential zones, identification of sites for water harvesting structures, water resource management, runoff and geographic characteristics of the drainage system(P T Aravinda et al.,2013) The use of GIS technique in morphometric analysis has emerged as a powerful tool in recent years particularly for remote areas with limited access. GIS techniques are the proven efficient tools in the delineation, updating and morphometric analysis of drainage basin. The term 'Morphometry' literally means measurement of forms introducing quantitative description for landform (Horton, R.E 1945). Morphometry is defined as the measurement and mathematical analysis of the configuration of the earth's surface and of the shape and dimension of its landforms (Clarke, 1966). Morphometric methods, though simple, have been applied for the analysis of area-height relationships, determination of erosion surfaces, slopes, relative relief and terrain characteristics, river basin evaluation, watershed prioritization for soil and water conservation activities in river basins (Kanth, 2012). The Morphometric analysis consists of linear, Aerial and relief aspects. Various important hydrologic phenomena can be correlated with the physiographic characteristics of drainage basins such as size, shape, slope of drainage area, drainage density, size and length of the tributaries etc. (Rastogi et al., 1976). The morphometry of the river basins relates to the hydrological and geomorphic response of processes like runoff, soil erosion, floods and droughts, river sedimentation, changing river flows and branching habit of the streams, flow characteristics of the drainage lines, and on the performance and sustainability of the associated dams and reservoirs if available within the basin (Garde, 2005; Mohd et al., 2013). Over the past 15 years, research has demonstrated the viability of techniques for automatically deriving a wide variety of topographic and topologic watershed information directly from Digital Elevation Models (DEM). Digital Elevation Model (DEM), is a grid of elevation values at constant horizontal grid spacing. Such a DEM can be used to determine topographic attributes such as slope and aspect, and watershed attributes such as basin area and stream network topology. Delineating catchment areas by employing geographic information systems (GIS) and digital elevation model (DEM) is being preferred to manual techniques due to the improved accuracy, less duplication, easier map storage, flexibility, and simplicity in data sharing, timeliness, greater efficiency and higher product complexity (Nahida Hameed Hamza Alqaysi,2016). According to Strahler (1964), systematic description of the geometry of a river basin and its stream channel requires measurement of linear aspects of the drainage network, areal aspects of the drainage basin, and relief (gradient) aspects of the channel network and contributing ground slopes. As linear aspects, stream order, stream length, stream number, and bifurcation ratios are considered. Areal aspects are basin area, basin perimeter, length of overland flow, drainage density, stream frequency, drainage intensity, circularity ratio, elongation ratio and form factor. The basin relief, relief ratio and relative relief are commonly evaluating relief aspects (Horton, 1945; Melton, 1957; Miller, 1953; Schumm, 1956; Strahler, 1964). Based on their ideology, similar work has been emerged throughout the world by different researchers using different techniques. In India, morphometric studies of various drainage basins have been carried out by Rastogi and Sharma 1976, Nautiyal 1994, Magesh et al. 2012a, b, John Wilson et al. 2012, and Magesh and Chandrasekar 2012. Application of remote sensing provides a reliable source for the preparation of various thematic layers for morphometric analysis. The digital elevation data can be used for generating the elevation model of a landscape to any extent. The resolution of the image may vary with respect to the satellite sensors. The processed DEM can be used for generating the stream network and other supporting layers (Mesa 2006; and Magesh et al. 2011). Identification of drainage networks in a basin can be achieved using traditional methods such as field observation and topographic maps alternatively by advanced methods like remote sensing and extracting features from digital elevation models (D.R. Maidment 2002).

Objectives

The main objective of this study is - To calculates Morphometric parameters viz; stream order, mean stream length, stream length ratio, bifurcation ratio, mean bifurcation ratiio etc. of the Vaippar River Basin through QGIS.

Study Area

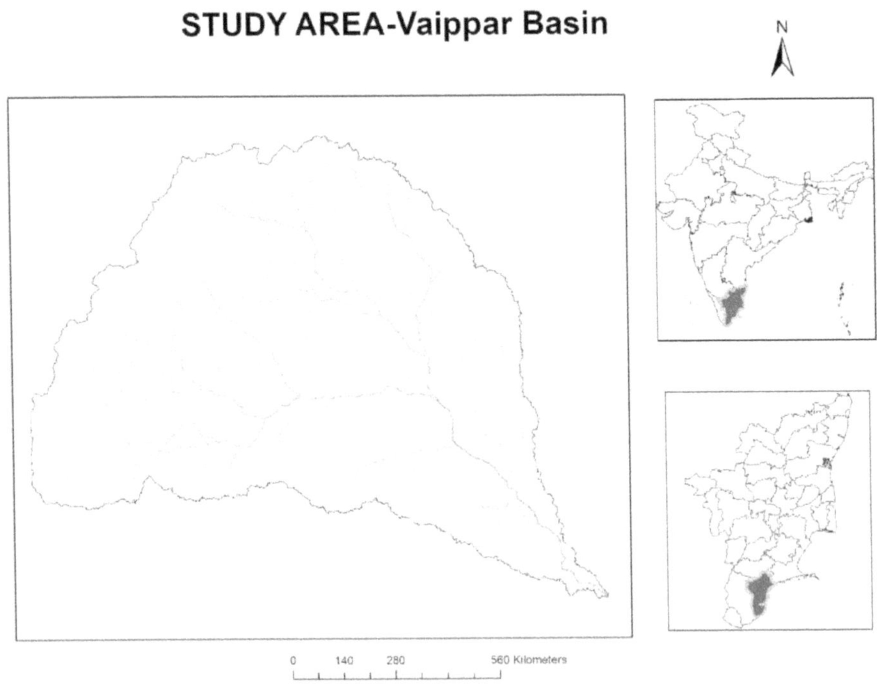

Figure 1

The Vaippar river basin is located in the southern part of Tamilnadu, India. The Vaippar basin lies ($8^{0}52$ to $9^{0}45$'N to $77^{0}1$' to $78^{0}15$'E) with an area 5339 sq.km. This basin has variation in the climatic as well as physiographic aspects such as soil texture, soil depth, rainfall, land use/landcover. Vaippar basin covers parts of Virudhunagar, Thoothukudi, Madurai and Tirunelveli districts of Tamil Nadu. On the basis of physiography, the basin can be divided into two broad sections, namely the hilly tracks with altitude above 100 meters and the vast stretch of black cotton soil plains. The basin is located on the eastern side on the eastern side of the Western Ghats. Though the basin extends up to the Bay of Bengal, the direction of monsoon winds restricts the rainfall considerably. The area has been selected because of its under developed nature and also for its varied lithological conditions such as geomorphology, hydrological characteristics, consolidated nature of rock etc.

Material and Methods

The morphometric analysis was based on GIS technique. The delineation of Vaippar River basin, watershed and drainage network was done from downloading the earth explorer SRTM DEM model data and by using QGIS 2.18 software. The morphometric parameters such as stream order (U), stream length (Lu), mean stream length(Lsm), stream length ratio (RL), bifurcation ratio (Rb), mean bifurcation ratio (Rbm) were identified .The methodology for the calculation was based on the formula suggested by Horton [1945], Strahler [1964], Hardly [1961], Schumn [1956], Nookaratanm et. al. [2005] and Miller [1953] which are mentionedin Table 1.

Table 1

Morphometric Parameters	Formula	Reference
Stream Order (Su)	Hierarchical rank	Strahler [1964]
Stream Number(Nu)	Total order wise stream segments	Horton [1945]
Stream Length (Lu)	Length of the stream	Horton [1945
Mean Stream Length (Lum)	Lum = Lu/Nu	Strahler [1964]
Bifurcation Ratio (Rb)	Rb = Nu/Nu+1	Schum[1956]
Mean Bifurcation	Rbm=Average of bifurcation ratio	Strahler [1964]

Where; Lsm = mean Stream Length, Lu=Total Stream length of order u, Nu = Total number of stream segment of order u, Nu+1=Number of stream segment of next higher order, , Rb = Bifurcation Ratio.

Results and Discussion

Morphometric parameters of watershed of Vaippar River Basin is extracted from the SRTM digital elevation model of 30m were processed using QGIS. The basis for assuming the limiting threshold value is obtained and presented.r.watershed was used to obtain the a watershed basin. The characteristics features streams and its order were

Table 2

Stream Order	Stream Number	Mean Stream Length	Stream Length Ratio	Bifurcation Ratio	Mean Bifurcation Ratio
1	4647	847.23	0.002	2.17	
2	2137	898.15	2.048	1.83	
3	1167	851.49	1.946	1.65	
4	706	777.51	2.122	2.08	
5	338	682.60	3.90	3.34	
6	101	678.84	1.228	1.16	
7	87	648.15	0.843	1.06	
8	82	711.29			
					1.89

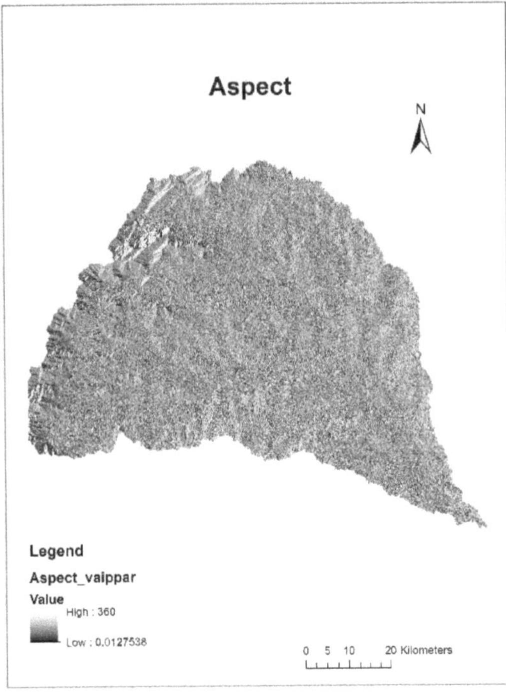

Figure 2

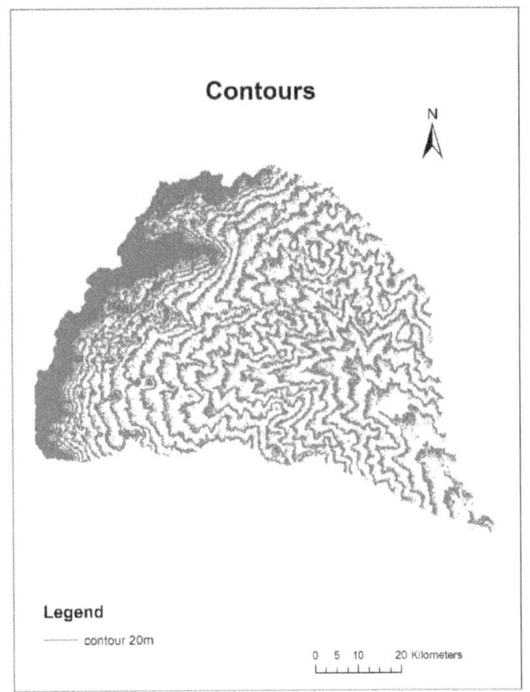

Figure 3

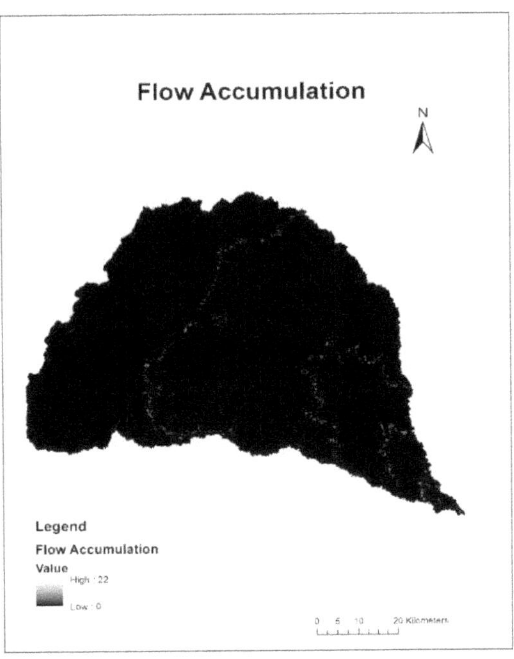

Figure 4

Figure 5

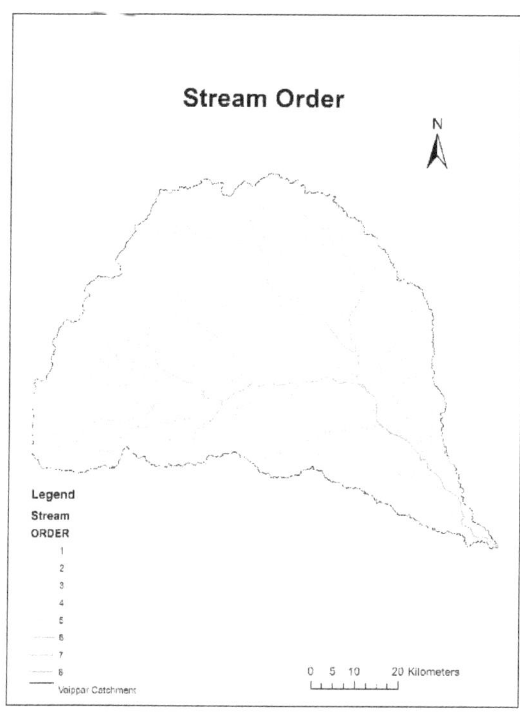

Figure 6

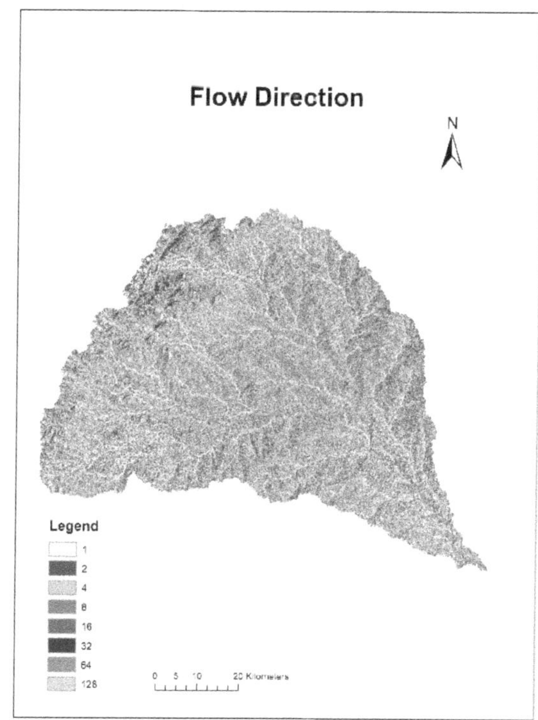

Figure 7

1. **Stream Order (u):** Stream order is defined as a measure of the portion of a stream in the hierarchical tributaries' (L.B. Leopold et. al., 1969). A.N. Strahler (1969) law is used for the determination of stream order to find the bifurcation rate in the Pair River Basin. There are 8 orders of streams covering an area of 1305.82 km2 throughout basin.
2. **Stream Number (Nu):** After assigning stream orders, the segments of each order are counted to get the number of segments of the given order (u). In first order streams there are 4647 segments. The second order stream have 2137 segments. Third order stream segments are 1167. Fourth order stream segments are 706; fifth order stream segments are 338; sixth order stream segments are 101 seventh order stream are 87 and eight order has 82 segments.
3. **Stream Length:** According to Horton (1945), Stream length is the total length of stream segment of each of the consecutive order in the basin tends approximate a direct geometric series in which the first term is the average length of the first order. It is the quantification of hydrological characteristics of bedrock and the drainage extent. When bedrock is of permeable character then only subtle numbers of relatively longer streams are formed in a well-drained basin area (Kulkarni, 2015, Sethupathi et al. 2011). It is clearly identified that the stream length is higher in first-order streams and decreases as the stream order increases i.e., first order stream length is 41394550 m.,
3. **Stream Length Mean (Lum):** Stream length Mean reveals the size of component of drainage network and its contributing surface (Strahler, 1964). It's directly proportional to the size and topography of drainage basin. It has been computed by dividing the total stream length of order 'u' by the number of stream segments in the order. Strahler (1964), indicated that the Stream length means is a characteristic property related to the size of drainage network and its associated surfaces. Stream length means values for the Vaippar river basin 847.23 km to 711.2 with a mean Lsm value of 761.9075 km.
4. **Bifurcation Ratio (rb):** Bifurcation ratio is related to the branching pattern of the drainage network and is measured by the ratio between given order (Nu) to the next higher order (Nu+1) to find out the bifurcation pattern of the stream related with geomorphology of the basin. 'Mean bifurcation ratios vary from about 2.00 for flat or rolling basins to 3.00 - 4.00 for mountainous, hilly dissected basins' (R.E. Horton, 1945). The mean bifurcation ratio of the Vaippar River Basin is 1.89 which indicate the basin comes under mainly plain area.

Conclusion

The morphometric analysis of Vaippar river basin using Geographic Information System retrieved that this tool helps to analysis the drainage basins easily and accurately in short time duration. The analysis of linear aspects of drainage basin result shows that the basin has a dendritic pattern with eight order streams. The values of basin are 1.89 which is derived from the bifurcation ratio in which the geologic structures do not distort the drainage pattern.

References

R. E. Daffi1, I. I. Ahuchaogu (2017), Delineation of River Watershed and Stream Network Using ILWIS, Asian Journal of Environment & Ecology

Nahida Hameed Hamza Alqaysia,b, Mushtaq Abdulameer Alwan Almuslehia(2016), Delineation of the Watersheds Basin in the Konya City and Modelling by Geographical Information System, J. Int. Environmental Application & Science, Vol. 11(3): 303-311(2016)

Lokesh Shrivastava, Shashikant Nag and Ritu Rani(2017), GIS-BASED MORPHOMETRIC ANALYSIS OF PENCH RIVER BASIN, International Journal of Recent Scientific Research Vol. 8, Issue, 5, pp. 16953-16959, May, 2017

Sharma H.S. 1981. Perspectives in geomorphology, Naurang Rai, Concept publishing company, New Delhi110015, pp 109-140.

Tribhuvan, P.R. and Sonar, M.A.(2016), Morphometric Analysis of a Phulambri River Drainage Basin (Gp8 Watershed), Aurangabad District (Maharashtra) using Geographical Information System, Cloud Publications International Journal of Advanced Remote Sensing and GIS 2016, Volume 5

Clarke, J.I. (1966). Morphometry from Maps. Essays in geomorphology. Elsevier Publ. Co., New York, 235 - 274.

Kanth, T.A. (2012). Morpho.metric analysis and prioritization of watersheds for soil and water resource management in Wular catchment using Geo-spatial tools", Inter. J. of Geo., Earth and Environ. Sci., 2 (1), 30 - 41.

RASTOGI, R.A. and SHARMA, T.C. (1976) Quantitative analysis of drainage basin characteristics. Jour. Soil and water Conservation in India, v.26 (1&4), pp.18-25

Garde, R.J. (2005). River Morphology, New Age International (Pvt) Ltd. Publishers, New Delhi.

Mohd, I., Haroon, S. and Bhat, F.A. (2013). Morphometric Analysis of Shaliganga Sub Catchment, Kashmir Valley, India Using Geographical Information System, International Journal of Engineering Trends and Technology, 4 (1).

Horton, R.E. (1932). Drainage-basin characteristics. Trans. Am. Geophys. Union, 13, 350

- 361. 12)Horton, R.E. (1945). Erosional Development of Streams and Their Drainage Basins, Geolo. Soc. of Am. Bull. 56 (3), 275 - 370.

P T Aravinda, H B Balakrishna, Morphometric Analysis of Vrishabhavathi Watershed Using Remote Sensing and GIS, IJRET: International Journal of Research in Engineering and Technology, Volume: 02.

Miller,V.C. (1953). A quantitative geomorphic study of drainage basin characteristics in the Clinch Mountain area, Varginia and Tennessee, Project NR 389042, Tech. Rept. 3., Columbia University, Department of Geology, ONR, Geography Branch, New York.

Melton, M.A. (1957). An Analysis of the relations among elements of climate, Surface properties and geomorphology, Project NR 389042, Tech. Rep. 11, Columbia University.

Schumm, S.A. (1956). The Evolution of drainage systems and slopes in badlands at Perth Amboi, New Jersey. Geological Society of America Bulletins, 67(5).

Strahler, A.N. (1957). Quantitative analysis of watershed geomorphology, Trans. Am. Geophys. Union, 38, 913 - 920. Strahler, A.N. (1964). Quantitative Geomorphology of drainage basins and channel networks, Handbook of Applied Hydrology, Ed. Ven Te Chow, 04, 9 - 76.

Watershed and Stream Network Delineation using QGIS in Noyyal River Basin, Tamil Nadu, India

P. Rishitha[1] and M.V.S.S. Giridhar[2]

[1]*M.Sc. Department of Geography, Central University of Tamilnadu, Tiruvarur.*

[2]*Assistant Professor, Centre for water Resource, JNTU, Hyderabad.*

ABSTRACT

GIS (Geographical Information System) and SIT (Spatial Information Technology) brought a revolutionized inventions in management of water resources and drainage basins. Morphometric analysis is one among of which can quantitatively evaluate characterstics of any earth surface feature or composition of any stream segment of a drainage basin. Various components such as stream segments, basin length, basin parameters, basin area, altitude, volume, profiles of land which indicate the nature of development of basins can be quantitatively calculated. In this study, Noyyal river, a tributary of Kaveri river in TamilNadu is taken for morphometrical analysis. The basin morphometric parameters such as radial and linear are computed. The 5th order streams of Noyyal are mainly dendtric and radial type. From the parameters calculated, Noyyal river basin drainage is not affected by geological structures.

Introduction

GIS(Geographical Information System) means "A powerful set of tools for collecting, storing, retrieving at will, transforming and displaying spatial data from the real world" (Burrough. 1987) QGIS functions as GIS software, allowing users to analyze and edit spatial information, in addition to composing and exporting graphical maps. QGIS supports both raster and vector layers. Geographical Information system (GIS) and Remote sensing techniques using satellite images are used as a convenient tool for Morphometric analysis. Digital Elevation Model (DEM) and Shuttle Radar Topography Mission (SRTM) widely used in drainage basin analysis. Morphometric analysis of watershed is the best method to identify the relationship of various aspects in the area. It is a comparative evaluation of different watersheds in various geomorphological and topographical conditions (Chandrashekar H, Lokesh K V, Sameena M, Roopa J and Ranganna G 2015). Watershed is a natural hydrological entity from which surface runoff flows to a defined drain, channel, stream or river at a particular point (Vittala S S, Govindaiah S and Gowda H H 2004). Drainage basin watershed analysis based on morphometric parameters is very important for watershed planning since it gives an idea about the basin characteristics regarding slope, topography, soil condition, runoff characteristics, surface water potential, etc. The morphometric analysis of watershed aids to know the aspects of linear, areal, and relief parameters (Abboud I A and Nofal R A 2017).

Study Area

Noyyal originates in the Vellingiri and Poondi hills and passes through Coimbatore, Trippur, Erode and Karur districts of western Tamilnadu. The boundary of the river Noyyal is between north latitude 10°54'00" to 11° 9' 03" and east longitude 76°39'30" to 77°55'25" The average width of basin is 25km. The total area of the basin is 3510 square km. The Noyyal is a seasonal river which has a good flow only for short periods during the north-east and south west monsoons. Occasionally, flash floods occur when there is heavy rain in the catchment areas. Apart from these periods, there is only scanty flow of water of the year. The Noyyal River has three distinct topographic segments: the slopes and valleys in the Western Ghats, a well-drained region; the lower catchment area in the flat plains; and the eastern Noyyal river basin which is fairly flat or gently undulating. The principle climate type of the river basin is semi-arid subtropical monsoon. In the western part, the climate is pleasant due to the winds rushing through Western Ghats pass, while the eastern part is comparatively hotter. The maximum temperature is in the range of 29.1 to 36.6 degree Celsius and the minimum is 17.3 to 24.4 degree Celsius recorded as mean temperature for the Noyyal river basin on an average estimation of various stations.

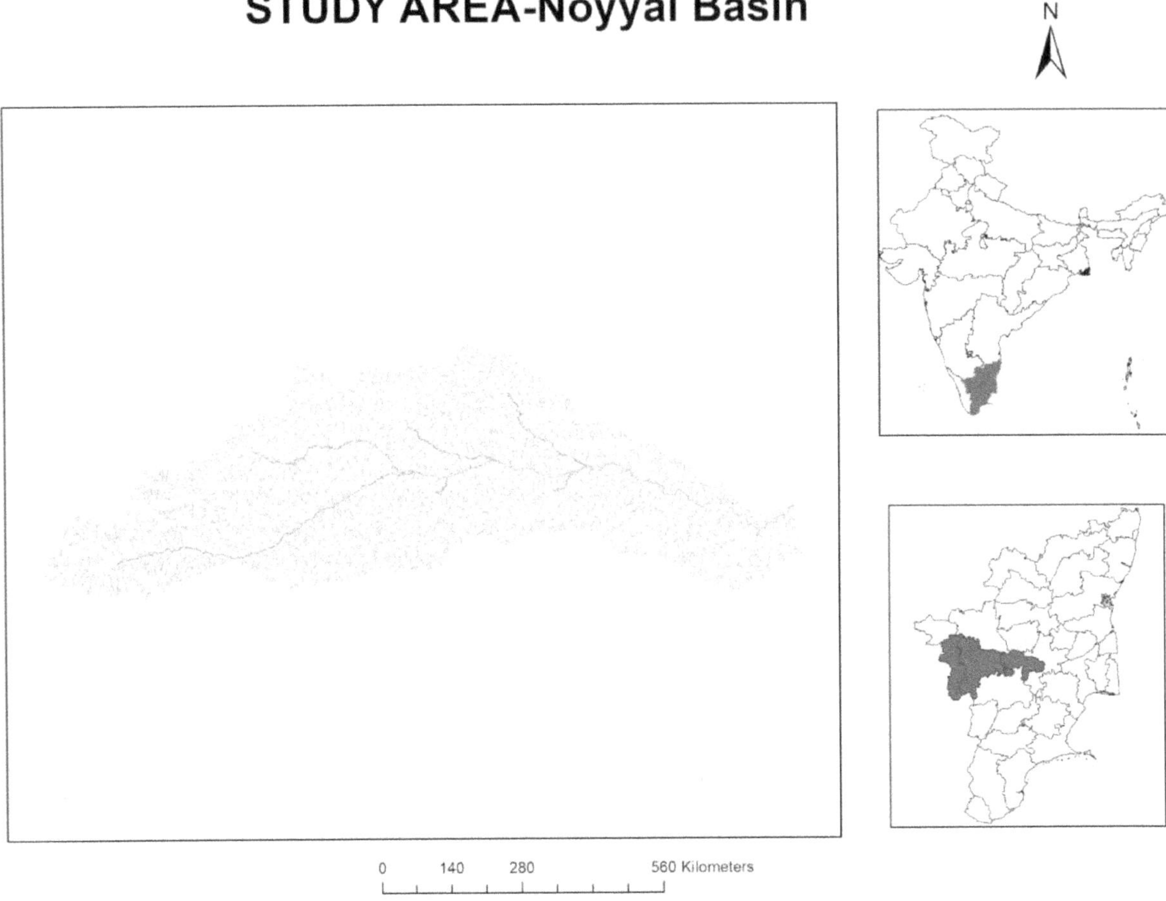

Figure 1

Methodology

The data of study area is from Dem downloaded from USGS Earth explorer. The delineation of study area is done by geo-processing in QGIS. The output is generation of Flow direction, Aspect, Contours, Flow accumulation, slope and stream order. The smallest, unbranched fingertip streams are designated as 1st order, the confluence of two 1st order channels give a channel segments of 2nd order, two 2nd order streams join to form a segment of 3rd order and so on. When two channel of different order join then the higher order is maintained. The trunk stream is the stream segment of highest order (Strahler1964). The parameters computed using QGIS are Stream order, Stream Number, Mean Stream Length, Stream Length ratio, Bifurcation ratio, Mean bifurcation Ratio.

Result

The present study area Noyyal river basin morphometrical analysis is done using QGIS. The following parameters are results which are quantitatively analyzed:

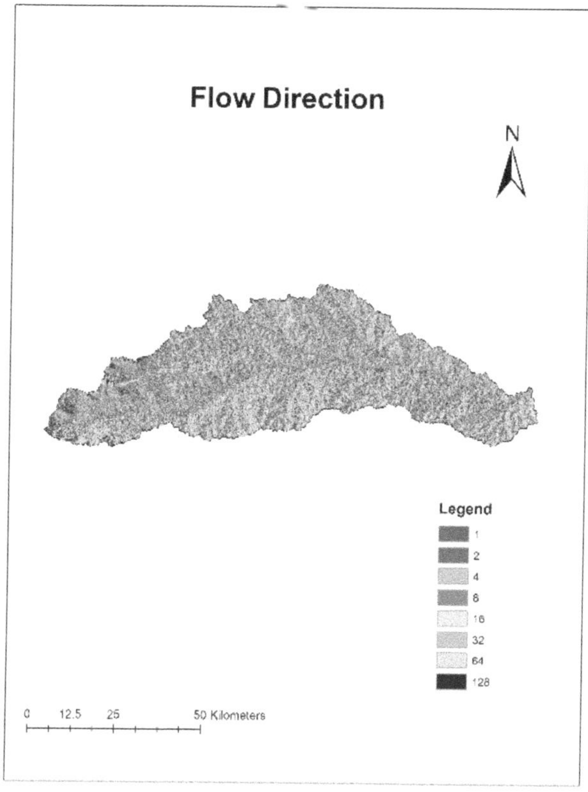

Figure 2

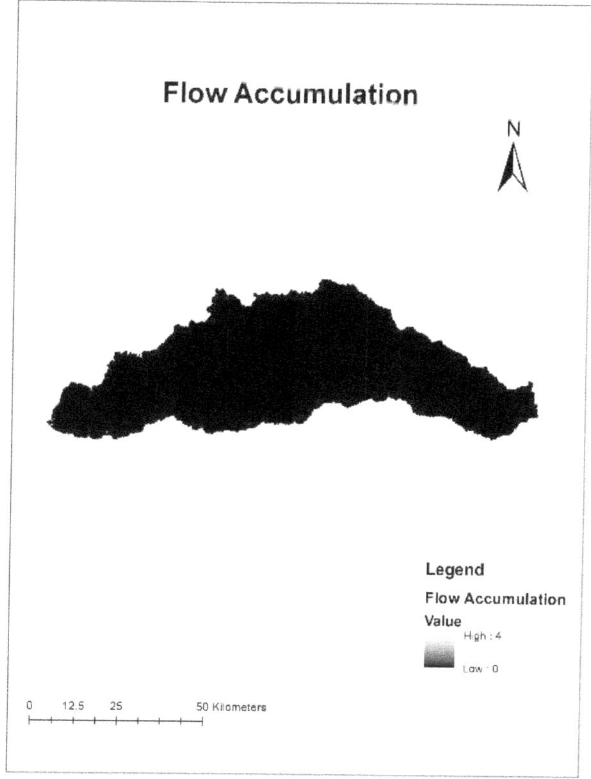

Figure 3

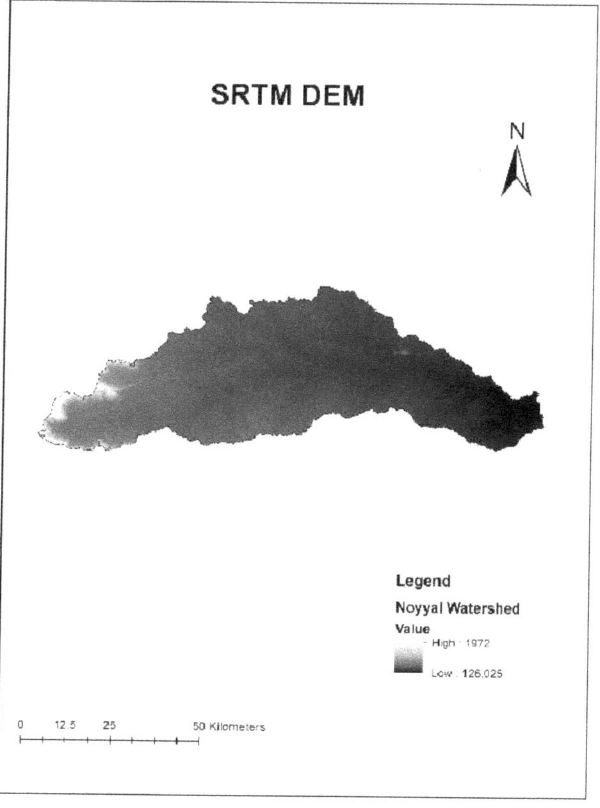

Figure 4

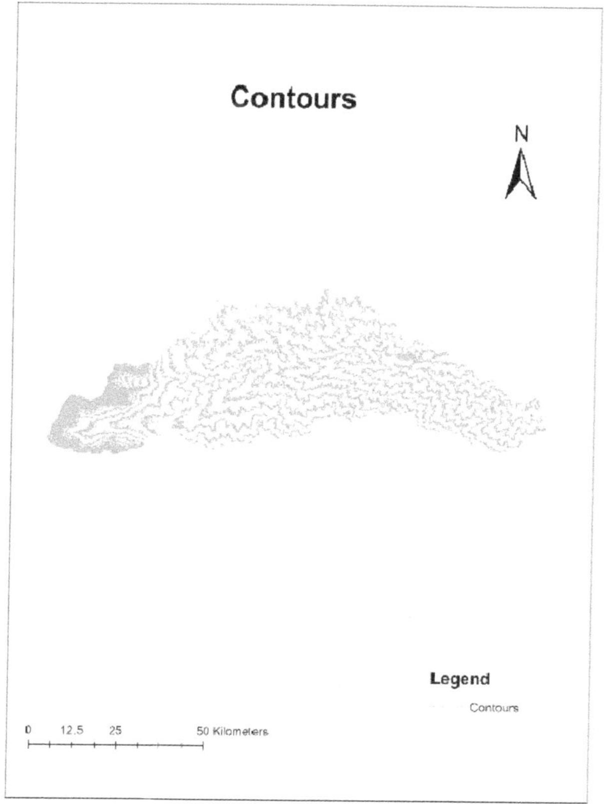

Figure 5

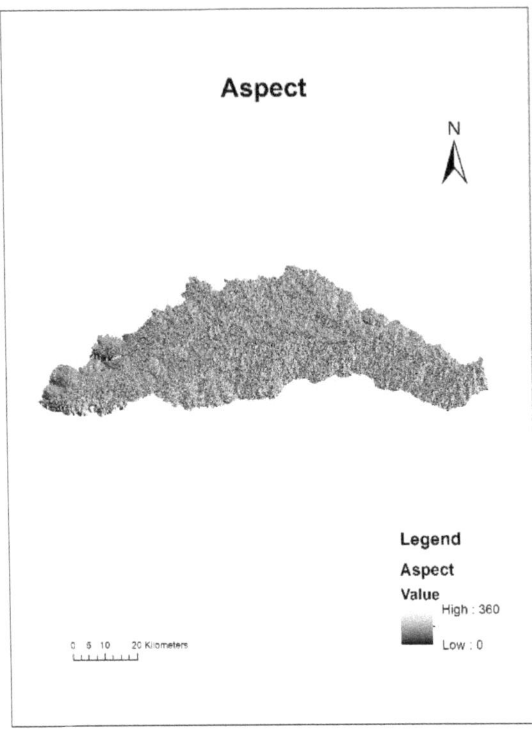

Figure 6

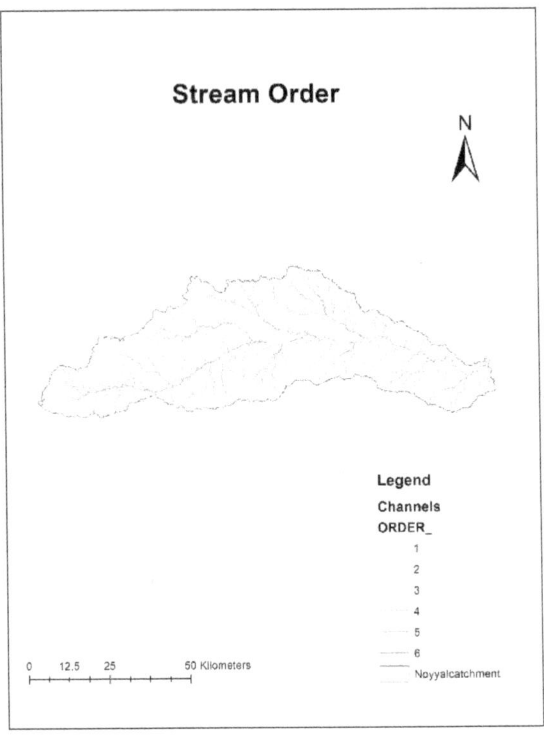

Figure 7

Stream Order

The present study area is ordered using Strahler stream ordering. Stream ordering is a widely applied method for stream classification in a river basin. Stream ordering is defined as a measure of the position of a stream in the hierarchy of tributaries (Leopold 1964).

Stream Number

Strahler's stream ordering starts in initial links which assigns order one. It proceeds downstream. At every node it verifies that there are at least 2 equal tributaries with maximum order. If not, it continues with the highest order; if yes, it increases the node's order by 1 and continues downstream with the new order (Strahler 1964). First order streams for Noyyal river are 3580.

Mean Stream Length

Stream length is one of the most significant hydrological features of the basin as it reveals surface runoff characteristics streams of relatively smaller lengths are characterstics of areas with larger slopes and finer textures. Longer lengths of streams are generally indicative of flatter gradients. Generally, the total length of stream segments is maximum in first order streams and decreases as the stream order increases. The number of streams of various orders in the basin are counted and their lengths from mouth to drainage divide are measured with the help of QGIS software. The mean stream length for Noyyal River basin is 664.90km.

Stream Length Ratio

Stream length ratio (Rl) is the ratio of the total stream length of the one order to the next lower order of stream segment. An increasing trend in the stream length ratio from lower order to higher order indicates their mature geomorphic stage (Vinutha D N and Janardhana M R 2014).

Bifurcation Ratio (RB)

Bifurcation ratio is the ratio of a number of the stream segments of specified order and a number of streams in the next higher order (1–6). The Rb is dimensionless property. Basically, there are two classes of Rb value; low and high. Low class means the drainage pattern is not affected by the geologic structures (1,3–6), whereas the high class means the drainage pattern is controlled by the geologic structures.(Chandrashekar H, Lokesh K V, Sameena M, Roopa J and Ranganna G 2015). For Noyyal river the Bifurcation ratio is between 1 and 2. Highest is 2.28km. This is low so, it is not affected by geologic structures.

Mean Bifurcation Ratio (RBM)

Low Mean Bifurcation Ratio indicates the negligible impact of geological structures on drainage pattern. For Noyyal river the Mean bifurcation ratio is 1.84km.

Table 1

Morphometric Parameters	Formula	Reference
Stream Order (Su)	Hierarchical rank	Strahler [1964]
Stream Number(Nu)	Total order wise stream segments	Horton [1945]
Stream Length (Lu)	Length of the stream	Horton [1945
Mean Stream Length (Lum)	Lum = Lu/Nu	Strahler [1964]
Bifurcation Ratio (Rb)	Rb = Nu/Nu+1	Schumm[1956]
Mean Bifurcation Ratio (Rbm)	Rbm=Average of bifurcation ratio	Strahler [1964]

Where; Lsm = mean Stream Length, Lu=Total Stream length of order u, Nu = Total number of stream segment of order u, Nu+1=Number of stream segment of next higher order, , Rb = Bifurcation Ratio.

Table 2

Stream Order	Stream Number	Mean Stream Length	Stream Length ratio	Bifurcation Ratio	Mean Bifurcation Ratio
1	3580	774.20	2.10	2.16	
2	1656	796.17	1.89	1.68	
3	984	707.15	2.60	2.28	
4	431	620.56	1.55	1.45	
5	296	582.9	1.87	1.63	
6	181	507.9			
					1.84

Conclusion

The quantitative analysis of morphometric parameters is found to be of immense utility in river basin analysis, watershed prioritization for soil and water conservation, and natural resources management at micro level (Noyyal river is a small tributary of Kaveri). The morphometric parameters evaluated using QGIS helped us to understand various parameters such as stream order, stream length, bifurcation ratio. The study area Noyyal river drainage basin is not much effected by geological structures. QGIS in conjunction with high resolution satellite data(DEM SRTM data) helped in better understanding the landforms and their processes and drainage pattern demarcations for basin area planning and management.

References

Gayen S., Bhunia G.S. and Shi P.K. 2013. Morphometric analysis of Kangshabati-Darkeswar Interfluves area in West Bengal, India using ASTER DEM and GIS techniques. Geol Geosci 2(4):1-10

Reddy G.P. Obi, Maji A.K. and Gajbhiye K.S. 2004. Drainage Morphometry and its Influence on Landform Characteristics in a Basaltic Terrain, Central India- A Remote Sensing and GIS approach. Int. J Appl Earth Obs Geoinfo. 6:1-16.

Chorley, R.J., 1972: Spatial Analysis in Geomorphology. Methuen and Co. Ltd., London.

Chorley, R.L., 1967: Models in Geomorphology. In Chorley, R.J. and Haggett, P. (eds.) Models in Geography, London. 59-96.

Dutta, D., Sharma, J.R., and Adiga, S. (2002). Watershed characterization, development planning and monitoring remote sensing approach, technical report ISRO- NNRMS-TR-103- 2002 Bangalore: ISRO.

Rastogi R.A. and Sharma T.C. 1976. Quantitative analysis of drainage basin characteristics. J Soil Water Conservation in India 26(1 & 4): 18-25. 17. Reddy G.E. Obi, Maji A.K. and Gajbhiye

K.S. 2002. GIS for morphometric analysis of drainage basins. GIS India 4(11): 9-14.

Krishnamurthy J., Srinivas G., Jayaram V. and Chandrasekar M.G. 1996. "Influence of rock types and structures in the development of drainage networks in typical hardrock terrain", ITC Journal, 3 (4), Pp. 252259.

Jenson SK, Domingue JO, (1988) Extracting topographic structure from digital elevation data for geographic information system analysis. Photogrammetric engineering and remote sensing 54(11): 1593-1600.

Jordan, D. L. (2004). "An introduction to GIS applications in hydrology." Southwest Hydrology 3: 1416.

Karadağ AA, (2013). "Use of Watersheds Boundariesin the Landscape Planning."

A Morphometric Analysis of Vellar River Basin of Tamil Nadu using Geospatial Technology

Swetha. S[1] and M.V.S.S.Giridhar[2]
[1]*M.Sc. Geography, Central University of Tamil Nadu*
[2] *Professor, Centre for water Resource, JNTU Hyderabad.*

ABSTRACT

Water resource degradation is a perilous hurdle all over the world especially in a drier region like Tamil Nadu.Water resources always has a prime importance among human beings as they are lifeline of all the living organisms in this planet. So the sustainable management of these water resources is a necessity. Therefore it is paramount to know the actual characteristics of a river basin which is done through morphometric analysis. Hence this study intends to assess the morphometric parameters of the river basin using geospatial technology. In this study Vellar River basin has been analysed to know its morphometric characteristics which will further help to understand the basin topography and for the sustainable management and protection of the river basin.This study is based on the SRTM digital elevation model from which the stream network and drainage basin has been delineated and analysed.From this study it is observed that the basin has an area of 7009.51 with stream orders ranging from 1 to 5 which has a mean bifurcation ratio of 2.95. This study primarily uses geospatial technology which proves to be a great tool for hydrological investigations and so that this method could further more used to analyse theintense problems of water resources.

Introduction

The main purpose of this paper is to find out the morphometric characteristics of the river basin taking into account its prime importance on the hydrological investigations. The significance of this morphometric analysis of the drainage basins is obvious because of its noteworthiness in water resources planning and management as the raising population and their increasing dependency over the water resources is proliferating day by day. Urban sprawling and population growth in countries like India, leads to increasing stress on water resources, because of growing demand for drinking, irrigation and industrial needs so as the need to protect those is of more concern .The drainage basins are the land areas that are drained by a river and its tributaries which holds a rich biodiversity within it. Morphometric analysis literally means the quantitative description or evaluation of the particular parameters of a land surface or a drainage basin based on certain criteria like linear, areal and relief aspects and using different mathematical equations. In this technological era the analysis of the characteristics of the water resources has become easier and faster due to the emergence of computer aided mapping of natural resources using geospatial technology. The geospatial technology has crude potential in hydrological studies. Remote sensing derived DEM's has been potentially utilized in water resource studies and GIS is serving as a great tool for this purpose and made everything simpler and uncomplicated. Furthermore many geospatial portals and web services provide us with lot more information on the inputs and methods. Therefore the use of geospatial technology in water resources analysis, monitoring and management is given a preeminence.

Objective

The major objective of this study is to investigate and identify various drainage parameters to understand the geometry of the basin through the use and application of GIS by delineating the basin from remotely sensed DEM from the same for the conservation and management of water resources in a sustainable manner.

Study Area

The study area lies between 78°34″ and 79°75″ E longitude and 11°94″ and 11°18″ N latitude. The study area Vellar river basin is one of the seventeen river basins of Tamil Nadu with a total geographical area of 7009.51Sq.Km covering 22 taluks, 40 blocks falling in parts of 8 districts, namely, Dharmapuri, Salem, Namakkal, Trichy, Perambalur, Ariyalur, Villupuram, and Cuddalore districts of TamilNadu. The Vellar River has it's originate from three rivers Anaimaduvu, Thumbal and Singipuram river. (i)Anaimaduvu River originates from Velanguttu hills at an altitude of 1122m and flows from west to southeast direction in Salem district (ii) Thumbal river originates from Thumbal hills at an altitude of 772m (iii) Singipuram river originates from Tengal hills, Jambuttu hills and Perumal hills of Attur taluk of Salem district, and joins the confluenced river of Kallar and Anaimaduvu at Vaittikavundan pudur travels as Vasista Nadhi up to Kalpaganur and thereafter the river is called as Vellar river. The major tributaries, viz., Swethanadhi, Chinnar river, Anaivari odai, Gomukhi river, Manimuktha river and Periyaodai are joining with Vellar river and the river flows through Dharmapuri, Salem, Namakkal, Trichy, Perambalur, Ariyalur, Villupuram, Cuddalore districts and finally confluences with the Bay of Bengal. The Basin is divided into three divisions based on the physiography like Western hilly terrain, Central hill valley complex terrain and Eastern deltaic plain and coastal region. The average annual rainfall of the Upper Vellar sub- basin is 847mm and the Lower Vellar sub-basin receives 1177mm annual average rainfall.

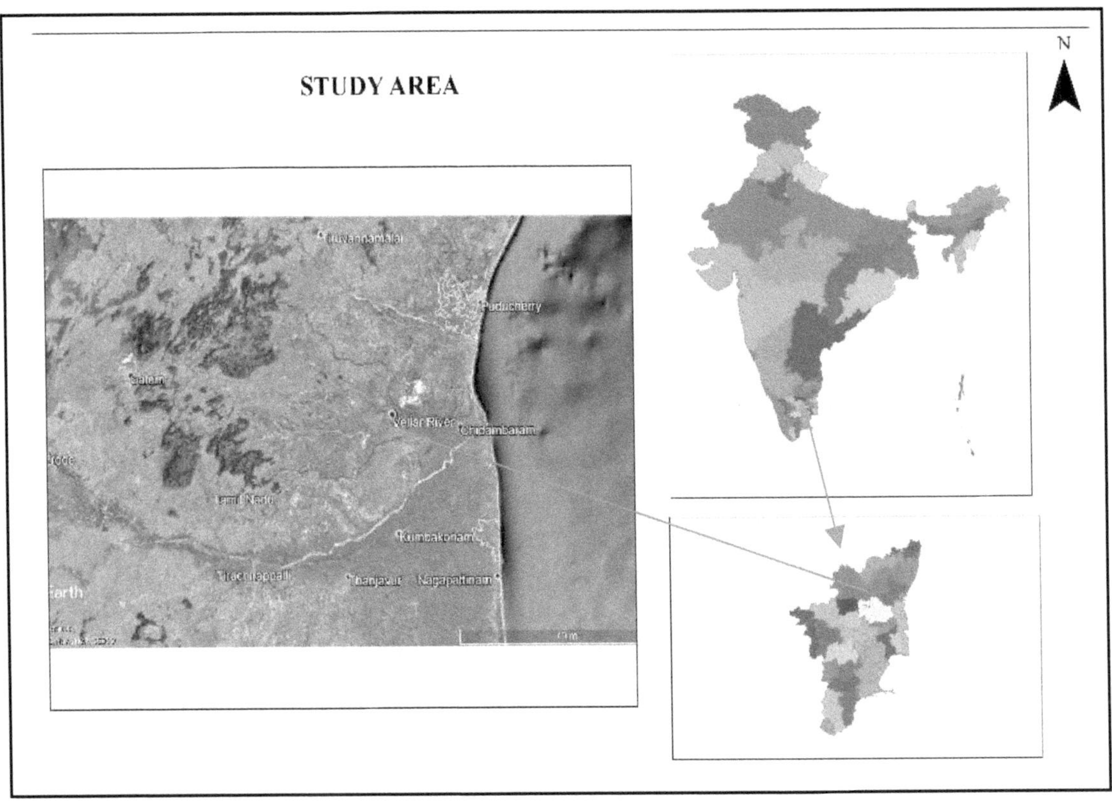

Figure 1

Methodology

The present study is based on the Digital Elevation Model of the River basin. These Digital Elevation Models (DEM) provides the 3 Dimensional evaluation of the earth surface. Here in this study the Shuttle Radar Topographic Mission (SRTM) data of Digital Elevation Model of 90m resolution has been used. From this DEM from SRTM the hydrological analysis and extraction of Stream Networks and Drainage Basins has been carried out. The SRTM data has been downloaded from the USGS Earth Explorer website which is freely available. The extraction of Channel network, drainage basin and further hydrological analysis of the drainage basin is carried out from a series of tools and techniques of QGIS (2.18.14) which is a open source GIS software .Finally the morphometric parameters has been calculated using standard formulae.

Table 1

Morphometric Parameters	Formula	Reference
Stream Order (Su)	Hierarchical rank	Strahler [1964]
Stream Number(Nu)	Total order wise stream segments	Horton [1945]
Stream Length (Lu)	Length of the stream	Horton [1945
Mean Stream Length (Lum)	Lum = Lu Nu	Strahler [1964]
Bifurcation Ratio (Rb)	Rb = Nu Nu−1	Schum[1956]
Mean Bifurcation Ratio (Rbm)	Rbm=Average of bifurcation ratio	Strahler [1964]
Basin Relief (Bh)	Vertical distance between the lowest and highest points of watershed	Hadley and Schumm (1961)
Relief ratio (Rr) R=basin	□□ = R/L Where, relief L = Basin length	Schumm (1963) Harton (1932)
Form factor (Rf) = area	□□ = A/(Lb)2 Where, A of the basin, km2; and Lb = length of the basin, km.	
Length of Overland Flow (Lof)	Lof= 1/2 Dd Where, Lof= Length of Overland Flow Dd= Drainage Density	Horton (1945)
Drainage density (Dd) (in km.)	Dd= Lu/A where, Lu=Total length of streams; A=Area of watershed	Horton (1945)

Result and Discussion

The Following Subsections illustrate the various morphometric parameters of the Vellar river basin. The values of these parameters are obtained various methods by different authors. The linear aspects has been calculated by using the methods of Strahler (1964), Horton (1945) and Schumm (1956). The areal aspects has been evaluated using methods of Horton (1945 and 1932), Schumm (19560 and Miller (1953). The relief aspects of the basin has been figured out by the methods of Schumm (1956).

Stream Order (Su): The stream order in this paper has been derivate based on the method proposed by Strahler (1964). Strahler Stream Order in hydrography deals with the hierarchy of streams from the source (or headwaters) to downstream. The headwaters are the first order and downstream segments are defined at confluences where two streams running into each other. Here in Vellar river basin the stream orders has been calculated using this method and as the results we got streams of 1 to 5 orders (Fig1.. The frequency of the streams is increasing as the decreasing order. The 5th order stream is the oldest stage of the river where it drains to Bay of Bengal.

Stream Number (Nu): After determining the stream order the number of streams in each order is counted to get the number of segments of the given order (u).In the study area there are totally 97 stream segments in which 69 are of first order streams, 19 belongs to second order streams ,6 to third order streams, 2 to fourth order streams and 1of fifth order stream. In this the highest percentage of stream segment is holded by first order streams.

Stream Length (Lu): The stream length (Lu) has been calculated on the basis of the Horton's law. According to Horton (1945), Stream length is the total length of stream segment of each of the consecutive order in the basin tends approximate a direct geometric series in which the first term is the average length of the first order. Here the Table.1. 2 shows the stream order wise length of the Vennar River Basin. The total stream length is greater in the 1st order streams which is 644153.18 m whereas for the 2nd order stream is 184326.63 m, 3rd order stream is 45077.56m, 4th order stream is 28519.02th order is 41062.78 m.

Mean Stream Length (Lum): Stream length Mean reveals the size of component of drainage network and its contributing surface (Strahler, 1964). It has been computed by dividing the total stream length of order 'u' by the number of stream segments in the order. The mean stream length as calculated based on this method are 9335.55m for 1st order streams, 9701.40m for second order streams, 7512.93 for 3rd order streams, 14259.5m for 4th order streams and 41062.78 for 5th order stream (Table.1.2).

Stream Length Ratio:

In this study the stream length ratio has been calculated based on Horton's law. The Stream length ratio of Vellar river basin varies from 1.44 to 0.24. The Stream length ratio of 5th order stream is 2.44, 4th is 0.63, 3rd is 0.24 whereas for 2nd is 0.28(Table.1.2).

Bifurcation Ratio (Rb): Bifurcation ratio is the ratio of the number streams of an order to the number streams of next higher order (Horton, 1945, Strahler, 1964). It is a dimensionless number denoting the ratio between the number of streams of one order and those of the next-higher order in a drainage network. It may be a useful measure of proneness to flooding: the higher the bifurcation ratio, the greater the probability of flooding. The Bifurcation ratio for the study area varies from 3.63 to 2. It is 3.63 for 1st order, 3.16 for 2nd order, 3 for 3rd order and 2 for 4th order (Table.1.2).

Mean Bifurcation ratio (Rbm): The mean bifurcation ratio is the average of bifurcation ratio of all orders according to Strahler (1957). The mean bifurcation ratio (Rbm) characteristically ranges between 3.0 and 5.0 for a basin when the influence of geological structures on the drainage network is negligible (Verstappen 1983).The mean bifurcation ratio for Vellar river basin is 2.95.

Table 1

Stream Order	Stream Number	Total Stream Length	Mean Stream Length	Stream Length Ratio	Bifurcation Ratio
1	69	644153.18	9335.55		3.63
2	19	184326.63	9701.40	0.28	3.16
3	6	45077.57	7512.92	0.24	3
4	2	28519.02	14259.51	0.63	2
5	1	41062.78	41062.78	1.44	

Basin Area: Drainage basin, also called Catchment Area, area from which all precipitation flows to a single stream or set of streams. The total area of the Vellar river basin is 7009.51 sq. km.

Basin Length: The length (L) is the longest length of the basin from the headwaters to the point of confluence Gregory and Walling (1973). The basin length determines the shape of the basin. High basin length indicates elongated basin. The length of this basin is found to be 240122.84m.

Basin Perimeter: The perimeter of a drainage basin is defined as the horizontal projection of its water divide. The perimeter of this drainage basin is 1001.83km.

Basin Relief (Bh): Basin Relief is defined as the ratio between the highest and lowest points of the watershed. The ratio can be positively correlated with the rate of sediment loss from a basin and it plays an important role in landforms, drainage development, and surface and sub- surface water flow, permeability and erosion properties of the terrain. The basin relief of Vellar river basin is 1355.98m

Relief Ratio (Rr): It is the ratio of the total relief of the basin to it the longest basin length measured along the main drainage line (Schumm 1956). The low value of relief is due to hard rock and low degree of slope. The highest value indicates steep slope and high relief. The relief ratio found for Vellar river basin is 0.0056.

Drainage Density (Dd): Drainage density is the stream length per unit area in region of watershed. Drainage density (D_d), expressed as a ratio of the total channel length of streams of all orders in a basin to the basin area, was introduced as an important morphometric parameter by Horton (1932). It indicates the closeness of spacing of channels, providing a quantitative measure of the average length of stream channel for the basin. The drainage density of the Vellar river basin is found to be0.134km

Form Factor (Rf): According to Horton (1932), Flow factor is the ratio between the basin area to the square of the basin length. It is a dimensionless ratio. Form factor is an indicator forflood formation and move, degree of erosion and transport capacities of sediment load in a watershed. The value of form factor varies from 0 to 1 where 0 means highly elongated shape whereas 1 means perfectly circular. The form factor of the Vellar drainage basin is found to be 0.12 which indicated its elongatedbasin.

Length of Overland Flow (Lof): Length of Overland flow is the distance covered by surface runoff before turning into a stream channel. In the development of drainage basins, length of overland flow (Lg) is usually equal one half reciprocal of the drainage density. The length of overland flow of velar river basin is 0.06km.

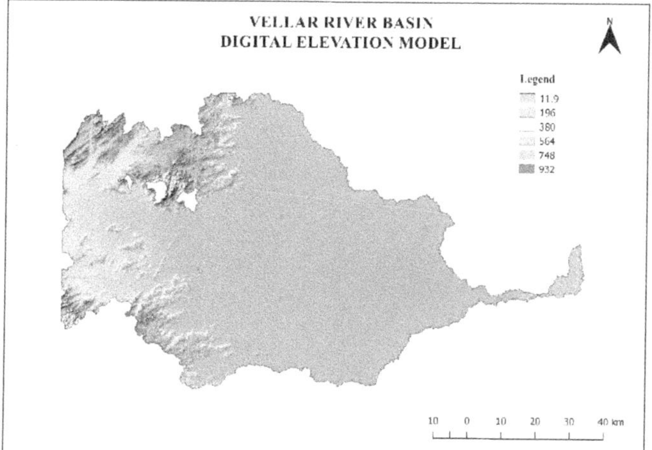

Figure 2

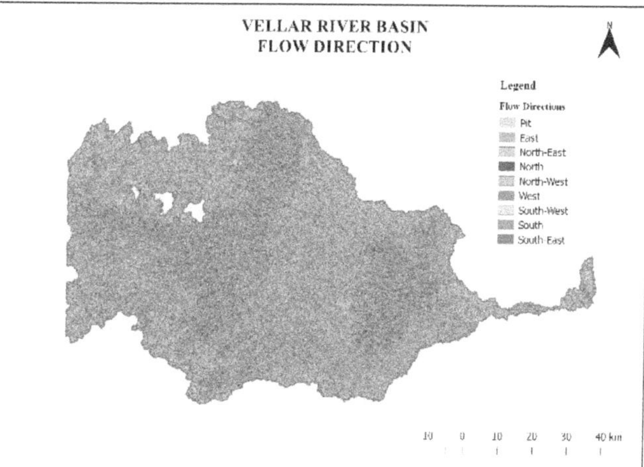

Figure 3

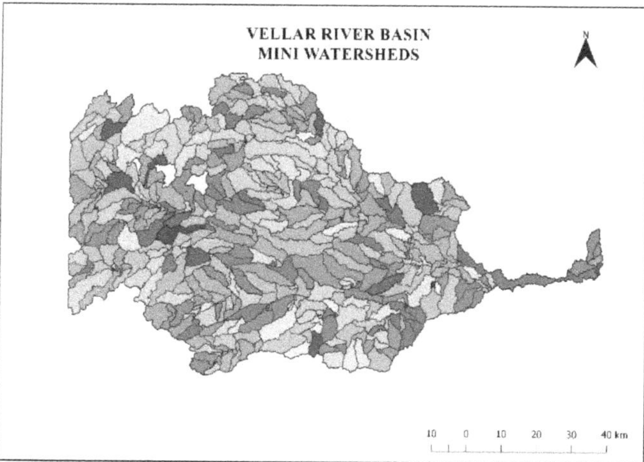

Figure 4

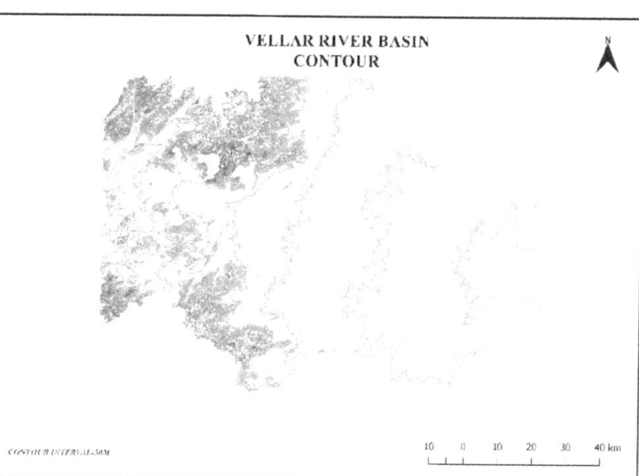

Figure 5

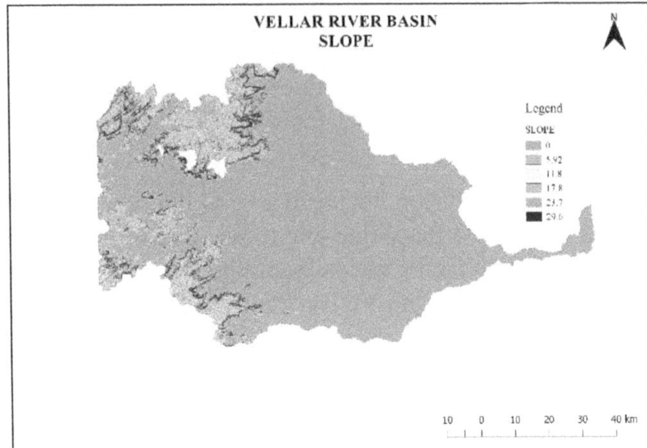

Figure 6

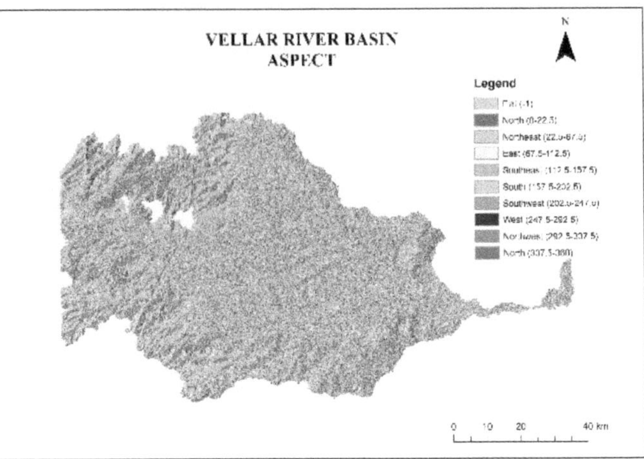

Figure 7

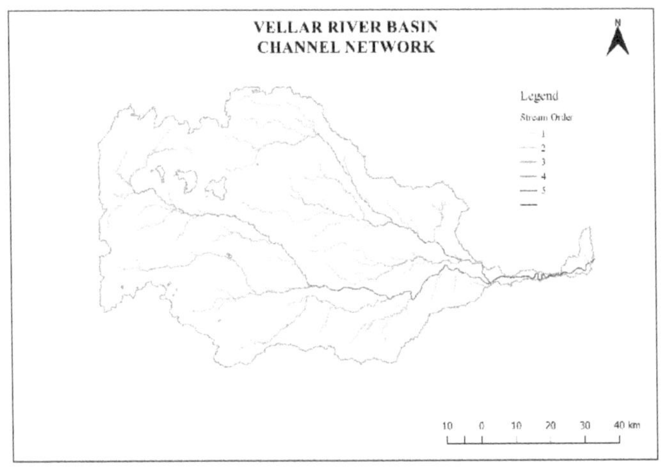

Figure 8

Conclusion

In this study the geospatial technology has been used which proved to be an effective tool for the analysis of Morphometric parameters. In the study area the stream order varies from 1 to 5U.The bifurcation ratio of the river basin which depicts that the drainage pattern is not affected by the geologic structures. From the analysis we could able to see that the river basin has a dendritic drainage pattern. In a state like Tamil Nadu where the water resources are dwindling gradually day-by-day, this geospatial technology proved to be a great tool for analyzing, monitoring and management of water resources and which is to used more efficiently.

References

Lokesh Shrivastava1, Shashikant Nag, Ritu Rani.2017. International Journal of Recent Scientific Research Vol. 8.

Singh, Prafull & Thakur, Jay Krishna & Singh, Umesh. (2013). Morphometric analysis of Morar River Basin, Madhya Pradesh, India, using remote sensing and GIS techniques. Environmental earth sciences. 68. 1967–1977. 10.1007/s12665-012-1884-8.

S Sukristiyanti, R Maria, H Lestiana. 2018. IOP Conf. Series: Earth and Environmental Science 118 (2018) 012028.

Jatoth Veeranna, Konda Gouthami, Perugu Balachandra Yadav and Mallikarjuna, V.R. 2017. Calculating Linear and Areal and Relief Aspect Parameters Using Geo-Spatial Techniques (ArcGIS 10.2 and SWAT model) for Akkeru River Basin Warangal, Telangana, and India.Int.J.Curr.Microbiol.App.Sci. 6(10): 1803-1809.

Tribhuvan, P.R. and Sonar, M.A.2016. Morphometric Analysis of a Phulambri River Drainage Basin (Gp8 Watershed), Aurangabad District (Maharashtra) using Geographical Information System. International Journal of Advanced Remote Sensing and GIS 2016, Volume 5.

Magesh N.S., Jitheshlal K.V. and Chandrasekar N. 2013.Geographical information system- based morphometric analysis of Bharathapuzha river basin, Kerala, India, Appl Water Sci 3: 467-477.

Mohd, I., Haroon, S. and Bhat, F.A. (2013). Morphometric Analysis of Shaliganga Sub Catchment, Kashmir Valley, India Using Geographical Information System, International Journal of Engineering Trends and Technology.

Schumm, S.A. (1956). The Evolution of drainage systems and slopes in badlands at Perth Amboi, New Jersey. Geological Society of America Bulletins, 67(5).

Strahler, A.N. (1957). Quantitative analysis of watershed geomorphology, Trans. Am. Geophys. Union, 38, 913 - 920. Strahler, A.N. (1964). Quantitative Geomorphology of drainage basins and channel networks, Handbook of Applied Hydrology, Ed. Ven Te Chow, 04, 9 - 76.

R. E. Daffi, I. I. Ahuchaogu.2017. Delineation of River Watershed and Stream Network Using ILWIS 3.7.1 Academic. Asian Journal of Environment & Ecology 4(4): 1-8, 2017; Article no.AJEE.35972ISSN: 2456-690X

Morphometric Analysis of Kodayar River Basin using Geospatial Techniques

Hemalatha I[1] and M.V.S.S. Giridhar[2]

[1]*M.Sc. Department of Geography, School of Earth Sciences, Bharathidasan University*

[2]*Professor, Centre for Water Resource, JNTU Hyderabad.*

ABSTRACT

Morphometric analysis is a quantitative description and analysis of landforms as practiced in Geomorphology that may be applied to a particular kind of landform or to drainage basin and large regions generally. This study has been made to understand the Morphometric characteristics of the Kodayar river basin with an aim to compute the detailed Morphometric analysis and their bearing on the hydrological condition of the region. The drainage area of the basin is 716.128 km^2. The stream order of the basin is mainly controlled by Physiographic and Lithological condition of the area. The present study makes to prioritize the Kodayar basin based on Morphometric parameters using Geospatial techniques. Various Morphometric parameters, namely linear parameters and Shape parameters have been determined using SRTM DEM for this basin. Morphometric analysis of any drainage basin can be very much effective in determining the landform characteristics and the processes and the components behind its development. The Morphometric parameters of the study area have been discussed with respect to linear, areal and relief aspects.

Introduction

The drainage basin is the rudimentary unit in fluvial geomorphology within which the relationships between landforms and the processes that reform them have been studied.]. The term 'morphometry' manifest the meaning of 'measurement of form' derived from morpho (form) and metry measurement. Morphometry litrelly means measurement and mathematical analysis of landforms may be applied to a distinct type of landforms or to a larger drainage basin generally. It provides important parameters about the basin characteristics. The study of morphometric analysis also helps to delineate the groundwater potential zones in a watershed. The diverse hydrologic phenomena can be correlated with physiographic attributes of drainage basin like size, slope, drainage density etc. Remote Sensing and GIS techniques are the proven systemized tools in the delineation, updating and morphometric analysis of drainage basin. The morphometric parameters are calculated by using Geographic Information system (GIS). GIS was used in estimation of linear and areal aspects of morphometric parameters. The remote sensing technique is an authentic method for basin study as the satellite images provide the opportunities to get a concise view over a large area. The DEM is used assuming that the water will flow from higher to lower elevation using steepest descent, will produce a stream extraction model with a thematic layer of aspect, slope, relief, and drainage density, stream frequency.Kodayar River is the major river system in Kanyakumari district The present study is an attempt to evaluate morphometric aspects of Kodayar River basin and to find out the recent changes in the hydrology of the Kodayar River.

Objectives

The aim and objective of the study is to delineate the morphometric characteristics of the Kodayar river basin using DEM then to study and evaluate the morphometric characters of the basin and to integreye the thematic maps through QGIS.

Study Area

Kodayar River is a major river system in Kanyakumari district originates from the Kodayar Lake (Muthukuzivayal plateau in Kalkulam taluk) and joins to Pahrali River at Moovattumugam, Chittar River I and Chittar River II. The basin extends between 8°57"N and 8°24" N latitude to 77°12"E and 77°49" E longitude. Pechiparai Dam is built across the River Kodayar. The Kodayar River drops 50 feet down forming the Thirpparappu waterfalls. Kothai is name of a female in Tamil. In Tamil it means "Beautiful girl of no mistakes"(Kuttram illadha Azhagiya pen). Kodayar Lake which is created as a result of the Pechiparai Dam. Surrounding the area of this reservoir is a dense green forest, which is noted for its variety of flora and fauna. The construction of this dam and reservoir was done a mile below the meeting point of Kuttiyar, Kallar and Chittar – the three tributaries of the Kodayar River.

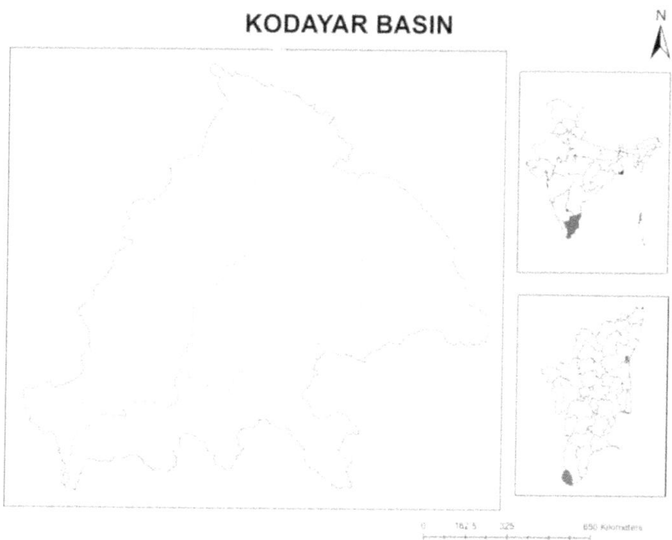

Methodology

This paper Shuttle Radar Topography Mission (SRTM) DEM to study the morphometric parameters. Delineation of sub- watershed boundaries, drainage network and extraction of terrain features like slope are done using SRTM DEM. Geographic Information System (GIS) is used for the hydro-spatial analysis. The drainage characteristics of Kodayar basin are studied to describe and evaluate its hydrological characteristics by analysing SRTM DEM data. The use of SRTM DEM data and GIS enables rapid, precise and inexpensive alternative for morphometric analysis. Data extraction and analysis is carried out using QGIS 3.6. Length and frequency of the stream of various orders are recorded for the analysis. Basic morphometric parameters such as drainage area (*A*), perimeter (*P*), basin length (L_b), and stream order (N_u) and mean stream length are estimated. Derived parameters like bifurcation ratio (R_b), mean stream length (R_l), drainage density (D_d), elongation ratio (R_e), circulatory factor (R_c), form factor (F_f) and RHO coefficient are calculated from the basic parameters. Relief-related parameters like basin relief, slope, relief ratio, gradient ratio, hypsometric integral and ruggedness number are estimated from DEM

Table 1

Morphometric Parameters	Formula	Reference
Stream Order (Su)	Hierarchical rank	Strahler [1964]
Stream Number(Nu)	Total order wise stream segments	Horton [1945]
Stream Length (Lu)	Length of the stream	Horton [1945
Mean Stream Length (Lum)	Lum = Lu Nu	Strahler [1964]
Bifurcation Ratio (Rb)	Rb = Nu Nu+1	Schum[1956]
Mean Bifurcation Ratio (Rbm)	Rbm=Average of bifurcation ratio	Strahler [1964]
Basin Relief (Bh)	Vertical distance between the lowest and highest points of watershed	Hadley and Schumm (1961)
Relief ratio (Rr) R=basin	□ □ = R/L Where, relief L = Basin length	Schumm (1963) Harton (1932)
Form factor (Rf) =area	□ □ = A/(Lb)² Where, A of the basin, km2; and Lb = length of the basin, km.	
Length of Overland Flow (Lof)	Lof= 1/2 Dd Where, Lof= Length of Overland Flow Dd= Drainage Density	Horton (1945)
Drainage density (Dd) (in km.)	Dd= Lu/A where, Lu=Total length of streams; A=Area of watershed	Horton (1945)

Results and Discussion

Morphometric parameters of watershed of Kodayar River Basin is extracted from the SRTM digital elevation model of 30m were processed using QGIS.The basis for assuming the limiting threshold value is obtained and presented. r.watershed was used to obtain the a watershed basin. The characteristics features streams and its order were

Table 2

Stream Order	Stream Number	Mean Stream Length	Stream Length Ratio	Bifurcation Ratio
1	464			
2	227	2.0311	2.0	2
3	26	2.1218	2.1	2
4	5	2.9928	2.9	2
5	2	1.1309	1.1	2.5
6	1	1.4343	1.4	2

Stream Order (u)

Stream ordering is the first step of quantitative analysis of the watershed. The stream ordering systems has first advocated by Horton (1945), but Strahler (1952) has proposed this ordering system with some modifications. Author has been carried out the stream ordering based on the method proposed by Strahler,. It has observed that the maximum frequency is in the case of first order streams. It has also noticed that there is a decrease in stream frequency as the stream order increases.. The ordering of the basin have been carried out by the method suggested by Strahler. The Kodayar basin is 6th order basin.

Stream Number (Nu)

Whole Kodayar basin has 725 streams, of which 64% are the first order streams having 464 Segments. The second order stream segments are 227 and account for 31.31%, third order stream segments are 26 and accounted 3.58%, fourth order stream segments are 5 and account for 0.68%, fifth order stream segments are 2 streams and accounted 0.27% and sixth order stream segment is 1 and account for 0.13%. The total number of stream segments is found to decrease as the stream order increases in all the sub basins.

Stream Length (Lu)

The stream length characteristics of the sub-watershed confirm Horton's second law (1945)Horton RE. 1945 "law of stream length", which states that the average length of streams of each of the different orders in a drainage basin tends closely to approximate a direct geometric ratio. Stream length of the basin indicates surface runoff characteristics. The total stream length in Kodayar river basin is 4.872km. The total stream length decreases with the increase in stream order.

Mean stream length

The mean stream length is a characteristic property related to the drainage network and its associated surfaces (Strahler, 1964). The mean stream length (Lsm) has been calculated by dividing the total stream length of order by the number of stream. The mean stream length of first order stream is 895.9m, second order stream is 901.6m, third order stream is 747.9 Km, fourth order stream is 947m fifth order stream 647m, and sixth order stream 735m. The mean stream length of stream increases with increase of the order.

Stream length ratio

Horton (1945, p.291) states that the length ratio is the ratio of the mean (Lu) of segments of order (So) to mean length of segments of the next lower order (Lu-1), which tends to be constant throughout the successive orders of a basin. Changes of stream length ratio from one order to another order indicating their late youth stage of geomorphic development (Singh and Singh, 1997). Stream length ratio shows a significant relation with surface flow discharge and erosion stage of the basin. The mean stream length of the second, third, fourth, fifth, sixth, is 2.0, 2.1, 2.9, 1.1, 1.4 respectively.

Bifurcation ratio

Bifurcation ratio (Rb) may be defined as the ratio of the number of stream segments of given order to the number of segments of the next higher order (Schumn, 1956). Horton (1945) considered the bifurcation ratio as an index of relief and dissections. Strahler (1957) The bifurcation ratio varies from sixth order has 2, fifth order has 2.5, fourth order has 5.2, third order has 8.3, and second order has 2 of bifurcation ratio.

Mean bifurcation ratio

The mean bifurcation ratio (Rbm) characteristically ranges between 3.0 and 5.0 for a basin when the influence of geological structures on the drainage network is negligible (Verstappen 1983). Geological and litho-logical development of the drainage basin may be the reason for these variations. The Rbm of Kodayar river Basin is 2.1.

Basin area (a)

Basin area is the direct outcome of the drainage development in a particular basin. The area of Kodayar basin is about 716.128sqkm. which indicates that rainwater will reach the main channel more rapidly, where the water has much further to travel.

Basin length (lb)

The length (L) is the longest length of the basin from the headwaters to the point of confluence Gregory and Walling (1973). The basin length determines the shape of the basin. High basin length indicates elongated basin. The length of the basin is 6.663Km

Basin perimeter

The perimeter of a drainage basin is defined as the horizontal projection of its water divide. The perimeter of the Kodayar basin is 242634 m

Basin relief

The relief (R) is defined as the differences in elevation between the highest and the lowest points on the valley floor of a study area. Basin relief is an important factor in understanding the denudation characteristics of the basin and plays a important role in landforms, drainage development, surface and sub-surface water flow, permeability and erosion properties of the terrain. From the morphometric study, it should be noted that the maximum relief value of Kodayar River Basin is 1792 m.

Relief ratio

Difference in the elevation between the highest point of a watershed and the lowest point on the valley floor is known as the total relief of the river basin. The relief ratio may be defined as the ratio between the total relief of a basin and the longest dimension of the basin parallel to the main drainage line (Schumm, 1956). In the study area, the value of relief ratio is 0.2689.

Drainage density (dd)

Drainage density is defined as a ratio of total length of all streams to the total area of the basin. Drainage density of the basin reveals the terrain configuration that is property of rock of area. The overall drainage density (Dd) of Kodayar basin is 0.0183km

Form factor (ff)

According to Horton (1932), form factor may be defined as the ratio of basin area to square of the basin length. The value of form factor would always be less than 0.754 (for a perfectly circular watershed). Smaller the value of form factor, more elongated will be the watershed. The value of form factor of the Kodayar River basin is 0.061sqkm.

Length of overland flow

It is the distance covered by surface runoff before turning into a stream channel. In the development of drainage basins, length of overland flow (Lg) is usually equal one half reciprocal of the drainage density. The length of overland flow is 0.00915km in the Kodayar river basin.

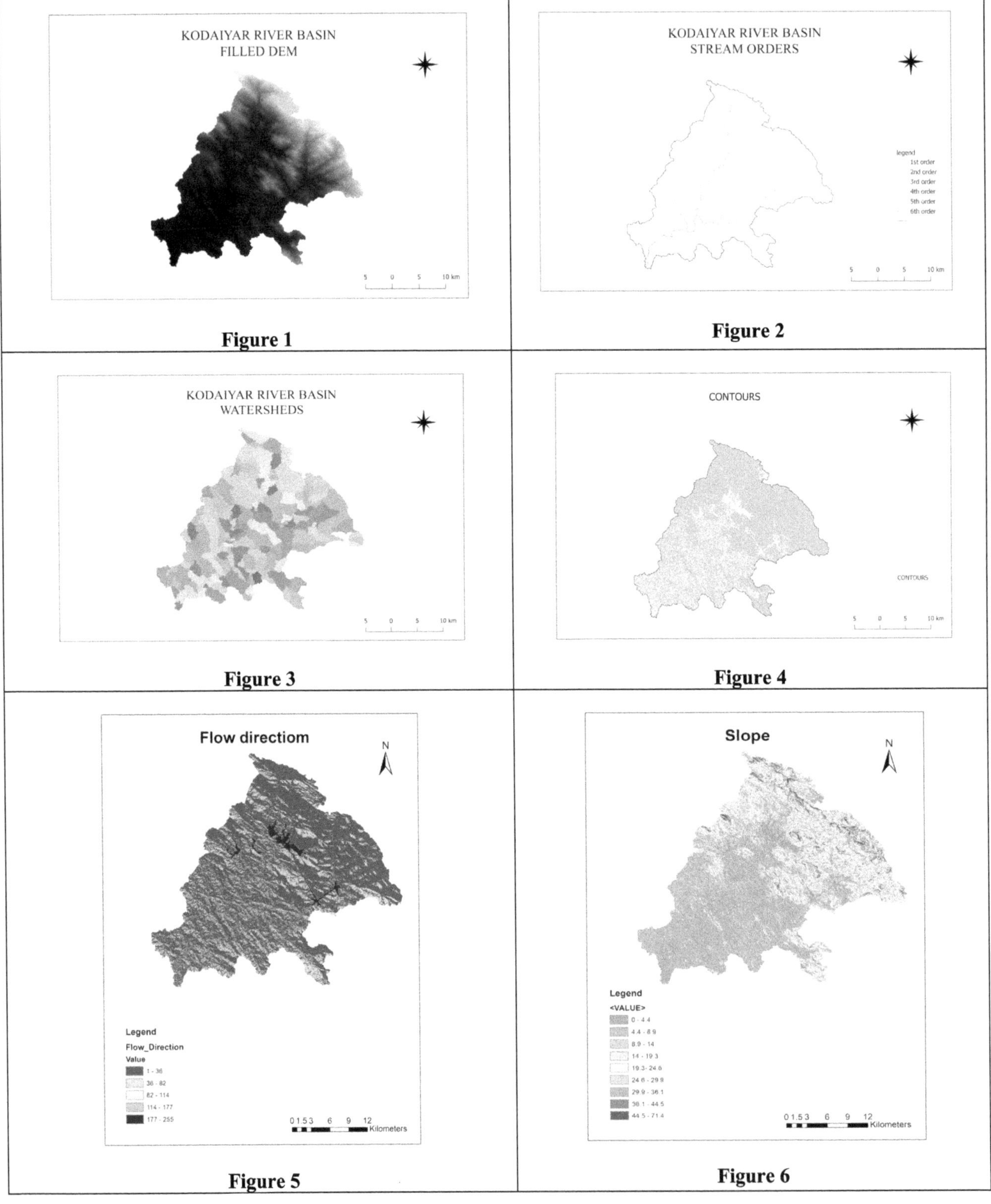

Figure 1

Figure 2

Figure 3

Figure 4

Figure 5

Figure 6

Conclusion

Analysis of morphometric parameters in the study area Kodayar can helps to find the river basin's highly eroded areas, vulnerability to the floods, water conservation, natural resource management. The studied basin has a dendritic drainage pattern with high drainage texture showing a 6th order stream network. The morphometric analysis of the River Kodayar was carried on the SRTM-DEM with 90 meter spatial resolution. QGIS3.6 software was used for analysis of the Perimeter (P), Basin Area (A), Basin length (Bl) and Stream Length (Lu).The bifurcation ratio ranges from (2.5 to 2) so there is less chances of flood risk. The morphometric parameters evaluated helped us to understand various terrain parameters such as nature of the bedrock, infiltration capacity, runoff, etc.

References

Kanth, T.A. (2012). Morphometric analysis and prioritization of watersheds for soil and water resource management in Wular catchment using Geo-spatial tools", Inter. J. of Geo., Earth and Environ. Sci., 2 (1), 30 - 41.

Mohd, I., Haroon, S. and Bhat, F.A. (2013). Morphometric Analysis of Shaliganga Sub Catchment, Kashmir Valley, India Using Geographical Information System, International Journal of Engineering Trends and Technology, 4 (1).

Volume: 02 Issue: 08 | Aug-2013, Available @ http://www.ijret.org 514 MORPHOMETRIC ANALYSIS OF VRISHABHAVATHI WATERSHED USING REMOTE SENSING AND GIS P T Aravinda1 , H B Balakrishna2 1 Assistant Professor, Department of Civil Engg, Govt Engg College Ramanagara, Karnataka, India,

Schumm, S.A. (1956). The Evolution of drainage systems and slopes in badlands at Perth Amboi, New Jersey. Geological Society of America Bulletins, 67(5).

Strahler, A.N. (1957). Quantitative analysis of watershed geomorphology, Trans. Am. Geophys. Union, 38, 913 - 920. Strahler, A.N. (1964). Quantitative Geomorphology of drainage basins and channel networks, Handbook of Applied Hydrology, Ed. Ven Te Chow, 04, 9 - 76.

Miller,V.C. (1953). A quantitative geomorphic study of drainage basin characteristics in the Clinch Mountain area, Varginia and Tennessee, Project NR 389042, Tech. Rept. 3., Columbia University, Department of Geology, ONR, Geography Branch, New York.

Pandurang Anandrao Pisal(july 2015). Morphometric analysis of Laxmi odha basin, Sangli district, Maharastra, India.

Garde, R.J. (2005). River Morphology, New Age International (Pvt) Ltd. Publishers, New Delhi.

Lokesh Shrivastava1., *Shashikant Nag1 and Ritu Rani2(2017) GIS-BASED MORPHOMETRIC ANALYSIS OF PENCH RIVER BASIN.

Magesh N.S., Jitheshlal K.V. and Chandrasekar N. 2013.Geographical information system-based morphometric analysis of Bharathapuzha river basin, Kerala, India, Appl Water Sci 3: 467-477.

Determination of Dielectric Constants of Different Materials by using GPR with 400 MHZ Antenna

[1]M. Sevu, [2]B. Venkateswara Rao and [3]D. Ajay Kumar

[1]Student of M.Tech, [2]Professor of Water Resources, [3]Student of M.Tech
Centre for Water Resources, Institute of Science and Technology,
JNT University Hyderabad, Kukatpally, Hyderabad-500 085, Telangana, India.

ABSTRACT

The main aim of this thesis is to estimate the dielectric constant value of different over burden materials such as bitumen pavement, unconsolidated soil, concrete bed and sand layer for the purpose of ground penetrating radar (GPR) surveys. Dielectric constant plays a vital role in estimating the depths in the GPR surveys. In the present thesis GPR surveys were conducted by using 400 MHz antenna at JNTU Hyderabad campus by taking seven profiles over the bitumen pavement, one profile on trench filled with soil after burying objects, one profile on concrete bed and one profile on sand layer. In all these profiles anomalies such as sewer pipe, Pvc pipe, electric cable and concrete pipe were detected using GPR with known depths, that is, depth of burial. While processing the GPR data with RADAN 7 software this known depth of burial of anomaly is substituted to get the travel time of the EM wave there by velocity and dielectric constant. From the experimental investigation of ten GPR profiles in the study area and the results obtained from the survey. It is found that the dielectric constant has increased due to decreases of EM wave velocities as it penetrates deeper depth in the presence of moisture in the sub surface. The dryer is the over burden material lesser the dielectric constant. The values of various overburden materials are 6.08 for bitumen pavement, 10.74 for unconsolidated soil, 5.28 for concrete bed and 4.40 for sand layer.

Keywords: Ground penetrating radar (GPR), Electromagnetic(EM), Radan 7, dielectric constant, geological.

Introduction

Ground-penetrating radar (GPR) is a new geophysical method that uses radar pulses to image the ground. This geophysical non-destructive method uses Electromagnetic Radiation in the range of microwave band of the Radiofrequency spectrum, to produce a cross-section profile of subsurface without any drilling, trenching or ground disturbances. The GPR profiles are used for evaluating the location and depth of buried objects and to investigate the presence and continuity of natural subsurface conditions and its features. GPR has been used successfully in a variety of media, including rock, soil, ice, fresh water, pavements and structures. GPR is also used in constraining problems of diverse fields such as archaeology, environmental site characterization, utility detections, glaciology, hydrology, landmine, unexploded ordinance detection and structural geology

In this paper, the present study was carried by GPR 400 MHz antenna the following objectives.

1. To determination of dielectric constant of different materials.
2. Subsurface anomalies such a sewer lines, metals pipe, cables, PVC pipes and any other buried objects may be detected. And also evaluated the location and depth of buried objects.

Location of the Study Area

The study area is located in the campus of Jawaharlal Nehru Technological University Hyderabad, kukatpally, Telangana, India which is bounded by latitudes 17°29’24” N and 17°29’50” N, and longitudes 78°23’30” E and 78°23’41” E. The areal extent of the study area is 89.19 acres. The topography of the study area is sloppy and well drained.

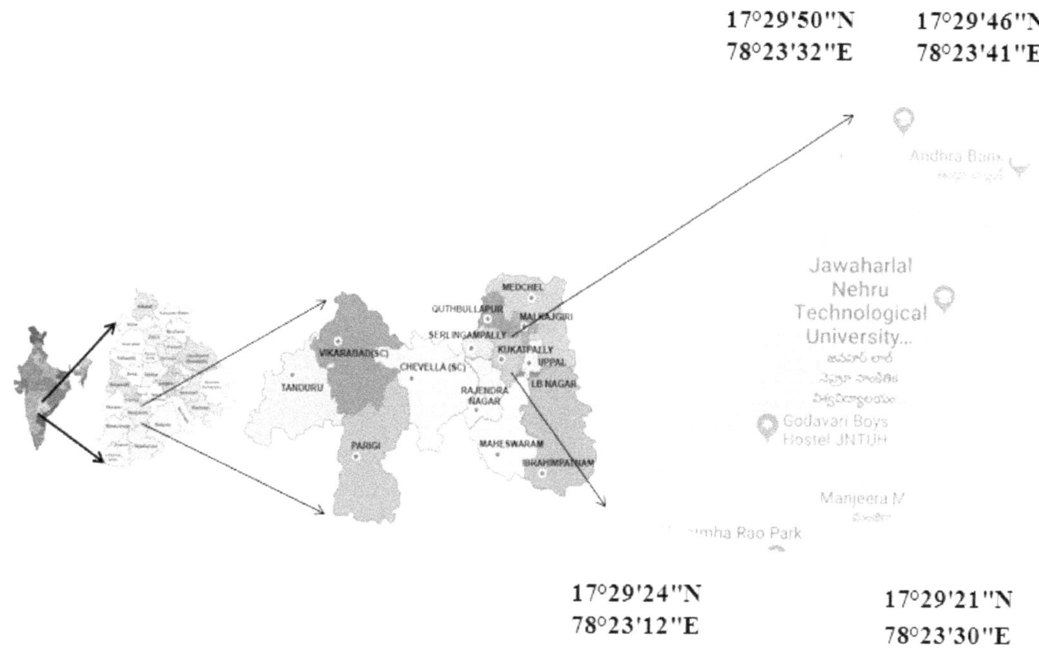

Figure 1 Location of the Study area

Geology of the Area

The geology is relatively homogeneous and consists of Archean granites. The campus is underlain by massive and hard grey and pink granites of Archean age. These rock formations are about 2,500 million years old. The ground water occurs in weathered formation under unconfined conditions, in joint and fractures in semi-confined conditions below the weathered mantle.

Research Methodology

The basic methodology in Ground penetrating radar (GPR) is based on the propagation and reflection of high-frequency electromagnetic waves into the subsurface. Then EM wave propagates in the ground at a velocity that is primarily a function of the relative dielectric permittivity of the subsurface. The EM wave encounters the interface of two materials having dissimilar dielectric properties, a portion of the energy is reflected back to the ground surface, and received by a receiver antenna however, it transmit data to the control unit display and store the data. Pulses are sent at regular intervals, which allow to rapidly generating image of the subsurface. The dielectric constant of the subsurface layers is created by plotting the differing travel time and depth of anomaly. The dielectric constant is a material property which affects the speed and reflection amplitude of electromagnetic GPR pulses. The strength, amplitude of the reflection is determined by the contrast of dielectric constants and conductivities of the two properties of those materials.

Materials having with higher dielectric constant values the radar energy penetrate slower and less distance. The water saturation of the material dramatically raises the dielectric constant of the material. Hence before conducting the survey area should be carefully inspected for signs of water. Metals are considered to be a complete reflector and do not absorb any amount of energy pass through it. Few Materials such as metal sheet, fine metal mesh, and pan deck can't be visible. In general the greater the dielectric constant of a material the slower the Radar energy will move through it.

Procedures for Data Acquisition

1. To start the GPR Survey on the green start button is pressed.
2. All the settings present in the control unit like depth range and sample per scan, Gain type and soil type are adjusted to the field condition. In the field the survey wheel is calibrated for 10 meters distance.

3. While moving antenna over the surface of the ground pulses of high-frequency EM waves are transmitted into the ground and the transmitted energy where encounters any anomalies, EM waves are reflected back and are received by antenna and are stored in the control unit.
4. The GPR control unit stores the reflections against two-way travel time in nanoseconds (ns) and then amplifies the signals.
5. To continue the data collection one can choose to save or discard a file. Then the instrument is positioned at the beginning of the next profile. Then the green Start button is to be pressed one time to begin the next data file to start.
6. During the field data acquisition itself, one can conclude that what type of anomaly exists in sub-surface, in location and depth of the anomaly. Once field data acquisition is completed, then stop the instrument. Raw data is collected from control unit and the post data processing is done with the RADAR 7 Software.

The equipment used to carry out GPR survey:

1. control unit
2. antenna
3. power supply (battery)
4. a measuring tape

Complete GPR Process

Following are the broad state in GPR survey which is shown in a flow chart

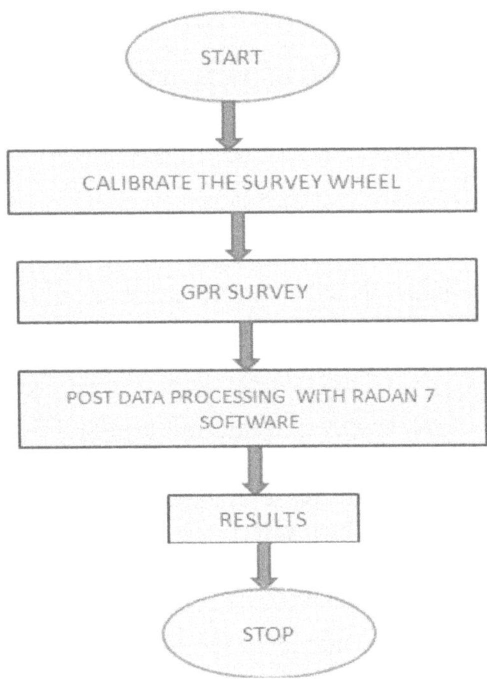

Figure 2 Flow chart of complete GPR survey

About Softwares

In this present work, RADAN 7 Software is one of the Most Advanced GPR Data Processing Software & interpretation with multi-layer velocity estimation and migration. RADAN 7 is a windows based system, providing a familiar and easy to use environment for all level of experience and easy processing feature automated processing wizard that recommends the best processes for cleaning up and improving data.

Data Processing

The sequential post data processing steps of GPR data are shown in the flow chart.

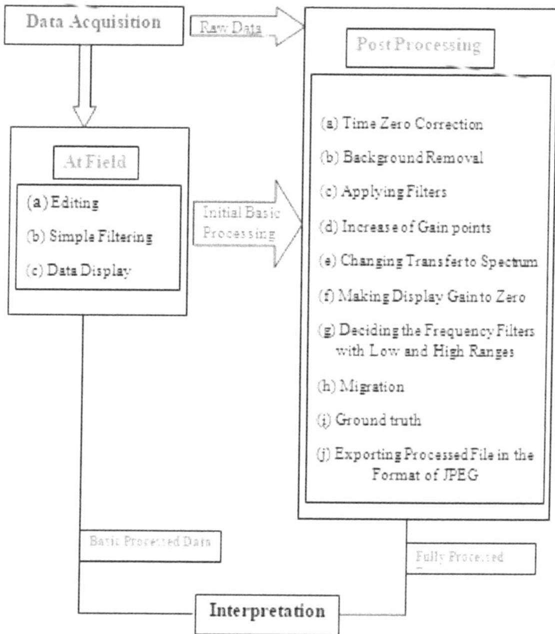

Figure 3 Flow Chart for Data Processing Steps

Results and Discussions

Discussion of Ground Penetrating Radar Profiles in the Study Area

The different types of utilities and its dielectric constant variations observed while conducting the GPR survey in site. Survey was conducted in ten locations in the campus in which three profiles were discussed in this paper.

GPR profile 1

Profile1 was carried out in the JNTUH campus near the main entrance gate over the highly compacted bitumen's pavement. The type of anomalies and its actual depths are verified physically in the site. In this profile four different anomalies (A, B, C and D as shown) were detected during survey. The anomaly present at 'A' is observed as concrete pipe with hyperbola having strong reflections. The Anomaly is iron pipe again which is present at 'B' and also checked physically in site. The Anomalies identified at 'C' and 'D' are as sewer pipe are close by. Based on the obtained profile it can be observed that the road substructure is well compacted clearly indicating layers. Once a depth measurement is obtained from the field, the signal velocity and dielectric can be calculated using the two- way travel time (2WTT) from the RADAR data.

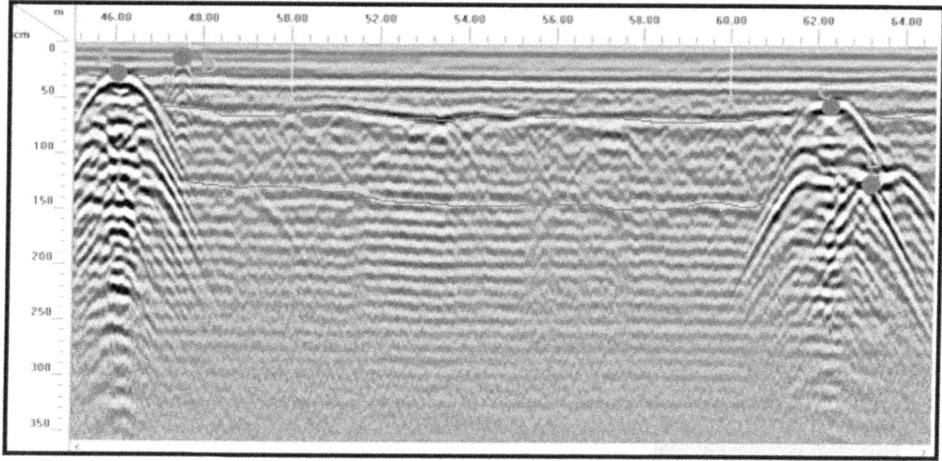

Figure 4 Processed GPR data of profile 1

Table 1 Dielectric constant of Bitumen- aggregate-soil cover at different depths

Anomaly Type	Anomaly present	Depth (cm)	Time (ns)	Velocity (m/ns) $v = \frac{2d}{t}$	Dielectric constant $\varepsilon_r = \left(\frac{c}{v}\right)^2$
Concrete pipe	A	29.80	5.11	0.11663	6.616
Iron pipe	B	10.34	1.57	0.13719	4.781
Sewer pipe	C	52.26	10	0.1045	8.141
Sewer pipe	D	116.5	23.2	0.1013	8.770

From the table 1 it can be observed that the lesser the target depth more is the velocity of the EM wave and lesser the dielectric constant. The EM wave is faster in the smaller depths because it is travelling only high compacted bitumen dried or any moisture. In the deeper depths along with compacted bitumen layer it is also travelling in the gravel and soil which is not as compacted as bitumen and also contain some moisture.

GPR profile 2

The profile was carried out in front of Godavari boy's hostel over bitumen's pavement in the JNTUH campus. The type of anomalies and its actual depths are verified physically in the site. In this profile four different anomalies were detected during survey. The anomaly present at 'A' is PVC pipe, which is present at lesser depth and the anomaly located at B in the profile in cement pipe, this anomaly has produced very strong reflection of hyperbola. The anomaly name is iron pipe present at 'D', the strength of a metal reflection increases with pipe size. On the other hand, it decreases with depth and/or presence of corrosion. It will also decrease when surrounded by material with a high dielectric constant.

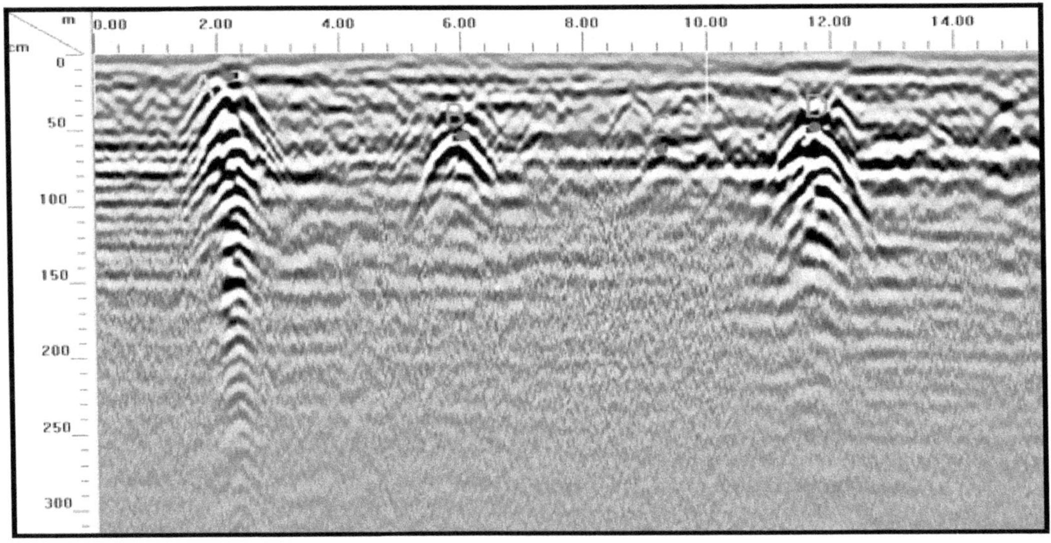

Figure 5 Processed GPR data of profile 2

Table 2 Dielectric constant of Bitumen- aggregate-soil cover at different depths

Anomaly type	Anomaly present	Depth (cm)	Time (ns)	Velocity (m/ns) $v = \frac{2d}{t}$	Dielectric constant $\varepsilon_r = \left(\frac{c}{v}\right)^2$
PVC with water	A	14.24	2.33	0.1222	6.04
Cement pipe	B	55.51	9.07	0.1224	6.007
Rock	C	62.71	10.3	0.1276	5.527
Iron pipe	D	51.58	8.43	0.1223	6.017

From the table 2 it can be observed that the velocity is more or less the same throughout the depth infact there is a little bit increase of velocity with depth. This may be due to the dryness of the subsurface.

GPR profile 3

The Profile was carried out in the JNTUH campus near the main entrance gate. The Anomaly detected at A is drainage pipe, this anomaly produced stronger reflection of hyperbolas whenever the radar energy move through it. The Anomalies present at 'B' and 'C' are iron pipes. The contrast in electrical conductivity between the materials scanning through and the target will affect the brightness of the reflection. Metal targets are showing bright reflections because they are conductive. In addition to the reflected RADAR wave, metal targets will return a small extra signal resulting from them becoming charged. The anomaly detected at 'D' is big drinking water main made up with metallic materials. The strength of a metal reflection increases with pipe size and it decreases with depth and/or presence of corrosion.

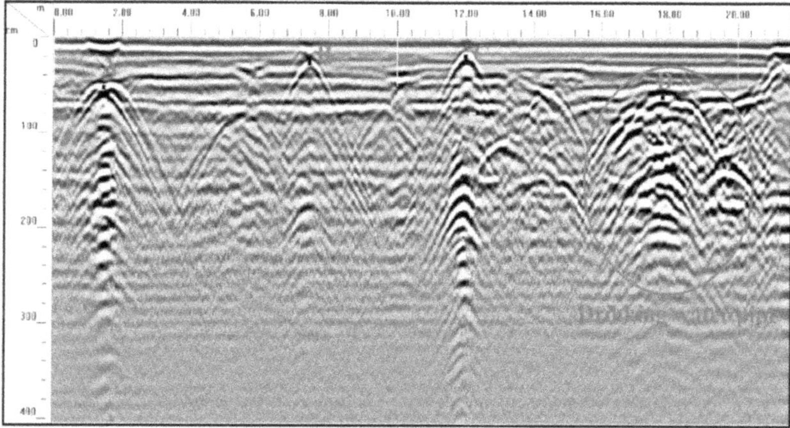

Figure 6 Processed GPR data of profile 3

Table 3 Dielectric constant of Bitumen- aggregate-soil cover at different depths

Anomaly type	Anomaly present	Depth (cm)	Time (ns)	Velocity (m/ns) $v = \frac{2d}{t}$	Dielectric constant $\varepsilon_r = \left(\frac{c}{v}\right)^2$
drainage pipe	A	52.87	8.58	0.1246	5.79
Iron pipe	B	26.77	4.28	0.1307	5.26
Iron pipe	C	19.19	3.13	0.1330	5.03
Drinking water pipe	D	62.14	10.02	0.1251	5.74

From Table 3 also indicate that EM wave velocity has decreased with depth consequently the dielectric constant has increased with depth.

Presentation of Results of Ten GPR Profiles

From the tables of Dielectric constants of various overburdens it can observed that

1. The average dielectric constant value of bitumen's pavement is 6.084.
2. The average dielectric constant value of unconsolidated soil is 10.74.
3. The average dielectric constant value of concrete bed is 5.28.
4. The average dielectric constant value of sand layer is 4.40.

It appears from the above investigations that dry sand has the lowest dielectric constant followed by concrete bed, bituminous pavement compared to this unconsolidated soil. The differences can be attributed due to presence of moisture level in the each bed the more the moisture level the greater is the dielectric constant. Dry sand is dryer than the concrete which is dryer than Bitumen's pavement and it is dryer than unconsolidated soil. Presence of moisture content in the beds results in the reduced velocity of EM wave and increased dielectric constant.

Conclusion

Based on the experimental investigation of ten GPR profiles in the study area and the results obtained from the survey the following conclusions were drawn. In general it is found that the dielectric constant has increased due to decreases of EM wave velocities as it penetrates deeper depth due to presence of moisture in the sub surface the dryer is the overburden the lesser the dielectric constant.

Recommendations

The dielectric constant values obtained from the study may be used in future investigations to determine the exact depths of anomalies and recalibrates instrument for dielectric constant.

Acknowledgment

I take this opportunity to express my sincere gratitude and whole hearted appreciation to my guide, Dr. B. VENKATESWARA RAO, Professor of Water Resources, and Director of Institute of Science and Technology, Jawaharlal Nehru Technological University Hyderabad, for his help, encouragement, inspiration during the whole thesis, and for his valuable suggestions in successfully completing and in writing of this paper. I thank my parents and all my family members, for their everlasting encouragement, love, cooperation and moral support in all my endeavors.

References

Steven C. Fisher, Robert R.Stewart, (1992). "Processing of Ground Penetrating Radar (GPR) data" CREWES Research Report Vol. 4, pp. 11.1-11.22.

P. M. Reppert, et al., (2000). "Dielectric constant determination using ground-penetrating radar reflection coefficients," Journal of Applied Geophysics, vol. 43, no. 2–4, pp. 189–197.

S. Feng and P. N. Sen, (1985). "Geometrical model of conductive a dielectric properties of partially saturated rocks," Journal Applied Physics, vol. 58, no. 8, pp. 3236–3243.

Antoine Robert, (1998). "Dielectric permittivity of concrete between 50 Mhz and 1 Ghz and GPR measurements for building materials evaluation", Journal of Applied Geophysics, Volume 40, Issues 1-3, October 1998, Pages 89-94.

Al-Qadi and S.Lahouar, (2005). "Measuring layer thickness with GPR: theory to practice". Construction and building materials 19(10), pp. 763-772.

Bevan, Bruce, and jeffrey Kenyon, "Ground penetrating Radar for Historical Archaeology. MASCA Newsletter vol 11, no. 2, December 1975, p 2 – 7.

Christina Salat, Andreas Junge, (2010). "Dielectric permittivity of fine grained fractions of soil samples from eastern Spain at 200MHz." GEOPHYSICS, 75(1), J1-J9.

Geophysical Survey Systems Inc, (2015). "GSSI SIR 4000 Manual" www.geophysical.com.

Richard J. Yelf, (2007). "Application of Ground Penetrating Radar to Civil and Geotechnical Engineering" Electromagnetic Phenomena Journal, Volume 7, pp. 103-117.

F. Bengal, J. B Garvin, (1990). "Microwave dielectric properties of dry rocks: Institute of Electrical and Electronics Engineers" Transactions on Geosciences and Remote Sensing, Vol. 28, no. 3, pp. 325–336.

J. R. Wang, (1980). "The dielectric properties of soil-water mixtures at microwave frequencies: Radio Science, Vol. 15, pp. 977–985.

J. L. Davis, A. P. Annan, (1989). "Ground-penetrating radar for high-resolution mapping of soil and rock stratigraphy", Geophysical Prospecting 37, pp. 531-551.

Design of Hydraulically Balanced Water Distribution Network

[1]Borkani Rajesh and C. Sarala[2]
[1]Student of M.Tech, [2]Professor of Water Resources
Centre for Water Resources, Institute of Science and Technology,
JNT University Hyderabad, Kukatpally, Hyderabad, Telangana, India.

ABSTRACT

Water is the most essential thing on this Earth for the survival of living species. Due to rapid urbanization, there is increase in demand of water supply all over the world. So there is a need of efficient water supply system in order to satisfy their demands. The main objective of the present study is to design the water distribution network by using EPANET software in the view of future demand of the people present in Adibatla village. The data collected from Adibatla village area, population data, distribution network data such as pipe length, diameter of the pipes of 100mm, 150mm, 200mm and 300mm. the Adibatla area digitization was done using google earth to simulate the network in EPANET software. The present study area is Adibatla village located in Ranga reddy district its population calculated by the arithmetic increase method and was found to be 17342. The distribution network is designed for 30 years with the required quantity and quality of water. The network consists of 81 junctions and 84 pipes, total length of pipes in the network is 6.09 km and reservoir capacity is 11.3 MLD, the pipe material is of ductile iron with roughness coefficient 130. By comparing values of hydraulic parameters the EPANET and manual calculated values haves got nearly same by obtaining some error which is negligible.

Keywords: EPANET, Analysis and Water Distribution Networks, GOOGLE Earth

Introduction

Water distribution system is a hydraulic framework consisting of aspect such as pipes, tanks, reservoirs pumps and valves etc. It is necessary to supply water to the public and provide drinking or potable water to the end users; Efficient water supply is of foremost importance in designing a new water distribution network and also spanning the existing one. In water distribution systems, water often needs to be pumped to a higher elevation with adequate pressure. For example, hydraulic pumps transport water from a treatment plant into an elevated storage tank. It is also necessary to investigate and establish a good network ensuring sufficient head. Determination of flows and pressure head in network pipes has been of great concern for those dealing with designs, construction and conservation of public water distribution systems. Analysis and design of complex piping networks can be tedious, especially if the networks consist of large number of pipes and system appurtenances. Manual calculations may not be practical to obtain results simultaneous and repetitively. By considering these reasons, the use of EPANET a computer software for designing and analyzing the system is appropriate. Hydraulic modeling technology involving pipeline system has rapidly evolved into an essential tool to facilitate design and optimized management of reliable water distribution systems. Such a study is necessary for changing the arrangement of consumption, delivery added features such as, an extension of booster pumps, pressure regulating valves or storage tank.

In this paper, the present study was carried out with the following objectives.

1. To forecast the population growth rate for the future demand.
2. To estimate the required quantity of the water to be distributed depending on the growth rate and future demand.
3. To design the water distribution pipe network using EPANET software.

Location of the Study Area

The study area is located in Adibatla Village of Ranga Reddy district in Telangana State between 17° 13' 53 to 17° 14' 16 " North Latitude and 78° 32' 34" to 78° 32' 39" East Longitude. The location map of the study area is shown in fig 4.6, covering an areal extent of 13.47 km^2. According to 2011 census, the population in the study area is 2582 and it is estimated that the population will increase to 17342 by 2048. The climate in the study area is semi arid type. During the summer months, the maximum temperature is around 42 °C while in winters the minimum temperature may come down to as low as 12 °C, Heavy rain from the south west monsoon falls between June and

September and the average annual rainfall is 833 mm. The major soil found here are black cotton soil, red soil and laterite soils.

Figure 1 Location of the Study area

Statement of the Problem

In today's world there is rapid increase of population. So, as a part of this, basic needs of the people are also increasing, water is one of those basic and important needs that human need. The water is supplied to the consumers from the reservoir and river through a proper distribution system in order to avoid losses and thefts, and to satisfy the demand of the consumer in the aspects like quality, quantity and pressure

The study are of thesis adibatla has no distribution system, and the Population in the area is increasing year by year. In order to satisfy the demands of the people living in the area, a distribution network system is designed by using EPANET software considering the future population for the next three decades.

Literature Review

Arjun Kumar et al. (2015) designed water distribution system using EPANET for Kathgarh village. In order to fulfil the water demand of the continuously growing population, it is essential to provide the sufficient and uniform quantity of water through the designed network of pipes. The general features of the area like information about the main water source, population of the area, demand of water, requirement of the pumps, distribution network and water tanks are essential for efficient design of water distribution system. The population of study area is 990. The type of distribution employed is tree or dead end system with design period of 20-40 years. The design of the water supply scheme is to supply efficient water to meet the daily requirement of people in this area.

Bhagvat Zolapara et al. (2015) carried out the design of Water Supply Distribution Network Using EPANET for Zone-1 of Kherali village, this network worked on EPANET based on public demand on population growth rate. EPANET software is used for simulation of hydraulic and water quality behaviour within pressurized pipe networks. It tracks the flow of water in each pipe, the pressure at each node, the level of water in each tank. In this study EPANET was used to carry out the hydraulic analysis of the distribution network in the study area. The results obtained are verified such that, the pressures at all junctions and the flows with their velocities at all pipes are feasible enough to provide adequate water to the network of the study area.

Rasooli Ahmadullah and Kang Dongshik (2016) designed Hydraulically Balanced Water Distribution Network based on GIS and EPANET. Their study showed the design of water distribution network using GIS, which is imported to EPANET software to analyze and to simulate model of KABUL CITY. The physical components are pipes, nodes, reservoir and different types of valves and the non-physical components are behaviour and operational aspect of system. Satellite images and infrastructure information in vector data are taken as input data. EPANET tracks the flow of water and aims to nullify the head loss and increase the discharge.

Philip et al. (2016) carried out smart optimization and sensitivity analysis in water distribution system using EPANET. Parameter uncertainty in water pipe network models are studied using newly developed simplified mathematical notions. These enable studies to be done using public domain software, including EPANET. The results obtained can be easier to use and interpret than those obtained from more general mathematical notions. The general idea is to study how a flow- and pressure-related quantity varies as a set of state parameters are varied. The quantity considered here is the average pressure, enabling smart optimisation of a water distribution system by keeping the average pressure unchanged as water demands change, by changing the speed of the pumps. Another application area considered, using the same mathematical notions, is the study of the sensitivity of parameters. The various parameter categories have very different sensitivities to a given change in the average pressure that can be tolerated. The critical state parameters to determine accurately in the models depend on the network. For the combined schemes studied as examples, variation of the pressure with reservoir depths is only related to the reservoir depths, and the pressure does not vary with the tank diameters. Pipe diameters are the most sensitive, pipe roughness coefficients are medium sensitive, and pipe lengths are the least sensitive

Theoretical Background

In this chapter the various methods of population forecasting, types of water supply, layout of distribution system and EPANET software are described.

Population Forecasting Methods

The population forecasting is the methodology which involves the prediction of the population in future years by means of having past census data. It is to be forecasted because it governs the future growth of population and development of the city in all aspects such as industrial, commercial, social and administrative

Population forecasting is based on the following methods.

(i) Arithmetic Increase Method
(ii) Geometric Increase Method
(iii) Incremental Increase Method
(iv) Decreasing Growth Rate Method

Arithmetical Increase Method

This method is suitable for large and old communities with considerable growth. In this method, the average increase in growth of population per decade is calculated from the past census reports. Arithmetical increase method is also called as constant growth method.

Therefore, Population after n^{th} decade will be $P_n = P_0 + n.x$

Where, P_n is the population after 'n' decades, 'P_0' is latest known population, 'n' is the number of decades and 'x' is average increases in population.

Geometrical Increase Method

In this procedure the percentage growth rate increases in population from decade to decade is assumed to be remains constant. This method gives more values and hence should be applied for young and rapidly growing cities. This method is also called as logarithmic growth rate method or exponential growth method.

The population at the end of nth decade 'P_n' can be estimated as

$$P_n = P\,(1 + I_G/100)^n$$

Where, 'I_G' is geometric mean (%), 'P' is Present population and 'n' is number of decades.

Incremental Increase Method

This method combines two methods namely arithmetical increase method and logarithmic growth method. This method is modified of arithmetical increase method and it is suitable for an average size city under normal condition where the population growth rate is found to be in increasing order.

Hence, population after n^{th} decade is $P_n = P + n.X + \{\frac{n\,(n+1)}{2}\}.Y$

Where, 'P_n' is Population after n^{th} decade, 'X' is Average increase and 'Y' is Incremental increase

Decreasing Growth Rate Method

This method is applicable to the areas where the population is decreasing with the time. This method is an extension of geometric increase method applicable to the communities which are rapidly approaching towards the saturation.

Types of Water Supply

Continuous supply: In such a system, there is a continuous water supply for 24 hours per day

Intermittent supply: In this system, the supply of water is either done in a entire village or town for fixed hours or supply of water is divided into number of zones and each zone is supplied with water for fixed hours in a day or as per specified day

Water Distribution Systems

The main aim of distribution system is to deliver water to consumer with appropriate quality, quantity and pressure. Distribution system is used to describe collectively the facilities used to supply water from its source to the point of usage. According to IS 1172-1993 code recommended the pressure for single storey building is 7 meters for two storey building is 12 meters of water for three storey building is 17 meters of water. The fire hydrants pressure should not be less than 1 kg/sq.cm

(a) Gravity system
(b) Direct pumping system
(c) Combined system or dual system

Layouts of Distribution System

The distribution system consists of pipes which are generally laid below the pavement of roads, and hence their layouts, generally follow the layouts of roads. Depending upon their layout and direction of supply, the distribution systems are generally classified into four types. They are:

1. Dead end or Tree system
2. Grid iron system or Reticulation system or interlaced system
3. Circle or Ring system
4. Radial system

Methodology Adopted

The population forecasting for the study area is carried out based on the population variation of the study area in previous decades by means of application of various methods of population forecasting and suitable one is adopted. Based on the forecasted population of the study area, the design of water distribution network using EPANET software is carried out in order to supply water adequately. The study area for the design of water distribution network is the Adibatla village. Based on the past population trends the Arithmetic increase Method is best suited for this study area.

EPANET Software

EPANET is a Water distribution system model have become widely accepted both within the water utility industry and the general research area for simulating both hydraulic and water quality behaviour in water distribution systems. In 1991, Lewis Rossman started the development of EPANET and released the first version in 1993. Though there were commercial water quality models available at that the time, the importance of EPANET was its open source and its public domain status. EPANET is a freely available computer program that performs extended period simulation of hydraulic and water quality behaviour within pressurized pipe networks. EPANET has two capabilities to perform extended period simulation such as hydraulic modelling and water quality modelling. EPANET is designed to be a research tool for different kinds of applications in distribution system analysis like sampling program design, hydraulic model calibration, chlorine residual analysis, consumer exposure assessment. EPANET can be used to assess alternative management strategies for improving water quality throughout a system. EPANET can be used for both steady-state and extended period simulation (EPS) hydraulic simulations. In addition, it is designed to be a research tool for modelling the movement and fate of drinking water constituents within distribution systems. For the simulation purpose, the water distribution network is represented in a hydraulic model as a series of links and nodes. Links represent pipes whereas nodes represent junctions, sources, tanks, and reservoirs, valves and pumps are represented as links.

Programming of EPANET Software

EPANET is public domain software invented by the Water Supply and Water Resources Division of the U.S. Environmental Protection Agency's National Risk Management Research Laboratory. EPANET software provides an integrated environment for editing network input data, running hydraulic and water quality simulations, and the results are obtained in different formats. The hydraulic simulation performed by EPANET delivers information such as flows and head losses in links (pipes, pumps and valves), heads, pressures and demands at junctions, levels and volumes for water storage. The network hydraulics solver employed by EPANET software uses the Gradient Method, first proposed by Todini and Pilati which is a variant of Newton-Raphson method. EPANET is a computer program that executes an extended period simulation of hydraulic and water quality behaviour within the pressurized pipe networks. This distribution network contains pipes, nodes (pipe junctions), pumps, valves and storage tanks or reservoirs. EPANET software tracks the flow of water in each pipe, the pressure at each node, the height of water in each tank, concentration of the chemical added that and the entire the network during a simulation period.

1. Create Project with Default settings
2. Drawing the Network Map
3. Setting the Reservoir, Junction and Pipe Properties
4. Saving and printing the project

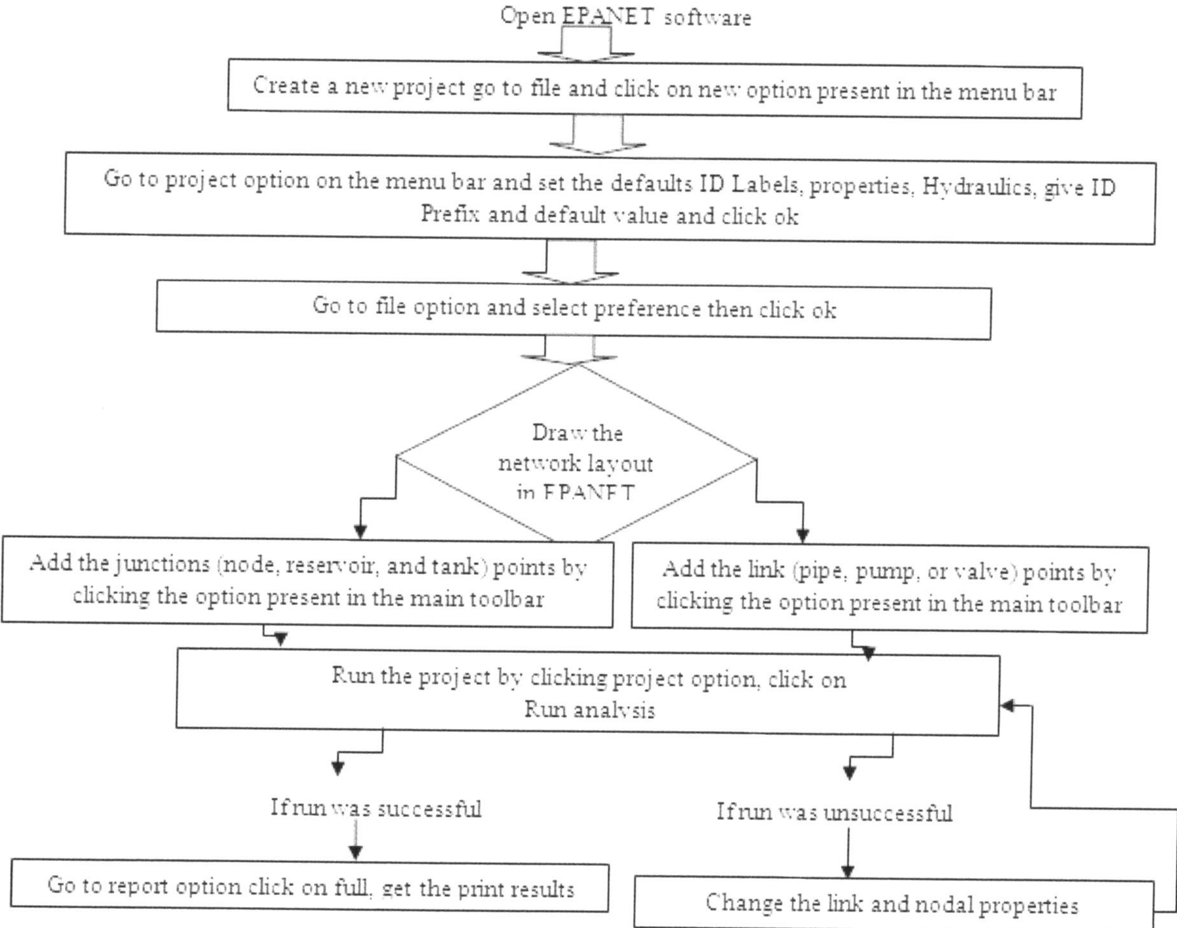

Figure 1 Flow chart of EPANET software for Water Distribution Network

Analysis of Epanet Results

Population forecasting for the study area

Design of pipe line network the population forecasting is most important. Population date has taken past three decades and we estimate future 30 years design. The population data is collected from the mandal office and is represented in table 1. The population forecasting is carried out by arithmetic increase method

Table 1 Population Data of Study Area

Year	Population	Per decade increased in population
1981	2167	-
1991	4125	1958
2001	6380	2255
2011	8962	2582

x = 2265

Population after 'n' decade can be determined by the formula

$$P_n = P_0 + n.X$$

Where,

P_n = Population after 'n' decade

P_0 = latest known population

n = No. of decades

X = Average increase in population

$$X = \frac{1958+2255+2582}{3} = 2265$$

P_{2018}=8962+0.7(2265)

=10547

P_{2048} =10547+3(2265)

=17342

Expected population after 30 years = 17342

Average rate of water supply = 135 LPCD

Total quantity of water required = 17342×135

QAvg = 2.34 MLD

Running Stsatus of Distribution Network

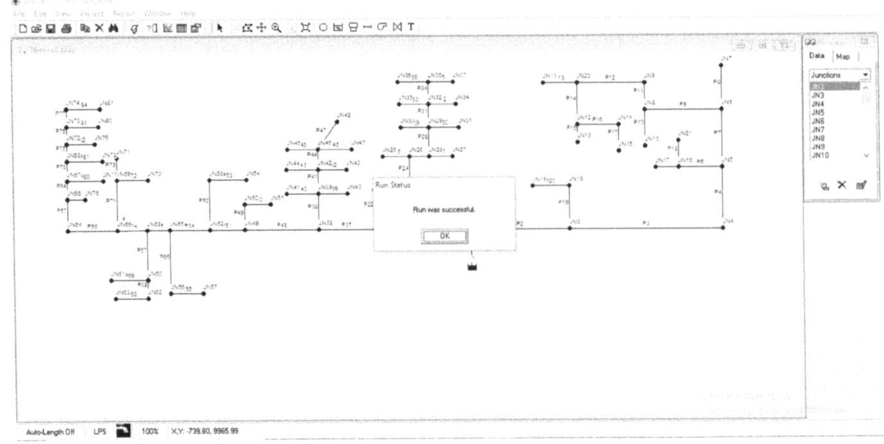

Figure 1 EPANET Software run status

Analysis of Junction in the Study Area

The main source of water supply to the study area is from Adibatla Village reservoir with a capacity of 11.35 MLD. There are 81 junctions and 84 pipes in the study area. At each junction the elevation values are recorded by using differential global positioning system and represented figure 2

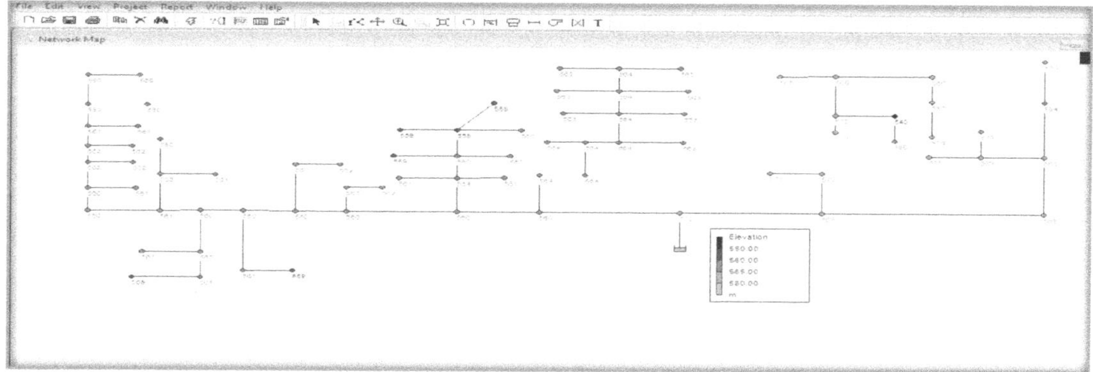

Figure 2 Network diagram for Elevations at Each Junction

The base demand at each junction is calculated by using the formula

Base demand (lps) = percapita demand * populations

At each junction, elevation and base demand values are given as an input in the EPANET software in order to obtained actual demand, total head and pressure head. The actual demand, total head and pressure head are calculated for one hour each junction in the study area which are shown in Table 2.

Table 2 Output Parameters Values at Each Junction

Junction No	One Hour Water Supply			Junction No	One Hour Water Supply		
	Actual demand LPS	Total head m	Pressure head m		Actual demand LPS	Total head m	Pressure head m
2	0.72	580.00	8.00	41	0.06	579.82	18.82
3	0.50	579.96	10.96	42	0.17	579.82	19.82
4	0.16	579.95	11.95	43	0.06	579.82	16.82
5	0.16	579.95	13.95	44	0.06	579.82	20.82
6	0.12	579.95	15.95	45	0.19	579.82	20.82
7	0.09	579.94	9.95	46	0.07	579.82	20.82
8	0.19	579.94	12.94	47	0.04	579.82	19.82
9	0.12	579.94	13.94	48	0.01	579.82	21.82
11	0.01	579.70	14.94	49	0.14	579.81	19.81
12	0.16	579.69	7.70	50	0.06	579.81	18.81
13	0.10	579.69	8.69	51	0.03	579.80	17.80
14	0.07	579.93	39.69	52	0.13	579.81	19.81
15	0.08	579.75	19.75	53	0.08	579.81	18.81
16	0.10	579.92	10.93	54	0.02	579.81	17.81
17	0.99	579.92	8.97	55	0.12	579.80	19.80
18	0.18	579.81	9.81	56	0.05	579.81	18.81
19	0.09	579.80	8.80	57	0.02	579.81	20.81
20	0.16	579.94	11.94	58	0.18	579.81	19.81
21	0.04	579.93	9.93	60	0.08	579.79	17.79
22	0.69	579.84	19.84	61	0.03	579.81	18.80
23	0.06	579.84	15.84	62	0.05	579.79	18.79
24	0.07	579.84	15.84	63	0.08	579.81	20.81

Junction No	One Hour Water Supply			Junction No	One Hour Water Supply		
	Actual demand LPS	Total head m	Pressure head m		Actual demand LPS	Total head m	Pressure head m
26	0.02	579.83	15.83	65	0.20	579.81	18.81
27	0.06	579.83	16.83	66	0.02	579.80	19.80
28	0.06	579.83	15.83	67	0.07	579.80	17.80
29	0.11	579.83	15.83	68	0.02	579.80	17.80
30	0.07	579.83	16.83	69	0.17	579.81	17.81
31	0.02	579.83	16.83	70	0.03	579.80	18.80
32	0.13	579.83	15.83	71	0.01	579.80	19.80
33	0.03	579.83	16.83	72	0.17	579.80	18.80
34	0.09	579.83	16.83	73	0.20	579.80	19.80
35	0.15	579.83	15.83	74	0.06	579.76	19.76
36	0.09	579.83	16.83	76	0.03	579.80	18.80
37	0.02	579.83	16.83	77	0.12	579.80	17.80
38	0.15	579.82	19.82	78	0.08	579.80	17.80
39	0.17	579.82	15.82	79	0.12	579.80	18.80
40	0.06	579.82	18.82	80	0.07	579.80	19.80

Analysis of Pipe in the Study Area

The study area consists of 84 pipes with varying lengths. The material used in the pipes is ductile iron and the length is measured by using Google Earth pro. The length of the pipes varied between 800 m to 30 m and these lengths are represented in figure 3. The diameter of the pipes ranged between 300 mm to 100 mm in the study area which is shown in Table 3

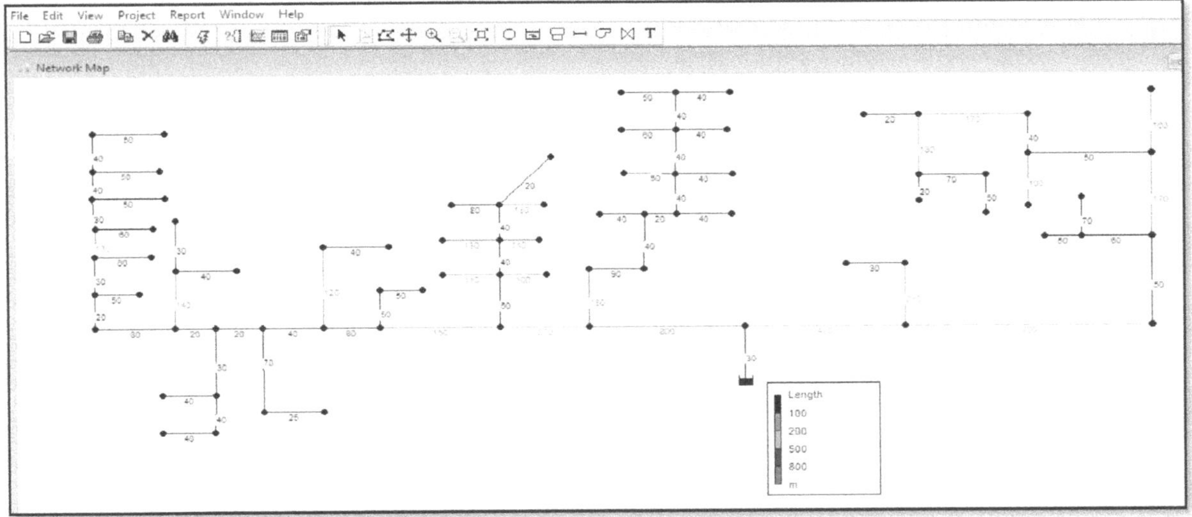

Figure 3 Network diagram for lengths at Each Pipe

The roughness coefficient for each pipe is 130 as the material is ductile. For each pipe, the length, diameter and roughness coefficient are given as an input to the EPANET software so as to obtain the flow, velocity and unit headloss for one hour and two hour time period in the study area. The output parameter values are determined when the pipe is open which are represented in Table 3

Table 3 Output parameters Values at Each Pipe

Pipe No	One Hour Water Supply			Pipe No			
	Flow LPS	Velocity m/s	Unit Headloss m/km		Flow LPS	Velocit y m/s	Unit Headloss m/km
1	9.61	0.14	0.08	41	0.61	0.02	0.00
2	3.41	0.11	0.08	42	0.06	0.01	0.00
3	2.65	0.08	0.05	43	0.06	0.01	0.00
4	2.49	0.08	0.05	44	0.31	0.01	0.00
5	1.13	0.14	0.32	45	0.04	0.02	0.02
6	0.99	0.13	0.25	46	0.07	0.01	0.00
7	1.19	0.04	0.01	47	0.01	0.01	0.00
8	0.09	0.01	0.00	48	2.57	0.08	0.05
9	0.98	0.06	0.03	49	0.09	0.01	0.00
10	0.06	0.01	0.00	50	0.03	0.01	0.01
11	0.70	0.04	0.02	51	2.34	0.07	0.04
12	0.58	0.03	0.01	52	0.10	0.01	0.00
13	0.01	0.01	0.00	53	0.02	0.01	0.01
14	0.41	0.21	1.37	54	2.12	0.07	0.03
15	0.10	0.05	0.10	55	1.93	0.06	0.03
16	0.15	0.02	0.01	56	1.09	0.02	0.01
17	0.08	0.04	0.07	57	1.04	0.03	0.01
18	0.04	0.02	0.02	63	0.08	0.01	0.00
19	0.27	0.08	0.26	64	0.99	0.04	0.01
20	0.09	0.05	0.09	65	0.07	0.01	0.00
21	5.48	0.17	0.20	66	0.02	0.01	0.00
22	1.17	0.04	0.01	67	0.24	0.01	0.00
23	1.11	0.04	0.01	68	0.12	0.01	0.00
24	1.04	0.03	0.01	69	0.03	0.02	0.01
25	0.02	0.01	0.01	70	0.80	0.03	0.00
26	0.84	0.11	0.18	71	0.21	0.01	0.00
27	0.06	0.01	0.00	72	0.03	0.01	0.01
28	0.72	0.02	0.00	73	0.01	0.01	0.00
29	0.07	0.01	0.00	74	1.51	0.05	0.02
30	0.02	0.01	0.01	75	0.71	0.02	0.00
31	0.52	0.02	0.00	76	0.42	0.01	0.01
32	0.09	0.01	0.00	77	0.15	0.02	0.01
33	0.03	0.02	0.01	79	0.03	0.01	0.00
34	0.26	0.01	0.00	80	0.12	0.02	0.00
35	0.02	0.01	0.01	81	0.08	0.01	0.00
36	0.09	0.05	0.08	82	0.12	0.02	0.01
37	3.62	0.12	0.09	83	0.07	0.01	0.00
38	0.91	0.03	0.01	84	0.09	0.01	0.00
40	0.06	0.01	0.00	58	0.03	0.01	0.00

Determination of Pipe Parameters Using Formulae

The velocity of flow in the pipe is determined by using Hazen-Williams equation

$$V = 0.85 \times C \times R^{0.63} \times S^{0.54}$$

Where, C = Roughness coefficient

R = Hydraulic radius (A/P)

A = cross sectional area of pipe

P = perimeter of the pipe

S = water surface slope (h_f/L)

L = length of given pipe

h_f = Elevation difference between respected nodes.

The discharge in the pipe is determined by using the continuity equation

$$Q = A*V$$

Where, Q = flow rate

A = cross sectional area of pipe

V = velocity of the liquid

$Q = \pi r^2 * V$ (where $A = \pi r^2$)

The head loss in the pipe is calculated by using Hazen-Williams equation

$h = A* Q^B$

Resistance Coefficient $A = 4.727\ C^{-1.852}\ d^{-4.871}\ L$

Flow Exponent B =1.852

Where, C = Roughness coefficient

d = Pipe diameter

L = Pipe length

Q = Flow rate

Comparison of Output Parameters in Pipe

The pipe output parameters namely velocity, flow and headloss are calculated by using formulae as well as by EPANET software. The difference in the output parameters values are shown in Table 4

Table 4 Comparison of Output Pipe Parameters in the Study Area

Pipe No	Flow (lps)			Velocity (m/s)			Headloss(m/km)		
	EPANET Software Values	Calculated Values	Difference	EPANET Software Values	Calculated Values	Difference	EPANET Software Values	Calculated Values	Difference
2	3.41	3.45	0.04	0.11	0.11	0.00	0.08	0.04	0.04
13	0.01	0.02	0.01	0.01	0.06	0.05	0.00	0.01	0.01
8	0.09	0.10	0.01	0.01	0.05	0.04	0.00	0.01	0.01
25	0.02	0.06	0.04	0.01	0.02	0.01	0.01	0.01	0.00

Conclusion and Recommendations

The capacity of the reservoir located in the study area is 11.35 million litres and the total amount of water that is needed to be supplied to the study area is about 2.34 million litres per day. The maximum actual demand is observed at junction 17 is 0.99 lps and the minimum actual demand of 0.01 lps is observed at junctions 11, 48 and 71. The total maximum head observed at junction 2 is 580 m and minimum total head at junction 13 and 14 is 579.69 m. The maximum pressure head of 39.69 m is observed at junction 14 and minimum pressure head of 7.70 m is observed at junction 12. Thc flow varied between 9.61 to 0.01 lps. in the study area. The maximum flow is observed at pipe 1 and minimum flow is observed at pipes 13, 47 and 73. The velocity of flow varied between 0.21-0.01 m/s. The maximum velocity is observed in pipe 14 and minimum value is observed in 31 numbers of pipes. The unit headloss values varied between 1.37-0.001 m/km. The maximum value is observed at pipe 14 and minimum value is observed in 28 pipes.

References

A Saminu, Abubakar, Nasiru, L Sagir (2013), "Design of NDA water distribution network using EPNET" International Journal of Emerging Science and Engineering (IJESE), ISSN: 2319-6378, Vol. 1, Issue:9.

Bhagvat Zolapara, Neha Joshipura, Jayadeep Patel (2015), "Case study on Designing Water Supply Distribution Network Using EPANET for Zone-1 of village Kherali", Indian journal of Research Vol. 4, Issue: 7.

RasoolI Ahmadullah, Kang Dongshik (2016), "Designing of Hydraulically Balanced Water Distribution Network Based on GIS and EPANET", International Journal of Advanced computer Science and Applications, Vol. 7, No. 2.

Darshan Mehta,Krunal Lakhani, Divy Patel, Govind Patel (2014). "Study of water distribution network using EPANET", International Journal of Advanced Research in Engineering, science and Mangement" ISSN:2394-1766.

G.Venkata Ramana, Ch.V.S.S.Sudheer, B.Rajasekhar (2015),"Network analysis of water distribution system in rural areas using EPANET" Procedia Engineering 119,496-505.

G.Anisha, A.Kumar, J.Ashok Kumar, P.Suvarna Raju (2016), "Analysis and Design of Water Distribution Network Using EPANET for Chirala Municipality in Prakasam District of Andhra Pradesh" International Journal of Emerging Science and Engineering and Applied Sciences (IJEAS) ISSN: 2394-3661, volume-3, issue-4.

Arjun Kumar, Vivek Thakur, Kankesh Kumar, Bharanidharan B, Neha Matial, Eshita Dey, Mahan Singh, Sarit Sharma, Neeraj Malhotra (2015), "Design Of Water Distribution System Using EPANET", International Journal of Advanced Research (2015), Volume 3, Issue-9.

Shivalingaswami.S.Halagalimath, Vijaykumar.H, Nagaraj.S.Patil (2016), "Hydraulic Modeling of Water Supply Network Using EPANET", International Research Journal of Engineering and Technology (IRJET), e-ISSN: 2395-0056, p-ISSN: 2395-0072, Volume: 3 Issue: 3.

Kakadiya Shital, Mavani Krunali, Darshan Mehta, Vipin Yadav (2015), "Simulation Existing Water Distribution Network by Using EPANET: A Case study of Surat city", GRD Journals, (Global Research and Development Journal for Engineering), Recent Advances in Civil Engineering for Global Sustainability, e-ISSN: 2455-5703.

D.Savic, G.Ferrari (2014),"Design and performance of district metering areas in water distribution systems" 16th conference on water distribution system analysis, WDSA 2014,doi:10.1016/j.proeng.

A.Ayad, H. Awad, A. Yassin (2016), "Integrated approach for the optimal design of pipeline networks"Alexandria Engineering Journal 57, 87-96

M.Prudhvi Kanth, S.Maanasa, T.Naga Rupesh (2016), "Design of water distribution network by using epanet software" ELK Asia Pacific Journals – Special Issue ISBN: 978-81-9304 11-5-4.

Impact of Ground Water due to the Solid and Liquid Dump and Evacuating Physicochemical Parameters in the Leachate

S. Venkat Charyulu[1], G.K. Viswanadh[2] and M.V.S.S Giridhar[3]

ABSTRACT

Dumping of solid waste, garbage is very new science to local places and surrounding areas which It effects the ground water and nearby places . In this paper, important of leachate effects and the it rate pollutions interm sof chemical tests. Here it is discussed about the Jawaharnagar local area which is situated in medchal village, Hyderbad city Telangana state India. In this venture, the nature of groundwater around the Jawahar nagar GHMC dumping yard, fully effected which is situated in Medchal-Malkajgiri locale, close Kapra, India., the rate of plootion is has been explored via completing Physico-Chemical investigation on groundwater and leachate tests gathered from the site, done to BIS standard. The technique incorporates computing the dissolved solids, pH, electrical conductivity, total disintegrated solids, Alkalinity, all out hardness, chlorides and nitrates in ppm (aside from pH), along these lines it is trailed by depicting ventures for structuring the reasonable landfill according to standard rules alongside different methods to neutralize supporting issue. finally it conclude the pollution intensity which has effected surrounding area with different chemical effects and measures to carried for the same.

Introduction

Rapid industrialization, growing population and changing lifestyle are the root causes for increasing rate of solid waste generation which leads to health hazards and an environmental burden. Not only the waste has increased in quantity, but the characteristics of waste have also changed tremendously over a period, with the introduction of so many new gadgets and equipment. The quantum of municipal solid waste generated in India is approximately 62 million tons annually out of which less than 60% is collected and around only 15% is processed. Therefore management of solid waste and related environmental impacts presents a huge challenge to both developing and developed countries.

Waste Generation in India

According to the Press Information Bureau, 62 million tones of waste is generated annually in the country, out of which 5.6 million tones is plastic waste, 0.17 million tonnes is biomedical waste, hazardous waste generation is 7.90 million tones per annum and 15 lakh tonne is e-waste. The percapita waste generation in Indian cities ranges from 200 grams to 600 grams per day

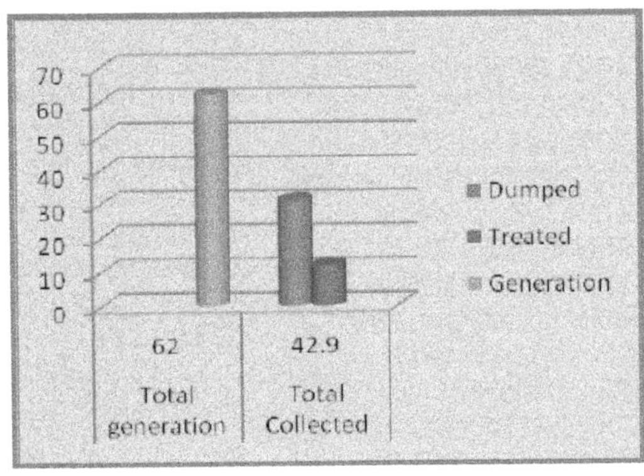

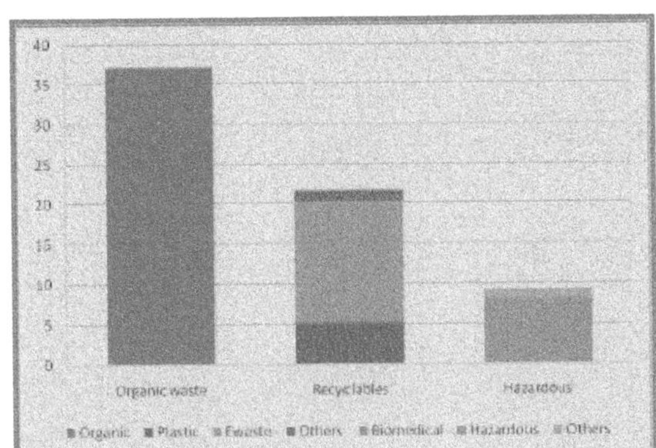

(Source: PIB, Government of India) PIB 2016

Figure 1 & 2 Collection vs Dumped Statistics (numbers in million MT per annum)

At present, 42.9 million TPA is collected, 11.9 million is treated and 31 million is dumped in landfill sites, which means that only about 75-80% of the municipal waste gets collected and only 22-28 % of this waste is processed and treated

Figure 3 Leachate lake near Jawahar nagar dump yard, Hyderabad

If a confining barrier beneath or surrounding the waste disposal site is absent, this leachate can migrate and contaminate subsurface and surface waters surrounding soils, vegetation, livestock and ultimately groundwater, which is a major source of water supply for drinking and domestic purpose in India.

- Several agro chemicals like DDT, fluorine, arsenic, lead compounds and organ phosphorus compounds found in polluted groundwater are super toxic and cause symptoms like nausea, vomiting, diarrhea, sweating, salivation and muscular tremors.
- Location of Jawahar Nagar dump yard, Jawahar Nagar village, Hyderabad, R.R.Dist.,Telangana, India.

Methodology

Very initial step was to select the location and visit the site, which in our project is Jawahar nagar dump yard located in Jawahar nagar village, Ranga Reddy district of Telangana, India.Leachate sample was collected in 1 litre pre-cleaned high density polyethylene bottle (HDPE) during premonsoon. Similarly, ground water samples were collected from 4 stations during pre-monsoon within 5kmradius from the dumpsite employing random sampling method.The groundwater characterization has been carried out for the parameters like pH, alkalinity, total dissolvedsolids (TDS), total hardness (TH), chloride (Cl-), and nitrate (NO3 -) by following the standard methodsprescribed as per Bureau of Indian Standard 10500 (BIS 2012).The physico-chemical characteristics like TDS, TH, CH, Cl- and NO3 - of collected leachate around the dumpsite during pre-monsoon were analyzed.WQI (Water Quality Index) for the water samples collected was calculated by considering abovementioned parameters. (Weighted Arithmetic Index method has been used).

Objectives of Present Study

- To analyze various physicochemical parameters including heavy metals in the leachate.
- Effect of the impact of leachate percolation on groundwater quality.
- Scope of design and operate the suitable landfill for the selected site.

Literature Review

A number of researches have been conducted on MSW management and Designing of Landfills during the recent years. The work of various researches have been studied and presented below:

1. Reducing the over-all cost of planning, design, operations andmaintenance of landfill facilities while ensuring the protection of public health and the environment.-Rakesh Kumar Dutta 1,V. Gayathri2 et.al
2. Building appropriate institutional framework along with policy-level directions will help facilitate the necessary change. - Mathangi Swaminathan et al
3. The high concentration of Total Dissolved Solids, Electrical Conductivity,

Hardness, Nitrates, Chlorides,Sulphates, in ground water near landfill deteriorates the quality of water. The principal threat to groundwater comes from inadequately controlled landfills where leachate generated from the fill is allowed to escape to the surrounding and underlying ground should strictly implement integrated waste management approach to handle large volume of wastes and protect environment.-P. Vasanthi 1 & S. Kaliappan 2 & R.Srinivasaraghavan 3 et.al

Physico-Chemical Parameters

Leachate sample was collected in 1 litre pre-cleaned high density polyethylene bottle (HDPE) during pre-monsoon. Similarly, ground water samples were collected from 4 stations during pre-monsoon within 5km radius from the dumpsite employing random sampling method.

Table 1 Geographical Details of the Study Area

S. No.	Sample	Station	Latitude	Longitude	Distance
1	GW1	Malkaram	17 31 38	78 34 52	1km
2	GW2	Y.S.R.Nagar	17 31 02	78 34 57	1km
3	GW3	Gabbi Lalpet	17 31 01	78 34 45	2km
4	GW4	Dammaiguda	17 30 12	78 35 27	2km

Each of the leachate and groundwater samples were analyzed for 6 parameters viz., pH, TDS, TH, TA, Cl- and NO3 -using standard procedures. The pH was recorded on site at the time of sampling with digital pH meter. The physicochemical parameters like Total Dissolved Solids (TDS), Total alkalinity (TA), Total Hardness (TH) and Chlorides (Cl-) of leachate, and ground water samples were analyzed titrimetrically. Chloride was included in the water quality assessment because of its measure of extent of dispersion of leachates in groundwater body. Nitrates determination was carried out using spectrometer. The main physical properties of water are color, taste, odor, turbidity, pH. These values should be within the permissible limits otherwise it is harmful. So tests have been conducted to find the quality of water present in and around Jawahar Nagar

PH

In chemistry, pH (potential of hydrogen) is a numeric scale used to specify the acidity or basicity of an aqueous solution. It is approximately the negative of the base 10 logarithm of the molar concentration, measured in units of moles per liter, of hydrogen ions. More precisely it is the negative of the base 10 logarithm of the activity of the hydrogen ion. Solutions with a pH less than 7 are acidic and solutions with a pH greater than 7 are basic. Pure water is neutral, at pH 7 (25 °C), being neither an acid nor a base. Contrary to popular belief, the pH value can be less than 0 or greater than 14 for very strong acids and bases respectively. It is measured using pH meter.

Total Hardness (TH) Hard water is water that has high mineral content (in contrast with " soft water "). Hard water is formed when water percolates through deposits of limestone and chalk which are largely made up of calcium and magnesium carbonates. Hard drinking water may have moderate health benefits, but can pose critical and grievous problems in industrial settings, where water hardness is monitored to avoid costly breakdowns in boilers , kettles and water heaters etc. Wherever water hardness is a concern, water softening is commonly used to reduce hard water adverse effects.

Total Dissolved Solids (TDS)

Total dissolved solids (TDS) is a measure of the dissolved combined content of all inorganic and organic substances present in a liquid in molecular, ionized or micro-granular (colloidal sol)suspended form. The principal application of TDS is in the study of water quality for streams,rivers and lakes, although TDS is not generally considered a primary pollutant (e.g. it is notdeemed to be associated with health effects. This method is used as it is the best yettime-consuming.

Total Alkalinity (TA)

Alkalinity is the ability to buffer acids. The measure of alkalinity is crucial in identifying the degree of buffering water has undergone against abrupt pH changes. Although corrosion is generally a result of low pH in water, increased alkalinity reduces calcium carbonate solubility, resulting in scaling.

Chloride ions (Cl-)

Chlorides are salts resulting from the combination of the gas chlorine with a metal. Some common chlorides include sodium chloride (NaCl) and magnesium chloride (MgCl2). Chlorine alone as Cl2is highly toxic and it is often used as a disinfectant. In combination with a metal such as sodium it becomes essential for life. Small amounts of chlorides are required for normal cell functions in plant and animal life. hlorides can corrode metals and affect the taste of food products. Therefore, water that is used in industry or processed for any use has a recommended maximum chloride level. Chlorides can contaminate fresh water streams and lakes. Fish and aquatic communities cannot survive in high levels of chlorides.

Nitrate ions (NO3-) Nitrate-nitrogen (NO 3 -N) in groundwater may result from point sources such as sewage disposal systems and livestock facilities, non-point sources such as fertilized cropland, parks, golf courses, lawns, and gardens.. The nitrite oxidizes the iron in the hemoglobin of the red blood cells to form methemo globing, which lacks the oxygen-carrying ability of hemoglobin. This creates the condition known as methemo globinemia (sometimes referred to as "blue baby syndrome"), in which blood lacks the ability to carry sufficient oxygen to the individual body cells causing the veins and skin to appear blue. The amount of nitrate ions present

Experimental Results and Conclusion

Table 2 Physico-Chemical Analysis of Groundwater and Leachate samples during pre-monsoon with comparison to World Health Organisation standards.

STATION	pH	TDS	TH	TA	Cl-	NO3-
GW1	7.8	384	205	298	78	28
GW2	7.3	640	1000	427	126	**174**
GW3	7.2	512	275	222	138	165
GW4	7.1	960	465	394	355	**196**
LEACHATE	9.6	35200	2900		45319	1012
WHO Standards	7 to 8.5	500	200	200	250	**50**

Leachate

The physico-chemical characteristics of collected leachate around the dumpsite during pre monsoon were analyzed and presented in the (Table number 2). From the results, it can be observed that pH is highly alkaline in nature. Alkaline pH is normally encountered at landfills, 10 years after disposal. Other analyzed parameters like TDS, TH, Cl- and NO3 - were found to have higher concentrations in the leachate collected during pre-monsoon season when compared to post monsoon leachate sample.

Groundwater

The results of physico-chemical analyses of groundwater samples collected during pre monsoon of 2019 were compared with the World Health Organization (WHO: 2006) as shown in (Table 2). The pH values of all the ground water samples collected around dumpsite fall within the WHO (2006) limits indicating the alkaline nature. The TDS value of the ground water samples ranged from 384-1408mg/l durin pre-monsoon season and from 528-1280 mg/l which exceeded the permissible limits of WHO(2006). Total hardness values of water samples ranged from 205-1000 mg/l during pre-monsoon were also above the permissible levels of WHO (2006). The chloride values ranged from 78-1100mg/l during pre-monsoon were above the permissible levels of WHO (2006). Highest chloride concentration was observed in GW4 station. The high chloride content in groundwater is from pollution sources such as domestic effluents, fertilizers, septic tanks, and leachates. Nitrate values ranged from 13-196 mg/l during pre-monsoon. Most of the water samples exceeded the WHO (2006) limits. It is further require to calculate the water quality index for the same area. And need of design and operate the suitable landfill for the selected site.

Spatio-Temporal Drought Analysis using GIS and SPI

Bhutham Santhosha[1] **and C. Sarala**[2]
[1]*Student of M.Tech,* [2]*Professor of Water Resources*
Centre for Water Resources, Institute of Science and Technology,
JNT University Hyderabad, Kukatpally, Hyderabad, Telangana, India.

ABSTRACT

Drought is defined as "severe water shortage". Low rainfall and fall in agricultural production has mainly caused droughts. In recent years, Geographic Information System (GIS) has played a key role in studying different types of hazards either natural or man-made. This study stresses upon the use of SPI and GIS in the field of drought risk evaluation. In the present work an effort has been made to derive drought risk areas facing agricultural as well as meteorological drought by use of temporal images based on normalized difference vegetation index and meteorological based standardized precipitation index. This study focused on assessing and analysing meteorological drought characteristics of Mahaboobnagar district based on rainfall, standardized precipitation index and geographic information system. SPI and monthly rainfall time series dataset for the period 1988–2015 were used to define the drought years and severity. GIS techniques, along with inverse distance weighted interpolation, were used to determine the spatial pattern of drought. Drought occurrences with severity were analysed based on average annual rainfall received during pre monsoon and post monsoon periods. Drought occurrence maps were generated in GIS environment by summarizing the percentage of drought occurrence for each category and for each time scale. The results obtained provide objective information on prevalence, severity level and persistence of drought conditions, which will be helpful to the resource managers in optimally allocating scarce resources. Thus, these parts of the district require urgent intervention on a priority basis to mitigate drought impacts.

Introduction

A Geographic Information System (GIS Software) is designed to store, retrieve, manage, display, and analyze all types of geographic and spatial data to produce maps and other graphic displays of geographic information for analysis and presentation, a management tool for drought analysis by mapping the mahabubanagr area of Mahabubnagar district. The main aim of this research is to study and evaluate the severity of drought in Mahabubnagar district. Drought is broadly defined as "severe water shortage". Low rainfall and fall in agricultural production has mainly caused droughts. Droughts cause losses of life, human suffering and damage to economy and environment. Droughts have been a recurring feature of the Indian climate therefore study of historical droughts may help in the delineation of major areas facing drought risk and there by management plans can be formulated by the government authorities to cope with the disastrous effects of this hazard. In recent years, Geographic Information System has played a key role in studying different types of hazards either natural or man-made. In the present work will be made to derive drought risk areas facing agricultural as well as meteorological drought by use of temporal images based Normalized Difference Vegetation Index (1988-2015) and meteorological based Standardized Precipitation Index.

In this paper, the present study was carried out with the following objectives.

- To Identify and assess drought severity in the study area by using Standardized Precipitation Index.
- To carry out normalized difference vegetation index analysis for 2001, 2007, 2015 satellite imaginaries are required.
- To generate drought occurrence maps by using Geographic Information System

Location of the study Area

A topographic map is a detailed and accurate two-dimensional representation of natural and man-made features on the Earth's surface. These maps are used for four number of applications, from encamp, hunting, fishing, and urban planning, resource department, and surveying. mahabubnagar district toposheet with scale of 1:80, 00000 scales it is georeferenced map. By using Arc GIS software version 9.3 the area is digitized. Georeferencing points are taken from the ground control points (GCP's) in the Google earths.

The latitude and longitudes for above diagram

makthal	16^{0}30'20.60''N	77^{0}31'43.72''E
Alampur	78^{0}15'00.36''E	15^{0}57'56.42''N
Nagarkarnool	16^{0}29'52.62''N	78^{0}19'31.33''E
Amangal	78^{0}31'45.45''E	16^{0}50'57.05''N

The Mahabubnagar District lies between 16°and 17° North latitude and 77° and 79° East longitude. Mahabubnagar is the largest in the area among the districts of Telangana region, with 18,432 sq.kms areas.

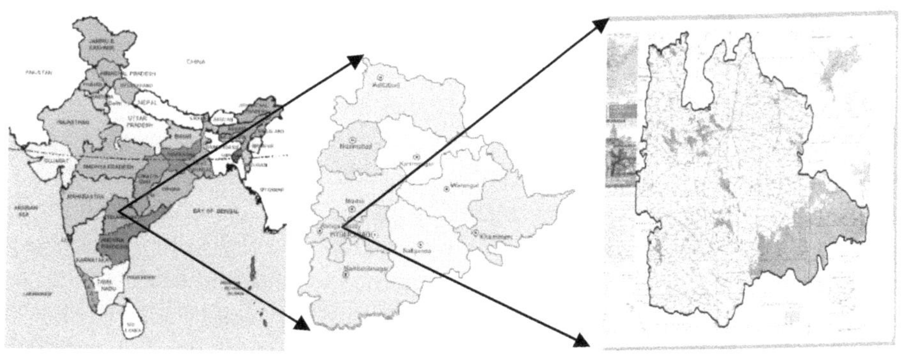

Figure 1 Location of the study area map

Statement of the Problem

Mahabubnagar district of telangana state is frequently hit by droughts. This is, however, not new to the drought-prone Mahabubnagar. With an average rainfall of just about 600mm a year, it is undisputedly the most backward and poorest of the districts in the State. It is also known for migration of laborers in search of work every year. Due to lack of adequate water, paddy cultivation is low. Mostly, maize and cotton apart from jowar and vegetables are cultivated. As for horticulture, it includes sweet oranges, Quinoa and mangoes. Past years, due to meagre rainfall, most of the maize has shown stunted growth, the cotton crop too has been hit badly, with pod size shrinking.

Drought is broadly defined as "severe water shortage". A droughts impact constitutes losses of life, human suffering and damage to economy and environment. In recent years, Geographic Information System has played a key role in studying different types of hazards either natural or man-made. This study stresses upon the use of GIS in the field of Drought risk Evaluation. In the present work will be made to derive drought risk areas facing agricultural as well as meteorological drought by use of temporal images based Normalized Difference Vegetation Index (1988-2015) and meteorological based Standardized Precipitation Index (SPI).

The standardized precipitation index (SPI) is a widely used index to characterize meteorological drought on a range of timescales. On short timescales, the SPI is closely related to soil moisture, while at longer timescales, the SPI can be related to groundwater and reservoir storage.

Literature Review

This chapter emphasis on the applications of Standardized precipitation index (SPI) and Geographic Information Systems (GIS) across various domains of Civil & Environmental Engineering by various Authors. Drought is a disastrous natural phenomenon. It differs from other natural hazards by its slow accumulating process and its indefinite commencement and termination. Even though it has scores of many definitions, it originates from a deficiency of precipitation over an extended period of time, usually a season or more. This deficiency results in a water shortage for some activity, group or environmental sector. Drought is also related to the timing of precipitation. Other climatic factors such as high temperature, high wind and low relative humidity are often associated with drought.

Review of literature on standard precipitation index

Yingxin Gu, et al. (2007) Carried out a live-year (2001 – 2005) history of moderate resolution imaging spectroradiometer (MODIS) normalized difference vegetation index (NDVI) and normalized difference water index (NDWI) data was analyzed for grassland drought assessment within the central United States.Initial results show strong relationships among NDVI, NDWI, and drought conditions. NDWI values exhibited a quicker response to drought conditions than NDVI. Analysis revealed that combining information from visible, near infrared, and short wave infrared channels improved sensitivity to drought severity. Analysis revealed that combining information from visible, near infrared, and short wave infrared channels improved sensitivity to drought severity.

Qiang Zhang et. al. (2009) are to classify anomalously wet and dry conditions by using the standardized precipitation index (SPI) and aridity index (I) for the rainy season (April–September) and winter (December–February). Trends of the number of wet and dry months decided by SPI were detected with Mann-Kendall technique. Different dry or wet tendencies can be identified in the Pearl River basin in the rainy season and in winter. The Pearl River basin tends to be dryer in the rainy season and to be wetter in winter. However, different parts of the basin show different patterns of dry and wet conditions— a general dry tendency can be observed in major parts of the basin in the rainy season, and a wet tendency in winter can be identified across the entire basin. Most of the stations show decreasing SPI index of rainy season but some are not, which may be due to the extremely inhomogenous spatial distribution of precipitation as a result of stronger convective precipitation and typhoon rain storms which are very common in the rainy season in the Pearl River basin. In terms of the number of dry or wet months in the rainy season, major parts of the Pearl River basin are characterized by a decreasing frequency of severe and moderate wet months. However, an increasing number of moderate wet months can be observed mainly in southeast parts of the Pearl River basin. Increasing frequency of severe dry months can be observed in some regions,and an increasing number of moderate wet months can be found in regions.

Ahmed Mustafa, Ghani Rah man (2018) to investigate the spatio-temporal variability of precipitation based on long-term data using Standardized Precipitation Index (SPI) to assess the trend and intensity of drought. SPI-3, SPI-6 and SPI-12 have been calculated to assess short- and long-term drought events, and MannKendall and linear regression tests were applied to detect the drought trend in the study region. The results of this study show that Jordan is facing an increasing trend of drought events. Various time scale SPI (12, 6 and 3 months) were used for better understanding of drought occurrence, magnitudes, and severity in the country. The analysis reveals that the amount and duration of rainfall are decreasing. This decreasing trend of rainfall is further exacerbating the issue of water scarcity as well as creating the issue of drought. It is forecasted in the study that the duration of drought events is increasing and the severity will be rather higher in future. The SPI results and spatial maps for whole study period indicate the presence of both local and national-level droughts. The probabilities of occurrence of local droughts are once every 2–3 years.The location of severe to extreme local drought is imposing the need to have a monitoring program to investigate the direct and indirect impacts on all sectors and to develop a proactive risk management approach and preparedness plans for various physiographic regions of the country.

Ghani Rahman et al. (2018) carried out the various indices like standard deviation, skewness, and kurtosis were calculated from annual rainfall of 15 meteorological stations during period of 1971–2015. Mann-Kendall technique was applied to both 1-month and 12-month SPIs for trend detection. All stations showed a positive skewness except Balakot. The highest variability was found in Parachinar and Balakot. The results of 12-month SPI showed two distinct dry periods, i.e., 1984 to 1989 and 1998 to 2002. Mann-Kendall trend test for 1-month SPI reveals more significant positive trend in Parachinar and there was variation in the results of other met-stations.

Review of literature on geographic information system

Muhammad Usman Liaqatl et al. (2001) suggested the satellite or remote sensing techniques can be used to monitor the current situation- before, during or after disaster. They can be used to provide baseline data against which future changes can be compared while the GIS techniques provide a suitable framework for integrating and analyzing the many types of data sources required for disaster monitoring.In recent years, the ever-increasing population and overstress on natural resources, soil degradation, decrease in water resources, and future projected climate change scenarios have become important areas of concern. Therefore, management of natural resources in developing as well as developed countries requires information on the state and changes in a range of biophysical variables. Droughts has been viewed as such a disaster where in a shortfall in precipitation has led to substantial reduction in production levels thereby leading to conditions which causes large scale migration and death of men and animals.

Muhammad Bilall et al. (2002) suggested the mitigation of the effects of disasters requires relevant information regarding the disaster in real time. Also the possible prediction and monitoring of the disaster requires rapid and continuous data and information generation or gathering. Since disasters that cause huge social and economic disruptions normally affect large areas or territories and are linked to global change, it is not possible to effectively collect continuous data on them using conventional methods. The space technology or remote sensing tools offer excellent possibilities of collecting this vital data. This is because the technology has capability of collecting data at global and regional scales rapidly and repetitively and the data is collected in digital form. The technology further provides an excellent communication medium.

Md. Rejaur Rahman (2004) suggested Beginning with a discussion of drought definitions, this review paper attempts to provide a review of fundamental concepts of drought, classification of droughts, drought indices, and the role of remote sensing and geographic information systems for drought evaluation. An understanding of the relationships between these two sets of parameters is necessary to develop measures for mitigating the impacts of droughts. Droughts are recognized as an environmental disaster and have attracted the attention of environmentalists, ecologists, hydrologists, meteorologists, geologists, and agricultural scientists. Temperatures; high winds; low relative humidity; and timing and characteristics of rains, including distribution of rainy days during crop growing seasons, intensity, and duration of rain, and onset and termination, play a significant role in the occurrence of droughts.

M.V.R.Sesha Saib and P.S.Royb (2014) suggested in recent years, Geographic Information Science (GIS) and Remote Sensing (RS) have played a key role in studying different types of hazards either natural or man-made. This study stresses upon the use of RS and GIS in the field of Drought risk Evaluation. In the present work an effort has been made to derive drought risk areas facing agricultural as well as meteorological drought by use of temporal images from NOAA-AVHRR (8km) based Normalized Difference Vegetation Index (NDVI) (1981-2000) and meteorological based Standardized Precipitation Index (SPI).Correlation and regression analysis was performed between NDVI, SPI, Rainfall anomaly and Food grain anomaly. SPI values were interpolated to get the spatial pattern of meteorological based drought. Food grain yield trend was plotted and an equivalent NDVI threshold was identified to get the agricultural drought risk. Similarly rainfall anomaly and NDVI were correlated and a threshold defined by IMD for meteorological drought was used to derive meteorological drought risk.

Theoretical Background

In this chapter we discussed basic concepts of standardize precipitation index (SPI), geographic information system (GIS). Drought studies are important because of their influence on the society and the economy of any nation. Extensive literature is available on droughts with respect to definition, methods of analysis and management procedures.

Theoretical Background on GIS (Geographic Information System)

A geographic information system (GIS) is a Software used to capture, manipulate, analyze, manage, store and present spatial or geographic data. In general, the term describes any information system that integrates stores, edits, analyzes, shares, and displays geographic information. GIS applications are tools that allow users to create interactive queries (user-created searches), analyze spatial information, edit data in maps, and present the results of all these operations. Geographic information science is the science underlying geographic concepts, applications, and systems.

1. **Geographic Information System**
2. **Components of GIS**

The important components of GIS consist of computer hardware, Set of application software module, Trained personnel.

(i) Hardware
(ii) Software
(iii) Data
(iv) People
(v) Methods

GIS Data Types

1. **Raster:** Raster image store information in a cell based manner. It can be aerial photo, satellite image, Digital Elevation Model (DEM). Raster images normally store continuous data.
2. **Vector:** Vector data are discrete. It store information in x, y coordinate format. There are three types of Vector data: Lines, Points and Area.

Theoretical Background on SPI (Standardized precipitation index)

Standardized Precipitation Index (SPI) expresses the actual rainfall as a standardized departure with respect to rainfall probability distribution function and hence the index has gained importance in recent years as a potential drought indicator permitting comparisons across space and time. The Standard Precipitation Index (SPI) shows the actual precipitation compared to the probability of precipitation for various time frames. The SPI is an index based on precipitation only. It can be used on a variety of time scales, which allows it to be useful for both short-term agricultural and long-term hydrological applications. A drought event occurs any time the SPI is continuously negative and reaches an intensity of -1.0 or less. The event ends w hen the SPI becomes positive. Each drought event, therefore, has a duration defined by its beginning and end, and intensity for each month that the event continues. The positive sum of the SPI for all the months within a drought event can be termed the drought's "magnitude". To Geo-referencing of toposheet first open arc GIS and in that go to Add toposheet by clicking on (+) Add Data option. Then next click yes and then ok. Now go to Customize —>Toolbars——>Georeferencing.Now zoom the map and check the coordinates printed on the corner of the toposheet. Now go to georeferencing toolbar and click Add Control Points. Now click on the corner of the toposheet and add Input X & Y. X will be the Longitude (higher value) and Y is latitude.Now enter the coordinates. But before entering convert degree minutes seconds to decimal degrees.Now move to the left bottom corner and add control point. It is a chance that your toposheet will suddenly disappear. To bring it back, click on toposheet and right click on Zoom to layer.Now repeat this all process for the remaining two corners on the right side. After that goes to the georeferencing toolbar and click on View Link Table.Check out the residual values, if they are quite less that means you have added the control points in the right manner. After this click on Georeferencing and choose Rectify which will create a whole new georeferenced raster image. And also choose Update Georeferencing which will add a world file to your existing toposheet and it will get georeferenced. Choose the desired Format & Output Location and then click on Save.Toposheet is Georeferenced now. And give the projection system by right clicking on the image —> Data Frame Properties —> Coordinate System. Choose the desired Coordinate System. Mosacing images first Go to raster then click on mosaic. The mosaic dialogue box will open in that now Click on Edit then Add images which are rectified. From the Add images window click on the Image Area options tab. After checking compute active area radio button, then click ok button. Click set mode for intersections Then Click edit &Set overlap functions. Click the automatically generate cut lines to intersections. Click yes. Check radio buttons of cutline exists, Feathering, smoothing Go to apply Go to edit then colour corrections check the Use colour Balancing, Use Histogram matching Click the process & Run Mosaic Give out put file name click ok.

Methodology Adopted

Based on the literature review and objectives identified for the study suitable methodologies are identified to carry out the present work. They are discussed in the following sections. Further, the area chosen for the study and its various details are also presented in this chapter.

The study area is delineated from the survey of India Toposheet number EM4407 on scale 1:50,000 scale. The area of interest is prepared for the study area by using the survey of India Toposheet.

Annual average fall

Table 1 Annual average fall

District mahabubnagar	Normal rainfall in mm	Actual rainfall in mm
2010-11	604	688
2011-12	604	536
2012-13	604	621

Collection of Data

We need data for execution of the work they are

I. Satellite data
II. Top sheet data
III. Rainfall data

Collection of Satellite Data

This satellite data was collected from the USGS earth explore bhuvan website. USGS earth explore is providing users the ability to query, search and order ,satellite images ,it started with first land sate satellite's launch in 1972 and is continuing with land sat 8 ,still operational. for almost 40years ,the land sat program has continuously collected spectral information from earth's surface. This unparalleled data archive gives scientists ability to accesses changes in earth's landscape. Mahabubnagar district falls under 4 grids.

The below table shows the Mahabubnagar district path/row, land sat name and spatial resolution.

Table 2 The satellite name path/row

S. No.	Date of image	Satellite/Sensor	Reference Sytem/Pah/Row	Spatial Resolution
1	13-11-2015	Landsat8/TM	WRS-2/144/48	30M
	30-05-2015	Landsat8/TM	WRS-2/143/48	30M
2	14-02-2015	Landsat8/TM	WRS-2/144/49	30M
	23-02-2015	Landsat8/TM	WRS-2/143/49	30M
3	03-03-2001	Landsat7/ETM+	WRS-2/144/48	30M
	24-02-2001	Landsat7/ETM+	WRS-2/143/48	30M
4	03-03-2001	Landsat7/ETM+	WRS-2/144/49	30M
	24-02-2001	Landsat7/ETM+	WRS-2/143/49	30M
5	01-05-2008	Landsat5/ETM+	WRS-2/144/48	30M
	02-11-2008	Landsat5/ETM+	WRS-2/143/48	30M
6	09-11-2008	Landsat5/ETM+	WRS-2/144/49	30M
	02-11-2008	Landsat5/ETM+	WRS-2/143/49	30M

Overall, NDVI is a standardized way to measure healthy vegetation. Having high NDVI values, there is health vegetation.

Method adopted in study area

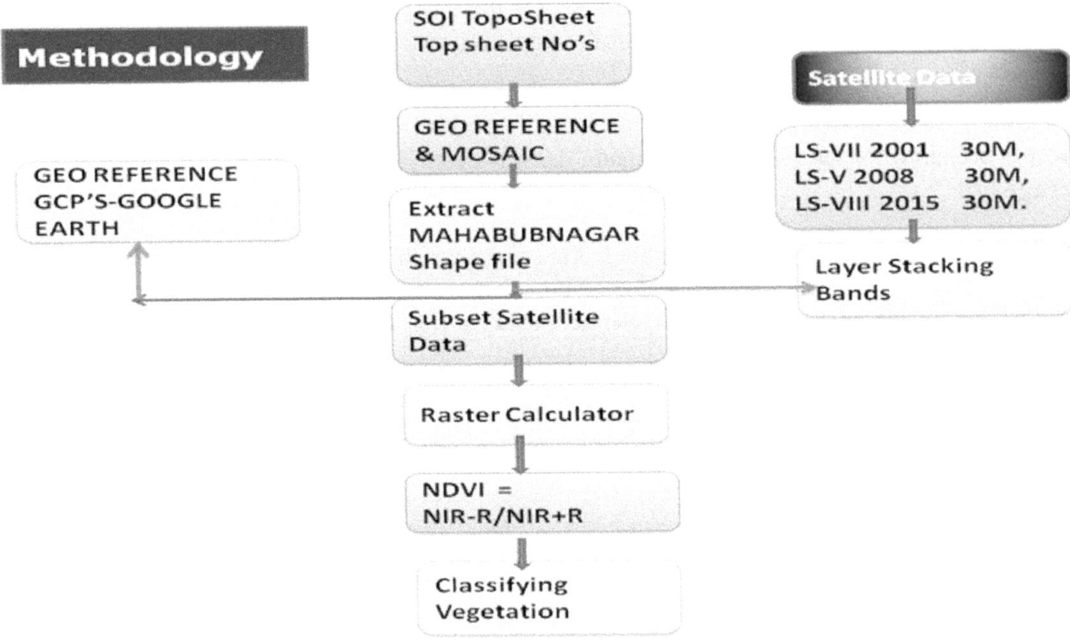

Normalized Difference Vegetation Index (NDVI) Calculation

Uses the NIR and red channels to measure healthy vegetation. Calculate it manually, this is the formula.

NDVI = ((IR - R)/ (IR + R))

- IR = pixel values from the infrared band
- R = pixel values from the red band

NDVI always generates a value between -1 and +1. It's really just a standardized way to measure healthy vegetation.

SPI method adopted in study area

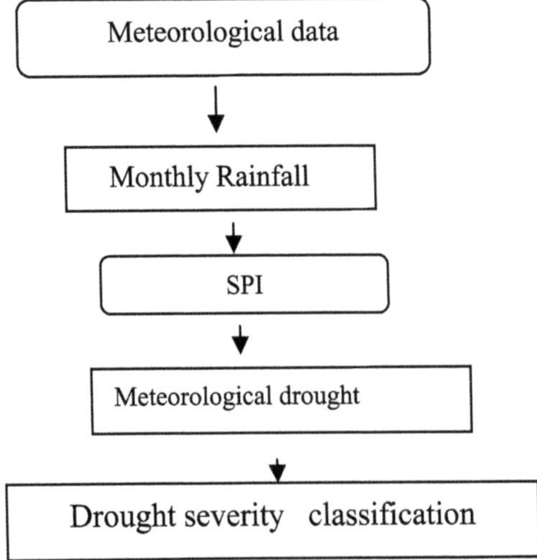

Applications to Study Area

Area of interest

Toposheet shows a topographic fcatures on the map either natural or man-made features. A topographic map is a detailed and accurate two-dimensional representation of natural and man-made features on the Earth's surface. These maps are used for a number of applications, from encamp, hunting, fishing, and urban planning, resource department, and surveying.

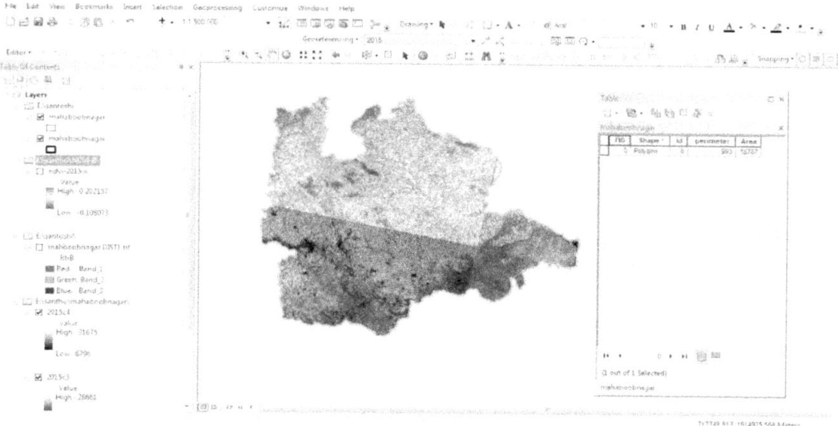

Figure 2 Area of interest is calculation

The area of interest (AOI) is the feature that is depicted on the product and represents the specific geographic area. The area of interest is based polygon feature. The above fig shows area of interest extracted from the 4bands and the area is calculated. The area of Mahabubnagar district 18787KM^2

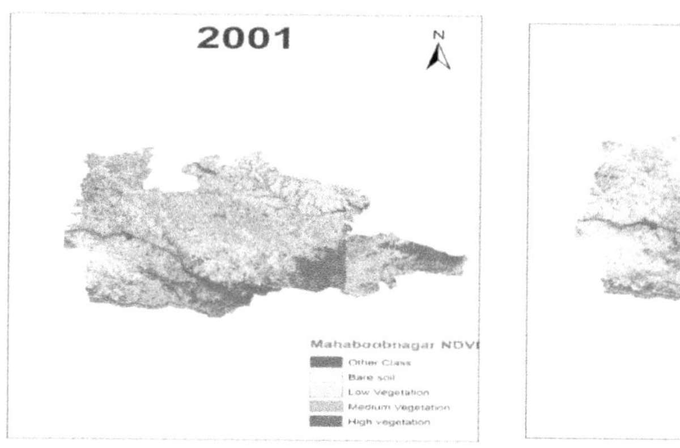

Figure 3 2001 NDVI analysis map

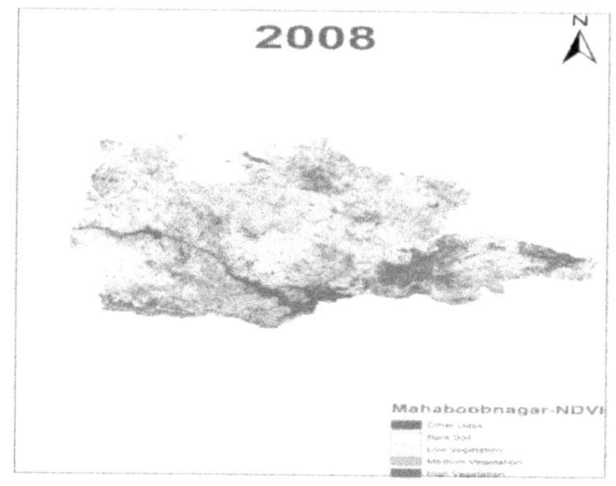

Figure 4 Show 2008 NDVI analysis map

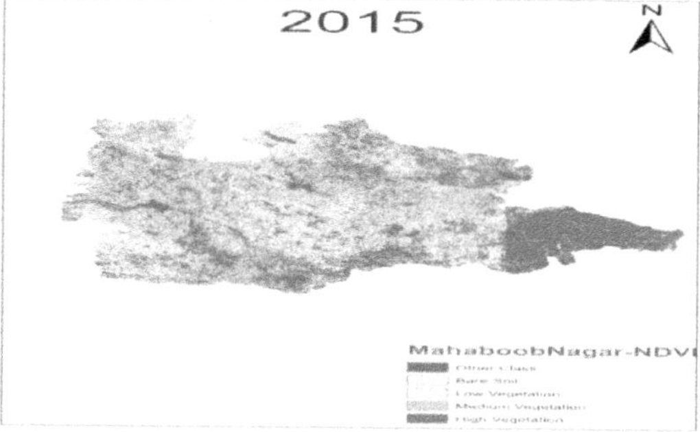

Figure 5 2015 NDVI analysis map

Image Pre-processing

(a) Layer Stacking: layer stacking of bands was done in the ERDAS Imagine -14 software.

(b) Mosaicking is the layer stacked image tiles were mosaicked and clipped with area of interest (AOI) shape file.

(c) Image rectification was done.

(d) Projection: The image was downloaded from the Universal Transverse Mercator projection reprojected in Geographic WGS 84, spheroid and datum Everest.

NDVI Calculation

Normalized Difference Vegetation Index (NDVI) is calculated by the values of reflectance of the visible and near inferred (NIR) bands captured by the senior of the satellite (NIR and red channels in its formula).

$$\text{NDVI} = \frac{(\text{NIR} - \text{Red})}{(\text{NIR} + \text{Red})}$$

Table 3 NDVI values

land sat	NDVI values	
	Highest	Lowest
2001 land sat: 7	0.991	-0.985
2008 land sat: 5	0.9	-0.89
2015 land sat: 8	0.992	-0.992

Rainfall Data

The rain fall data was collected from India Meteorological Department (IMD). From the year of 2001 to 2015 year to find out the metrological drought by using SPI software.

Monthly Rain Fall Data

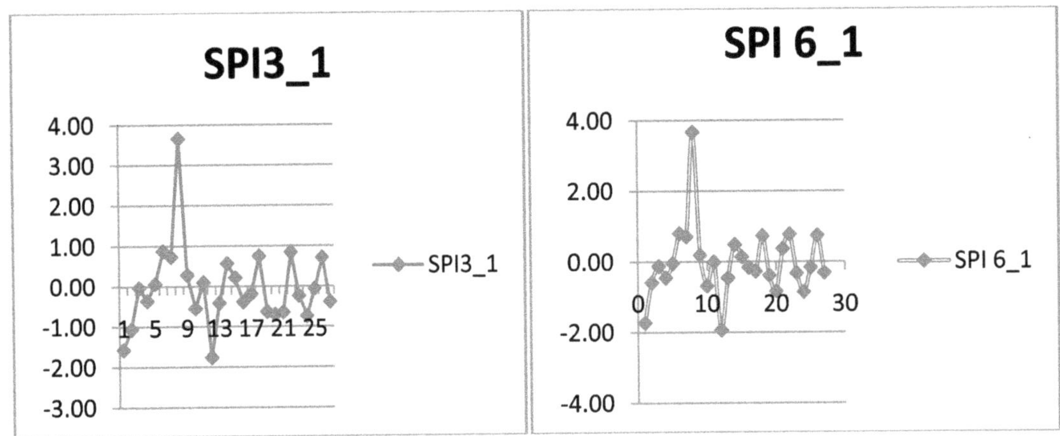

Figure 6 SPI 3 TIMESCALE 1988-2015

Figure 4 SPI 6 TIMESCALE 1988-2015

Calculation of SPI

To calculate the SPI in DrinC software we required Rainfall data the data files are selected in the File Management window (menu>Data>File Management).

3-month SPI

The 3-month SPI provides a comparison of the precipitation over a specific 3-month period with the precipitation totals from the same 3-month period for all the years included in the historical record. In other words, a 3-month

SPI at the end of February compares the December–January- February precipitation total in that particular year with the December–February precipitation totals of all the years on record for that location. 3-month SPI reflects short- and medium-term moisture conditions and provides a seasonal estimation of precipitation. In primary agricultural regions, a 3-month SPI might be more effective in highlighting available moisture conditions than the slow-responding Palmer Index or other currently available hydrological indices. It is important to compare the 3-month SPI with longer timescales. A relatively normal or even a wet 3-month period could occur in the middle of a longer-term drought that would only be visible over a long period. As with the 1-month SPI, the 3-month SPI may be misleading in regions where it is normally dry during any given 3-month period. Large negative or positive SPIs may be associated with precipitation totals not very different from the mean

6-month SPI

The 6-month SPI compares the precipitation for that period with the same 6-month period over the historical record. The 6-month SPI indicates seasonal to medium-term trends in precipitation and is still considered to be more sensitive to conditions at this scale than the Palmer Index. A 6-month SPI can be very effective in showing the precipitation over distinct seasons. For example, a 6-month SPI at the end of March would give a very good indication of the amount of precipitation that has fallen during the very important wet season period from October through March for certain Mediterranean locales. Information from a 6-month SPI may also begin to be associated with anomalous stream flows and reservoir levels, depending on the region and time of year.

Trend Analysis Using SPI

Values Analyzing monthly data the moderate drought occurred in 2001-02, 2003-04, 2005-06, 2007-2010 and 2014-15 in Mahabubnagar District. There was no severe or extreme drought occurred.

According to SPI values of 3, 6, and 12 months, the drought severity of the Mahabubnagar district has been categorizes by three classes.

Table 4 SPI Classification

SPI values	Class
>2	extremely wet
1.5 to 1.99	very wet
1.0 to 1.49	moderately wet
-0.99 to 0.99	near normal
-1 to –1.49	moderately dry
-1.5 to-1.99	severely dry
< -2	extremely dry

The SPI is calculated for every month in the Mahabubnagar district to find out the severe drought in that area. By using SPI software the above graphs are drawn. To draw the time scale the rainfall data has required. In this study the timescale are drawn monthly wise from 2001 to 2015 and also drawn the 3month scale for 2001, 2008 and 2015 as show above figures. The vales are taken on the graph is: on x-axis spi values and on y-axis years are taken to draw the graph. Finally the entire study gets moderate drought in the Mahabubnagar district. The SPI values are ranges in between -0.99 to 0.99 or -1 to -1.49 in my study area.

Conclusions and Recommendations

Conclusions

In present study, three images (2001 2008 and 2015) have been classified on the basis of normalised difference vegetation index (NDVI), a vegetation index calculated by NDVI is an index based on spectral reflection factor of the bottom surface feature. Every feature has its own characteristic reflectance varying according to the wavelength. NDVI value ranges between -1 to +1.A Higher price of NDVI infers the presence of healthy vegetation within the space whereas its lower price is that the indicator of distributed vegetation. In the year 2001 land sat 7 has used with the 30m resolution and NDVI Highest value is 0.991, and lowest value is -0.985. In the year 2008 land sat 5 has used with the 30m resolution and NDVI Highest value is 0.9, lowest value is -0.89.In the year 2015 land sat 8 has used with the 30m resolution and NDVI Highest value is 0.992, lowest value is -0.992 Standardized Precipitation Index (SPI) expresses the particular downfall as a regular departure with regard to downfall chance distribution operate and thus the index has gained importance in recent years as a potential drought indicator permitting comparisons across different rainfall zones.

According to the results for all years in spi 2001 2008 and 2015 the drought periods are shown (3 month time scale), 2008(6 month time scale) and 2015 (9 month time scale) are shown as drought years. Rainfall Data The SPI was initially designed to assess the meteorological drought, so that the first scale referred only to negative precipitation anomalies, although McKee and his colleagues acknowledged that the index might monitor each wet and dry periods (McKee et al., 1993).Nowadays, the SPI has become the most powerful and widely used of the simple drought Indices), Most studies refer to a single type of drought (meteorological, agricultural or hydrological). Thus, the results of the present study suggest that SPI as a stand-alone indicator needs to be interpreted with caution for drought intensity assessment particularly in low rainfall districts which are more vulnerable to droughts. The average rainfall of Mahabubnagar district is 604 mm, most of it being received during south west monsoon period (June – September).

References

Guttman, N. B., 1999: Accepting the Standardized Precipitation Index: A calculation algorithm. J. Amer. Water Resources Assoc., 35(2), 311-322.

W. C. Palmer, "Meteorological Drought", *U.S. Weather Bureau Manuscript* No. 45 58pp February 1965

Moumita Palchaudhuri, Sujata Biswas, Analysis of Meteorological Drought Using Standardized Precipitation Index – A Case Study of Puruliya District, West Bengal, India, International Journal of Environmental, Chemical, Ecological, Geological and Geophysical Engineering Vol:7, No:3, 2013

Livada and V. D. Assimakopoulos, Spatial and temporal analysis of drought in Greece using the Standardized Precipitation Index (SPI), Received June 2, 2004; revised February 10, 2005; accepted December 1, 2005 Published online December 28, 2006 #Springer-Verlag 2006

M. Naresh Kumara, C.S. Murthy, M.V.R. Sesha Sai, and P.S. Roy, On the use of Standardized Precipitation Index (SPI) for drought intensity assessment

Guttman B Nathaniel, 1998, Comparing the Palmer drought index and the standardized precipitation index, *Journal of the American Water Resources Association*, 34, 1, 113-121.

Ajay D. Malakiya, T. M. V. Suryanarayana, Assessment of Drought Using Standardized

Ahmed Mustafa, Ghani Rahman (2018). Assessing the Spatio-temporal Variability of Meteorological Drought in Jordan.Earth Systems and Environment,Volume 2, Issue 2, pp 247–264.

Yue S, Pilon P, Cavadias G (2002) Power of the Mann-Kendall and Spearman's rho tests for detecting monotonic trends in hydrological series. J Hydrol.

Eskinder Gidey,Oagile Dikinya ,Reuben Sebago · Eagilwe Segosebe Amanuel Zenebe(2018) Modeling the Spatio-Temporal Meteorological Drought Characteristics Using the Standardized Precipitation Index (SPI) in Raya and Its Environs, Northern Ethiopia. Earth Systems and Environment.

Ghani Rahman& Atta-ur-Rahman & Samiullah & Muhammad Dawood (2018). Spatial and temporal variation of rainfall and drought in Khyber Pakhtunkhwa Province of Pakistan during 1971–2015 Arabian Journal of Geosciences.

Jinyoung Rhee, Jungho Im, Gregory J. Carbone (2010).Monitoring agricultural drought for arid and humid regions using multi-sensor remote sensing data. Remote sensing of environment.

Repair and Rehabilitation of RCC Structures

Rathod Ravinder[1], A.Vittalaiah[2] and Akula Prakash[3]
[1,2,3] *Assistant Professor GRIET, Bachupally Hyderabad*
rathod506ravinder@gmail.com, vittal1559@gmail.com and akulaprakash93@gmail.com

ABSTRACT

The purpose of the project is to gain fundamental and practical understanding on concrete repair and rehabilitation of the structures. Large number of reinforced concrete (RC) structure are deteriorating, often prematurely, and need remedial measures to reinstate their safety and/or serviceability. Consequently, the need for repair and protection has grown considerably in recent years. While costs associated with repair of deteriorating concrete structures can be substantial, costs resulting from poorly designed or executed repairs may be even higher. Repair methods need to be designed with consideration for the anticipated or desired remaining service life of the structure. A distinction must be made between repairs intended to stop deterioration fully and those merely aimed at slowing down deterioration processes for a limited period of time. During the research for our project relevant repair methods for damaged concrete structures will be discussed, focussing on design methods. The project will be initiated with various sites nearby Nizampet inspection and repairs will examined. Then the respective repairs will be studied and classified into cracks, corrosion of concrete reinforcement, seepage and deterioration of surface coating. With the help of journals and publications a detailed study will be done on the causes for each repair and a suitable rehabilitation method will be suggested for each repair site by comparing various methods. This paper will consist of studies of various repairs along with pictures, referred case studies and other references.

Introduction

Concrete is the most widely used and versatile construction material possessing several advantages over steel and other construction material. Very often one comes across with some defects in concrete they are in the form of cracks, spalling of concrete, exposure of reinforcement, excessive deflections or other signs of distress. Corrosion of reinforcement may trigger off cracking and spalling of concrete, coupled with deterioration in the strength of the structure such situations call for repairs of affected zones and sometimes for replacements of entire structure.

Repair is the process of restoring something that is damaged or deteriorated or broken to good condition. Repairs are performed on damaged buildings to restore the strength after disaster.

Rehabilitation is the process of restoring the structure to service level, once it involves the upgrading or changing of a building's foundation in support of changes desired, its use, design goals or regulatory requirements. Assessing the existing condition of the structure and deciding which component of the structure should be repaired or restored based on all future requirement of the structure. Need for repair and rehabilitation of structures:

- Faulty design of the structure
- Improper execution and bad workmanship
- Extreme weathering and environmental conditions
- High degree of chemical attack
- Ageing of structures.

Methods and Materials

CRACKS: Cracks in building are of common occurrence. A building component develops cracks when stress exceeds strength. Cracks are of two types:

1. **STRUCTURAL**: Structural ones are due to faulty design, faulty construction or overloading which may endanger safety or buildings.
2. **NON-STRUCTURAL:** The non-structural cracks are due to internally reduced stress. Depending on width of crack they are classified into THIN (<1mm), MEDIUM (1mm to 2mm), WIDE (>2mm)

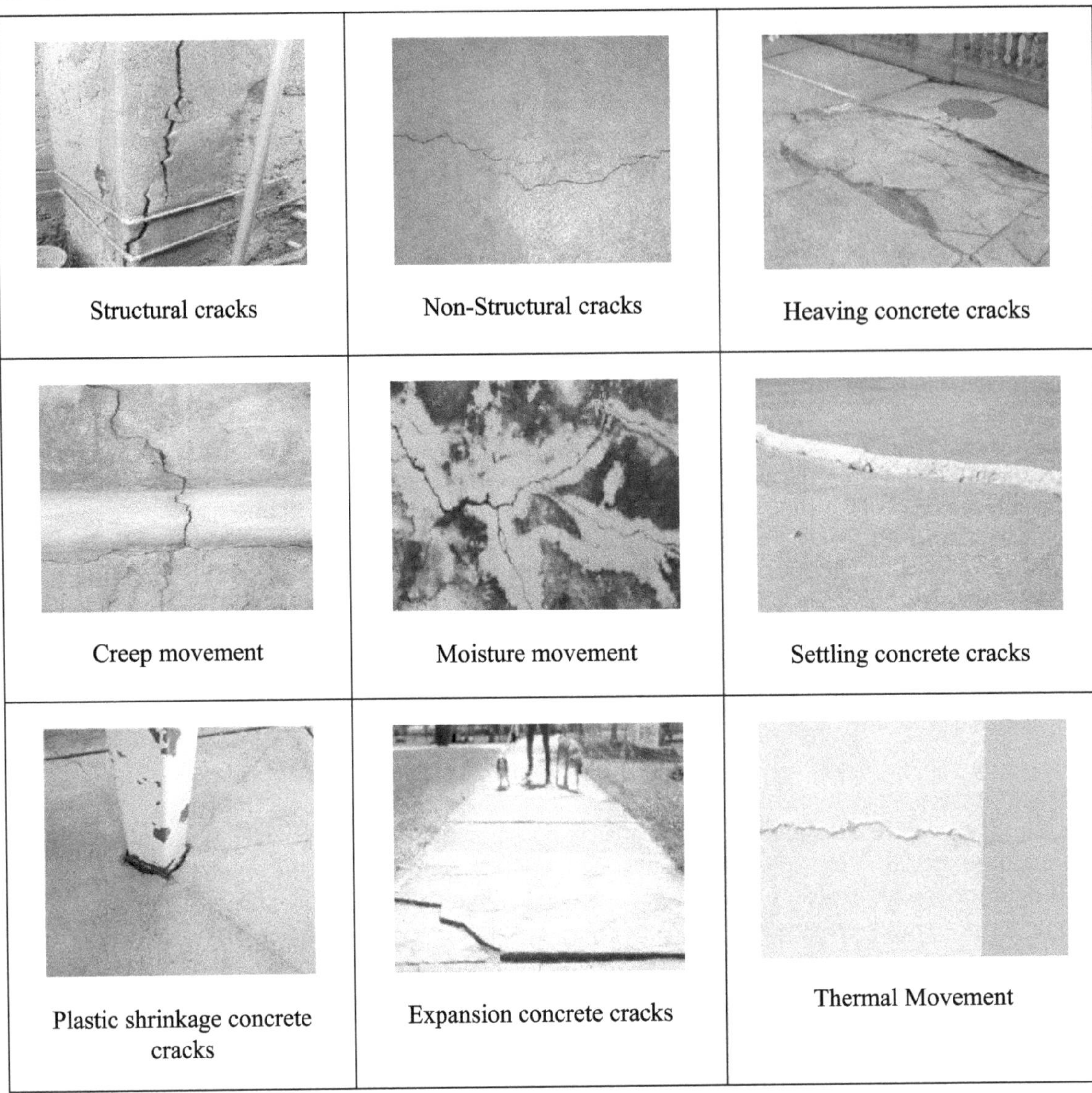

Structural cracks | Non-Structural cracks | Heaving concrete cracks

Creep movement | Moisture movement | Settling concrete cracks

Plastic shrinkage concrete cracks | Expansion concrete cracks | Thermal Movement

How do Cracks Develop?

Cracks indicate the deterioration of concrete due to uncoverable environment around concrete structures. Environment constitutes PHYSICAL, CHEMICAL AND MECHANICAL.

Physical agents causing cracks are: temperature, frost action, concrete cover, porosity of concrete.

Mechanical agents causing cracks are: abrasion, vibration, impact loads.

Repair work for Cracks in Beams and Columns

- Firstly, detect the damaged area by hitting with hammer, hollowness sound indicates the damaged area shoring to distributed load.
- Remove corroded area of cement.
- According to IS code the thickness of cover for columns and beams should be 40mm.
- The chemicals used are ARMATECH-108 an anti-corroding chemical is to be applied on steel reinforcement and SIKA LATEX to ensure proper bonding with old and new cement.
- Plasticizer is added to decrease permeability and increase in strength.

Corrosion

Steel reinforcement which is used in RCC, though on one side complements the concrete for its weakness in tension (tensile stress), it also impairs the durability and longevity of concrete, due to its proneness to corrosion. Repairs and rehabilitation of concrete structures, which of late has become an activity comparable to construction itself worldwide, is mostly because deterioration of concrete due to corrosion of embedded steel.

In the earlier part of this century when the Reinforced Concrete Construction began to be widely used, replacing almost completely the hitherto used construction materials viz. Timber, (stone) Masonry and steel sections etc., the life expectation of the R. C. structures was of the order of 100 years. However, at the turn of century we find these expectations belied and ironically newer constructions say 20 to 25 years old, show serious deterioration and distress.

The durability of concrete has become a highly discussed topic in global development. Even though several factors are responsible for early distress in reinforced concrete structures it is observed that in majority of cases, it is because of the corrosion of steel. The corrosion seems to be an all-pervasive phenomenon causing widespread destruction of all types of structures in all countries across the world and has come to be termed as 'Cancer' for concrete. Ideally speaking, a good concrete is supposed to provide adequate protection to the embedded steel. This is due to the protective alkaline environment (pH value as high as 12.5) provided by fresh concrete resulting in formation of protective coating on the surface of the steel, which passivates it from further corrosion.

However, over the passage of time, due to carbonation or ingress of chloride ions, pH value starts declining slowly and alkaline surrounding of the reinforcement bar is lost, heralding the corrosion process, which in turn causes cracks and spalling of concrete. It would thus be realized that the crucial factor giving quality and durability of concrete appears to be its impermeability, which can be ensured by providing sufficient cement content, low w/c ratio, complete compaction and curing. The same can be further improved by using proper admixtures and providing increased concrete cover.

However, these measures are firstly difficult to realize in practice fully and secondly the same are not found good enough in aggressive environment. Thus, it becomes necessary to provide additional protection to reinforcement steel, especially because of chloride induced corrosion (worse than carbonation corrosion) which can develop even in good quality concrete.

Most of the deterioration of RCC structures is mainly due to corrosion of reinforcement. Therefore, basic understanding of corrosion technology for both durability of structure and rehabilitation work is a must.

Steel embedded in hydrating cement paste rapidly forms a thin passivity layer of oxide which strongly adheres to the underlying steel and gives it complete protection from reaction with oxygen and water, that is from formation of rust or corrosion. This state of the steel is known as passivation.

Maintenance of passivation is conditional on an adequately high pH of the pore water in contact with the passivating layer. Thus, when the low pH front reaches the vicinity of the surface of the reinforcing steel, the protective oxide film is removed and corrosion can take place, provided oxygen and moisture necessary for the reactions of corrosion are present.

Causes of Corrosion of Steel Reinforcement in Concrete

Corrosion of steel in concrete is an electrochemical process. The electrochemical potentials to form the corrosion cells may be generated in two ways:

(a) Composition cells may be formed when two dissimilar metals are embedded in concrete, such as steel rebars and aluminium conduit pipes, or when significant variations exist in surface characteristics of the steel.

(b) Concentration cells may be formed due to differences in concentration of dissolved ions near steel, such as alkali's, chlorides, and oxygen.

The differences in electrochemical potential can arise from differences in the environment of the concrete. Electrochemical cells form also due to a variation in salt concentration in the pore water or due to a non-uniform access to oxygen.

Thus, one of the two metals (or some parts of the metal when only one metal is present) becomes anodic and the other cathodic. The fundamental chemical changes occurring at the anodic and cathodic areas are as follows.

When there exists a difference in electrical potential along the steel in concrete, an electrochemical cell is set up: there form anodic and cathodic regions, connected by the electrolyte in the form of the pore water in the hardened cement paste.

The positively charged ferrous ions Fe^{++} at the anode pass into solution while the negatively charged free electrons e^- pass through the steel into the cathode where they are absorbed by the constituents of the electrolyte and combine with water and oxygen to form hydroxyl ions $(OH)^-$. These travels through the electrolyte and combine with the ferrous ions to form ferric hydroxide which is converted by further oxidation to rust. The reactions involved are as follows:

Anodic reactions:

Fe -------> $Fe^{++} + 2e^-$

$Fe^{++} + 2(OH)^-$ ------> $Fe(OH)_2$ (ferrous hydroxide)

$4Fe(OH)_2 + 2H_2O + O_2$ -----> $4Fe(OH)_3$ (ferric hydroxide)

(Water) (Air) $2Fe(OH)_3$ ----> $Fe_2 \times O_3 \times H_2O + 2H_2O$ — Hydrated ferric oxide (rust).

Carbonation reactions

$4e^- + O_2 + 2H_2O$ ----> $4(OH)^-$

It can be seen that oxygen is consumed and water is regenerated but it is needed for the process to continue. Thus, there is no corrosion in, dry concrete, probably below a relative humidity of 60 percent; nor is there corrosion in concrete fully immersed in water, except when water can entrain air, for example by wave action.

The transformation of metallic iron to rust is accompanied by an increase in volume, which depending on the state of oxidation, may be as large as 600 percent of the original metal. This volume increase is believed to be the principal cause of concrete expansion and cracking. It should be noted that the anodic reaction involving ionization of metallic iron will not progress far unless the electron flow to the cathode is maintained by consumption of the electrons at the cathode; for this the presence of both air and water at the surface of the cathode is absolutely necessary.

In the absence of chloride ions in the solution, the protective film on steel is reported to be stable if the pH of the solution stays above 11.5. Normally there is sufficient alkalinity in the system to maintain the pH above 12. In exceptional conditions (e.g., when concrete has high permeability and alkali's and most of the calcium hydroxide are either carbonated or neutralized by an acidic solution), the pH of concrete near steel may be reduced to less than 11.5, thus destroying the passivity of steel and setting the stage for the corrosion process.

In the presence of chloride ions, depending on the $C1^- / OH^-$ ratio, it is reported that the protective film may be destroyed even at pH values considerably above 11.5.

For corrosion to be initiated, the passivity layer must be penetrated. Chloride ions activate the surface of the steel to form an anode, the passivated surface being the cathode. The reactions involved are as follows:

$Fe^{++} + 2Cl^-$ -----> $FeCl_2$

$FeCl_2 + 2H_2O$ --------> $Fe(OH)_2 + 2HCl$

The factors influencing the depth of carbonation are:

- Depth of cover
- Permeability of concrete
- Grade of concrete
- Time
- Whether the concrete is protected or unprotected
- The environmental influences.

The ultimate result cracking, spalling and corrosion.

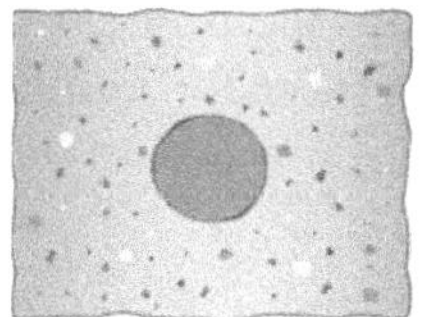

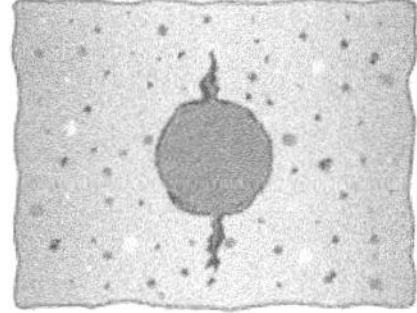

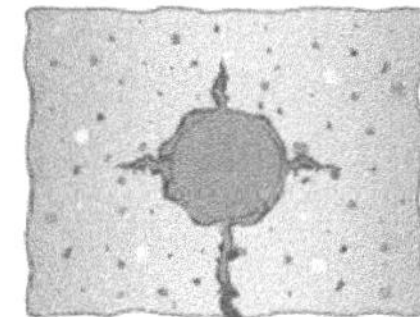

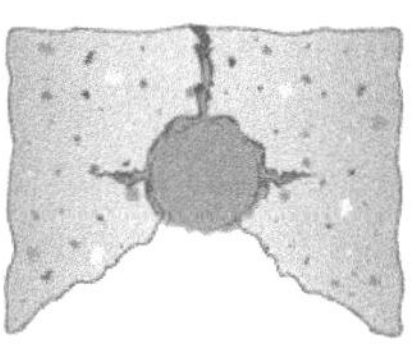

The corrosion cycle of steel begins with the rust expanding on the surface of the bar and causing cracking near the steel/concrete interface. As time marches on, the corrosion products build up and cause more extensive cracking until the concrete breaks away from the bar, eventually causing spalling.

Factors Influencing Corrosion of Steel Reinforcement

The factors which generally influence corrosion of reinforcement in RC structures are:

pH value, Moisture, Oxygen, Carbonation, Chlorides, Ambient temperature and relative humidity, Severity of exposure, Quality of construction materials, Quality of concrete, Cover to the reinforcement, Initial curing conditions, and Formation of cracks.

Repair of Concrete Columns for Cracks and Damages: Before starting the repair of a concrete column, the axial dead load, axial live load, horizontal load and its associated moments must be known.

Repairs to concrete columns can be divided into two categories. Surface or cosmetic repair only covers local deterioration and structural repair restores or strengthens the affected columns. If the deterioration does not significantly reduce the cross section, the conventional concrete repair can successfully be employed.

Methods of Strengthening Concrete Columns

Column strengthening is a process used to add or restore ultimate load capacity of reinforced concrete columns. It is used for seismic retrofitting, supporting additional live load or dead load that not included in the original design, to relieve stresses generated by design or construction errors, or to restore original load capacity to damaged structural elements.

There are several techniques which are used to strengthen reinforced concrete columns like reinforced concrete jacketing, steel jacketing, and FRP confining or jacketing.

Repairing Concrete Methods for Cracks and Damages

Unloading Columns: unloading the column is necessary. Entire cross-section of the repair column is capable of carrying the reintroduce design load. Without unloading, new repair does not carry any load. Drying shrinkage of new material reduces the share of the load.

Redistribution of the Load: In corrosion of reinforcement and concrete deterioration, Redistribute a load of column concrete with alternative supports for repair.

Supplemental Reinforcing Steel: The supplemental vertical bar to fix outside the original cage with extra ties. Provide adequate cover and Place Apartment's bars outside the tie bars to increase column dimensions. Use Hairpin ties, of stainless steel laterally to support the supplemental bars. Column ties cannot disturb at the repair of the longitudinal bars as it causes buckling.

Concrete Removal: Remove concrete within a column cage and unload the column. If not, the longitudinal bars are buckle and compression failure of column take place.

Corroded Reinforcing Steel: It's not necessary to remove the corrode reinforcing bar with the reduce cross-sectional area if the loss is supplement with additional reinforcing bars. The partially corrode reinforcing bars are thoroughly clean by sandblasting to obtain the bare metal. The bars with excessive corrosion are replacing with fresh reinforcement having full laps on both sides.

Corroded Ties: Replace the corrode ties by adding stainless steel hairpin ties that are anchor into the concrete. It is often necessary to deposit extra material around columns to provide an adequate cover over the supplemental ties.

Low-strength Concrete: Where the concrete strength is low, resulting in insufficient load-carrying capacity, several alternatives are available, Shore the column and remove and replace the in-place concrete. Shore column and increase the size of the column to reduce bending stresses, and increase confinement on placed weak concrete. Wrap the column with carbon- or glass-reinforced plastic. Install a supplemental column.

Conclusion

Every building has some life span after time passes certain problems arises like paint deuteriation, corrosion, seepage problems, deflections in beams etc. Buildings will become unstable due to all these problems. So, repair works should be done in order to gain the strength of the structure. Repair and Rehabilitation is necessary to save hazardous failure of structures. It is recommended for old buildings which have some signs like cracks, corrosion of embedded materials, etc. Therefore, timely maintenance of structures is required. Most of the olden structures are given strength by doing process of repair and rehabilitation like Charminar.The selection of technique is used as per cost, location of site and other factors. Thus, for proper maintenance, the techniques likewise

References

Critical chloride content in reinforced concrete—a review -Ueli Angst, Bernhard Elsener, Claus K Larsen, Øystein Vennesland Cement and concrete research 39 (12), 1122-1138, 2009

Recent Advances in Repair and Rehabilitation of RCC

Structures with Non-metallic Fibers -Abhijit Mukherjeel, Mangesh Joshi2

Structural rehabilitation of columns with reinforced concrete jacketing E S Ju´lio1, F Branco2 and V D Silva1 1 FCTUC, Coimbra, Portugal 2 IST, Lisboa, Portugal

Handbook of Repair and Rehabilitation of RCC Buildings.

Analysis of repairs and rehabilitation of R.C.C Structures

Chajlani Jeetendra (MTech scholar Civil Engg. Department RGPV Bhopal (M.P.)), Kushvaha Suresh (HOD Civil Engg. Department RGPV Bhopal (M.P.)), Hussain Aslam (Asst. Prof. Civil Engg. Department RGPV Bhopal (M.P.))

Water Quality Analysis using Ecoclean 2300

G. Sreenivasa Rao[1], B. Sreeja Reddy[2], P. Srividya[3] and M.V.S.S Giridhar[4]
[1]Assoc. professor, [2,3]Student, Department of Civil Engg, MVSR Engg College, Hyderabad,
[4]Professor, CWR, IST, JNTUH

ABSTRACT

Indian Subcontinent has abundant water resources in the form of perennial rivers and subsurface water. The present study has been carried out on surface water of Nallacheruvu under Greater Hyderabad Municipal Corporation (GHMC), Telangana state and ground water in the vicinity to assess its quality for drinking and irrigation. A study has been made using Eco clean 2300 which is a herbal reagent for waste water treatment. The main objective of this study is to assess the water quality parameters using Ecoclean2300 and also to improve the efficiency of waste water treatment process. Physical and chemical parameters were analysed like pH, hardness, total dissolved solids, chlorides, Alkalinity, EC, Dissolved Oxygen and BOD using suitable methodologies before and after treatment with Ecoclean2300.The result shows that Ecoclean2300 can be used in the places where sewage treatment plant is not available. It is cost effective and treated water can be used for small scale industries, secondary works like gardening, cooling towers and boiler plants.

Keywords: Ecoclean2300, Greater Hyderabad Municipal Corporation (GHMC)

Introduction

Fresh water lakes are vital resources for any country because they regulate the urban climate and also have a prominent effect on ground water quality and ground water table. Population growth, urban runoffs, sewage discharge and improper agricultural practices can disturb or disrupt aquatic ecosystems leading to eutrophication of inland water bodies causing the deterioration of water quality, which in turn interferes with most of its beneficial uses. Hyderabad, a capital City of Telangana state, geographically situated in land locked arid zone and no perennial river but a seasonal River Musi flowing through it. There are two dams built on the Musi river that are Osman Sagar and Himayat Sagar. Both of the reservoirs constitute the major drinking water sources for Hyderabad. For longer periods, it is the capital city of so many rulers and in long run expanded to the 8500 sq.km in Telangana southern Indian State. ECOCLEAN-2300 is a single Dose Herbal reagent for Sewage water Flocculation, Sedimentation & Disinfection of water. No smell No Bacteria.

ECOCLEAN-2300 is a unique combination of the extracts of Azadirachta indica (Neem), Moringa olifera (Drumstick), Strychnos potatorum (Cleaning nut), Occimum sanctum (Tulsi) and Melaleuca alternifolia (Tea Tree oil) blended together using organic solvents and stabilizers. Dosed at 40 ppm (40 ml for 1000 litres of sewage water), it not only induces flocculation and causes separation of both suspended and dissolved solids but also disinfects the water, freeing it from all harmful bacteria including fecal coliforms and E-coli and improves TDS, TSS,BOD,COD of water with no smell and side effects.

Materials and Methods

The present study has been carried out on surface water of Nallacheruvu under Greater Hyderabad Municipal Corporation (GHMC), Telangana state and ground water in the vicinity to assess its quality for drinking and irrigation. The following experiments are performed such as pH, Hardness Test, Total Dissolved Solids, Electrical Conductivity, Chlorides, Dissolved Oxygen, Alkalinity and Bio chemical oxygen demand. Twenty samples were analysed for each parameter.

Figure 1 Nalla cheruvu

Figure 2 Study area map using Google EarthPro

Results and Discussion

The collected water samples were analysed for pH, hardness, total dissolved solids, chlorides, Alkalinity, EC , Dissolved Oxygen and BOD using suitable methodologies before and after treatment with Ecoclean2300.

pH: The pH of ground water samples are comparatively more than the pH of surface water samples. The mean pH of the Surface samples before and after treating with ECOCLEAN 2300 7.25 and 7.44 respectively(Table.1 and Table 2). All the values are within the permissible limits. The probable cause for rise in pH is due to the presence of neem in Ecoclean 2300 which is alkaline in nature .

- **Hardness:** Most of the the ground water samples showed hardness within the permissible limits. After adding the ECOCLEAN 2300 to the samples, the hardness of all the samples has decreased and are within the permissible limits(Table.1 and Table.2).

- **EC & TDS:** All the surface and ground water samples the values within the above permissible limits for Electrical conductivity and Total dissolved solids. The high concentration may be due to leaching of solid waste from ground surface. Electrical conductivity and Total dissolved solids of all the samples after treating with ECOCLEAN 2300 are reduced and are within the permissible limits (Table.1 and Table.2).
- **Chlorides:** Chlorides of the ground water samples are within the permissible limits. Chlorides of the samples after treating with the ECOCLEAN 2300 showed increment but are within the permissible limits (Table.1 and Table.2).
- **Dissolved Oxygen:** The results showed that DO is nil in all surface water samples . A minimum of 4-5mg/l of DO is good for the survival of aquatic life. DO of all the surface water and ground water samples has increased after treating the water with ECOCLEAN 2300 which is essential(Table.1 and Table.2).
- **Alkalinity:** Alkalinity of the all surface water and ground water samples are above the permissible limits. The alkalinity values are reduced and are within the permissible limits after treating with ECOCLEAN 2300(Table.1 and Table.2).

The results shows that most of the parameters under consideration are with in permissable limits only after adding ECOCLEAN 2300 and are not in acceptable limits for surface and ground water samples. Hence it can be used for other secondary works such as water coolant, gardening and it can also be used for irrigation which are based on their sodium adsorption values. It is observed that the dosage of ECOCLEAN2300 is actually 0.5 ml for every 1 litre and also time allowed for settling of the particles is 10-15 minutes.

Table 1 Surface Water quality of Nalla cheruvu

S. No.	Parameters	Before adding Ecoclean 2300	After adding Ecoclean 2300
1	pH	7.25	7.44
2	TDS	914.2ppm@29.5°c	907ppm@31.1°c
3	EC	1681.8µs@29.6°c	1632.8µs@31.1°c
4	Alkalinity	850mg/l	450mg/l
5	Hardness	295mg/l	160mg/l
6	Chlorides	225mg/l	312mg/l
7	DO	-	2.6mg/l
8	BOD	80mg/l	216mg/l

Table 2 Ground Water quality of Nalla cheruvu

S. No.	Parameters	Before adding Ecoclean 2300	After adding Ecoclean 2300
1	PH	7.3	7.48
2	TDS	1090.4ppm@30.1°c	1063ppm@31.3°c
3	EC	2000µs@30.1°c	1900µs@31.3°c
4	Alkalinity	650mg/l	525mg/l
5	Hardness	278mg/l	235mg/l
6	Chlorides	225mg/l	363mg/l
7	DO	3.2mg/l	6mg/l
8	BOD	266mg/l	498mg/l

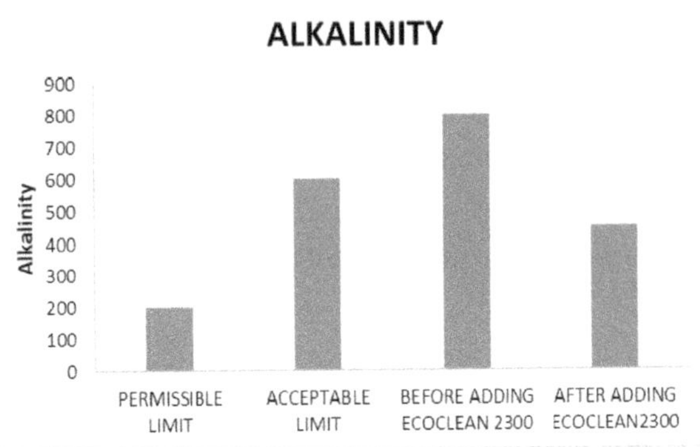

Figure 3 Alkalinity in Nalla cheruvu

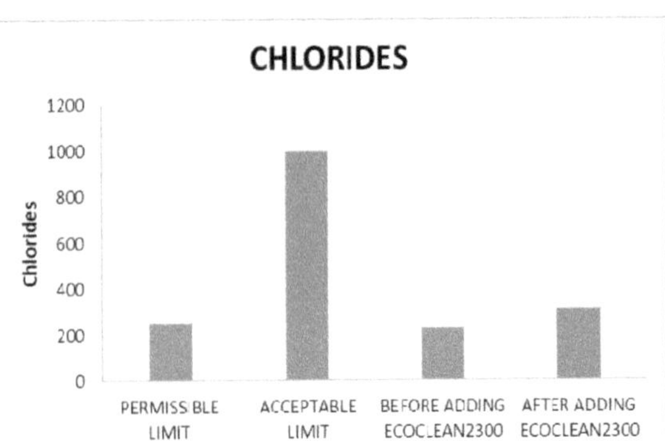

Figure 4 Chlorides in Nalla cheruvu

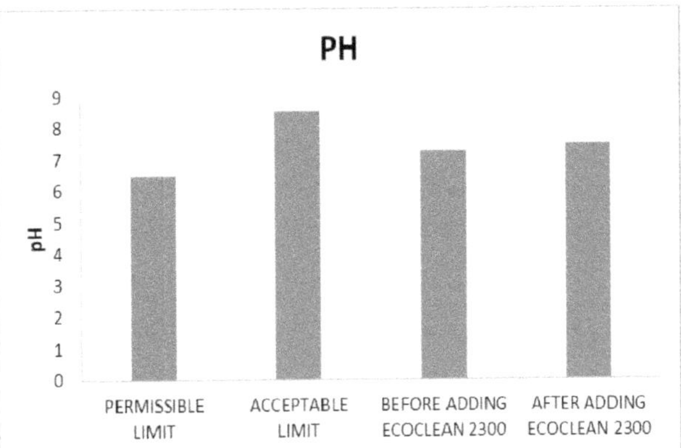

Figure 5 pH in Nalla cheruvu

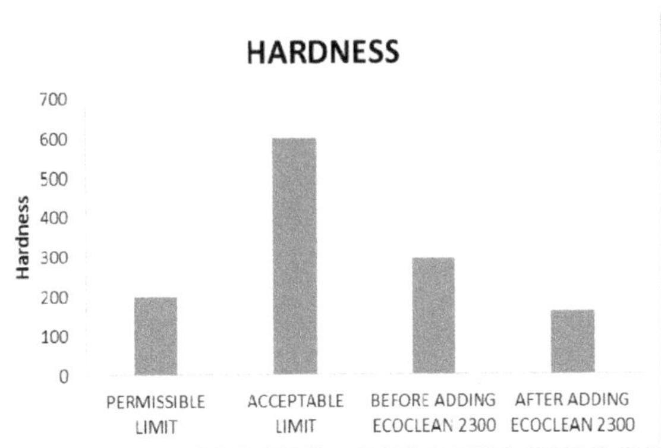

Figure 6 Hardness in Nalla cheruvu

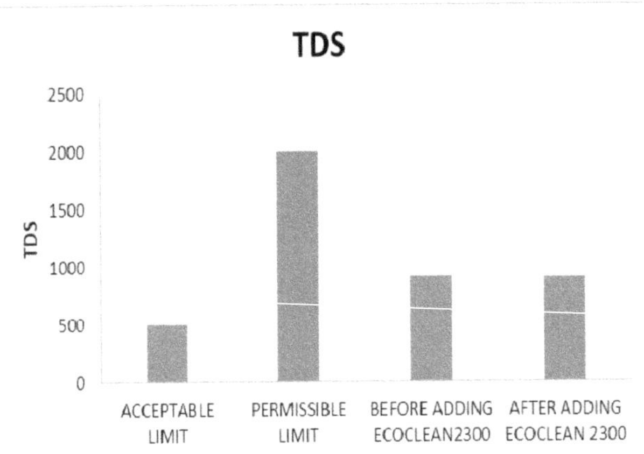

Figure 7 TDS in Nalla cheruvu

Conclusion

In the present study, water samples were collected from surface sources and bore wells located in the study area. The result shows that Ecoclean2300 can be used in the places where sewage treatment plant is not available. It is cost effective and treated water can be used for small scale industries, secondary works like gardening, cooling towers and boiler plants. Biological parameters must be studied to decide whether treated water is suitable for drinking purpose.

References

Applied Hydro geo chemistry by Fetter

BIS IS 10500(1983) Bureau of Indian standards, Indian standards specifications for drinking water.

"Preliminary investigations of ground water quality in Hyderabad city, Telangana, India ",by K Niranjankumar, A.G.S. Reddy, R Mallikarjunareddy and K.Srinivas Varma

Waste water engineering, B.C Punmia and Ashok Jain by Lakshmi publications.

Environmental Engineering laboratory manual.

www.gooogle.com website

www.wikipedia.org

Ground water hydrology author: David Keith Todd by John Wiley and sons publications.

Engineering Hydrology. K. Subramanya by Tata McGrawHill Publications.

Water supply in Sanitary Engineering author: G.S. Birdie by Dhanpat Rai and sons Publications.

www.wrmin.nic.in website of Ministry of Water Resources for BIS 10500(1991) standards.

Manual on Water supply and treatment published by Ministry of Urban Development New Delhi.

Author Index

P

R

S

T

U

V